RICE

Origin, Antiquity and History

RICE

Origin, Antiquity and History

Editor

S.D. SHARMA

Formerly Principal Scientist and Head
Genetic Resources Division
Central Rice Research Institute
Cuttack, India

CRC Press
Taylor & Francis Group
Boca Raton London New York

CRC Press is an imprint of the
Taylor & Francis Group, an **informa** business

Science Publishers
Enfield, New Hampshire

CRC Press
Taylor & Francis Group
6000 Broken Sound Parkway NW, Suite 300
Boca Raton, FL 33487-2742

First issued in paperback 2019

ISBN-13: 978-1-57808-680-1 (hbk)
ISBN-13: 978-0-367-38396-1 (pbk)

Library of Congress Cataloging-in-Publication Data

Rice: origin, antiquity and history/editor, S.D. Sharma.--
1st ed.
    p. cm.
  Includes bibliographical references and index.
  ISBN 978-1-57808-680-1 (hardcover)
  1. Rice--History. I. Sharma, S.D. (Shatanjiw Das), 1934-

SB191.R5R48 2010
633.1'8—dc22

                                                          2010001073

Visit the Taylor & Francis Web site at
http://www.taylorandfrancis.com

and the CRC Press Web site at
http://www.crcpress.com

Contents

Contributors

Ahuja, S.C.

CCS Haryana Agricultural University, V. & P.O. Kaul-136021, District Kaithal, Haryana, India

Ahuja, Uma

CCS Haryana Agricultural University, V. & P.O. Kaul-136021, District Kaithal, Haryana, India

Aidy, Ibrahim Rizk

Principal Rice Breeder, Agricultural Research Center, Field Crops Research Institute, National Rice Research Program, Egypt

Ali, Jabir

Assistant Professor, Centre for Food and Agribusiness Management Centre, Indian Institute of Management, Lucknow, India

Badawi, A. Tantawi

President, Agriculture Research Center, Aricultural Research Center, Giza (Egypt)

Coclanis, Peter A.

Associate Provost for International Affairs, and Albert R. Newsome Professor of History and Economics, University of North Carolina, Chapel Hill (USA)

Falvey, Lindsay

Formerly-Dean, Faculty of Land and Food Resources, University of Melbourne, Melbourne, Australia

Ferrero, Aldo

Dipartimento di Agronomia, Selvicoltura e Gestione del Territorio, Università degli Studi di Torino, Grugliasco, Italy

Guimarães, Elcio Perpetuo

Senior Officer Cereals/Crops Breeding, Viale delle Terme di Caracalla, FAO-AGP, Room C778, 00153 Rome, Italy

Hartmann, John

Presidential Teaching Professor, Department of Foreign Languages and the Center for Southeast Asian Studies, Northern Illinois University, DeKalb, IL

Heu, Mun-Hue

Professor Emeritus, Seoul National University, Republic of Korea

Kaneda, Chukichi

Technical Advisor, Japan Association for International Cooperation of Agriculture and Forestry (JAICAF), Tokyo, Japan

Khush, Gurdev S.

Former Principal Plant Breeder, International Rice Research Institute, Los Banos, Laguna, Philippines

Kumar, Sushil

Associate Professor, Centre for Food and Agribusiness Management, Indian Institute of Management, Lucknow, India

Luo, Wei

Professor, Geography Department, Northern Illinois University, DeKalb, IL

Maximos, Milad

Rice Breeder and Geneticist, National Rice Research Program, ARC, Egypt

Moon, Huhn-Pal

Visiting Professor, Kongju National University, Republic of Korea

Nesbitt, Mark

Ethnobotanist and Archaeobotanist, Royal Botanic Gardens, Kew

Olaoye, Raimi Adebayo

Associate Professor, Department of History, University of Ilorin, Ilorin, Nigeria

Pereira, José Almeida.

Rice Researcher, Embrapa Meio Norte Avenida Duque de Caxias, 5650, Caixa-Postal: 01, 64006-220, Teresina Piauí, Brazil

Pokharia, Anil K.

Birbal Sahni Institute of Palaeobotany, Lucknow, India

Sattar, Mofarahus

Independent Agricultural Researcher and Consultant, H-4, Red Castle Apartments, 584 Sheorapara, Rokeya Sarani, Mirpur, Dhaka 1216, Bangladesh

Sharma, S.D.

Principal Scientist and Head (Retd.), Genetic Resources Division, Central Rice Research Institute, Cuttack-753006, Orissa, India
Present address: 36, VIP Colony, IRC Village, Bhubaneswar-751015, Orissa, India

Simpson, St John

Curator, British Museum, London

Svanberg, Ingvar

Senior Lecturer, Dept of Eurasian Studies, Uppsala University, Uppsala, Sweden, and Södertörn University

Sysamouth, Vinya

Independent Scholar and Executive President of the Center for Lao Studies, San Francisco, CA

Tang, Shengxiang

Professor, China National Rice Research Institute (CNRRI), Hangzhou, 310006, Zhejiang Province, China

Vepa, Swarna S.

Visiting Professor, Madras School of Economics, Chennai (Tamil Nadu), India

Vidotto, Francesco

Dipartimento di Agronomia, Selvicoltura e Gestione del Territorio, Università degli Studi di Torino, Grugliasco, Italy

Virk, P.S.

Plant Breeder, International Rice Research Institute, Los Banos, Laguna, Philippines

Xuan, Songnan

Professor, China National Rice Research Institute (CNRRI), Hangzhou, 310006, Zhejiang Province, China

Preface

I was born in a small village of eastern India in mid-1930s. Rice was grown as the main crop in my village. I remember an incident when I was a young boy of about seven years old. My father, who was a farmer and came from a priestly community, consulted the new almanac for the year. He found that astrologically it would be more auspicious if I did the ceremonial sowing of rice for that year. There was (and is) a special day in the lunar calendar for ceremonial sowing of rice. On that day, my father and mother woke me up early in the morning, took their bath and asked me too to take my bath. We put on new clothes. Then they arranged all the items for worshipping the Goddess of Wealth in a new bamboo basket along with a small iron share and a small sample of rice seeds. Then my father and myself proceeded to our rice field. It was just dawn when we reached the rice field.

My father sprinkled water on a small piece of land in the field, installed the icon of the Goddess in the field and worshipped Her. He covered the head of the deity with a new small red scarf, applied sandal paste and vermillion on her forehead, offered her flowers and incense, lit a lamp and offered fruits and coconut to the Goddess. Then he muttered some prayers in Sanskrit and asked me to join him. Though I could not understand a word of it, I knew, he must have meant, "O Goddess of Wealth, bless us with a rich crop of rice this year so that we have enough to eat and enough to sell to meet all our needs. Bless us with happiness and prosperity." Then he asked me to "plow" a small patch of land of the field with the share, sow the rice seeds, level the ground and water it. Then we bowed before the Goddess once again and left for home.

That year, if I remember vaguely, rainfall was sufficient (there were no irrigation facilities in my village), there were no pests or diseases (there were no insecticides in those days) and the plants grew lush green (there were no chemical fertilizers in those days; farmyard manure was the only source of manuring). When the rice plants were filled with grains in the month of September, there was the festival of "eating new rice" in the whole village. There was special prayer in the village temple; special dishes were prepared in every house; people were dressed in new clothes and, as per customs, younger people sought the blessings of their elders. In the month of December, when the newly harvested crop was brought home, as per custom, the Goddess of Wealth was worshipped on every Thursday. The floor was decorated with rice paste, the doors were decorated with rice earheads and special dishes of rice were prepared in every house. Every year, I used to wait anxiously for the month of December because my mother used to prepare special dishes of rice on every Thursday.

Rice was everything to us. It was our breakfast in the morning, lunch at noon, snacks in the evening and dinner at night. On festive occasions, special dishes were made of rice only. Our festivals revolved round the rice crop. It was a food crop and also a cash crop for us. Rice was our language too. If a boy was uncontrollable, we used to call him a "wild rice" meaning that he is as nuisance as a wild rice plant. If a girl was very beautiful, we said that she is "made of a single grain of rice". Rice was our God. If somebody had to swear, he would hold a handful of rice in his hand and swear by the Goddess of Wealth. In fact, rice was our life.

I have described the life of the people in a small remote village of eastern India in mid 1930s but the pattern of life was not much different in any other village in the whole of eastern, southeastern or southern Asia. If one draws a line from Kerala State of India to Honshu Island of Japan on the map of Asia and visits any village in the east of this line, he would find the same life style. In fact, it was more or less the same one or even two thousand years ago.

To be sure, let us visit the Shinto shrine, Sumiyoshi Taisha, near Osaka in Japan in the month of June. The rice planting ceremony is held here every year on June 14 and the ceremony is more than 1,700 years old.

> Sumiyoshi Taisha is the nation's leading shrine boasting its old history and vast size. The shrine preserves a deity rice field at the corner of its premises to hold the annual rite for the planting of seedlings. The ceremony features a mixture of entertainment, prayers and amusement. Beautiful professional entertainers from the nearby geisha quarters serve as women planting rice seedlings. Shrine maidens attired in red and white clothes and children wearing dance clothing and pretty makeup also participate. Farming cattle adorned with costumes show up. Some participants emerge on the stage laid over the paddy to perform the Noh play and interlude and chant a Noh text. The whole area where the event takes place gives an air of an ancient or medieval Japan.[1]

Or let us visit the Ifugao people in the northern part of Luzon Island of the Philippines where rice has been traditionally grown in terraces since 2000 years.

> The Ifugaos are a tribe with a rich culture. Numerous centuries old rituals are performed in their rice growing. The rice culture leader called *tomona* determines the start of the planting season. Rice planting season is ushered in by the *tunod* ritual in the *tumun-ok* or *payoh* which is the main terrace selected hundreds of years ago from among those belonging to the *kadangyan* (nobility). A statue of a rice god called *bulol* in a sitting position is touched by the hand with blood from butchered chicken or pig. Seeding starts in the months of November and December and planting of rice seedlings commences in January and February.[2]

Throughout the history of mankind, man was a hunter-gatherer except for the last 10,000 years or so when he started domesticating plants and animals of his surroundings. Wild rice (*Oryza rufipogon* and *Oryza nivara*) grew abundantly in South and Southeast Asia and also in the southern part of China. In fact, it grows like that even today. The plant did not miss the attention of early man (rather of early woman). He started collecting its seeds for his subsistence and propagating the plant near his habitat.

In course of time, man has changed the plant a lot. Today he has rice varieties which can be grown in countries of the equatorial belt such as in Java Island and also in high latitude countries such as Hokkaido Island of Japan. It is grown near sea coast and also in high Himalayan valleys. It is grown in uplands and also in 30 feet deep water. There are types which hardly yield one ton per acre and also types which yield more than 3 tons. Man has also changed the shape, size, color and quality of its grains. There are long fine grains which cook fluffier and have aroma. And there are types which stick together when cooked. Man has learnt to use almost all parts of the plant. Mats, shoes, hats, boards, etc are made from its straw. Husk is used for packing, as pulverizers, etc. Bran is used for producing oil. The grain is used in various ways besides being used as staple food, for noodles, dumplings, cakes, pop rice, rice flakes, wines, etc. Today it is the staple of one third of the mankind.

And with the passage of time, man has carried the crop far and wide. China passed it on to Korea, Japan, Taiwan and the Philippines before the start of the Christian era. In the fourth century BC, Greeks carried it to Central and West Asia and to Greece. In the eighth century, Arabs introduced it into Spain and in the next century in Sicily. From Spain, rice was carried to Po valley of Italy and from there to France. Spaniards,

Portuguese and British carried it to the New World. The most recent introduction of rice was into Australia. Today rice is cultivated in all the continents of the world except, of course, Antarctica. However, ninety percent of world's rice is still grown and consumed in Asia only.[3]

Rice has changed man's life and life-style. Once man domesticated rice, he gave up his nomadic life and became sedentary to grow this crop. In the Neolithic age, he made tools of bones, stones, bamboos and woods to sow, weed, harvest and thresh. Today he makes tractors, harvesters and combines to do the same job. He terraced the hills so that water remains deposited in his rice fields, built tanks and canals to irrigate his rice fieds and marveled in engineering skills. Let us visit some of these tanks built in the first five centuries of Christian era in Sri Lanka.

> Some of them were of considerable size, great artificial lakes and many of these 'tanks' . . . . were skillfully connected with each other to form a vast irrigation system. Modern irrigation engineers have evinced much admiration for the way in which the ancient Sinhalese succeeded in their irrigation schemes which are far from easy, even to their modern counterparts . . . . The construction of these early irrigation systems was a remarkable practical feat. They required constant attention and their construction and maintenance must be regarded as the leading feature of early Sinhalese economic life.[4]

The same thing happened in Cambodia when Jayavarman II and his successors were ruling the country in the 9th century AD. They built a sophisticated irrigation system that included giant man-made lakes and canals that ensured three crops of rice in a year.

> The Angkorian system of aquatic management was so sophisticated that architectural historian Sumet Jumsai na Ayudhya has argued that it actually represented a new stage of civilization in Southeast Asia . . . . [*The emperor*] was literally, the lord of life, the master of the waters—it was thanks to his hydraulic craft that the land could produce three or even four rice crops per year. Later, when the Angkorian Empire finally collapsed, it was largely due to the breakdown of its system of aquatic management and the falling into obsolescence of its vast system of dikes, canals and reservoirs.[5]

The same was the case of Vijayanagara Kingdom of India which flourished in the fifteenth century. Its prosperity was based on a vast number of tanks which irrigated the crops and on which the prosperity of the kingdom depended.

The importance of irrigation was well understood from early times; dams were constructed across streams and channels taken off from them wherever possible. Large tanks were made to serve areas where there were no natural streams, and the proper maintenance of these tanks was regularly provided for. The extension of irrigation was encouraged at all times by granting special facilities and tax concessions . . . .[6]

The Asian rice has a cousin that was born and nurtured by the people of West Africa. A different kind of wild rice (*Oryza barthii*) grows wild in sub-Saharan Africa. The people around the inner delta region of river Niger domesticated it about 3000 years ago and created a different type of rice which we now know as *Oryza glaberrima*. Its cultivation spread from inner delta region of River Niger down the stream in the valley of River Sokoto and then to area around Lake Chad. Its cultivation also spread upstream of River Niger and then to the valleys of River Senegal and River Gambia. The people had not only developed the plant but also the technology to grow this crop.[7] When the people of this region were carried to the New World, they not only carried this plant to the new World but also provided the technology for its successful cultivation there.[8] When the Asian rice was introduced in the homeland of this African rice, the Asian rice was gradually replacing the local one because the Asian one's better yielding capacity though the local one has better intrinsic merits. Now, thanks to efforts of agricultural scientists, the merits of both the types have been combined and new types of rice have been developed.[9]

This book is an attempt to tell the story of rice from the time when it was first domesticated in the river valleys of Yangtze about seven thousand years ago and in the valleys of Mekong and Ganga soon thereafter and how the people of various countries have shaped the plant, how they have improved the techniques of its cultivation, how they have found various uses of its grain and other plant parts and in turn how the plant has shaped cultures of the various peoples and helped build their civilizations. It is an attempt to tell how the plant has influenced their languages, beliefs, customs, social habits, festivals, etc. It attempts to tell how man has carried the crop to different regions of the world where topography, soil, climate, vegetation and ethnic and social conditions differ. In each of these regions of the world, people went through their own trials and tribulations for its cultivation, their own successes and failures, their own experiences and learning. And "this is what the human story is, not the emperors and the generals and their wars, but the nameless actions of people who are never written down, the good they do for others passed on like a blessing . . . ."[10]

A book narrating all these aspects of this magnificent crop in all the parts of the world is difficult to be narrated by a single person. The story can be told better by persons who have first hand knowledge and experience of this crop in their own countries. The editor acknowledges with gratitude the contributions of all the authors who willingly agreed and contributed chapters for the book.

I would be ungrateful to my wife (Usha), daughter (Surabhi), sons (Shamik, Sujagya) and daughter-in-laws (Charu, Reena) if I did not acknowledge their continuous support and encouragement during preparation of the book. I am especially thankful to my grandson (Shaunak) and grand daughters (Shansita, Suhani) for diverting my attention and keeping me refreshed and entertained with their playful activities.

S.D. SHARMA

References

1. Tanaka, Junzo (2003).
2. Peñafiel, Samuel R. (2006).
3. Nanda, J.S. (2003).
4. Pakeman, S.A. (1964), Quoted by Toynbee (1972:113).
5. Barnhart, James (2003).
6. Nilakanta Sastri, K.A. (1955:298).
7. Porteres, R. (1956).
8. Carney, Judith A. (2002).
9. NERICA varieties are the hybrids between the African rice, *Oryza glaberrima* and the Asian rice, *O. sativa*. The hybrid derivatives are better than either of the parents in one respect or the other over their parents, either through superior weed competitiveness, drought tolerance, pest or disease resistance or simply through higher yielding potentials. See website: http://www.warda.org.
10. Robinson, Kim Stanley (2002:333).

1 | Domestication and Diaspora of Rice

S.D. Sharma

Until about ten thousand years ago, mankind was roaming in the forests, hunting animals and gathering fruits for his survival. Then he started domesticating some of the wild plants that were growing in his surroundings. That was the beginning of agriculture. Almost during the same period, he started domesticating some of the wild animals that were roaming around him and that was the beginning of animal husbandry. He gradually started selecting better kinds of these plants and and that is how he started plant breeding though he was not aware of it. He did the same thing for his animals as well.

There are two species of rice that man has domesticated, one in Asia called *Oryza sativa* or the Asian cultivated rice and the other *O. glaberrima* or the African cultivated rice. The first one (*O. sativa*) domesticated in Asia has now spread to different parts of the world but the other one (*O. glaberrima*) domesticated in West Africa is still confined to its native land. Outwardly, the two species look similar so much so that an ordinary farmer cannot differentiate the two. In fact, it took quite some time even for the botanists to recognise them as two distinct species. They are considered as different species because they do not cross easily in nature and do not interchange their genes easily. The Asian cultivated rice (*O. sativa*) was introduced in West Africa in the 16th century only. Before this time, only the African rice (*O. glaberrima*) was being grown there. In course of time, the Asian rice became more popular in West Africa. As a result, *O. glaberrima* is now grown in a much smaller area in its own native land.

For more than a hundred years, scientists have been trying to ascertain when, where and how the cultivation of these two rice species started. And who were the people responsible for starting their cultivation? A lot of information has been gathered by now by botanists, geneticists, archaeologists, linguists, historians, sociologists, etc but the picture is far from complete. This chapter tries to review the findings with regard to the questions: how did these two cultivated species of rice originate, where were they domesticated, who were the people responsible for their domestication, how was the technology of rice cultivation developed and how did the Asian rice (*O. sativa*) spread to different parts of the world?

ASIAN RICE

Variation within the Cultivated Rice

The variation within the Asian cultivated rice (*O. sativa*) is enormous. The morphological variation is quite impressive. There are dwarf varieties which grow up to a height of 120 cm only and there are tall plants that grow as high as 1000 cm in deep water. The leaves are generally green but there are varieties which have completely purple leaves. The husk colour could be fawn or yellow or golden or purple. There are awned and awnless types. The grain types could be short and roundish or long and slender, its colour could be white or chalky or red or even black. The grains may cook fluffy and get separated or may be sticky; some are aromatic, others lack any aroma. There are physiological differences as well. There are some varieties which mature in 70 days and some others in 180 days; some flower only when days are short and nights are long, some others do not depend on day length for their flowering. There are types which can grow at high altitudes in cold weather of the Himalayas; there are also types adapted to hot weather of Sri Lanka. They also differ with regard to their grain characteristics.

But the major difference recognised by rice researchers is the genetic difference among them. The type that grows in the northern part of China, Korea and Japan is called the *japonica* type and the other one that grows in warmer regions of China, southeast Asia (mainland) and India is the *indica* type. The *japonica* type has spread to high latitude and colder contries of the world such as Italy, Hungary, etc whereas the *indica* type has spread to the warmer countries of the world. In fact, there is a third type known as *bulu* (also called *javanica*) type that is grown in Java (Indonesia) and some adjacent islands. The rice geneticists now consider it as a tropical form of *japonica* type only. There are many morphological and physiological differences between *indica* and *japonica* rices but all these characters overlap and even many intermediate types exist.

Studies of isoenzymes of rice cultivars have offered a better insight into the genetic differentiation within *O. sativa*. These isoenzymes are better suited for phylogenetic studies as they have no outward expressions and hence have been less affected by human selection. Glaszmann[1] studied such enzymes at 21 polymorphic loci in 1968 rice varieties representing various ecotypes and concluded that the rice varieties of the world could be grouped into six clusters. His findings not only supported the traditional classification of rice varieties into two main groups (*japonica* and *indica*) but also brought out finer details regarding ecotypic differentiation within the Asian cultivated rice.

There is variation for many invisible traits such as resistance to various races of pathogens and biotypes of insects, various isoenzymes and biochemical characters. And a variant could prove very valuable for the breeders, agronomists, entomologists or pathologists or biochemists. As a consequence, geneticists have collected thousands of rice varieties from the farmers' fields and stores and conserved them in the national and international genebanks in living form for breeding new varieites in future. For example, the International Rice Research Institute (IRRI) in the Philippines now conserves "more than 102,547 accessions from the Asian cultivated rice (*Oryza sativa*), 1651 accessions from the African cultivated rice (*Oryza glaberrima*), and 4508 accessions from 22 related wild relatives" which "enables analysis of traits undergoing selection in the course of domestication."[2] The other genebanks which hold sizeable collections are that of International Institute of Tropical Agriculture (IITA) at Ibadan (Nigeria) and Centro Inernacional de Agricultura Tropical (CIAT) at Cali, Colombia and the national genebnks of China, India, Japan and USA.

Wild Relatives of the Cultivated Rice

There are 20 different species of wild rices that grow in the tropical and subtropical regions of the world; all of them grow in the tropical regions only. Of these wild rices, there are two species that are close relatives of the cultivated rices and produce large quantities of seeds. One of these, called *Oryza rufipogon*, grows wild in China, southeast Asia and south Asia. The other one, known as *O. barthii* (syn. *O. breviligulata*) grows wild in northern part of tropical Africa. These two wild rices attracted the attention of Neolithic man for his subsistence especially when he had not enough wild animals to hunt or wild fruits to gather. In fact, some of the aborigines still continue to collect seeds of these wild rices in India, Southeast Asia and Africa.[3]

The Asian wild species (*O. rufipogon*) has also much variation within itself. The main two types within this species are: a swamp type and an upland type. The swamp type grows where there are permanent deposits of water, elongates as and when water level in the pool rises and sets seeds when the day length is short. In other words, it is a photoperiod-sensitive species. Its panicle is lax, grains are slender and awns are slender. It is a perennial plant and survives by its stubbles in swampy places. It is distributed in the coastal delta regions of the rivers of south and southeast Asia and south China. The other type is an annual type that grows in seasonal swamps in the plateau regions of south China, southeast Asia (mainland) and south Asia. It is about 150 cm tall. It does not elongate like the swamp form. It has a robust panicle and bolder seeds and is gregarious in the sense that all the plants set their seed at the same time. The taxonomists have been debating if these two types should be considered as two distinct and separate species or they should be clubbed as a single species having two different forms. Those who treat them as two different species call the annual type as *O. nivara* and the perennial type as *O. rufipogon*. These wild rices cross with cultivated rices in nature and produce various intergrades. The Chinese scientists treat the perennial as well as the annual forms as a single species and call it *O. rufipogon*. In recent years, the molecular geneticists have detected differentiation similar to *japonica-indica* differentiation within the wild species, *O. rufipogon* as well as *O. nivara*.[4]

Progenitors of the Cultivated Rice

The cultivated rices do not grow wild in nature except sometimes as escapes. So the cultivated rices must have originated from their wild relatives as a result of human efforts. In other words, man must have domesticated their wild relatives so as to evolve these cultivated types. In the early years of rice research, many species were suspected to be the progenitors of the Asian cultivated rice but the search ultimately narrowed down to two species, *O. rufipogon* and *O. nivara* that man must have domesticated. As mentioned already, the two species are so closely related that some scientists treat them as a single species and call it *O. rufipogon* only. These two species grow wild in coastal region and in the eastern part of India, in southeast Asia and south China. According to palaeobotanists, the climate of the middle and lower Yangtze basin was warmer in the Neolithic age than what it is today and *O. rufipogon* was growing wild in that area also.

Before domestication, man must have 'harvested' the grains of wild rice

(*O. nivara* and *O. rufipogon*) for centuries and in the process must have selected for non-shattering of spikelets (grains) on maturity, absence of awn, white pericarp (outer layer of the rice grain), synchronous flowering of its panicles, more number of grains per panicle and heavier grains as and when these traits were available to him for selection. There were many other characters that man selected for, such as grain shape and size, sensitivity to photoperiod, duration of growth, various colors for the husk, glabrous leaves and husk, aroma in grains, etc. but these traits were not essential in making the plant a true domesticate.

According to the one school,[5] the cultivated rice was first domesticated in the delta regions of rivers. The primitive cultigens were the lowland types from which the upland types were developed later. This view has been elaborated in Chapter 4 (Domestication of Rice in China and Its Cultural Heritage).

According to the other school,[6] the upland types of *O. sativa* originated from the annual species *O. nivara* in upland where it was cultivated for quite some time. Later, when it spread to lowland areas, it crossed with the *O. rufipogon* and the lowland types were developed. In India, the eastern India (the present-day Chhattisgarh, western Orissa, Jharkhand, eastern Uttar Pradesh) was the probable area where the upland (*aus*) types were first domesticated. When these upland types spread later to middle and lower basins of Ganga and came in contact with *O. rufipogon* in the lower valleys, the upland rices crossed with *O. rufipogon* and the lowland (*aman*) types were developed. The ecological situation and the distribution of *O. rufipogon* and *O. nivara* in southeast Asia (mainland) is similar to that of India. It appears that the upland types originated there first and the lowland types were developed later. According to this view, the annual type (*nivara*) was the probable progenitor because it has already acquired the anuual habit, self-pollination, gragarious habit, synchronous flowering and seed setting, higher seed productivity and bolder seeds. According to Sang and Ge, "although we still do not have evidence to reject either hypothesis, it is more parsimonious to consider the origin of *indica* rice from the annual ancestor."[7]

In recent years, Oka school has tried to accommodate the above two contrasting hypotheses by proposing that perennial-annual intermediate type could have been the probale ancestor of the modern cultivated rices. This has been summarised by Hiroko Morishima as follows:

> It has been a subject of discussion whether the perennial or annual type is the ancestor of *O. sativa*. Sano *et al.* (1980) inferred that the perennial-annual intermediate type is most

probably the immediate ancestor. "Intermediate type" implies the population that is habitually clonal and partially outbreeding but can propagate sexually if seed propagation is advantageous. Such populations are now mostly the secondary products of natural hybridization between perennial wild and cultivated plants. When the primitive perennial population was exposed to disturbed or dry conditions, the population genotype probably shifted toward wild annuals in natural habitats or a primitive cultivated type in man-made habitats through the "intermediate type." A trade-off which constrains energy allocation into sexual and asexual reproduction might have moved the plants toward higher seed yielders.[8]

Socio-cultural Evidences

Social traditions such as myths, folklores, rituals and festivals have deep roots in the society and have their origin in antiquity. So is the case with languages; its words, idioms and proverbs have their origin in antiquity and even indicate association with other languages and cultures. The people of east Asia, southeast Asia and south Asia are rich in traditions, rituals, myths and festivals related with rice. Their languages have many words for rice and rice products. People of no other part of the world have such long traditions of rice; nor are their languages full of words, idioms, phrases and proverbs associated with rice. This indirectly proves that rice must have been domesticated in this part of the world only. This has been discussed in detail in Chapter 3 (Rice in Social and Cultural Life of People) of this book.

Archaeological Evidences

In the first half of the 20th century, there were different views about the place where rice could have originated. Some favoured China as the centre of origin of *O. sativa*; some others were of the opinion that it originated in India. Still others thought that it was domesticated in southeast Asia. According to Harlan,[9] rice could have had a diffused origin. In other words, it might have originated in more than one place and more than one people could have played their role. It also implies that it might have originated at different places at different times and even the process of its evolution might have been different at different places.

The middle and lower Yangtze River valley in China, the Mekong River valley in southeast Asia and the middle Ganga valley are the most probable places where rice could have originated. In pre-historic times, the South China Sea was a land mass which geologists call Sundaland. During 8000-4000 BC, the Arctic ice melted and this land mass (Sundaland) got inundated. The people of Sundaland must have fled in different directions: mainland southeast Asia, Island southeast Asia that have escaped inundation and to Pacific islands. Later, the sea somewhat receded so that the areas that were earlier close to the sea are now farther from the sea. It is possible that the people of the southeast Asia along with the people of this larger land mass that is now inundated might have played an important role in the origin of *O. sativa*.

During the last fifty years, the Chinese scientists have excavated many sites of Neolithic culture especially in the middle and lower basins of the Yangtze River. They have recovered carbonised rice grains along with the primitive tools of bones, stones and ceramics dated seventh millennium BC. Among these, the Pengtoushan and Bashidong are the oldest sites and hence are noteworthy.

> The oldest directly-dated rice grains have been found in two areas: the Yangzi River drainage basin (6500 BC) and to the north in Henan at Jiahu (6000-7000 BC). Some of the best evidence for early rice is from the Pengtoushan and Bashidang sites on the Liyang Plain near Dongting Lake. Both sites belong to Pengtoushan Culture (7500-6100 BC). Village life was well established at the time. Bashidang is surrounded by the earliest combination of defensive walls and ditches in China. Nearly 15,000 rice grains were recovered from a 100 square meter area of water-logged deposits. The rice from Bashidang has considerable variation, so cannot be assigned to a rice sub-species and the grains are slightly smaller than the modern domesticated rice.[10]

It is difficult to determine if the rice samples discovered from these sites were domesticated. But the rice samples discovered at Hemudu, a site near the mouth of River Yangtze, definitely belonged to domesticated type as they lack awns in their glumes—a trait not seen in the wild progenitor species. The earliest rice samples of this site could be dated 5000 to 4500 BC.[11]

The earliest evidence of rice cultivation in Thailand comes from impressions of rice in pottery discovered at the archaeological sites Non Nok Tha and Ban Chiang in northeast Thailand. The samples of rice discovered at these sites dated as fourth millenium BC have been identified as having some primitive characters and are intermediate between wild and cultivated rice. North Vietnam had rice cultivation as back as 2000

BC but even by 111 BC agriculture technology of Vietnam was primitive by Chinese standards. Since southeast Asian forests were covered with hardwood forests, they had to wait until the advent of iron tools for their largescale clearance and rice cultivation. Transplanting in Vietnam began about 1000 AD.[12] Therefore, archaeology does not provide support to the idea that peninsular southeast Asia was the place of origin for the domestication of rice.

There have been evidences of rice cultivation in the middle Ganga valley in the sixth millenium BC but such early dates for the excavated samples have been questioned because systematic sampling methods were not followed. Recently, new evidences for domestication of rice in Ganga basin have come up[13] but these evidences have yet to gain acceptance by the archaeologists in general. According to Peter Bellwood, there are evidences that rice cultivation was established in the Ganga valley by 3000 BC. By 2000 BC, rice appeared as a summer crop in Harappan civilization when this civilization was approaching the end of its mature phase (2500-1900 BC). According to him, there is no evidence so far that rice originated as an independent crop in south Asia or if it was an introduction from southeast Asia. According to him,

> Eastern and northeastern India must have played roles in either an independent development of rice cultivation or its successful introduction into the subcontinent from the east perhaps around 3000 BC but we have no clear archaeological evidence for this as yet from countries such as Bangladesh and Burma.[14]

It must be noted that the probable areas of domestication of *O. sativa* according to the botanists and geneticists do not match with the areas suggested by archaeologists. According to Te-Tzu Chang, a rice geneticist, rice was domesticated in the sub-Himalayan belt stretching from Nepal and middle Ganga valley to the border regions of China and Vietnam. It includes Sikkim, Assam, Yunnan, northern Myanmar, northern Thailand and Laos that lie in between.[15] In fact, Chang was summarising the findings of many other rice botanists and geneticists. For example, wild rice is most frequent and has maximum diversity in this region. According to Morinaga, rice cultivars of sub-Himalayan region around Sikkim have wide compatibility with all other ecotypes of rice.[16] Similarly, Nakagahara found maximum variation for certain isozymes in the border regions of south China and southeast Asian countires.[17]

Linguistic Evidences

The linguists have tried to ascertain the language(s) spoken by the people residing in the middle and lower Yangtze valley in the Neolithic period when rice was supposed to have been domesticated and their subsequent movements in the pre-historic times especially during 6000 BC to 3000 BC. According to linguists, five phyla of languages were spoken in or around middle and lower basin of Yangtze river around 6000 BC. They were Austronesian, Austroasiatic, Hmong-Mien, Tai-Kadai and Sino-Tibetan. Some linguists recognise a much larger group (a macrophylum) of languages called Austric and include the first four phyla of the above-mentioned five phyla. According to them, the speakers of Austric langua-ges originally lived in the middle and lower valleys of Yangtze River and practised rice cultivation. One of these linguistic groups played the major role in the domestication of rice in the Yangtze valley. The speakers of other linguistic groups acquired rice cultivation quite early and played impor-tant roles in the spread of rice cultivation along with the technolgy of its cultivation.[18]

At present, the Austro-Asiatic (AA) languages are spoken over a large area stretching from south China to eastern part of India but such areas are now not contiguous. It is presumed that the Austro-Asiatic speakers were spread over this entire area until the speakers of other languages made a dent in their area. According to one view, the Austro-Asiatic speakers moved along the river courses of Red River, Chao Phraya, Mekong, Brahmaputra and Ganga during the third and second millenium BC. They were agriculturists and during their movement they carried rice cultivation wherever they went and settled. The Austronesian (AN) languages are now spoken in a few pockets of south China and also in Taiwan, islands of southeast Asia, Polynesia in the east and Madagascar in the west. The Austronesian people adopted the sea route and carried rice cultivation to Taiwan, Philippines and Indonesian islands.[19] The speakers of Tai-Kadai languages have moved from southeast China south-westward during the pre-historic period and are now settled in Thailand, Laos, Vietnam and Yunnan. Miao-Yao group was probably closest to the middle Yangtze valley during neolithic time and has moved least southward. It is now spoken in pockets in the areas bordering China and Southeast Asia. The migrations of people speaking Austro-Asiatic (AA), Austronesian (AN) and Tai-Kadai group of languages in different direc-tions was probably caused due to southward "push" by the Sino-Tibetan speakers from the north. The presence of Han speaking Chinese in south China and Tibeto-Burmese speaking people in Myanmar is the result of such subsequent movements. All these people acquired the technology of rice cultivation quite early in pre-historic times, developed their own technologies for rice cultivation and participated in the spread of rice cultivation in the east, southeast and south Asia. For example, the Tai people started moving from the east of the Guangxi-Guizhou region in a southwesterly direction around 2000 years ago into what is today northern region of mainland southeast Asia. They moved along with their technoculture of growing rice and are now settled in Vietnam, Mekong valley, Chao Phraya of Thailand, Shan States of Myanmar, Assam State of India and Yunnan Province of China.[20] Chapter 2 (Tai Participation in the Spread of Rice Agriculture in Asia) discusses in detail the technology developed by the Tai speaking people and its spread in southeast Asia during the last two thousand years.

Evidences from Molecular Genetics

Molecular genetics has now fairly established that the wild species *O. rufipogon* (*sensu lato*) was already differentiated in the line similar to *japonica-indica* before its domestication and hence the two sub-species could have originated from different populations of wild rices.[21] It has also been suggested that the *japonica* could have originated from the perennial rice (*O. rufipogon*) and the *indica* from the annual wild rice (*O. nivara*).[22]

Besides the nuclear DNA, mitochondrial DNA (mtDNA), chloroplast DNA (cpDNA) and ribosomal DNA (rDNA) have also been studied and have helped studying domestication of rice. Based on studies on the DNA sequence variation in three gene regions, two nuclear genes and one mitochondrial gene of *O. rufipogon*, Jason Londo and his associates have concluded

> India and Indo-China may represent the ancestral center of diversity for *O. rufipogon*. Additionally, the data suggest that cultivated rice was domesticated at least twice from different *O. rufipogon* populations and that the products of these two independent domestication events are the two major rice varieties, *Oryza sativa indica* and *Oryza sativa japonica*. Based on this geographical analysis, *O. sativa indica* was domesticated within a region south of the Himalaya mountain range, likely eastern India, Myanmar and Thailand, whereas *O. sativa japonica* was domesticated from wild rice in southern China.[23]

On the other hand, basing their studies on three FNPs (functional nucleotide polymorphisms) for domestication-related genes of rice (*Wx*, *qSH1* and *qSW5*), local origin information and RFLP variation patterns at

179 loci distributed throughout the rice genome, Konishi, Ebana and Izawa have proposed that *japonica* rice could have originated in Indonesia, the Philippines or a southern part of Indo-China.[24]

Molecular geneticists have studied the domestication-related genes i.e. the genes that were basically responsible for making rice a domesticated species. One of the main characters that helped domestication of rice is the absence of its shattering. Shattering of spikelets in wild rice is due to formation of an abscission layer between the rachis and the spikelet. In cultivated rice, this layer is not formed and hence the grains do not shatter. The origin of this trait in populations of wild rices and subsequent exploitation of this trait by man has played the key role in domestication of rice. The two genes that are most important in relation to shattering in rice are *sh4* and *qSH1*. *sh4* activates the abscission process while *qSH1* regulates abscission-layer formation. Sequence analysis of *sh4* has revealed that a single base-pair mutation is responsible for non-shattering and this change is the same in both *indica* and *japonica* rice varieties.[25] This would imply that rice was domesticated only once, a conclusion that would contradict the hypothesis that rice was domesticated more than once. Vaughan, Lu and Tomooka[26] have tried to explain this apparent contradiction by proposing "a cycle of introgression, selection and diversification from non-shattering domesticated rice, importantly in the initial stages in its center of origin in the region of Yangtze valley."

White pericarp is another gene related to domestication and has been investigated. The wild rice has red pericarp that is governed by *Rc* gene. The mutaion of this gene has caused the white pericarp in cultivated rices. According to Sweeney and his associates "the predominant mutation originated in the *japonica* subspecies and crossed both geographic and sterility barriers to move into the *indica* subspecies." They also remark "Our finding provides evidence of active cultural exchange among ancient farmers over the course of rice domestication coupled with very strong, positive selection for a single white allele in both subspecies of *O. sativa*."[27]

Glutinous rice is a special type of rice. This character is not met with in the wild rices and hence is worth investigating like the domesticating genes. Glutinous rice has culinary importance and cultural significance for the people of southeast and east Asia. Generally rice grains contain two types of starch: amylose and amylopectin but glutinous rice has no amylose. It consists of amylopectin only. Glutinous rice becomes sticky and forms a lump when cooked. It is used as staple food by some people of mainland southeast Asia and also for desserts on festive occasions. In southeast Asia, it is especially grown by Tai-speaking people who migrated to this area about 1100 to 1500 years ago. It is generally grown in Laos, Cambodia, Vietnam, Myanmar, northeast India and China. It is also grown in Korea and Japan for special preparations and in festival days. The gene that controls this character is *Wx*. In recent years, the DNA of this gene has been analyzed in detail by Kenneth Olsen and Michael Purugganan. According to them, this character originated only once in tropical *japonica* type in mainland southeast Asia many thousand years ago and has spread to other varieties of rice by natural hybridization and introgression.[28]

Spread of Rice Cultivation in Prehistoric Times

From the middle and lower Yangtze valley, the neolithic communities proliferated to the north, south and most probably by boat to the east to Taiwan.[29] By 5000 BC rice cultivation had expanded along the south China sea coast as far as Guangdong and by 4000 BC perhaps upto Thailand and northern Vietnam. By 3000 BC rice cultivation was established as a crop in Southeast Asia. Rice is recorded in Taiwan in the third millenium BC, in Luzon (Philippines) and Kalimantan in the second millenium BC and in Sulawesi in the first millenium AD. According to Pejros and Shnirelman,

> We are quite convinced that the spread of rice through the Southeast Asian islands was connected with the southward migrations of the early Austronesian languages. The earliest rice is dated in Taiwan to the second half of the third millenium BC. In Kalimantan, the dates for rice grain in shred are given as 3850 ± 260 BP. In northeastern Luzon (Philippines) rice has been discovered at a site dated to the second half of the second millenium BC while rice was introduced in Sulawesi (Indonesia) no later than the middle of the first millenium AD. These scant but reliable archaelogical data accord well with modern theories on the routes of ancient Austronesian migrations. They also disprove the hypothesis of some botanists that rice cultivation spread from south to north in Southeast Asia.[30]

Rice was introduced into the middle part of the Korean peninsula around third millennium BC along the riversides of the Han-river. Afterward, rice culture was extended toward the southern parts of Korea where techniques of rice cultivation were improved. When the Han state of China conquered Gochosun, a prehistoric state of Korea, in BC 109, Korea was exposed to an advanced civilization along with an advanced technology of rice cultivation.

Around 350 BC wet-rice agriculture was introduced into Japan. It was first inroduced into the northwest part of Kyushu Island. Whether it was introduced from Korea or China is still a debated subject; it could have been introduced through both the routes.

> From there it spread northeastward in three successive waves, reaching the northeastern Tohoku region by the beginning of the Christian era. Arguably, from its sociopolitical ramifications, no other historical event was as significant for the development of what is now known as the Japanese nation.[31]

Technology of Rice Cultivation

Selection of genotypes adapted to different ecosystems, development of improved methods of crop cultivation and innovation of agricultural implements have played significant roles in establishing rice as a major cereal of the world. The ancient writings of China provide a lot of information about the history of cultural practices of rice; the Indian records in this regard are poor though they are rich in spiritual and philosophical contents. Te-Tzu Chang[32] has summarised much of the information available in Chinese records for the benefit of English readers:

The climate of the Yangtze valley of China was relatively warmer and more humid in the Neolithic period. Later, as the climate became drier, rice cultivation receded eastward as well as southward. Rice was probably a minor food during Zhou dynasty (1122-255 BC). It was merely a supplement to the diet of millets and legumes. Rice was probably seeded by broadcasting and the varieites were generally awned. Spades and hoe blades made of wood, stones and bones of large animals were used in preparing fields. Clay pots were used as cooking utensils. Use of iron in making farm implements, use of water buffalo as draft animal, development of irrigation systems and transplanting were the major breakthrogh in rice cultivation in subsequent periods. According to Chang, rainfed rice cultivation was well established during the Chou dynasty. After Yü initiated flood control measures in the second millenium BC, the area under rice began to expand. Water works began around 700 BC, initially for controlling floods. Irrigation projects are recorded in Honan as early as 563 BC. By the 3rd century BC, 'rice men' were appointed to supervise planting and irrigation of rice fields.[33] Transplanting is first mentioned in Eastern Han (AD 23-220) writings.

Water buffalo was used as a draft animal since Shang dynasty (1570-1045 BC). The iron plough, spade and scythe came into use during the Warring States (403-221 BC). By the eastern Zhou dynasty (255-249 BC), rice was already the staple food crop of the people of the middle and lower Yangtze basins. By the 3rd century AD, rice had become so productive a crop in Yangtze valley and south of it that there was mass migration of people from north to south during the Chin and Song dynasties (317-1279).

The introduction of early maturing Champa rice from Vietnam in the 11th century AD and its largescale cultivation in central and south China made double cropping possible and led to population increase.

The distribution of spike-tooth harrow drawn by the water buffalo in southeast Asia probably reflects the spread of Chinese technology of rice cultivation.

> The spike-tooth harrow drawn by the water buffalo greatly facilitates puddling of the heavy clayey soils and levelling of the field for transplanted rice. This practice originated in China and is popularly used in a long arc that extends from Burma to Malaysia and the Philippines. It is less popular, however, in Indonesia where Indian influence probably is relatively greater. The less widespread of the harrow in India is partly reflected by the relatively large area of the direct-seeded crop.[34]

Spread of Rice Cultivation to Other Regions

From India, rice cultivation was introduced into Iran quite early. During the Achaemenid period, the reign of Darius the Great (522-486 BC) extended up to the Indus river; he could not have remained unaware of rice cultivation in India. Speaking about the rice cultivation in Iran during this period, Peter Christensen[35] says:

> Rice was introduced no later than in Achaemenid times, and even though we do not know the details, it was certainly widespread by Sassanian times. In the 10th century, rice was the most cultivated crop after wheat and barley and bread baked fom rice flour formed a typical element of the Khuzi's basic diet.

In Khuzistan and Susi (southeastern parts of Iran which are very close to the present day Iraq), rice cultivation was already being practised in the 4th century BC. According to Diodorus Siculus, when Eumenes and his troops were in Susiana *ca.* 318/7 BC 'marching through the country . . . . he was completely without grain, but he distributed to his soldiers rice, sesame, and dates since the land produced such fruits as these in plenty.'[36] During Achaemenid period, Mesopotamia was a part of the Achaemenid Empire.

When Alexander the Great invaded India in 327 BC, the Greeks became better aware of rice as a crop and tried to introduce it in the Middle East. By the end of the 1st century AD, the Romans occupied Egypt and they were sailing from Egyptian ports on the Red Sea to south Indian ports. They were importing spices, pearls and ivory and paying with their gold coins. The Romans introduced rice cultivation in Syria, Palestine and Asia Minor and most probably in Egypt as well.[37]

Arabs occupied Egypt in 642 AD and Baghdad in 750 AD. The Arabs promoted rice cultivation in Egypt and Mesopotamia. The Fayyum region of Egypt and southern part of Iraq became the main rice growing areas. Rice and fish were the staple diet of Khazars who formed the state of Khazaria in the region of Black Sea and Caspian Sea during the seventh and 10th century.[38]

In the 8th century, Moors (Arabs of Morocco) occupied Spain and ruled this country for five hundred years. During their reign, they introduced rice cultivation in the Valencia province where it is cultivated even to this day. They also planted almonds, figs, apricots, lemons and oranges. They also occupied a large part of Portugal where they introduced rice cultivation. Moors got a foothold in Sicily in 827 and by 903 they became the masters of the island for more than 150 years. Here too, they introduced rice cultivation. In the 10th century, Sicily was exporting rice.[39]

Rice was not a new cereal for Italians. It was known to them even during the Roman period when Roman physicians used it as a medicine. In the medieval period, the Italian merchants were importing rice from Egypt, Sicily and Spain. There are unconfirmed reports that rice cultivation was introduced in Piedmont from Portugal in the 10th century. There are, however, historical evidences that rice cultivation started in Po valley in the last part of the 15th century with the capital investment by the merchants of Lombardy. They invested on clearing forest land and developing canal system for irrigation. The merchants wanted to maximise their profit and used to employ women and children so much so that an ordinance was issued in 1590 in Lombardy banning this practice. Rice cultivation was banned near towns because it was considered to cause "malaria".[40]

When Ottomans became the masters of West Asia, north Africa and east Europe, they promoted rice cultivation in their domain. They imported rice from Egypt to their capital Istanbul. Busbeq, the ambassador (1554-1562) of the Austrian Emperor Ferdinand I to the Sultan of Ottoman Empire noted, "There are two things from which, in my opinion, Turks derive the greatest advantage and profit, rice among cereals and camel among beasts of burden, both are admirably adapted to the distant campaigns which they wage . . . . When the Sultan went on campaign, he was accompanied by 40,000 camels loaded with rice." In the sixteenth and 17th century, when the Ottomans had to keep a large army in Hungary, they promoted rice cultivation in that country so that rice need not be transported all the way to Hungary to feed their army garrisoned there. "The Ottomans brought cultivators specialised in rice growing and rice cultivation spread in countryside."[41]

Around 200 to 400 AD, Polynesians from Borneo travelled to Madagascar in their out-rigger boats and settled there with their rice and rice culture.[42] It is not clear whether the Asian cultivated rice was introduced in West Africa by the Arabs who reached there from east (Egypt and Sudan) as well as from north (Maghreb) in the 11th century and introduced their religion, their writing system and many of their favourite crops. There is, however, no firm evidence so far that they introduced Asian rice in West Africa.

The 15th century was the era of exploration by Portuguese navigators. The Portuguese reached west coast of Africa (by sea route) in the 15th century and in 1498 Vasco da Gama touched an Indian port that is now in the present-day Kerala State of India. The Portuguese sailors carried Asian rice from south and southeast Asia on their homeward journey and distributed it among farmers in the west coast of Africa. According to Olga Linares,[43]

> Although it is not known with certainty when and where the first varieties of Asian rice O. sativa were first introduced into West Africa, the general consensus is that, beginning in the 16th century, the species spread and was adopted by people living in the Upper Guinea Coast who had previous experience growing the local African species.

Around the same time, the Portuguese introduced maize and cassava from the new World into West Africa. The Asian rice, once introduced, was so well received by the African farmers that at present only 5% of the rice varieties grown in Sierra Leone are O. glaberrima varieties.[44] The situaton in other West African countries is no better in this regard.

When the Europeans started colonizing Americas, they introduced Asian crops like rice and sugarcane and carried out large scale plantation of these crops. Early Spanish explorers introduced Asian rice to the Caribbean islands and South America. Rice first arrived in Mexico in the 1520s at Veracruz which was selected for its warm, wet, climate. Rice cultivation started in Colombia in the Madalena River valley around 1580 and was first cultivated as an upland crop like wheat. Rice cultivation started much later (in 1761) in the Mainas province of Brazil.[45]

Rice arrived in South Carolina around 1685 when sea captain John Thurber's ship had to land in Charleston for repair. Thurber gave a sack of "Gold Seede" rice from Madagascar either to Dr. Henry Woodward or Thomas Smith who was a landgrave (governor of a major land grant). However, a bushel of rice had been sent to the colony on the supply ship William and Ralph as early as spring 1672. By 1700, Charles Town (Charleston) was exporting 330 tons of rice to England and English Caribbean colonies. By the time America gained its independence in 1776, rice had become one of the country's major export crops. Early in the 1800s, rice spread to Louisiana but not until the 20th century was it produced in California.[46]

In Australia, the first rice cultivation started probably during the gold rush of 1850-1860 when the Chinese propsectors tried to grow rice for their use in Queensland. And in 1891, the New South Wales Department of Agriculture officially started experiments for growing upland rice. But the real foundation of rice cultivation in Australia was laid by the Japanese parlimentarian, Isaburo Takasuka. The Victorian government allotted him 200 acres of land on the Murray River basin to demonstrate rice growing. After persevering through floods and droughts, he produced a crop of rice for commercial sale in 1914.[47]

AFRICAN RICE

Before the Asian rice was introduced into West Africa in the mid-16th century, the people of West Africa were growing their own rice which is a different species (*O. glaberrima*), not the same one as the Asian rice (*O. sativa*). The two species look outwardly so similar that it took quite some time even for the botanists to realise that they are two different species. Before the introduction of the Asian rice, the African rice was being cultivated from west coast of Africa up to Lake Chad. Its cultivation was limited in the north by the Saharan desert and in the south by the tropical rain forest. Even in this narrow belt, its cultivation was further limited to certain pockets only; in and around Lake Chad, in the valleys of River Niger and its tributary, River Sokoto, Inner Niger Delta and in the valleys of river Senegal and Gambia. It was also grown in Casamance, Guinea, Guinea-Bissau and Sierra Leone which receive comparatively better rainfall than other parts of sub-Saharan Africa. Besides, it was not the staple food of the people of West Africa except in the pockets of its cultivation. It was always grown for subsistence and hardly ever for commerce.

The African rice (*O. glaberrima*) does not offer as much variation as does *O. sativa*. However, African rice has different ecotypes that are grown in different ecological conditions such as inland wetlands of Niger Inner Delta, tidal floodplain of Senegal, mangrove swamps of Sierra Leone and rainfed uplands of Guinea, Liberia and Ivory Coast. There is also variation in husk colour, grain (pericarp) colour and grain type as in the case of *O. sativa* though the variation is not as great as in the case of the latter. The morphologists call the variation between the two species as 'parallel variation'. Geneticists could recognise only 'feeble' differentiation like *japonica* and *indica* in this species.[48] This is because nature did not offer much variation in altitude and latitude to *O. glaberrima* as it did to *O. sativa*. Besides, the population pressure in West Africa was not as intense as in Asia to cause "cultivation pressure" leading to selection pressure for developing diverse ecotypes and varieties. There were no iron implements and no water buffalo or oxen; hence the area under cultivation could not be extended with limited human labour. The genetic diversity of its progenitor species, *O. barthii* was also limited as compared with that of its Asian counterpart (*O. nivara*).[49]

Compared with *O. sativa*, the African rice (*O. glaberrima*) has many shortcomings: its grains often shatter on the ground even before reaching the threshing floor, the rice is red-skinned and, when cooked, tastes gritty. Its per acre yield is low and it has not become a trade commodity. The merits of the African rice are that it spreads so fast in its young stage that it can smother weeds; it is better adapted to adverse soil conditions of Africa, is more resistant to drought, diseases and insect pests of Africa and is also tolerant to iron toxicity. It is also more adapted to low management and late planting, situations that prevail in Africa. It is, however, treated as a luxury food for the chiefs and is used in religious functions and rituals of African natives.[50]

There are two wild species of rice in Africa that are closely related to the African cultivated rice (*O. glaberrima*). One of them is perennial (*O. longistaminata*) and the other one is annual (*O. barthii*), these two entities are distinct and different. These wild species have awns and shed their spikelets (grains) on maturity. On the other hand, most of the varieties of *O. glaberrima* lack hairs on their leaves and husk (lemma and palea) and do not shatter.

There is general agreement among botanists and geneticists that the African cultivated rice *O. glaberrima* originated from its annual wild relative *O. barthii*. This annual wild rice grows naturally in the swamps of tropical West Africa in the sub-Saharan belt. It is evident that this wild relative of African rice attracted the attention of the early man for his subsistence. Even to this day some people of Africa harvest the grains of this wild species for their subsistence.[51]

The People Who Domesticated *O. glaberrima*

There was no domestication of cereals in West Africa until 4000 BP.[52] Pastoralism, extensive grain gathering and fishing were the dominant food economy in West Africa up to that time. With the increasing desertification of Sahara, the people of Sahara gravitated toward the Senegal and Niger rivers and Lake Chad. Agricultural land clearance around Inner Niger Delta began around 3000 BP. The Bozo people were then the autochthonous inhabitants of Inner Niger delta area and fishing in the river was their main occupation. The present-day Bozo people along with other inhabitants of Inland Delta area speak Mande language. The Bozo people also share their myths and rituals with many of these groups. According to Kevin MacDonald,[53]

> Could not an incoming food producing people fleeing drought in the north have had a decisive impact on fishing peoples trying to survive in a similarly declining environment? . . . . The Ndondi Tossokel people (upon encountering stands of wild grains in and around the delta more vast than they had ever encountered in the north) began to practice rice cultivation in the centre of the delta and continued millet cultivation along its margins (with varying degree of livestock husbandry persisting)? In doing this, they interacted with the regions indigenes, intermarrying and absorbing others into their ranks by their economic success?

Ceramic imressions of rice have been found from archaeological sites at Ganjigana in northeast Nigeria. These go back to 1800 BC and continue up to 800 BC. Many charred grains of rice have also been found at a neighbouring site, Kursakata dating from 1200 BC up to the beginning of the Christian era. It is, however, not known if these were the cultivated African rice or its wild relative. The African wild species that man exploited in this part of the world was *O. barthii* (syn. *O. breviligulata*).

In recent years, a large quantity of rice grains was recovered from Dia in the Middle Niger Delta. These have been dated eighth to 4th century BC and seem to be domesticated rice because their size and shape match with the cultivated African rice.[54] Similarly, domesticated rice dated 300 to 200 BC have been discovered from Jenne-Jeno in Mali.[55] By the beginning of the Christian era, African rice (*Oryza glaberrima*) was being cultivated on the rich silt at the confluence of the Niger and the Bani rivers where the Jenne Jeno was situated, in the midst of a string of large villages.[56]

According to Ronald Porteres, the Inner Niger Delta is the primary centre of origin of African rice (*O. glaberrima*); the secondary centres of domestication being the river basins of Gambia and Casamance where the non-floating wetland types suitable for cultivation in the flood plains were developed. The upland types were developed in the upper basin of Niger River i.e. in the uplands of Fouta Jalon in Guinea.[57] According to Carpenter, rice was cultivated widely in West Africa before the Portuguese introduced the Asian cultivated rice in the west coast of Africa in the 16th century.[58]

Spread of Cultivation of African Rice

According to Porteres, African rice was domesticated in the Inner Niger Delta region of Niger River and from there it spread westward as well as eastward. Its cultivation spread down the River Niger and also in the valley of River Sokoto which is a tributary of River Niger. Its cultivation spread to the basin of Lake Chad. Its cultivation spread upstream along the River Niger and then downstream along the Gambia and Senegal rivers where the non-floating types were developed. It spread to Casamance where tidal swamps were developed. It also spread to the source of River Niger in Fouta Jalon in Guinea. The upland types were developed in the mountainous region of Fouta Jalon. The upland types then spread to Sierra Leone, Liberia and Ivory Coast. The river Bandama in Ivory Coast marks the eastern limit of rice cultivation in West Africa. To the east of the river, yam is cultivated.[59]

In West Africa, the people of two ethnic groups—Mande and Jola—are traditionally associated with rice cultivatiom. The Mande people lived in the upper Niger Basin when the Mali Empire flourished in the 13th century. From there they moved in south and southwest directions and now are spread from Gambia to Ivory Coast i.e. within the western part of West Africa. The Mande people follow a rice-based cropping system. It is believed that southward movement of Mande people has gone hand in hand with the expansion of rice-based cropping system. When they moved southward, they cleared the forests for cultivation of rice.[60]

The lower Casamance has a tropical climate and a marshy, coastal landscape. Numerous creeks (*marigots*) filled with brackish water and lined with mangrove vegetation, branch out from the Casamance River, criss-crossing the low-lying areas and creating an amphibian landscape. The Jola people have been living here since the 16th century if not earlier.

They probaly migrated there from the neighbouring Guinea. They cultivate rice in the land between the creeks of the river Casamance. They build dykes on a new land, fill it with the rainwater and then leach out the salt. Then they transplant rice seedlings already grown elsewhere. Traditionally, they do not sell rice, instead they store it in granaries often for years with pride.[61]

The other people who have been growing rice in coastal region are Baga and Temne. The Baga people migrated from the middle Niger and settled in the estuaries of Gambia and Senegal River and have become excellent rice farmers in the brackish water area. Most of the Temne people live in Sierra Leone. The Temne farmers migrated to the estuaries of the Great and Little Scarcies at the end of the 19th century. They fell dense mangrove forests spread over an area of about 20,000 hectares and brought the land under rice cultivation. Their language is akin to that of Baga people.

The Bandama River of Ivory Coast has been the dividing line between rice-based cropping system on the western side of this river and yam-based cultivation on its eastern side. There has been much discussion about social, cultural and ethnic reasons for it. Becker and Diallo have discussed the views of various authors and conclude that it is due to "southerly dispersal of Mande people who, either themselves or through the actions of people they displaced, brought rice-growing technology into the Ivorian savanna and forest. This dispersal dates firmly to the late 16th century but began gradually in the 14th and 15th centuries with the breakup of the large empires of the Sudan along the Niger River and the dry period."[62]

Ever since the introduction of Asian rice by European powers in West Africa in the mid-16th century, the area under its cultivation in this region has been increasing and, at the same time, the area under the African rice has been receding so much so that the African rice now hardly occupies 5% of the total area under rice in its homeland.

When the colonisation of America started in the 18th century in a big way, there was a heavy demand of labour in the New World and the European settlers found the sub-Saharan Africans as the best source of cheap labour. Africans were transported as slaves to Brazil, West Indies and United States particularly to work in the farm lands.[63] As Judith Carney has pointed out, it is the African labour and West African technology of rice cultivation that laid the foundation of rice cultivation and rice industry in South Carolina of United States as well as in Brazil in the 18th century.[64]

Present Position

The main interest of the European powers in West and East Africa was, however, to promote plantation/commercial crops; improvement of rice, therefore, did not receive their serious attention. During the British rule, some rice research was done at Moore Plantation in Nigeria and at Rice Research Station, Rokupr in Sierra Leone. The French conducted rice research for their West African colonies through Institut de Recherches Agronomiques Tropicales et des Cultures (IRAT) and Office de la Recherche Scientifique et Technique Outre-Mer (ORSTOM) at Bouake in Ivory Coast. Serious attention to this crop was paid only after 1960 when the International Rice Research Institute at Los Banos (Philippines), International Institute of Tropical Agriculture (IITA) at Ibadan (Nigeria) and the West Africa Rice Development Association (WARDA) and now the African Rice Centre under the aegis of WARDA were established. Together, these institutes have collected and conserved about 2800 accessions of the African rice before its disappearance. In fact, the African rice has not disappeared; the cultivation of Asian and African rice closeby has led to natural hybridization between the two species and genes of Asian rice have already introgressed into the African rice.[65] This has enriched the genepool of African rice.

In recent years, the scientists at the African Rice Center have crossed the two species to combine the stress-tolerance and disease resistence of *O. glaberrima* with the yield potential of *O. sativa*. From the progeny of the crosses, they have selected varieties adaptable to different African ecosystems and capable of much better yield. These new types of plant, named 'Nerica' (New Rice for Africa) hold great promise to meet the growing demand of rice in African countries.

References

1. Glaszmann (1987).
2. McNally *et al.* (2006).
3. McIntosh (2005:94).
4. Yamanaka *et al.* (2003).
5. Oka (1974, 1988), Morishima (1984).
6. Chang (1976a), Sharma *et al.* (2000), Sharma (2003).
7. Sang and Ge (2007).
8. Morishima (2001).
9. Harlan (1992).
10. Crawford (2005).
11. Crawford (2005).

12. Chang (1976).
13. Saxena (2006), Tiwari (2006).
14. Bellwood (2004:87).
15. Chang (1985b).
16. Morinaga (1968).
17. Nakagahara (1978).
18. Pejros and Shnirelman (1999).
19. Higham (1996, 1999).
20. See Bellwood (1992), Higham (1999), Pejros and Shnirelman (1999).
21. Second (1982), Vitte *et al.* (2004), Garris (2005), Londo *et al.* (2006).
22. Yamanaka *et al.* (2003).
23. Londo *et al.* (2006).
24. Konishi, Ebana and Izawa (2008).
25. Li *et al.* (2006).
26. Vaughan, Lu and Tomooka (2008).
27. Sweeney *et al.* (2007b).
28. Olsen and Purugganan (2002).
29. Higham (1996).
30. Pejros and Shnirelman (1999).
31. Ohnuki-Tierney (1994).
32. Chang (1976c, 1983).
33. Chang (1976c, 1983).
34. Chang (1976).
35. Christensen (1993:117).
36. Potts (1999:358).
37. Scheidel (2001:81).
38. Adshead (1997), Sato (1997).
39. See website: http://www.cliffordawright.com/history/risotto.html
40. Adshead (1997).
41. Inalcik (1994).
42. Collins and Burns (2007).
43. Linares (1994).
44. Paul Richards (1996).
45. See Chapter 14 in this book.
46. See Chapter 13 in this book.
47. Markus (2000).
48. Misro and Mishra (1969).
49. Chang (1976), Oka (1977).
50. Ruskin (1996).
51. Murray (2004).
52. MacDonald (1997).
53. MacDonald (1997).
54. Murray (2004).
55. Sweeney and McCouch (2007).
56. Nieane and Ki-Zerbo (1997).
57. Porteres (1970).

24

58. Carpenter (1978:4-5).
59. Porteres (1976).
60. Olson (1996).
61. Hudgens, Trillo and Calonec (2004).
62. Becker and Diallo (1996).
63. Littelfield (1981).
64. Carney (2001).
65. Simon *et al.* (2005).

2 | Tai Participation in the Spread of Rice Agriculture in Asia

John Hartmann, Wei Luo and Vinya Sysamouth

INTRODUCTION

It has been said that history is basically the search for origins. The very title of this book on the *Rice: Origin, Antiquity, and History* indeed ties history to the search for beginnings, namely the search for the origins of domesticated rice. The pioneering archaeobotanical research that took place during the last quarter of the preceding century for candidate sites for the earliest domestication of rice pointed exclusively to the middle and lower Yangtze region as early as 8000 years ago.[1] A recent, closer and more critical look at these earlier findings by Londo *et al.*[2] suggests that "cultivated rice was domesticated at least twice from different *O. rufipogon* populations and that the products of these two independent domestication events are the two major rice varieties, *Oryza sativa indica* and *Oryza sativa japonica*. Based on this geographical analysis, *O. sativa indica* was domesticated within a region south of the Himalaya mountain range, likely eastern India, Myanmar, and Thailand, whereas *O. sativa japonica* was domesticated from wild rice in southern China." Sharma, Tripathy and Biswal[3] assert "that rice might have been domesticated at many sites simultaneously." Thus newer genetic findings have overturned the widely held presumption of a single origin for rice domestication in China. The lesson to be learned here is two-fold: pronouncements about origins require careful fieldwork and analysis; new analytical technologies such as the electron microscope, DNA studies, and GIS can be used effectively to come to less speculative and more scientifically based conclusions. In this paper, we champion the use of GIS technology.

Attempts have been made to employ linguistic tools to trace the origin and dispersal of rice in Asia by human agents but have met with limited success because the ethnological dimensions of human migrations in southern China and Southeast Asia are uncertain up until the present. Now that it appears that rice was domesticated in two separate locations —globally speaking, eastern India on the one hand and southern China on the other—the words for 'rice' in these two regions appear to support this finding as well. In Western languages words for 'rice' (*ris, riz, rice, oruza,* and *arrazz*) all descend from Dravidian *arisi*, and point to auto-chthonous origination of the *indica* varieties in the Indian region.[4]

There are three Old Chinese[5] terms for 'rice': *dao4, tu2,* and *can4*.[6] Chinese civilization, however, arose in the North, around the area where millets (*he2*) were domesticated. While rice appeared in the Huang He river valley in northern China rather early, it remained a minor cereal there well into the historical period.[7] The Han Chinese word for 'rice' in the north is *dao, dau,* or *tao*. Written with the 'grain' radical, it designates husked grain of some sort, husked rice perhaps, before later shifting its meaning to 'rice plant'. The term *tu2* is seen as synonymous with *dao4* in Old Chinese; and *can4* designated 'hulled rice'. Sagart points out that a problem with the term *dao4* is its near-homophony with terms for millet. This leads us to suspect that the early Chinese in the North viewed rice, a non-native grain, as some kind of millet, and so we turn our gaze southward toward the Tai in the South and closer to the region of early domestication of rice there.

In proto-Tai the word for rice is *khau_{C1}*; in languages directly descended from proto-Tai or distantly related we find variants such as *gao, hao, ?au,* etc., the geographic distribution of which will be detailed later, after a discussion of the Yue below. Among the Austroasiatics living in close proximity to the Tai in the Yue region were the Vietnamese. Not surprisingly, one of the Vietnamese words for 'rice' is *gao_4*, 'raw rice' clearly cognate with Tai *khau_{C1}*. Other Vietnamese words for 'rice' include *lu2a* 'unhusked rice'—the plant and the grain or grains in general, and *cơm* 'cooked rice'.[8]

THE YUE

The Yue (also spelled Yueh), purported "ancestors" of the Tai, whose ethnolinguistic composition is uncertain, but whose origins appear to be south of the Yangtze, most likely participated in the earliest domestication of *japonica* in the second zone of independent origins of domesticated rice in southern China. The prehistoric, neolithic Yue and the historic Tai have many affinities and are considered by Chinese and Western scholars to be the ancestors of present-day Tai. A small portion of them may have been ancestors of present-day Vietnamese as well. The fact that the Tai and Vietnamese agricultural vocabulary shares many items in common[9] further illustrates a combined history of agricultural development and raises enduring questions as to what ethnic groups can legitimately lay unequivocal claim to the origination and spread of wet rice farming in the zone of origins of both rice and identifiable peoples in southern China. What the Bai Yue "One Hundred Yue" (a term later used by the Chinese to separate the Yue from the "Man" and "Yi"—'barbarian' and 'foreigner', respectively) had in common with the Tai was, among other characteristics, that they were inhabitants of valleys and lowlands, had many customs relating to water or rivers, and were cultivators of rice. They also lived in pile houses and tattooed their bodies. All of these characteristics are found in Tai societies. More convincing of a suggestive link is linguistic evidence from a Chinese phonetic transcription of a Yue song and other Yue words written by Lu Xiang in 77-6 BCE that demonstrates phonological, semantic and syntactic similarities between Yue and Tai, especially the Tai-Zhuang language of Guangxi.[10] Barlow[11] cautions that the Yue, widely dispersed over much of southern China for millennia, were probably linguistically diverse and spoke a number of languages, not all related necessarily to proto-Tai. Moreover, the term Yue often referred to a region as well as to a collection of peoples, and the mix of language groups had to include Austroasiatic speakers as well. In the final analysis, it is not unreasonable to claim that the somewhat mythical Yueh and the clearly more historic Tai of southern China contributed to the domestication, cultivation and spread of rice in Asia. In the remainder of this paper, we will detail the techno-cultural dimensions of the spread of the language propelled by what we believe was a sudden technological and cultural leap forward that has its echoes in the transition to rice farming in the Japanese islands not coincidentally about the same time as the appearance of Proto-Tai—some 2000 years ago.[12]

THE TAI

Like the search for the beginnings of rice agriculture, the quest for the origins of an ethnolinguistic group categorized as Tai has been an ongoing enterprise for generations. The question of Tai origins and the emergence of rice agriculture in Asia are related because the Tai are inextricably associated with irrigated rice farming. The distribution of Tai languages today does indeed invite the question of a Tai homeland. We[13] have shown elsewhere that the proto-language emerged somewhere in the border region of present-day southern China and northern Vietnam, south of the middle and lower Yangtze River region where *japonica* was indigenous and east of the India-Myanmar-Thailand zone of *indica* origins.

The early Tai, then, were positioned geographically to participate in the development and expansion of "wet," i.e. lowland, irrigated rice agriculture from either direction. Based on our linguistic research, the Tai expansion was from the east out of the Guangxi-Guizhou region in a southwesterly direction into today's upper mainland Southeast Asia. Their migratory pattern points back to the Yangtze agricultural zone as the impulse for their eventual emergence as a participant in early rice domestication and cultivation. Colin Renfrew[14] has made the claim that Indo-European languages spread peacefully into Europe from Asia Minor around 7000 BC with the advance of farming. A similar claim can be made for the Tai languages: they spread out of the Guangxi-Guizhuo region along with their techno-culture for growing rice about 2000 BP. The explanations of Jared Diamond[15] about societies like the Tai who expanded because of the geographical advantages and interactions with other groups, such as the Han to the north, whose own agricultural inventions, especially irrigation techniques, likely prompted the Tai to develop their own potential as successful rice farmers. The notions espoused by both Renfrew and Diamond are controversial, but they nevertheless provide a way of looking at the somewhat sudden emergence and expansion of the Tai into areas that were previously peopled primarily by Austroasiatic peoples.

The Tai have been closely identified historically with the emergence of irrigated rice technology and culture. In Tai terms—comparative, historical and technical—the system of gravitational irrigation that employs dikes or weirs to direct water from streams and rivers through a series of man-made ditches or channels is called *"Müang fai,"* $müang_{A1}$ meaning 'irrigation channel, ditch, canal' and fai_{A1} meaning 'dike, weir, dam'. Several very interesting articles have appeared recently dealing with the early development by the Tai in southern China of a truly revolutionary technological and societal complex in an environmental niche that lends itself to this unique type of irrigation engineering—*"Müang fai"*. The extent and importance of this techno-culture is described by Tanabe,[16] who has written the best and most complete comparative study of Tai rice-growing ecology and technology to date.

> *Müang-fai* irrigation is thus a skillfully designed system from water input at the *fai* through to distribution by terminal networks. The farming system which has been established within

the intermontane basins depends upon this tremendously well organized system of water utilization. The existence of such systems and associated customary regulations is also recorded outside Thailand especially in Sipsong Panna of Yunnan (China), the Shan States of Burma, the Tonkin hills of Vietnam, and in Laos. This implies that an irrigation system essentially similar to *müang fai* is widely distributed among Tai-speaking populations and hill dwellers in intermontane basins and mountain valleys throughout the northern part of mainland South East Asia.

O'Connor[17] has written a brilliant essay synthesizing what he believes was a regional phenomenon—the revolutionary development of a new "agro-cultural complex" (p. 973). One can infer from his new cultural history that the early Tai taught the neighboring Burmese to the west and the Vietnamese to the east a new way to produce dependable crops of rice by using a system of water channels to carry water into leveled fields into which rice seedlings are carefully transplanted. From a comparative-historical linguistics standpoint, it is interesting to note that the Tai word *müang*$_{A1}$,[18] in fact, has been borrowed into Vietnamese ('gutter, ditch, canal')[19] and Burmese ('ditch').[20] In both cases, it is not unreasonable to assume that both the term and the technology were borrowed in a single act. The term *khan*$_{A2}$ 'small dike between or surrounding rice fields to impound water' has been borrowed by the Chinese, further supporting the claim to Tai primacy.[21] According to O'Connor, a village-as-unit cooperative social structure that was needed to bring manpower to bear on the digging of water channels and weirs and their maintenance year in and year out led, in turn, to political advantage and the eventual hegemony of the Tai (and their Burmese and Vietnamese neighbors) in what was previously a predominantly Mon-Khmer region. O'Connor's study shows that, by contrast, the independent, household-as-unit gardening style of agriculture favored by the earlier Khmer agriculturists did not lead to dependable rice production or the political consequences of communal cooperation. The Khmer were at the mercy of annual flood waters during an unpredictable monsoon season; the Tai, by contrast, controlled the flow of water from mountain-fed streams into fields or *naa*$_{A2}$ surrounded by water-retaining bunds. The result over time was a steady, inexorable expansion of irrigated rice production and a well-fed, expanding population. The Tai have been depicted by historians as warrior types; in the final analysis, rice may have been their best weapon and the hunger of their adversary, where the term is appropriate, their best expansionist strategy.[22]

In his article, O'Connor advocates a return to a regional anthropology so that similarities can be discovered across larger areas and away from the more current tendency to focus on ethnography of a village and its inhabitants. Perhaps no discipline is more regional in its scope than comparative-historical linguistics, where geographic boundaries serve as linguistic isoglosses in space, and temporal layers reveal language family ties over time. O'Connor's comparisons and histories focus primarily on the Tai of Yunnan proximate to the lands of the Burmese and the Vietnamese and the historically antecedent Pyu, Mon, and Khmer.

A Linguistic Geography of Tai Rice Terms

A linguistic study of terms associated with irrigated rice technology serves to enlarge the picture of origins of the Tai and their participation in the development of domesticated rice in southern China and its spread in a southwesterly direction into the upper reaches of today's mainland Southeast Asia. The research objective that is proposed here is one of reconstructing the past of Tai rice growing technology and culture in southern China as remote as the proto-Tai period, *ca.* 2000 BP. In his discussion of the origins of the Tai and their ubiquitous cultivation of glutinous rice (*Oriza glutinosa*), Terweil[23] says, ". . . . the idea that they (the Tai) might have been the first to develop and cultivate this particular type of rice is worthy of further investigation." Such sentiments too motivate our attempts at historical reconstruction by using the fieldnotes and articles of linguists and ethnographers, such as Gedney, Terweil, Tanabe, Phya Anuman Rajadhon, Wu Wenyi, Zhang Gongjin and others.

The locus of irrigated rice culture is the paddy field or naa_{A2}. The construction of the paddy field, the technical aspects of water engineering that brings water into the field, and the social aspects of the engineering and the cultivation—construction and maintenance groups, the use of indigenous tools and body movements in manipulating plants, pests, soil, and water, in short, as detailed an ethnography as can possibly be reconstructed from the published information available is ultimately aspired to but will see only a first attempt in this paper.

The strong connection between early Tai water engineering capabilities—$m\ddot{u}ang_{A1} fai_{A1}$—and socio-political organization—$m\ddot{u}ang_{A4}$—is spelled out in the text of a Tai Dam origin myth: $pang_1 nan_6 pen_1 s\ddot{u}ang_4 nan_6,$ $yet_2 naa_4 baw_2 mii_4 m\ddot{u}ang; yet_2 m\ddot{u}ang_4 baw_2 mii_4 caw_3,$ which translates as 'In those times, it was such that if one wanted to make wet rice fields there were no irrigation channels; if one wanted to make a chiefdom, there were no chiefs.' One reading of this assertion in the myth is that the overlords of the early, smaller $miiang_{A4}$, polity-cum-environmental cells comprised of a configuration of villages in an intermontaine valley, mobilized and managed the manpower of a moderately large rice-growing area watered by a common source controlled by the system of $miiang_{A1}fai_{A1}$—water channels and weirs—constructed and maintained by these cooperative labor groups. It was this techno-culture that was the foundation of what O'Connor calls an "agro-cultural complex."

Both O'Connor and Lansing are disciples of Condominas, who "reminds us, agriculture is at once a social and a technical process."[24] In his writing on the Tai Dam, Condominas[25] himself points out that among the many titled functions of the *caw miiang* (chief or overlord of the polity), one is "his role in the organization of production (the importance of the irrigation system for the Tay, developing valleys and mountain hollows into rice fields is well known)." The Tay *caw miiang*, far from being an oriental despot, was the center of a "network of personal relationships" of the techno-cultural $miiang_{A1}fai_{A1}$ complex.[26]

Objectives and Methodology

The geographic center of this historical advance in agricultural technology has been placed by some scholars in Yunnan Province.[27] However, the linguistic evidence for these suppositions has not been convincingly presented. Our approach is to employ the methodology of comparative-historical linguistics and the technology of GIS to see if the older theory of Yunnan origins of irrigated rice growing by the Tai of Yunnan can be supported or if the data point to the Tai of a different region. A secondary purpose is to outline the semantic field of the rice-growing complex for historical-comparative utility and for intrinsic interest revealed in the details of the technology and culture.

In approaching the task of devising a persuasive manner and methodology of presentation, we have plotted terms used in the $miiang_{A1}fai_{A1}$ techno-culture on a linguistic map of the areas that are candidates for the historical origin of the irrigated rice revolution in southern China. That such language origins can be approximated by this means is enunciated by Edmondson[28] who echoes Bailey,[29] namely:

Dialectologists tell us that we should find a wave-like pattern of forms that show an origin where a change began and a periphery, cf. Bailey.[30] There can, of course, be perturbations in the wave and cases in which the spread does not propagate according to a strict geographic pattern, e.g. near-identical word forms may not necessarily be found contiguously. There may also be other waves of change from different origins that interfere. Nevertheless, one expects to be able to discern a pattern.[31]

We can view the direction of linguistic change and the spread of Tai rice agriculture in terms of proto-Tai by citing synchronic forms using the proto-Tai tonal categories and their relationship to proto-Tai initials following a chart developed by W.J. Gedney.[32] The space limitations of this article, however, do not allow us to demonstrate the methodology implied in its use as a comparative tool.

GIS Mapping of Tai Linguistic Geography of Southern China

The task of constructing a preliminary GIS map of Tai linguistic geography concerning irrigated rice agriculture begins with a small, initial set of lexical items that might define the $müang_{A1}fai_{A1}$ techno-cultural complex: rice, rice seedlings, irrigated rice field, irrigation channel, weir, and so forth, and plot their Tai forms across the Tai region of southern China, citing cognates from the three branches of the Tai family: Northern, Central and Southwestern (See Fig. 2.1 for a diagram). Following the principle of "directionality of language change" enunciated by Bailey and Edmondson, the patterns that might show the greatest change from an

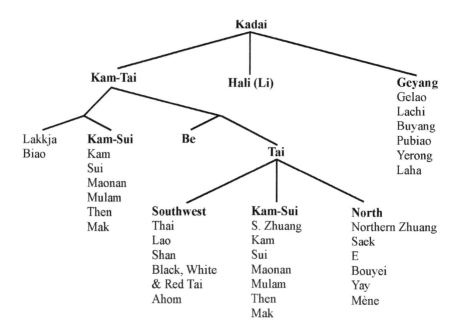

Fig. 2.1. Diagram of the Kadai languages showing the relationship between Kam-Tai and Tai.

earlier or proto-Tai form should generally be the point of origin, and forms closest to preserving the proto-Tai form at the periphery. There are numerous exceptions, of course.

Siamese (Thai of the Central Plains and one member of the Southwestern Tai branch of the Tai family) is a noteworthy example of a Tai language that preserves many of the older or proto-Tai features at the periphery. Conversely, languages or dialects of the Northern Tai branch—the reputed area of origin of proto-Tai, such as Po-Ai, Yay, and Saek, would expectedly reflect the greatest change, all things considered.

Table 2.1. The word "rice" spoken in different languages and location

Language Group	Language	"Rice"	Location spoken
	Siamese	κηααω3	Bangkok
	Isan	κηαω3	Roi Et
	S. Zhuang	κηαω3(	Lungming
	Lao	κηαω3	Vientiane
	Tai Don	κηαυ3	Lai Chau
Tai	Saek	⊗αω6((	Nakhon Phanom
	Tai Dam	ξαω3	Son La
	Tai Dehong	ξαυ3	Luxi
	Tai Lue	ξαυ3	Jinghong
	Yay	ηαω6	Lao Cai
	N. Zhuang	ηαυ4	Wuming
	Bouyei	ηαυ4	Wangmo
	Sui	/αυ4	Libo
Kam-Sui	S. Kam	θ↔υ4	Rongjiang
	Maonan	Ιυ4	Huangjiang
	Mulam	ηυ3	Luocheng

Notes: 1. The languages in the table above are sorted in the order of decreasing similarity to the proto-Tai word for "rice"- *khau$_{C3}$*. In other words, the descending order replicates the general locations of dialect pronunciations most similar to proto-Tai (preserved at the periphery) and point back to those dialects most changed over time and space (the region of Tai origins).[1]

2. See Fig. 2.1 of the Kadai language family tree and the Kam-Tai branch in particular.

---

1. We would like to thank Dr. Li Jinfang, Central University for Nationalities, Beijing, for assistance in creating this table.

As a first example of the series of changes, the direction of change from putative point of origin to the periphery, examine the linguistic geography of the word for 'rice' itself. The proto-Tai form has been reconstructed by Li[33] and others as $*khau_{C3}$, which is identical to the form in many Southwestern Tai languages, such as Yuan (Chiang Mai), Lue, and Tai Mau (Chinese Shan). Siamese differs from $*khau_{C3}$ only in vowel length: $khaau_{C3}$. Based on the principle of directionality, Buyei $?au_{42}$, which has undergone a loss of the initial kh- would appear to be at or near the point of origin; the Lue and Tai Dam forms $khau_{C3}$ are two examples of forms at the periphery, and they preserve the proto-form. The data are shown in Table 2.1 and illustrated in Fig. 2.1.[34]

Preliminary Findings—Comparative and Historical

The illustrative example of the changes of the pronunciation of the terms for 'rice' and handful of other lexical items used to plot a linguistic geography and spread of Tai $müang_{A1}fai_{A1}$ technology are only part of a preliminary attempt to see what conclusions can be drawn as a first approximation of the origins of the Tai and their participation in the spread of domesticated rice. We should be heedful of the remarks of Lass[35] and others about the tendency of historical linguists to use poor data to arrive at wished-for conclusions. In this paper, the data are reliable for the most part but incomplete. Further comparative research is needed. However, from this preliminary comparative study of the geographic spread of $müang_{A1}fai_{A1}$ technology, it appears safe to say that, like proto-Tai itself, the phenomenon had its origins to the east of the Southwestern Tai languages of Yunnan. Judging from the degree and direction of change, the origins are in the Guizhou-Guangxi area. At this stage in our investigation, it is not possible to tease apart the origins of $müang_{A1}fai_{A1}$ technology and the history of the domestication of rice from proto-Tai itself.

By the 12th century, the Tai were surely in Yunnan and most likely in northern Thailand. King Ram Khamhaeng of Sukhothai devised a system for writing Thai in 1283 AD, and in his (the earliest) inscription (No. 1) of 1293 AD there appears the oft-quoted phrase, nay_4 nam_6 mii_4 $plaa_1$; nay_4 naa_4 mii_4 $khaaw_3$ 'in the waters there are fish; in the (irrigated) rice fields there is rice', indicating that the practice of irrigated rice growing was established at that time and place in Tai history. In 1281 AD, under the rule of King Mengrai, a "trunk canal" for diverting water from the left bank of the Ping River was excavated for a length of 34 kilometers (20 miles).[36] Such a large-scale project would seem to indicate that the Tai had long since developed the technical refinements and political organization needed for their *müiang$_{A1}$fai$_{A1}$* techno-culture. The *Mangriasat*, a set of laws attributed to King Mengrai, detail regulations dealing with the management and maintenance of *fai$_{A1}$* and irrigation water usage, including fines levied for damages caused by boats or rafts passing through a space left for them in the weir.

CONCLUSIONS

From this first approximation, combining knowledge of historical events with comparative-historical linguistics, it would not seem unreasonable to offer a provisional two-part conclusion. First, *müiang$_{A1}$fai$_{A1}$* technology was relatively well developed by the proto-Tai period, that is, roughly 2000 years ago. Second, accepting the linguistic argument that proto-Tai origins can be traced back to the Guizhou-Guangxi area, Tai irrigation engineering of rice fields employing water channels and weirs originates there as well—not in Yunnan.[37] The use of GIS mapping to display the data used to arrive at this conclusion demonstrates that the competing theory of a Tai homeland in Nan Chao (Yunnan Province) is no longer convincing. Because Proto-Tai can safely be reconstructed back to 2000-3000 years BP, we cannot use the methods of comparative-historical linguistics to posit an earlier date for the rise of Tai rice agriculture without obtaining data for extra-Tai languages, such as the Kam-Tai languages that broke away from proto-Tai and even more remote Kadai forms. However, the direction of movement of the Tai out of their Guizhou-Guangxi homeland certainly indicates that the Tai at that point in time picked up some or all of the earlier techno-culture from a more northerly source and much earlier time (5000-7500 BC), namely, sites in south China near Lake Dongting and the middle reaches of the Yangzi River.[38]

In a final demonstration of the generally southwesterly movement of Tai irrigated rice technology, we have constructed a map (Fig. 2.1) based on a selected lexicon of 25 items (plants, animals, fish, simple technologies) associated with the Tai rice-centered culture. The map illustrates the two millennia dispersion of early Tai out of southern China and into Southeast Asia, moving from *müiang* to *müiang*, basin to basin, and eventually into larger river valleys, most notably the great Mekong and its tributaries. In an interesting discussion of the history of the early Tai and the antecedent Yue, the British historian, Chris Baker (n.d.), asks—and answers—a heuristically helpful question about "dispersion."

But what moved? We should not imagine "Tais" setting out from southern China around 0 BC, and "Tais" finishing up many generations later in Assam, the Shan States, the Chaophraya and Mekhong valleys, and the hills stretching between the Upper Yangzi and Red Rivers. What moved, first of all was the language. What moved also were some speakers of the language. What went with them were some technology, social habits, and ideas. But in the process of dispersion was not a simple population migration and cultural diffusion.

In this paper, we have attempted to show in some detail the complex process of the diffusion of the Tais that Baker talks about and the contribution of the Tais, a hybridized people to be sure, to the spread of hybridized rice agriculture in Asia. The migration of the Tai and concomitant spread of their techno-cultural irrigated rice ethos was not a linear process. Indeed, Saek, one of the languages of the Northern Tai branch, which is centered in the far northeastern corner of Guizhou province in China, was discovered in Nakorn Phanom province in northeastern Thailand by William Gedney. How this Northern Tai language figuratively leap-frogged such a great distance into the region where Southwestern Tai languages predominate is a mystery. As in many migrations, peoples pick up and move great distances and small, often driven by vagaries of the weather, famine, disease, war, population pressures, political unrest, and simple wanderlust. Many Tai villages in Northwest Vietnam, Laos, Burma, and Yunnan were forcibly moved in the First Reign of the Thai Chakri Dynasty, creating a new patchwork of Thai languages and cultures in the sparsely populated Central Plains and the new royal capital of Bangkok.[39] In most instances Tai farming populations would naturally have carried crop seeds, most notably rice, with them to plant in their new settlements. In times of famine too, areas of surplus would seize the opportunity to transport rice great distances, even in small quantities by mule train, to sell or barter to communities in need. In all of this, there was a certain inevitability that the Tai, or some agricultural group like them, would spread the cultivation of irrigated rice of one variety or another. Rice was indigenous to the north of their homeland as was the water buffalo, the requisite draft animal in Jared Diamond's equation for expanding farming societies. Their irrigation technology was simple enough: channels to direct gravity-fed streams from hills and mountains and contain and direct the flow with dikes improvised year in and year out by teams of village men from readily available bamboo, trees, reeds and nearby stone and sand. There was no need to wait for "Chinese civilization" to move in from the North to give the Yue and their Tai descendents instruction in irrigated rice cultivation. Similarly, no help was required to instruct women and children in selecting the best seeds from wild varieties for domestication. Tai participation in the early dispersal of cultivated rice in Asia was not inconsequential.

References

1. Smith (1995:123).
2. Londo *et al.* (2006).
3. Sharma, Tripathy and Biswal (2000).
4. Panikar and Gowda (1976).
5. The dates for Old Chinese vary from the 11th century BCE—771 to a later estimate of roughly the 1st millennium BCE. For a discussion, see Sagart (1999:5).
6. Sagart (1999:181).
7. Sagart (1999:176).
8. Nguyen, Dinh-Hoa (1966).
9. Hartmann (1998).
10. Leshan Tan (1994).
11. Barlow (1987).
12. Curry (2008).
13. Luo and Hartmann (2000).
14. Colin Renfrew (1987).
15. Jared Diamond (1997).
16. Tanabe (1994:27).
17. O'Connor (1996).
18. "A1" indicates the proto-Tai tone categories used by Li Fang-Kuei (1977:28) and others. In the body of the paper, IPA transcription has not been used in an effort to make the narrative more accessible to the general reader. GIS tables and maps, however, are constructed with IPA notation.
19. Nguyen Din-Hoa (1996:284).
20. Judson (1955:243).
21. Luo Meizen, personal communication.
22. In his discussion of serfs of non-Tai origin, Condominas (1990:64-65) comments on the observation of Henri Roux and Tran Van Chu (1927:195): "the prohibition, on religious grounds, against the Sa' making use for their own benefit of an advanced technique, in this case the plough, which as *pua* (serfs) they use in the rice fields of the Tai. It is true the non-Tay serfs had no rights to the communal rice-fields." Our interpretation of this situation, as reported by Condominas, is that the Tai (Dam) controlled the technology, production and surplus distribution of rice among non-Tai people under their control to a large extent.
23. Terweil (1981:2).
24. Lansing (1991:49).
25. Condominas (1990:55).

26. Condominas (1990) indicates what the constraints on his power were: ".... the house of the *tao* .... is simply the grandest and richest house in the village, but remains open to all, and Father Bourlet has well described the comings and going of people and the ease of visitors, who feel themselves at home in a house they themselves have built. The *tao*, despite his divine origins, has powers limited by the resources of a narrow social space; these limited dimensions still allow him to control his subjects under customary law. The villagers may leave him for another lord, or even depose him, as happened even in the *müong* which were principalities (at the instigation, it is true, of the aristocracy). (p. 86).
27. Watabe (1978:120).
28. Edmondson (1989:10).
29. Bailey (1973).
30. Bailey (1973).
31. Bailey (1973:36) says, ".... the patterns of a language are the cumulative result of natural, unidirectional changes, which begin variably and spread across the social barriers of age, sex, class, space, and the like in waves."
32. W.J. Gedney (1972).
33. Li (1997:344).
34. The data for administrative boundaries and most of the place locations come from Center for International Earth Science Information Network (CIESIN, http://www.ciesin.org/). Some place location data come from the GEOnet Names Server at National Imagery and Mapping Agency (NIMA, http://164.214.2.59/gns/html/).
35. Lass (1996).
36. Watabe, Tadao (1978:21).
37. For several generations, the notion that the proto-Tai homeland was the old Nan Zhao kingdom in Yunnan persisted. The American historian Frederick Mote was among the first scholars to refute the theory. The history of Nan Zhao has been detailed by David Wyatt (2002:10-13), who notes that in the *Man Shu* text, "the lists of Nan Zhao words are identifiable as Lolo [Tibeto-Burman] and untraceable as Tai."
38. In particular, the site of Pengtoushan dating 5000-7500 BC (Pearson and Underhill 1987:810; Yan Wenming 1991:120, Fig. 1) and around Hemudu to the south of the mouth of the Yangzi during the 6th millennium BC (Liu J., 1985).
39. Hartmann (2004).

3 | Rice in Social and Cultural Life of People

S.C. Ahuja and Uma Ahuja

It is little wonder that rice is revered next only to God. Through the ages, it has evolved to become an indispensable moiety of religious rites, rituals and festivals, ceremonies, *rites de passage (sanskara)*,[1] festivals,[2] vows (*vratas*) and offerings made to God. No other grain is so closely woven into social, religious and cultural fabric and can ever hope to achieve the inimitable relationship achieved by rice with humankind. Even in areas where rice is not mainstay of people, it is rice which is used in ritualistic observances.[3]

The great Indian sage *Parashara* has stressed precisely and wisely the essentiality of rice. In ancient Sanskrit text *Krishi Parashara*, he sang in its praise "Rice is vitality, rice is vigor too and rice is indeed the means of fulfillment of all ends of life. Gods, demons and human beings all subsist on rice."[4] Rice in Asia is not only considered as basic sustenance of body but also soul of mankind incorporating male and female creative forces as earth and water both unite to produce rice.

A strong rice culture holds over a vast part of Asia from India, Sri Lanka, Bangladesh, South East Asia, China, Korea and Japan. Rice in these parts of the world was not a mere cereal; it was the root of their civilizations. The ancient civilizations recognized rice cultivation as the basis of social order with strong moral and religious overtones. Rice not only nourished the world's oldest civilizations but also shaped them in ways which guided them for the hard work needed in its cultivation.

Names after Rice

Pervasiveness of rice is apparent in the nomenclature of countries, places, people, clans and some of the ceremonies.[5] The names of important people such as *Shuddhodana* (pure rice), father of Lord Buddha owe derivation to Sanskrit word *odana* that means rice. Similarly names of some important states and countries are based on rice i.e. Java (island of rice), *Denjong*/Sikkim (valley of rice) and *Mizomonokumme*/Japan means 'land of luxurious rice'. Surnames of some powerful Japanese owe their derivation to rice. Examples are: Toyoto (bountiful rice field), Honda (main rice field), Nakasone (middle rice field), Sawada (rice field), Tanaka (middle rice field) and Narita (developing rice field).

In India, *Nellore* means town of rice. Even names of many rites/ceremonies owe their names to this tiny grain. Sending paddy and betel/turmeric to brides' house by groom's family becomes *dhanpan*, the feast at the time of marriage offered by bride's father to bridegroom and party is *bhat* (literal meaning boiled rice) and ceremony of gifts performed by maternal uncles is called *bhat bharna*. The dance performed by groom's mother after departure of marriage party, performed with other women is termed *Tilchauli* (*til* = sesame; *chauli* = rice).[6]

The image of golden ears of rice stretching across the lord's domain expressed the power of a feudal lord in Japan and emblem of plenty as one of the twelve ornaments embroidered on vestment of state in imperial China. Rice is much more than just a food in Bali. It is an inseparable part of daily life as well as a part of the island's history and religion. It is a part of the culture, a part of the mythology and a part of the pride of individuals and of the race. It could be said that without rice there could not be a Bali or Balinese.

Innumerable myths, ceremonies, rituals, fairs and festivals, folklore, taboos, make penning these beyond the page restriction of this chapter. The authors humbly present some of these sans any claim of being exhaustive.

Myths Associated with Origin of Rice

A number of myths are prevalent on topics as diverse as origin of rice, its cultivation, domestication and introduction into specific areas. There are myths regarding grains being large like melons, grains coming to granary on their own, how the husk appeared on rice and development of different grain colors.[7] Tales, legends or myths associated with origin of rice in all the Asian countries can be grouped into three types: (a) Mother earth (b) Divine/heavenly origin, and (c) Origin in swamps, brought to human by various agents.

Mother Earth

Such myths are prevalent in India, Myanmar, Thailand, Indonesia and Malagasy. According to the Indian myth, when all food disappeared from earth, people appealed to Lord Brahma for help. He consulted Lord Vishnu and decided to awaken Lord Shiva who invited Goddess Annapurna to earth and begged for rice.[8] In north India, female deity associated with rice is Annapurna or Parvati, Lord Shiva's wife while in south India, it is rice Goddess Ponniamma, one of the seven virgin sisters i.e. Durga or incarnation of mother earth. Durga is considered to be an incarnation of Parvati.

Tribals in India too believe in the origin of rice from earth but in a way of their own. Konds of Kalahandi (Orissa) believe that in primitive days they used to eat *siali* seeds and lived in forest like monkeys. They lived promiscuous life. Mother earth, Nirantali, objected to it. The Kond chief replied that they had no food to serve as feast on marriage and without feast no relationship was possible. Mother Nirantali gave them rice, rice beer, pig, cocks and instructed the chief to marry only when the crops would ripen. Henceforth they started living civilized life.[9]

In Southeast Asia various myths related to either mother earth or the Rice Goddess are prevalent in different areas. Each myth relates a story of incest, rape, immolation, incarceration or sufferings endured by mother earth in delivering rice. Having originated directly from the body of the goddess, rice is considered sacred.

Folktales of east Java relate that Dewi Sri was raped and murdered by strangers while tale from Bali tells by Batar Guru. However, in Central Java, there is a different myth which goes as follows: 'Once there was a king, an incarnation of Batar Guru with super natural powers, he created a beautiful woman, named Retno Dumilah. One day, king getting impatient for her love, embraced her strongly and she died of the shock. From the place where Retno Dumilah was buried, there grew a rice plant which would produce foodstuffs forever.' This story has been handed down the Javanese society until now and, to celebrate her merit, people regularly perform thanksgiving ceremonies.[10] Hindu Balinese (Indonesia) believe that people in olden times had sugarcane to eat. Then Lord Vishnu raped and caused the mother earth to give birth to rice and directed the rain God Indra to teach people to grow it. Afterwards people started eating this tasty food. In South India, people believe that Mahavishnu married the earth and mother earth gave birth to rice. God gave rice to inhabitants of earth when humans prayed the god for food for improving their strength and fertility.[11]

In Thailand, people believe that rice came from the corpse of Rice Goddess Mae Kusok killed by gaze of an ascetic Lap Ta.[12] Similarly, in Malaysia, according to the Dusun, the primordial male deity buried the primordial female one to make foodstuffs for humans to eat. From her grave came rice, taro, sweet potatoes and onions as well as buffaloes, cattle and pigs.[13]

Divine Origin

Various legends relating to origin of rice recognize that rice came from heaven but differ in the way it was carried down to earth. Buddhist, Hindu, Shinto, Tao and Muslim,[14] all people believe in divine origin of rice.

The old Indian text *Aitareya Brahmana* traces the origin of rice to God Indra or rain god, a male deity. From Indra's bone, sweet drink flowed and became *nyagrodha* tree and from his marrow flowed 'soma' juice which became rice. Indra is the deity in *marutham* or wet agricultural lands.[15]

The *Kachins* of North Myanmar believe that they were sent forth from the centre of the earth along with the seeds of rice and were directed to a wondrous country where every thing was perfect and where rice grew well.[16]

In East Java, people believe that a pipit flying in the sky carried seed from heaven. The bird was carrying the seed sent by King Suroloyo for Princess Sri Sedono. On the way, the bird dropped the seed which grew vigorously and became a rice plant. She took care of the plant to be named *padi* until it produced grains. She then spread the grains over East Java.[17] The Balinese believe that Brahma (Sanghyang Keshum Kidul in Bali) gave rice to man, sending four doves with rice of three colours as well as turmeric. According to *Catur Bhumi* mythology, the God *Bathara Ciwa (Shiva)* sent a bird to earth carrying four seeds of rice with different colors: yellow, black, red and white. On the way, the bird ate yellow seed so that only three were left. Until now three different types of rice, namely black (glutinous), red and white rice have been the main foodstuffs of Balinese.[18]

In the Far East, Japanese people believe that the heavenly swan while flying donated few rice panicles to the Sun Goddess, Amaterasu Omi Kainc and to mythological ancestor of Japanese imperial family. The Sun Goddess sowed the grains which sprouted and gave good yields. The emperor sowed rice seeds in the fields of heaven and celebrated with a feast and by giving his grandson sprouts of rice from the sacred garden for him to plant.[19]

Hani people of Yunnan (China) believe that a young hero, Mamai, ascended to heaven on a winged colt to ask the Heaven God for cereal grains. During the nuptial ceremony of Mamai and Lady Rice Ear, the younger daughter of the Heaven God, the hero escaped alone from heaven. On the colt he had hoarded rice in its stomach. The furious bride pursued him and cut off a wing of the colt with her sword. The hero and his colt fell and the grains burst from the belly of colt. In this way, rice came into the land of Hani.[20]

Malagasy people believe that rice belonged to God who considered it as one of the symbols of his superiority and sovereignty. Then men stole it to earth with connivance of God's daughter. She came down to earth with her husband, bringing rice in her hairy clothes. God was displeased so he sent cardinal birds and small parrots down to earth to devour all rice grains grown by his son-in-law. Only one story collected in the Malagasy relates that rice originated on earth with the banana but God forbade men to use it and ordered the serpent, which at that time was able to stand up just like men, to guard it. However, the serpent betrayed God and presented this forbidden food to some men. This incurred the wrath of God who then cursed both the serpent and men. From that time onwards, the serpent became a reptile without paws and men became excessively fond of rice.[21]

Dayak in Malaysia believe that once upon a time people ate nothing but fungus, which grew on tree trunks, together with roots, fruits and pith. A fellow, Se Juru got up among the branches to gather some fruits of *sibau* tree; he continued there for a long time, climbing higher and higher. At length, he came upon a new beautiful place in the sky: the land of Pleiades. There Se Kera instructed him all the mysteries of farming, art to cut down and burn forest, reap and store rice. After giving Se Juru three kinds of rice, Se Kera let him down to earth by a long rope, close to his own village. Since that time, the Dayaks farm according to the instruction of Se Kera.[22]

The *Agganna Sutta*, the 27th of the longer dialogues of the Pali version of the Buddhist cannon, deals specifically and at length with the origin of human society. A story of rice is as fallows: First, there was the world of water and darkness when the earth formed itself and made *ghee* (butter oil) and honey. Some beings tasted it, became greedy and broke off pieces, which became sun and moon and the stars. After all the savories on earth were used up, mushrooms began to grow, were tasted and eaten. Then came the time of the creeping plants and when these had all been consumed, finally rice appeared. At each of these stages of nourishment, body became more solid.[23]

Origin in Swamps

In some parts of China, Vietnam and Borneo, it is intriguing to hear about the mythical origin of rice growing in swamps. Outside this zone, a dog or a rat is believed to have brought rice from heaven (the Lolo in Yunnan, the Muong in northern Vietnam and the Ngaju Dayak in southern Borneo). These myths probably indicate domestication of rice. The animals include not only dogs and rats but also ants (East Asia, Southern Asia and Melanesia).[24]

People of a mountain tribe Pa Theng, in Vietnam, believe that the dog, the cat and the pig stole rice seeds from sky for them; so after the harvest they now give the first bowl of new rice to these animals.[25]

The Rengma Nagas version in Nagaland tells: Finding rice growing in the pond in the primordial era, men sent a rat to fetch it and the rats have ravaged granaries ever since. The Chinese version in the standard form runs as follows: After a deluge, Heaven God sent animals to bring rice to humans and only the dog succeeded in doing so. Swimming across the water, the dog lost the grains except those stuck to his tail. Ever since then, rice bears ears only on the top of its stalk. A number of versions of this story are distributed over Sichuan, Hubei, Guangdong and Jiangsu while some variants from Guangdong, have a rat instead of dog as a hero.[26]

However, in the Philippines no animal is associated with domestication. According to the myth, a young girl was sitting dangling her feet in a mountain spring. She saw a golden sheaf flowing down the stream. It was laden with golden grain. She buried the grains in mud near the stream. The grains sprouted and grew. After a few years, there were enough grains to fill their hut. His father was able to pay his debt and became a rich man.[27]

In India, Gonds of Madhya Pradesh and Kabris of Assam remember people associated with introduction of rice in particular areas with reverence. A statue of a woman thought to have introduced rice can be seen in village temple in Arunachal Pradesh.[28]

The Vietnamese myths and legends written in 13 to 14th century include 11 legends related to Huang kings' period. A myth associates the origin of rice with that of Vietnamese people, tells that Lac Long (dragon) and Au Co (fairy) fell in love and married. Some time later Au Co delivered one hundred boys. Au Co settled with her fifty sons and taught them to grow sticky rice, sugarcane, mulberry trees. After the harvest, she taught them to make rice cakes, dig canals and other useful skills.[29]

Some other legends are prevalent among Akhas and Kachins in Myanmar and Ao Nagas in Nagaland (India). These legends tell of a widow obtaining the first rice seeds and getting instructions regarding the appropriate rituals for growing rice. At first rice grew continuously from the stubbles that remained in the field after harvesting. However, on one occasion, the widow whistled and clapped her hands and, from that time onwards, rice no longer grew again naturally.

Goddesses/Rice Mother

Most important, perhaps, is the widespread belief in rice-soul, a living principle that makes rice grow. In most rice-growing countries of Asia, spirit is believed to reside in the Rice Mother or the Rice Goddess. The Rice Goddess is known as Mae Posop or Mae Kusok in Thailand, Dewi Sri and Mbok Sri in Bali and Indonesia, Inari in Japan, Hmall Hngo among Mon Khmers, Mae Khau among Thai tribe in Vietnam, Semangat Padi in Malaysia and Kelah in Myanmar. In Malaysia, the rice spirit is personified as a little girl and referred by affectionate nicknames such as *anak sambilan bulan* (nine-month child) or as *anak maharaja cahaya* (sunray prince) *si dang rupa sari* (flower princess), *si dang sari tonghat* (tonghat flower princess), *si dang muri* (white princess), *si dang gemula* (crystal princess), *si dang yas* (beautiful princess), or *sii dangomala*, (crystal prince). In Manipur, rice goddess is *Phaoibi Warol* and worshipped whereas Garos of Meghalaya offer fowl as sacrifice to *rohimi*, the mother of rice.

In Japan, *Inari*, the god of rice, is a major Shinto deity, depicted as either male or female form and believed to protect rice. The rice grain embodies the Japanese deity and represents the peaceful deity. Most deities have dual qualities and powers. The Deity of the Rice Paddy has only the *nigimitama* or peaceful soul and no *ramitaama* (violent spirit).

In some South Asian countries, Rice Goddess is represented by a special sacred rice variety, specific for each rice family. This sacred rice is 'abode and sanctorum of soul' or spirit of rice plant. Cutting of sacred rice is not handled by men, never sold and cut by finger knife. At harvest, special stalks are taken home with respect, believed to contain spirit of rice goddess and for their rebirth in next farming season. A statue of rice goddess made of straw, bamboo, or some other materials is installed in fields.[30] In East India it is represented by an early maturing variety which is offered to Goddess *Lakshmi* at harvest. In South India, a rice variety 'ponni' represents *Lakshmi* whose idol is placed in paddy fields.[31] In East

Java, Mob Sri is represented by sacred rice strain while Sedna, her husband by black glutinous rice. Planting is done by men in a set sequence. Sedna is planted along short ends of fields to protect Mob Sri. In Sumatra, farmers create a 'navel' at the centre of the field where special plants make refuge for rice spirit. In Indonesia, farmers erect miniature granaries and conduct planting ritual at the centre of the field where felled forest trees are buried.[32]

In Vietnam, a married woman preferably one with many children, takes the special unhusked rice from the attic of her house to the field. These are never pounded as they are supposed to contain the essence of rice mother and sown in the sacred field adjacent to a sacred bamboo structure.[33]

Just as mothers, rice goddess is considered to be the protector of rice and hence is treated in a similar way: respectful and protective. The rice soul is highly venerated. Asian people, since ancient times, believe that rice grain has a soul like humans and must be nurtured with rituals to procure a bountiful harvest. Often seed is saved with special rituals to keep the spirit alive from year to year. Farmers in particular will hold various rituals that demonstrate their deep respect and gratitude to rice goddess at varying stages of growth throughout the growing season. Almost similar rituals are performed at six crop stages in Orissa (India) and South East Asian countries. Ritual observations for the spirits of the rice plant include the obligatory taboos as well as the most sacred rites practiced including ceremonial food offering and community feasts.

The majority of rituals conducted are for rice spirit/goddess/mother. Elaborate rituals are conducted in Thailand, Malaysia, Indonesia (Bali, Java and Sumatra) and India (Orissa, Bengal, Manipur, Mizoram and Assam). Rituals conducted in Java and Thailand for rice mother is most physically evident in the shape of field shrines in Java that are created and on which offerings to Sri Dewi are placed. Java farmers perform rituals on all the major stages of rice growth. Food offering ceremonies are conducted at all stages of growth.[34] Such ceremonies are performed in almost all Asian rice-growing areas (Table 3.1). When grains start forming, almost similar rituals are observed in Bali, Java, Thailand, Malaysia and Orissa (India). In central Thailand when rice grains begin to swell in the fields, the rice plants are called pregnant and a pregnancy ritual is held for Mae Phosop. Selected plants are dressed and offered things that are liked by pregnant women. At the conclusion, a bamboo marker is set at the boundary of the field, to warn people not to disturb the pregnant Goddess.[35]

Table 3.1. Rituals conducted at various stages of the rice plant in various Asian countries

Stage	Thailand	Indonesia	Malaysia	Java	India
Planting seeds				nyerbar wiji	
First ploughing	Rage Thai	Anak padi	wiwit ngluku	wiwit ngluku	Akshayamuthi
Planting	Raeg Dam Raeg	Dara biak	burbur bumi	burbur bumi	Batauli
Young seedling	Bun Khaaw	Indu Dara	buntoni	buntoni	Harihari
Pregnancy	Khaaw nooj	ngandong	Bunting kecil	kelemon	Garbha Sakranti
Grain formation	Hoo Khaaw		Bunting besar	Slametan metik	Saptapuri amavasya
Harvest	Raog kiaw	nanchang padi	jumput semangat		Dhanu Sankranti
Thanksgiving				sedekah bumi	Nabanna

Other Goddesses Related to Rice Cultivation

In addition to rice mother, other goddesses are also associated with rice. In India, Lakshmi is the Goddess of all crops as in pre-industrial period grain was the symbol of wealth and prosperity. *Annapurna* is pan-Indian deity for rice whose shrines are located in Varanasi in North and Nellore in South India. In addition to this, Lord Indra or rain god is also associated with rice crop.[36]

In Bali another goddess Dew Danu or irrigation, is worshiped. In Vietnam and China, gods of cloud, rain, thunder and lightning are associated with rice.

In the Shinto tradition of Japan, the mountain spirit is invited to irrigated fields and worshipped as the rice-field spirit (*ta no kami*). There she presides until the harvest when another ritual is held to release her back to the mountain. In western Japan, Ta-No-Kami is the God of agriculture, farming and paddies and is depicted as a wealthy rotund figure seated on two bales of rice with a sack of jewels over his shoulder and holding a hammer in his right hand.[37]

Rice Rituals and Rice Festivals

In most parts of South, Southeast and East Asian countries, the principal subsistence crop is rice. The people's lives depend upon their rice harvests and it is the central foodstuff of every meal. Not surprisingly, therefore, every major stage in the traditional rice cycle has become focus of the ritual attention. In this way, they endeavor to assure supernatural protection for their principal food as well as to buttress their own hard labor. Moreover, given the great significance of rice among these people, it is hardly surprising to find that, in traditional animistic worldviews, they have endowed their principal foodstuff with soul force, just as they have done to other major constituents of the environment: human beings, animals, forests, rivers, etc. Only rice is considered to possess spirit comparable to one possessed by humans. Rice becomes pregnant, gives birth and dies. Therefore, the fertility of the crop is allied with the fertility of females and rice is seen as female in gender.

The rituals and rites may differ in their external (physical) approach but are performed with the same aim of propitiating God, village deity, (*Isht devata, Thakur deo*) and rice spirit, Rice Mother or Rice Goddess. The agricultural rituals, unlike the personal rites, are invariably integrated with festivals as they are observed in community. Each festival has two components, the ritual and the festival, both going on together. While ritual consists of a series of conventional rites, the festival constitutes the joyous celebrations by the families, relatives and community.

Rice rituals are not restricted to a country, a land, tribe or community or type of rice cultivation. These are conducted with equal reverence, respect and honor in all rice growing Asian countries.

Rice cultivation is a hard work carried out at right time for sowing, transplanting, weeding, harvesting, threshing and processing. To organize the labor and to follow strict timetable in order to timely finish the work, people became accustomed to work in groups, leading to celebrations in form of festivals and turning the hard job of rice production into joyous affairs. Majority of Asian festivals are tied to agricultural cycles and in many countries, months are named according to activity in rice fields. In India, in different states, New Year starts with either sowing or harvesting of rice.

Japanese festivals of rice, *matsuri*, chiefly of sacred origin, are related to the cultivation and the spiritual well-being of local communities. Many such festivals feature a parade of *mikoshiw* (portable shrines) and contests or games that give opportunities for community members to play together and match skills.[38]

Pre-sowing Rituals

A number of pre-sowing rituals, ceremonies and festivals are celebrated in India, Japan and many other Southeast Asian countries to ward off evil spirits. Some have become of major importance or even of national importance.

The importance of sowing is reflected in the Japanese custom wherein the Emperor himself initiates the sowing of rice crop. In Thailand and Cambodia, 'royal plowing ceremony' is a major public ceremony marking the beginning of rice planting season. In olden times, the king used to perform it but nowadays a minister performs the ceremony. Local spirits are invoked. People take a few blessed seeds from the ground and mix it with their own seed for good luck. Once the main ceremony is over, local spirits and gods are invoked at the village level to start the new planting season.[39] This is reminiscent of *Akshaya Tritiya* ritual in Orissa, Assam and Bengal.

Pre-sowing festivals in India include *Akshyamuthi* or *Akshaya Tritiya* of Orissa, *Eruvaku Purnima* of Andhra Pradesh, *Damurai* of Chhota Nagpur, *Chapchar Kut* of Mizoram and *Mopin* of Arunachal Pradesh[40] (Table 3.2). In most parts of India, agricultural year is welcome on Akshaya Tritiya—

akshaya signifies that the effort (of sowing) will not go waste and *tritiya* means the third day of lunar month corresponding to the month of April-May in the Gregorian calendar. Every household in Orissa celebrates this festival by ceremonial sowing of paddy in the field. Oblation is offered to the goddess of Destiny, Shasthi while the tribal farmers observe *Dhanabuna Parva* (the festival for sowing paddy) to propitiate the village and household deities for reaping a good harvest. Various tribes celebrate the festival with different names. All tribes worship mother earth, offer sacrifice before performing the ritual of *akshyamuthi*.[41] Similar rite of blessing the seeds is performed in Malaysia on an auspicious day.[42] It is interesting that animal worships are done before sowing and after completing the harvest as part of thanksgiving in India, Japan, Bali, Thailand, Java and Philippines.

In Manipur, Meiteis start plowing on a sacred day for cultivation, *Phairel Panchami* (about January) after worshipping rice goddess *Phaoibi Warol*. The Nagas perform the *Rialongcchi* ceremony in which a black fowl is scarified and some seeds of rice and other crops are anointed with the blood.[43, 44]

In Yunnan province of South China, before the start of plowing, married women prepare liquor, meat, cooked rice and accompany their husbands to mountain lands to perform rituals. Standing on the side of the fields, the women call out the rice-souls, asking them to come back and guard the land:[45]

> Wo, Wo, Wo!
> Rice souls come back!
> Come quickly to guard the mountain fields.

In Japan, a country with strong traditions of rice planting, a rite imitating the preparation of a rice field with a hoe is performed everywhere, a rite which is meant to chase bad spirits away and bring luck for the year. In Oki Island, West Japan, on the rice-seed sowing day, offerings are made in the nursery and prayers are said for a good harvest.

Rice Sowing Ritual

Sowing is the most important step in raising any crop and was considered a risky affair. Therefore, to do proper sowing so as to receive good results is considered to be a job of the greatest responsibility and is generally done by the owner of the land himself or by the learned priest. There are more rites and rituals for sowing than for other operations. This may be a reflection of the feeling of unknown future combined with expected hard work ahead.

Sowing in some tribes is an open affair while it is a secret affair for others. In tribal areas of India, it is accompanied by sacrifices. Sowing being the premier step for rice cultivation, many pre-sowing rites, rituals, fairs or festivals are celebrated to ward off evil spirits and to invoke the blessings of the ancestors.[46] Where pre-sowing is done at community level, seed is blessed or purified and small parts of it are taken by individual farmers to be mixed with their seed. Pre-sowing community functions are held by tribals of Orissa, Bihar and Assam (India), Thailand and Malaysia. In Assam the sowing festivals of Bohag Bihu corresponds with New Year and is held in pomp and gaiety. The rituals last for three days.

Germination and Growth of Crop

After sowing, prayers and magical rites are performed for proper germination and growth of the crop. Garos of Meghalaya (India) offer fish tail on start of seed germination. In Andhra Pradesh, village deity Akipen is worshipped and a fowl and a goat are sacrificed. Mundas in Orissa celebrate *Hero Banga* in June for propitiation of village deity, Desauli, abiding in a sacred groove. In Borneo, a dance feast is given just after sowing, masks are worn to frighten away both evil spirits and rodents.

In Orissa, the people of Oraon tribe celebrate *Harihari* in June-July when paddy seeds have just germinated and new shoots have appeared. They celebrate marriage of paddy seedlings with a lot of ceremonies, to shed fear that paddy seedlings will either rot or fail to germinate.[47]

Transplanting Festivals

Transplanting is monotonous and tiresome but in most places it often is a celebration when people work in groups.

In Bengal, Assam and Orissa, tilling in rice fields for transplanting is done after worshipping Mother Earth. In June-July when first rainfall occurs, Earth is said to be in menses. Touching of earth is avoided so special sandals made of banana bark are used. Plowing of fields is strictly forbidden for three days during these days. People eat precooked special rice cakes. When pollution ceases, people start plowing after worship. *Raja Parba* fair in Orissa and *Ambuvaci* in Bengal and Assam are held. In Java, the ceremonial purification of rice fields with wooden planks is held with rituals accompanied by procession and invocation.[48]

In Yunnan, South China special propitiatory rites are performed at village shrines through prayers and sacrificial offerings to village spirits. In Lancang, North China before transplanting people hang a bamboo basket from house pole on the day of dragon with offerings and the ritual specialist chants prayers for protection of rice crop. Village gods, spirits and water spirits are invited to reside in the plot and rest of the work is completed.[49]

Rice planting in Japan was always considered the most important event in the agricultural calendar both by Shinto and Buddhist followers. Community transplanting rituals are organized in shrines and temples. Different shrines have their distinct ways of performing the ritual. At Mibu, the planting of the first rice of the year, Flower Rice Planting ritual was ritualized into the Shinto ceremonies and is still seen in rural communities throughout Japan to this day.[50] Another leading shrine, Sumiyoshi Taisha boasting its old history and vast size preserves a deity rice field at the corner of its premises to hold the annual rite for the planting of seedlings.[51]

At the Murone Shrine, Iwate Prefecture, Japan, Planting Ritual *nanagusa gomori* is performed. People retire into the shrine, make miniature hoes of chestnut or oak wood and beat with them a dancing rhythm on the veranda of the shrine and sing rice-planting songs.[52] In Tsushima Shrine, Japan, the planting of brown rice on the shrine-field occupies a central position among all festivals. No night-soil is used there as fertilizer. After the planting a sacred rope (*shimenawa*) is hung around the field. Once every month the Buddhist shrine minister comes and says purification prayers over the field. Ritually impure persons must not come near the sacred field. Rice grown on this field is used as seed rice in the coming year.[53]

In India, like sowing, transplanting in some areas is started by priest while at some places by the owner. In Orissa, transplanting is started by priest and carried out by women. Transplanting is a women's job in Orissa. Mundas celebrate Batauli as sacrificial feast to mark the beginning of transplanting of paddy in June. Rice beer is offered on the occasion. In Coorg (Karnataka State) transplanting is done by men who stand in knee deep muddy water and a competition is held for transplanting seedlings. Landlord offers a feast called *Keilmurtha*. Agricultural implements are cleaned and homage is paid to them.[54]

Fertility Ritual

Fertility rituals are performed at different times at different places. It is conducted at some places at the time of transplanting while, at other places, as a pre-harvest ritual. In Orissa before transplanting girls pat the earth to render her fertile and men, carry palm leaves fans as if coaxing her to wake and bear abundant crops.[55] Oraons and Gonds perform war dance. Nagas play tug of war between males and females and Meitei men pull a bamboo against women with same intention. In Japan, people sing transplanting songs with sexual overtones to promote fertility of crop while in Java, Vietnam and Indonesia fertility rites are conducted as preharvest ceremony. It is celebrated as marriage of Mbok Sri with Sedna in Java, as *Ma Bua Pare* ceremony in Indonesia,[56] men portraying as water buffalo and making advances to each other and as a midnight fertility ritual leading to procession of sacred seeds and performance in Vietnam.[57]

Rites to Invite Rains

Rain calling rites include welcoming rains, special rites if drought occurs or rains are uncontrolled. With a view to welcoming rains, many tribals in India celebrate Asadi Parva and worship Indra (rain God). People perform this ritual to supplicate the rain God to be kind and grant them rains for good harvest. All the tribes performing this ritual sacrifice at least a fowl. Farmers in Indonesia offer rice to rain God to ensure favouable weather and a good crop.

In case of failure of rains, rain-calling ceremonies are performed in many countries. Frog rites are conducted in different states of India. Newars of Himalayas worship frog in a pond. Konds of Madhya Pradesh and farmers in Tamil Nadu and Andhra Pradesh conduct frog rites to call rains but at each place in a different way. In Assam, frog and bamboo marriages are performed.

In Thailand and Malaysia, cat takes the place of frog in similar rites. Another custom in Thailand includes putting two dolls and a frog in holes and prayers are held until it rains. In Isan region, village monks fire off decorated rockets to "pierce" the sky to bring the monsoon rains. In Ghost Festivals of northeast Thailand, male and female papier-mâché figures are paraded with other symbols of fecundity. Elaborate masks are a hallmark of these festivals. Both these festivals are major tourist attraction.[58]

In Tamil Nadu, Selliamma deity is worshipped for rain. Three priests perform ceremonies; seven courses of rice and vegetable are offered in upward direction. Festivals are held to appease Indra, the lord of rain. Special rain songs are sung in *Varsha Raga* to bring in rain.[59]

In case of drought various types of *yajnas* are performed. However, Meiteis follow a unique custom. Meiteis headed by raja strip themselves of clothes and stand in roadway cursing one another. At night, women gather in a field outside the town, strip themselves and throw rice into the neighboring pool.[60]

Magical rites are performed to stop rains if uncontrollable. In Bombay, a leaf plate full of boiled rice and curd is warmed to be off. If this fails then a live goat is laid on a file in open. If hail storm threatens, then the Oraons of Orissa ask priest for worship. The priest in charge of a particular hill offers goat and chickens and spirits are propitiated by scattering rice on ground.[61]

Weeding

Before the advent of chemical herbicides, manual weeding was done after performing ceremonies in India and in many Asian countries. Weeding rituals of India include *chitau amavasya* in Orissa, *sankranti* of Sawan (July-August) and Bhadon (Aug-Sept) in Himachal and *chitla* ritual in Mizoram. Some of these rituals are followed even today. It is interesting to record that the farmers make offerings to snail and oysters and prey not to cut feet of farmers.[62]

In Bali, women perform a special ritual while weeding. In south China, it is customary to perform offering rites for souls before starting to weed. Every family places eatables in a special basket hung above the household head's sleeping place.[63]

Protection of Crop

When the crop is half grown, the farmer puts his heart and soul for the welfare of the crop as the failure of crop means death for them. In olden times when chemicals were not available for protection of crop, a variety of prayers and magical rites were performed for the protection of crop.

In Bali, rats are one of the evils and ritual ceremonies are held when they go out of control. In tribal belt of Orissa many festivals are performed to keep evil spirits away from rice fields where sacrifices are mostly done. In Assam, during *Kati Bihu* farmers light a clay lamp and utter *mantras* for protection of crop from attack of birds, rodents, insects and animals. In Kashmir, when blight attacks, the farmer begs for an amulet from a holy man and ties it on the post in his rice field and performs many magical rites.[64, 65]

In Madhya Pradesh, branches of wild rice are hung up in the house in August when the crop is growing. In Manipur, Meiteis perform ceremony of goddess *Laphurit Leima Wadagnu Chomkhoi-doisibi* (daughter of a Hillman, Hao) in case of caterpillars' attack in October-November.[66]

Grain Formation Stage

In addition to the rituals conducted for rice mother mentioned earlier, a number of other rituals are performed in other areas.

In Orissa conception is celebrated as *Garbhana Sankranti* (pregnancy period) in October-November. People place food offerings in the corners of rice fields with the hope of safe pregnancy and high yield.[67] When milky grains are set, farmers in Thailand prepare a sweet *khaaw maw* and serve monks. Similar rituals are performed in all South East Asian countries.

People of Orissa worship Goddess Annapurna on *Saptapuri amavasya* in the month of August-September in anticipation of a good harvest, when early maturing rice matures and rest of rice is in the fertilization stage.[68]

Pre-harvest Rites

Farmers observe a number of pre-harvest rites, rituals and festivals before the actual harvest of the crop. These include thanksgiving to ancestors, worship of harvest deities and offering them new rice, making a sacrifice, or placing rice figures in the field to guard the harvest and cutting of special stalks of sacred rice. Thanksgiving rites are included in pre-harvest rites in many areas whereas these form a part of harvest rites or sometimes are performed separately and independently as 'eating new rice' in some areas.

In Java, pre-harvest rites take the shape of wedding ceremony of rice Goddess, Mbok and Sedana. Witnesses are an essential part of the ceremony, *slametan metik,* but only men attend and eat, taking home with them the remains of the feast. The sacred rice is cut with care, without noise and with finger knives so as not to disturb rice spirit.[69]

Maiba (priest) of Meiteis in Manipur invokes Phaoibi, the rice goddess at the harvesting rite to increase rice yield. The people of Kerala celebrate *onam.* The night before the harvest, they chant *poli* (increase) *poli deva,* which echoes across the field. The families return their home with first sheaf of rice crop, invoking the Gods to increase their bounty.[70] In Mizoram, a sacrifice called *Leuhmathawna* is performed in front of each *jhum* house with chanting for bountiful crops, a red hen would be sacrificed and seeds would be anointed with blood after the sacrifice, people would go to field and collect paddy.[71]

Garos of Meghalaya, before harvesting, pluck some ears of rice, pound them between two stones and offer them to their gods on a piece of plantain stem.[72] In Bengal, farmers ceremonially worship a number of harvest deities (Bhanju, Kartik, Kojagari) and celebrate annual festivals.[73]

Kochs in Assam harvest the early crop and offer that to ancestors. In Andhra Pradesh, people worship the village deity Satti Pen and offer a small quantity of newly harvested crop.[74] At harvest time, Lhota Nagas make sacrifice of a pig and offer share of meat to Rangsi, the fields' spirit, who gives good crop.[75] The people in Tamil Nadu before the harvest upbraid the rice stalks into garlands and hang them outside the houses and symbolically share the first yield with birds and small animals.[76]

At Kyoto (Japan), in Kamo shrine, the main event is a procession of people dressed as court nobles of the Heian period. Participants wear headgear decorated with leaves of the *aoi* plant (*Asarum caulescens*); the leaves are offered to the deities at the shrine with the hope of a bountiful rice harvest.[77]

In Isan (Thailand), when farming is over, farmers observe a ritual. They tie a few reserved bundles of rice to a pair of bamboo poles and strike them on hard dry floor and perform blessing ritual.[78] The transfer of grains from field to storage house is done in a symbolic way in Vietnam. Six initial sheaves of rice are deposited in a four pillar 'granary'. Prayers are performed to invoke *phi* or spirits of mountains, forest, sky, land and river. People take ritual meal and rice mother is brought back to home till next year sowing.[79]

The senior woman of household in Bali makes rice figures which watch over the harvest until ceremonially installed in the granary, allowing the ancestral spirits of the rice to live on until the next planting cycle.[80]

Harvest Festivals

Harvest time is an occasion of festivity in all the rice-growing countries. It is celebrated by different names in various states of India such as Magha/Bhogali Bihu in Assam, Paus Utsav in Bengal, Dhanu Sankranti in Orissa, Pongal in Tamil Nadu, Huttari in Karnataka, Makar Sankranti in Andhra Pradesh, Onam in Kerala and Pawal Kut in Mizoram.[81] In most of the states, it coincides with New Year. Most of these festivals last for two to three days. Each day is devoted to different rituals, as for cattle, humans, god or ancestors. *Sankranti* means transition (of sun to a new zodiac sign) and *Makar* means the constellation of Capricorn. At this time, the harvest festival is celebrated under the same name in Andhra Pradesh and as Pongal in Tamil Nadu and Sri Lanka. Both these festivals and Dhanu Sankranti of Orissa (on December 14) are celebrated as sun worship and thanksgiving ceremony. In Pongal, newly harvested rice cooked ceremonially is partaken.[82] In Orissa, sun worship is followed by ritual of Merikuntha (central pole in threshing floor) Puja.[83]

Animal Worship

During the rice rituals, the bullocks and cows are not ignored and given due attention and share in celebrations. Animal worship is an integral part of ceremonies associated with pre-transplanting and harvest. During *Eruvaku Purnima* and *Makar Sankranti* of Andhra Pradesh, Pongal of Tamil Nadu, Nabanna of Madhya Pradesh and Bohag Bihu of Assam, a day is reserved for cattle worship ritual in each of these festivals. Cattle are washed, decorated and fed with pullagam specially prepared for them on Eruvaku, in Pongal; money bags are tied and they are set free.[84]

Post-harvest Operations

After harvesting, threshing, winnowing and storing are equally serious undertakings and special rites accompany these processes.

Threshing of rice is also done ceremonially. Various communities perform a number of different kinds of rituals at this time. Threshing festival is *Kharihari Puja* of Madhya Pradesh and *Phoukouba* rite of Manipur. Kurmis of Madhya Pradesh offer a goat/fowl to Thakurdeo before threshing. Mundas do not thresh their rice until *Kharihari Puja* or threshing floor rite is done and spirits are worshipped with sacrifice of a fowl and oblation of rice beer. Oraons don't thresh their paddy before the village priest has prepared his own threshing floor and performed the public *Kharihari Pooja*. After threshing, the priest keeps 3-5 handfuls of the rice for the threshing floor and it is presumed that these will watch the threshing floor up to the next crop.[85]

In Manipur, Meiteis celebrate the ceremony of calling the rice (*Phoukouba* rite) with the help of *Maiba* (priest) on the last day of threshing paddy before being carried to house for storing.[86]

In Isan of Thailand when threshing is over, farmers observe a ritual by tying few reserved bundles of rice stalks to a pair of bamboo poles, strike them hard on ground and perform the blessing ritual before transporting bulk of rice to storehouse in village.[87] Even today, people believe that the soul of rice, "ina-dama," which is called the god of rice paddy, dwell in the last rice stubble. The person who reaped the last rice would bring it back home and hold service. This custom is still practiced at the rice harvest rituals.[88]

Storage Festival

Putting rice into granary concludes the cycle of labor. Throughout South East Asia, special rituals are held when harvest is put into granary. In some communities, it marks the biggest festival of the year to indicate that rice spirits have taken their abode. Some figures are placed on the granaries as doll, etc who are supposed to guard the granary. Rice is brought into house only after milling as rice spirit is still alive in paddy. In Thailand, the granary guardian is called Taa Pook or Phii Lao and is made from rice straw into the shapes of sitting dolls.[89] In Toraya region, Indonesia special stones collected from river beds are placed on granary. In Indonesia, in every traditional family compound, there will be an elevated, bow-roofed granary near the kitchen called a lumbung (built off the ground on posts to deter rodents) with a storage area for rice (and rice-related offerings): this rice barn is the house of Dewi Sri.[90]

In Mizoram, *Sawa Awhthi* ceremony is conducted after the storage of grain in the granary. For the rice to last, a sacrifice, called *bei pariawthi* would be performed in which a fowl would be killed and blood would be sprinkled over the rice. Meiteis worship the paddy room by adoration of the Phaoibi in her stone form in granary called *Kotlai Khurumba* rite.[91]

The Lahus of Southwest China believe that rice soul hides under the last pile of un-threshed rice. They take a couple of stones and two hardened lumps of earth, said to contain the souls of the crop, from below this last pile of unhusked rice and on their way back home, scatter husk and chant prayers to call back the rice soul. On reaching home, they place the stones and clay under the storage basket.[92]

Thanksgiving Festivals and Tasting New Rice

When promise of abundance is fulfilled, the major task that remains is giving thanks to spirits for success of harvest. This ritual is of such importance that it becomes the focus of the entire harvest festival.

The use of newly harvested rice is carefully regulated. As all that is new is sacred, man cannot touch it until the ceremony of dedicating a part of the produce to high powers and ancestors is completed. In most states, harvest festivals include such offerings while in some tribes eating of new rice is a separate ceremony. Crops are harvested before the festival day but nobody dares to eat before the festival as doing so would cause misfortune. In southern states, harvesting coincides with New Year and it takes the shape of elaborate festivals. Such festivals are celebrated in Bengal, Kerala, Tamil Nadu, Karnataka and Orissa while in Assam and Andhra Pradesh, it coincides with the sowing time. Mundas of Orissa who eat new rice parched with milk, curd and molasses and celebrate *Jon Nava* in September, have another festival called *Kolon Singh Borga* in November on harvest of transplanted rice.[93]

The Lahus of Southwest China, believing that rice has a soul, conduct the ceremony 'New Rice Offering' and *zha ha ku ve* ritual, before and after the harvest time respectively to summon the soul.[94] In Java, puppet shows are held on harvest to remind humans of divine origin and sacred nature of crops and sacrifice to Dewi-Sri.

New rice is offered to Buddha in a ceremonial way in rural Thailand, later offered to ancestors and afterwards to monks and people eat in elaborate rituals.[95] The festival of Alut Sahal Mangalle in Sri Lanka occurs in January every year. When the harvest of paddy is complete, certain quantities are carried to the temples and *devales* and offered to Buddha and the respective divinities.[96] In Myanmar, Java and Malaysia the first harvested ears are offered to the sun god Suriya that reminds one of Pongal of Tamil Nadu, Dhanu Sankranti of Orissa and Makar Sankranti of Andhra Pradesh. In Orissa, Nabanna is popular and celebrated with funfair. Paddy is considered as Goddess Lakhsmi herself. Therefore, nobody can take it unless first offered to deities. After harvesting in November, village deity Satti Pen is worshipped and small quantity of newly harvested crop is offered to deity in Andhra Pradesh.[97] In Madhya Pradesh and Tripura, *Nabanna* is a thanksgiving festival. Gonds offer green paddy to *Sai* tree (*Terminalia tomentosa*) and then to deities.[98]

In Assam Gopeswar is thanked. Cultivators from villages offer paddy and a thanksgiving Bhog is held. Koch of Assam harvest the early crop and offer that to ancestors[99] and Garos of Meghalaya before harvesting pluck some ears of rice, pound them between two stones and offer them to their gods on a piece of plantain stem.

In Bengal *Khetra Thakur* is thanked who is caretaker of rice fields in *Baisakh*. Flattened and fried rice (new) is offered to deity along with sugar candy, sweets, boiled rice and cooked vegetables.[100]

Lohta Nagas offer a sacrifice: they harvest the rice sown in a special plot, cook some and pretend to eat it, praying that squirrels, rats and birds may find village rice bitter. A small quantity of new rice is wrapped in a leaf and kept in the bin into which a days' supply is put every morning. The priest and his wife thus run the risk of eating this taboo-ed grain and part of it is reserved as a means of promoting fertility.

In villages of Japan, the entire community celebrates autumn festival of the first fruits at a special altar and in many places floats carrying symbolic gods are paraded through the streets. At the Imperial Palace, the Emperor fulfils the role of presenting offerings of new grains and produce to the gods. In Japan, long ago, the new autumn harvest (of rice) could not be eaten until after a festival in honor of the rice spirit. There was dancing,

Box 3.1. Festivals associated with various stages of rice
cultivation in India and other Asian countries

Pre-Sowing Festivals

Akshaya Tritiya of Orissa, Akshyamuthi, Dhanbhumi puja, Hon ba Parav;
Hera Parab(seed sowing ritual); Jamolpur, Bihanbuna, Puja Irumamne,
Eroksim or Erosim. Kedu Parav; Jhagdi Parav; *Rialongcchi* ceremony in
Manipur.

*Royal ploughing ceremony, Japan, Cambodia, Thailand; Purification of rice fields, Java;
Rite of imitating rice field preparation, Japan; Inviting mountain spirit, Japan; Rite
of blessing seeds in Malacca.*

Sowing Festivals

Eruvaku Purnima, Andhra Pradesh; Bohag Bihu, Assam; Mopin, Arunachal
Pradesh; Damurai Jharkhand; Worship of Thakurdeo in Madhya Pradesh;
Ganesh by Bhils, MP; Household Deity in Himachal Pradesh; Phaiobi in
Manipur; Ancestors in Coorg.

Sowing ceremony in Malacca.

Pre-Transplanting

Raja Parba of Orissa, Ambuvaci fairs of Bengal and Assam.

*Village deity worship of Yunnan; God and water spirit worship in Lanching, China
on Dragon Day.*

Transplanting Festivals

Kati Bihu of Assam; Koda Yatra, Asadi Para, Chitalagi Amavasya, Karam
Puja, Garbha Sankranti, Satpuri Purnima of Orissa; Fertility Rite, Orissa;
Batauli Celebration by Mundas, Orissa; Fertility Rite—tug of war, males and
females, Oraons in Orissa; Kojagari and Kartik puja of Bengal; Pulling of
Bamboos, Meiteis, Manipur; Keilmutha Festival, Coorg.

*Flower rice planting in Cibu, Planting of brown rice in Tsushima; Rice planting
ritual called Nanagusa Gomori, Iwate Prefecture, Japan; Sumiyoshi Taisha; Annual
rite for the planting of seedlings.*

Weeding

Rituals of Chitau Amavasya in Orissa; Sankranti of Sawan and Bhadon in
Himachal; Chitla ritual in Mizoram; Mimkut festival of Mizoram.

Rain Calling Ceremonies

Frog Worship by Newars of Himalaya; Frog carrying Rites, Madhya Pradesh;
Bengal, Andhra Pradesh, Tamil Nadu; Frog and Bamboo Marriage in Assam;
Selliamma Deity Worship & Cursing Indra, Tamil Nadu; Khati Awan in
Mizoram; Farmers & Raja Stripping, Manipur; Ladies Strip; Worship of
Hindodev & Cursing Indra in Bengal.

*Cat washing in Malaysia; Rituals of cat carrying, putting 2 dolls and frogs in a hole,
rocket-firings, procession of male & female paper mache doll in Thailand.*

Contd.

Box 3.1 continued

Welcoming Rains

Asari Parva, Indra Pooja, Batauli (Mundas); Karma Pooja (Binjals); Haryari (Oraons); Asadhia, Utaras Balijatra (Parojas).

Offering rice to rain god in Indonesia.

Protection of Crop

Siju tree Ceremony in Assam; Autumn Bihu, Assam; Rialnggechi Ceremony, rite for Grub control, Nagaland; Fowl Sacrifice, Mundas; Planting sacred tree branches, Mundas; Karma Festival by Oraons of Orissa; Kadhotha Festival, Orissa; Hanging Wild rice, Hareli Festival, Madhya Pradesh; Ceremony of Laphurit Leima Wadagnu Chomkhoi-doisibi (daughter of a hillman, Hao) in Manipur; Hariyari Devi Worship, Uttar Pradesh; Amut Banding, Kashmir.

Grain Formation Stage

Grabha Sankranti in Orissa; Hanging Rice Garlands in Tamil Nadu; Lakshmi Worship in Orissa; Ceremonial Reaping, Assam.

Pregnancy ritual in Thailand; Slametlan Metik ritual & Kelaman food ceremony in Java.

Pre-harvest Rites

Bhanju Worship in Bhadon (August-Sept); Kartik worship (Oct-Nov), Kojagari Lakshmi worship (August-September in Bengal).

Harvest Festivals

Magh Bihu, Assam; Paus, Bengal; Huttari, Karnataka; Gurubar Osha, Orissa; Nabanna, Orissa, Bengal and Tripura; Pongal, Tamil Nadu; Onam, Kerala; Makar Sankranti, Andhra Pradesh; Pazusata ceremony, Mizoram.

Metik food ceremony in Java; Htamin in Myanmar, Kamo shrine harvest festival, Japan; harvest ritual, Korea.

Threshing Festival

Kharihari Worship (Threshing floor rite) of Mundas and Oraons; Calling the Rice, Phoukouba rite, Manipur.

Thanks-giving Festivals

Thakur deo Worship, MP; Gurubara Osha, Orissa; Ancestors Worship, Assam; God worship, Garos; Sawa Awethi Ritual, Mizoram; Ancestor & Village Deity worship, Nabanna Purnima in Uttar Pradesh; Eating New Rice, Worship of Akipen, Jona Nova (September), Kolon Singh Borga Festival; (transplanted rice), Mundas, Orissa; Sarna Burhi, Oraons; Janthar Puja, Santhals.

Ancestor festival in Ammi Behiwa, harvest festival in Ishigaki Island, Japan; 'New Rice Offering Ceremony', Lahus, China.

Storage Festival

Ritual of Sikisa Sacrifice of Mizoram; Khurumba rite of Manipur.

singing and waving of fans. Everyone joined in a great feast. Now, that day is a national holiday; it takes place on November 23. The name of the festival has also been changed; it is now called Labor Thanksgiving Day. At midnight, the Japanese emperor offers rice at the shrine[101] (See Box 3.1).

Rice Divination: Offering Rice to Deities/God

Ever since its domestication, rice has acquired a sacred status in Asia whether as food or decorative item or medium of communication. Rice has been attributed human and superhuman qualities. Auspicious quality of rice and its significance remains untarnished even in areas where rice is not the primary crop. From whichever angle one looks at rice, its immense significance in the Indian ethos simply cannot be missed.

Some of the tribes perform sacrifices to propitiate minor gods in order to save themselves from their wrath. The ancient scriptures forbade cruelty towards animals and replaced animal sacrifice with offering of rice. The sacrificial value of rice has been highly lauded in ancient scripture, *Aitareya Brahmana* and considered it as representing the most important among sacrifices. *Caru, purdosa* and *payasa* were ritual oblations. "Purdosa is the animal which is killed. The chaff and straw of rice are considered the hairs of the animal, husk its skin, its smallest particles the blood, all the fine particles to which the rice is ground represents flesh and whatever other substantial part left in the rice are the bones."[102] This explains why rice is used so elaborately and extensively in worships.

In addition to this, rice is one of the sacred grains associated with heavenly bodies, days of the week and deities. In Japan, rice grain is believed to be the peaceful soul of the deity. Humans and their communities must rejuvenate themselves by harnessing the positive power of the deities (*nigimitama*). This can be accomplished by performing rituals during which humans harness divine power or they can do so by eating rice. Therefore, rice, soul, deity and the *nigimitama* (peaceful/positive power of the deity) are all symbolic equivalents.[103]

Rice used in rituals at home in India include *akshat* (unbroken), *nelkade* and *naivedyam*. It is also used as a seat for the sacred jar (*kumbham* or *kalasam*), in offerings to the sacred fire (*homa*) and in preparation of *kolam*. The most frequent use of rice is as *akshat* for blessings when an elder sprinkles rice grains mixed with turmeric powder (*akshadai*) on a young person. So, revered is *akshadai* that it can be a substitute for flowers, clothing and jewellery when making an offering to the deity. *Naivedyam* is ritual offering of food and only rice is the fitting seat (*asanam*) for *kumbham* or *kalasa* (a sacred jar) that forms an integral part of all worships (*puja*) for invoking the gods and ancestors. *Havis* (raw and boiled rice) is offered to the god of fire (*homa*) which is essential in Hindu rituals. *Kolam,* the artistic ritual designs created on the floor in front of each house are made of rice flour. The rice flour keeps away the evil eyes and demons and forms a protective wall before the house. When birds and insects partake of the rice flour, the souls of ancestors are pleased. *Nelkade* is a large sheaf of paddy hung outside the home to keep away evil eyes and demons.[104]

The ancient *puranas* of Tamil carry account of rice clearly highlighting the specific varieties used in important religious ceremonies. In eastern India names of rice varieties such as *Krishnabhog, Gopalbhog, Govindbhog, Sitabhog, Nripatibhog, Rambhog, Narsingbhog, Lakshmibhog, Laksminarayan-bhog, Deobhog* indicate specific requirements in terms of offerings to specific gods. In ancient times, in the temple of Lord Jagannath at Puri, daily freshly harvested rice was used as offering to the Lord and the farmers were compelled to cultivate a number of varieties. In Bhandara (Maharashtra), two interesting festivals, *Rishi Panchami* and *Shravan Dwadasi* are celebrated during August. The people of Maharashtra depend upon wild rice grown in swamps, locally known as *deodhan* (god's rice). In the days of Maratha rule, *deodhan* was specially distributed from government store houses.[105, 106]

Forms of Rice Used in Ceremonies

Ceremonial use of rice is widespread. It is used as sacred offering in most of the South Asian countries. The basic form of rice used for this purpose varies in different countries. In India, rice grains and cooked rice are used in offering. Raw rice in China and cakes in Japan and Indonesia are offered to deities in an artful manner. A special art for making earthen, metallic or porcelain offering vessels was developed in China.

In Japan, mirror or *mochi* cakes are offered on altars at New Year to Shinto deities. These mirror rice cakes embody the spirits of rice. Even the process of making mirror is considered as a ritual procedure rather than a culinary one.[107]

In Vietnam, steamed cakes of glutinous rice called *banh chitng* stuffed with mung-bean and pork are placed on family's ancestral altar during *Tet*, the Vietnamese New Year period. These cakes from the altar are served to visitors, sharing the blessings of the family ancestors.[108]

Perhaps the most spectacular rice offering of all are the elaborate construction of brightly colored rice dough made for temple festivals in Bali. Many different kinds of rice cookies (*jaja*) are part of the most offerings. The term *jaja* encompasses a range of tasty cookies and cakes eaten as ordinary sweets but are also used as offerings. People do not eat such cakes every day but only as leftovers from offering. The most common edible offering, the crispy *jaja uli* and *jaja begina*, is always in combination of a white and red color. In small offerings, only a little bit of *jaja uli* and *jaja begina* is used but the offerings of a large family consist of many kinds of tasty *jaja*.[109]

The making of offerings in Bali never stops—at shrines in the family compound, at clan temples, *subak* temples and hundreds of village temples, at full moon, at new moon, on auspicious days and on every 15th day when possibly harmful spirits are especially prevalent—rice is invariably involved.[110] Every morning after cooking the daily rice, housewives put some on small pieces of banana leaf and set this down where spirits are likely to be i.e. on the kitchen stove, in the courtyard, at the water source, in the house temple. Such offering are called *banten joten* or *banten nasi* and are intended for the invisible inhabitants who help with processes of daily life.

Ceremonial use of rice in various rituals in India dates back to *Vedic* period as is evident from *Vedas, Upanishads, Puranas* and epic stories of *Ramayana* and *Mahabharata*; plenty of references are found of its use as offerings. Rice is offered as oblation to gods, deities and the village deity in religious ceremonies, as alms to brahmans and to needy people.

Offering is made of unhusked rice (paddy), white unbroken grains, colored yellow (with turmeric), sun dried, cooked, fried, parched or popped rice, rice flour, rice cakes of various kinds and rice beverages. These forms offered in various ceremonies depend upon the religious belief of the people and convey different shades of meaning. Colored rice is festive shorthand for auspiciousness while rice itself is shorthand for either food or body. Fried rice is the symbol of fertility. A pot of cooked rice is compared with a family.[111] Raw and cooked rice is, respectively, considered as hot and cool. Raw rice finds place in the mouth of corpse before cremation while plain cooked rice is used in funeral rite and during *pitripaksha* (fortnight meant for ancestors) ceremonies to appease restless spirits and ancestors. Cooked rice is believed to scare demons particularly those that check fertility of union.[112] Rice colored with turmeric or vermilion is symbolically used as invitation. Popped rice is considered favorite of lord Krishna and is served to Him on His birthday and parched rice is offered to Goddess Lakshmi.

Another form of rice used in religious ceremonies and tradition is rice flour and its paste. The paste forms part of the make up of Kathakali perfomer in Kerala and also in making of *dhuli chitra* (ritual diagram on floor) and *bhitti chitra* (drawing on walls). These drawings form an

Table 3.2. Rice offered to gods and deities in religious festivals

God/Deities	Deities of disease	Festivals	Seasonal festivals	Family festivals
Vishnu	Mariamma	Dushera	Teej	Ahoi
Shiv	Piriyapattandamma	Diwali	Lohri	Karva Chauth
Ganesh	Durgamma	Janmashtami	Holi	Raksha Bandhan
Lakshmi	Mastiamma	Shivaratri	Basant Panchami	Bhaiya Duj
Saraswti	Poleramma	Ram Navami	Navroj	24 Ekadeshis
Kali	Ankamma	Durga Puja	Shravan Shukla	12 Sankrantis
Shashti	Ral Durga	Kumari Puja	Anna Chaturthi	12 Amavasyas
Koda (T.N.)	Chandi durga	Janmashtami	Makara Sankranti	12 Purnimas
Sun	Chandi	Ganesh Chaturthi		
Moon	Shoba Chandi	Shivaratri		
Rivers		Guru Purnima		

essential part of various rituals and *vratas* (vows); the observance of *vratas* and the performance of rites were and are obligatory, particularly for Hindu women.

In addition, processed and fermented rice beverages are used as offering to God. The tribal people of various states offer rice beverages in order to propitiate the deities during various ceremonies and in rites of passage and rituals at various stages of rice crop.[113] However, the Hindu peasants generally don't employ rice beer or fermented grains in rituals. This perhaps reflects an aesthetic aversion to the rotting connotation of fermentation.[114]

A garland of rice is a special offering to Lord Govindaji during *rathajatra* (chariot festival), one of the greatest Hindu festival and to Lord Shri Bijoy Govinda during a boat race held on the 11th day of Langban in Assam. Each of the two-team leaders wears a garland of 108 grains of rice and a garland of seed from the *Embelica officinalis* plant. After the race these are offered to Lord Bijoy Govida and Rameshwari (Table 3.2).

Symbolism

Ceremonial use of rice has not only physical but also symbolic values. Fried rice is symbol of fertility. A pot of cooked rice as also a vessel of water is compared with a family. As food or as a decorative item or as an object of play or as medium of communication, rice has been a powerful medium throughout Asia since ancient times. Urbanites, peasants and tribals alike endow rice with magical equivalence of food itself.

Rice stands for fertility and women power. This idea is the basis of all contemporary post-wedding rituals. Various forms of cooked and un-cooked rice provide aesthetic guides and model for interpretations of life, birth, growth, maturation, death and life after death. However, various forms of rice represent different moods.

Rice functions as the key symbol for Hindu culture.[115] Its uses codify, clarify and focus attention on basic life processes. Offering of rice stands as symbol of existence, consciousness and experience in all things. In ancestral worship of Hindu India or Buddhist Japan or traditional China, one puts rice to one's ancestors with the same reverence in these countries. The fact that rice season is full of rituals while dry season is devoid of such rituals though other crops are grown shows importance of rice.

Decorating floors, a tradition followed throughout India, is called *alpana* in Bengal, *sathia* in Gujarat, *rangoli* in Maharashtra and parts of Uttar Pradesh, *mandana* in Rajasthan and Madhya Pradesh, *jhoti* in

Orissa, *kolam* in parts of South India, *mugli* in Andhra Pradesh, *likhnu* in Himachal Pradesh, *aria pan* in Bihar, *aipan* in Uttar Pradesh and *chowk purana* in Haryana, Punjab and parts of Uttar Pradesh. Women create patterns using rice flour paste, wheat or barley flour and rice powder mixed with turmeric, cow dung and soil. Often colored rice, pulses and saw dust are used. Rice flour is used in these designs with a motive to provide food to ants, insects and birds.

These symbols welcome the new bride; announce a marriage or a birth. These drawn on all ceremonies related to *rites de passage* from birth to death or after death during *shradh* (death anniversary) ceremony. Warlis of Bombay draw circular patterns with rice paste on the occasion of funeral and marriage. The pattern stems from their mythic knowledge that the death is not the end of human existence but another beginning.[116] Likewise Sanja wall motifs (drawn with rice paste and grain) of Madhya Pradesh, Rajasthan, Haryana and Uttar Pradesh is drawn at the time *shradh* ceremony of the departed soul and to appease the ancestors.[117]

Rice in *rites de passages* (*Sanskara*: Rituals of Human Life)

The importance of rice in rites of human passage can best be judged from the Iban's terms for rice farming. Rice farming is described as "nurturing (*inang*), looking after (*nguang*) or caring for rice (*nginta padi*), using the same language that describes the actions of a mother in nurturing her offspring. Similarly, the growth of rice is identified with the life stages of the women who cultivate it. Thus, young rice plants are referred as *anak padi*, rice children. Later *dara biak*, 'young maidens', then *dara tuai*, 'older maidens' and finally after the panicles of rice have formed and begin to swell either *indu dara* "maidens ready for marriage' or *indu ngandong* "pregnant women'. Rice rituals express this symbolism in manifold ways. In rite of harvest, three clumps of rice plants especially heavy with grain are tied together and ornamented by the senior most woman of the family, to farm *padi tanchang*, a ritual construction representing marriageable maiden (*indu dara*).[118]

In all traditional societies, the transition of life is observed and celebrated with ancient rites that predate the written history. The milestones of birth, adulthood, marriage and death gave course for major festivities which if neglected would incur the wrath of ancestral spirits. Rice is intimately associated with the various rites of human life in India, Indonesia, China, Malaysia, Thailand, Korea and Japan.

In India, rice is an indispensable item in all rituals and ceremonies be it marriage, birth or post-death ceremonies, rites performed during pregnancy (*god bharai*), on the event of birth, at the time of the start of feeding milk, to ward off evil spirits from the mother and the new born, in the naming ceremony of a child, at the time of *annaprashan* (the ceremony of giving first solid food), his/her initiation to writing, on *mundan* (shaving off hairs) ceremony of boys and on girls attaining puberty in Kerala, Bengal and Tamil Nadu. A Jain father performs *akshat tilak* and puts some rice grains on his hand and says "Stay pure like white rice, adjust with hot and cool persons as rice adjusts itself according to climate." In one and all ceremonies, rice is used in one or the other form. In addition to the rituals concerned with an individual's life, rice finds place of equal importance in every home, social or family ceremony.

In *raksha badhan* (tying thread by sister on brother's wrist expecting protection at times of some problem), saying good-bye while departing for an important work or on a long journey, in festival of *bhaiya duj* (sisters put rice and vermilion/turmeric on their brother's forehead) or while receiving a newly-wed bride at the door step of in-law's home, the welcome *tilak* or departing *tilak* (mark on the forehead) is always done with rice and vermilion or turmeric. In addition to personal rites, rice finds place in all types of *pujas* (worships) performed on the occasion of selection of a site for village, and *bhumi pujan* (sanctification of a site for the construction of a building), construction of a house, etc.[119]

Chinese offer monster god *Nieri Gao*, a rice cake during the lunar New Year to appease his rapacious appetite and spare their villages and residents from killing. Chinese valentine day or Yuen Siu is celebrated with rice soup, ball and dumplings to signify family unity and harmonious relationships. New Year turnip cakes of glutinous rice, dried meat and turnips are traditional offerings to guests.[120]

Rice finds place in almost all the ceremonies of life in Java. A dish called *nasi punar*, steamed rice colored yellow with saffron, is prepared to celebrate birthdays and weddings. *Nasi buceng* (*Buceng telu*, consisting of three cones, *telu* literally meaning three) is prepared for traditional ceremonies such as celebration of the third month of a woman's pregnancy and commemorations of the passage of three days after death. The following types of rice porridge are served on ceremonial occasions in Indonesia. *Jenang procot* is served to pregnant woman hoping safe delivery. *Jenang senegkolo* is used in any kind of ceremony to keep off misfortune and *jenang abang* and *jenang putih* are served to celebrate a woman's pregnancy in order to indicate that the coming baby is really a legal baby. These two dishes are also prepared for birthday celebration.[121]

Rice in Birth of a Baby

From birth to wedding to funeral, rice plays an important part in every Asian country. In traditional Thai rituals for childbirth, rice is scattered around the room when a woman is in final stages of labor. Thus, baby is born into a space already marked with the sacred grain that will become a lifelong companion. In northern Thailand, a Tai mother will give a taste of rice to her newborn infant before she breastfeeds it for the first time, acquainting it first with the food that will provide subsistence for a lifetime.[122]

In Chinese ritual, birth of a baby demands serving of rice to friends who on the third day reciprocate this with gifts of raw rice wrapped in red topped with two stones. The stones ensure that the baby will grow up strong as a rock. On the first birthday, mothers are expected to serve crispy rice biscuits.[123]

In the welcoming ceremony of a three-month old baby, Balinese present an offering. At the base of the offering, *baten jejanganan*, the figure of a baby is laid out in uncooked white rice grains, symbolizing the beginning of the infant's life cycle.[124]

In Manipur, India, rice is indispensable in Meitei *rites of passage*. On sixth day after a child is born, they observe a rite known as *ipanthaba*. For that rice and a wreath of dried fermented fish are needed. The villagers present a measuring basket of rice and an egg to the parents.[125] In Kerala the celebration of a young woman's first menses involves a display of a large quantity of unhusked rice as a fertility symbol invoking her future reproductive powers.[126]

Rice in Weddings

Marriage is the most important event in anybody's life. Rice in various forms appears repeatedly in engagement and wedding celebrations. For example, rice wine decorates a magnificent bridal robe in Japan. In India, raw rice covered with golden turmeric or vermilion (*akshat*) is offered as an invitation to a wedding or boy's initiation. A basket of unhusked paddy forms the resting-place for bride-to-be, while another basket of cleaned white grains, forms platform for bride and groom for a later stage of wedding ceremony. Rice also finds place in the wedding play of some tribes accompanied by four kinds of pulses. In this game raw rice is placed in a pot of water which bride and groom are required to dig for a gold ring.

In traditional Vietnamese wedding process, a ceremony called *le en hoi*, takes place after a formal proposal has been made but before the actual wedding. It is the most important event in the entire process because its purpose is to confirm the relationship between the two lineages. The groom's kin assemble with different kinds of gifts including wedding cakes.[127]

In China, engagement of a daughter is announced through a tradition of preparing 'Fragrant Rice Crackers'. In India, turmeric rice forms integral part of invitation for attending marriage. The engagement and marriage ceremonies are respectively initiated with *tilak* and tying an amulet containing turmeric rice. The latter ceremony ends with blessings of showering rice. As rice is symbol of fertility, the newly weds are showered with rice not only in India but also in China, Philippines, Europe and the United States of America. Yellow colored rice is sprinkled on newly weds in Myanmar. After rites are over, showers of rice grains or colored rice consecrate the marriage in most parts of India; not only Hindus but also Parsis follow this ritual. The custom of welcoming a person by showering rice grains can be traced back to this ritual. In fact the Western custom of showering rice during wedding ceremony comes from this Hindu tradition. The priest, after performing the marriage rites and reciting the prayers, consecrates the marriage by a shower of rice colored with saffron on the newly wed couple.

Nepalese use rice grains as offering to holy fire *(homa)* during weddings. In one of wedding rites, *lajahoma* (*laja* is parched rice) wherein the groom makes bride to offer fried grains of rice (symbol of fertility) to the fire.[128] Korean newly weds are seated in the middle of a long table loaded with a variety of rice cakes. In Japan, bride and groom are required to take three sips of rice wine sake. In China parents of bride are expected to serve numerous sweet rice dumplings made from glutinous rice, the couple should eat a bowl of red and white dumplings. In Punjab bride and groom are made to sit in a corner of their respective homes with three pitchers of water, jaggery and rice. She and her friends are asked to pick up rice grains, chew and throw with a belief that by doing so bride acquires power to control husband according to her wishes.

The Meitei marriage ceremony involves a form of prognostication known as *chengluk lubak kaiba*. On the day of marriage, the groom's party brings a decorative basket made of cane and bamboo that contains rice to the bride's residence that is to be opened on the morning of fifth day to observe and foretell the couple's fortune.[129]

A *Kodva* marriage in Coorg, India offers a panoramic view of the extent of rice used in rites relating to this *sanskara*. There is no step of marriage ceremony without involving use of rice. At least 13 times rice grains are used during the marriage process. A dish-rice-lamp stands witness to the whole process of marriage so is the case with all ceremonious rites.[130] In Karnataka, a basket of unhusked paddy forms the resting place for bride-to-be while a basket of cleaned white grains forms platform in the wedding play of some tribes accompanied by four kinds of pulses.[131]

When the Balinese bride has moved into her husband's household, the family members of the groom visit the family of bride, presenting them with a large number of different *jaja* of both wet and dry categories. The *bental* (pillow) long *jaja* (rice cookies and cake) steamed in leaves of coconut palm and dressed up as husband and wife attract special attention during such a visit.[132]

In Bengal, when a newly wedded couple arrives at the bridegroom's house, a unique ceremony is performed. The bridegroom places a small bowl containing the rice grain on brides' head and with a small pointed spoon allows the grains to drop from both sides along the way from courtyard to the main house. This again is symbolic of ploughing field, scattering seeds, inducing fertility in bride. In Maharshtra a large pot filled with rice grains is kept on the threshold of bridegroom's house which the bride touches with her right toe and tilts the pot so that the grains fall and rice spreads on the floor. The extent of the grains spilled across the floor is believed to denote the prosperity bride brings to the family. Rice is also used to ward off evil spirits from the newly wedded couple. After a *Dhodia* wedding in Bombay, a man blows evil spirit from bodies of the pair from head to foot with the grains of rice.[133] A vessel of rice symbolizing prosperity is kept during traditional wedding ceremony of Nairs of Kerala. The Sanskrit word *Shyala* denotes the custom of bride's brother scattering fried grains at the marriage ceremony.[134]

Not only grains but also rice flour is vastly used in various ceremonies. It is rice flour which is the base for the delicate and intricate white line drawings made by women on their floors on festive occasions. Or with rice flour made into paste, they model into tiny oil lamps.[135]

In Malaysia, a vital part of marriage ceremony is *pokok nasi*, the custom of presenting gifts to the newly married couple. An artificial tree of boiled eggs attached to leaves and stems that are painted gold is prepared. The stem is then planted in the bed of glutinous rice, colored yellow with turmeric. The glutinous rice and eggs and leaves are symbols of fertility. In central Sumatra, rice is thrown over bridegroom's head before he enters his bride's house, together with betel leaves and lime. Later there is custom of pulling the chicken under yellow rice. At the end of a wedding ceremony in Central Java, drinks and foods are served and the bridal pair is presented with a specially decorated dish of yellow rice. The bridegroom takes a handful of rice, which he kneads into a compact mass and gives to bride to convey that the bride should be content with everything the bridegroom possesses or earns, however little this may be.[136]

Rice in Funerals and Death Anniversaries

In most Asian countries, the concept of soul or spirit and rebirth after death is deep-rooted. It has transformed and evolved into elaborate ceremonies at death and after death to pay respect to the deceased person. The deceased person should be ritually separated from living and corpse must be separated from spirit. The common folk representation of this separation is white color of plain cooked rice symbolizing unadulterated soul. Rites in death also involve much use of rice in different forms. A unique after-death rite in South India highlights the importance of this tiny grain. In the event of death of any person away from home, a mock cremation is conducted with bones and ashes. A replica of figure is made with 360 *palas* [*Butea monosperma* (Lamk.) Taubert; Fire of Jungle] leaves over black antelope skull, 32 pomegranate seeds for teeth and so on and finally life into the figure is brought back by sprinkling rice grains over it and only then cremation is done.[137] Tribals of Assam, Orissa, Manipur and Nagaland use rice beer and cooked rice in funeral rites.[138]

Of 16 *Sanskaras*, sons and grandsons perform the last one for the dead. Ancestor worship in Asia is essential for the welfare and prosperity of family. The offerings include balls of rice called *pindas*, foodstuff, etc. for all the dead relatives. It is performed once a year during *pitripaksha*, a fortnight specially dedicated to ancestors. Clearly, it may be said that life in Asia starts and ends with rice. Meiteis of Manipur perform a separate offering, *tarpan*, for a period of 15 days beginning from the full moon day in the month of *Langban* (September-October). The offering includes rice, sesame, leaves of *Cynodan dactylon*, fruits and flowers.

An offering, *Khayom lakpa*, is presented to supernatural beings to appease them and protect people from them. This offering is made at many rituals and in course of treating the sick. In a *Khayom*, rice, gold and silver coins, eggs and leaves of *lengthrei* plants are packed with seven layers of plantain leaves and tied to a bamboo rope. In ancestor worship in Hindu India or Buddhist Japan or traditional China, one offers rice to one's ancestors with the same reverence in all the three countries.

Rice in Language

Rice in Folklore

Iban refer to rice cultivation of hill rice (*bumai*) as *jalai idup* literally 'path of life' or indeed as *peng idup* 'life itself'. The following prayer reflects the idea as also expresses the Iban people's reverence for the rice seed:[139]

> *Oh scared padi*
> *You the opulent, you the distinguished*
> *Our padi of the highest rank;*
> *Oh scared padi*
> *Here I am planting you*
> *Keep watch over your children*
> *Keep watch over your people*
> *Over the little ones, over the young ones*
> *Oh do not tire, do not fail in your duty*

People's dependence on rice as food, source of sustenance, is exhibited through the honor, respect and reverence it receives. No wonder then rice has a unique place in the folk literature of all Asian countries. It appears in folk songs, folk tales, riddles, sayings in almost all the languages of the rice growing areas. In addition, folk literature is also available on rice cultivation practices (traditional wisdom on cultivation), this being a subject of a separate study has been excluded.

Folk Songs on Rice Cultivation

Almost every tribe in India, Malaysia, Thailand, Philippines, Japan, China and Vietnam has such songs that make the arduous task of rice transplanting easier. There are special transplanting songs in most of the Indian languages; singing these songs makes the hard work of transplanting easy and joyful. *Romani* songs of Himachal Pradesh and *Ropani* songs of Bengal are quite familiar. Songs and dances are dominant aspects of rice cultivation of *Tharus* of India and Nepal. The workers in different states have composed songs in their native languages and sing during transplanting, weeding, harvesting and husking of paddy. Such work songs are available in almost all states of India but those in Bengal especially the husking and transplanting songs for women workers and harvesting songs for men are well known.[140]

These songs depict hard work involved in the whole process of rice cultivation; the songs break the monotony of the work when sung in group and make the task enjoyable. Subjects of such songs vary from jovial mood of workers and sometimes depict sorrows and hardships of the life of laborers.

It is universally acknowledged that transplanting rice is a difficult job as an old Filipino song conveys:

> *Planting rice is never fun*
> *Bend from morn till set of sun*
> *Cannot stand and cannot sit*
> *Cannot rest for a little bit*

And also:

A popular song from Vietnam captures graphically the arduous task of rice planting:[141]

> *In the heat of the mid-day, I plough my field*
> *My sweat falls drop by drop like rain on the ploughed earth*
> *Oh, you who hold a full rice bowl in your hands*
> *Remember how much burning bitterness there is*
> *In each tender and fragrant grain in your mouth*

On a lighter mode, this Vietnamese nursery rhyme makes a refreshing appeal for rain:

> *Sky! Let the rain fall down*
> *So that there is water to drink*
> *So, I can plough my field*
> *Sky! Let the rain fall down*
> *So we can eat the rice*
> *And chopped aubergine!*

Though tough to cultivate, rice did not spare and has touched the heart and feelings of Urdu poets. Ahmed Nadim Kasami considers rice cultivation as an art and recites:

> *Dhan ki phasal ki tasvir hai meeraje-kamal*
> *Dhan ki phasal uthana bhi phankari hai*

Urdu poet describes the reaping and harvesting as an art. Vietnamese leader and Prime Minister, Ho Chi Min compares the pounding of rice to the adversity on to human soul.

> *"How it suffers, the rice under the pounding of the pestle*
> *But once that is over, how white it will be*

So it is for a man, living in this world
To be a man, one must be pounded by misfortune"

Domain of rice in folk literature is not limited to sayings and songs on cultivation, proverbs and riddles but is spread over rituals and love songs in Himachal Pradesh and Bihar, Western Uttar Pradesh and Kumaon, *Ukhanas* of Maharashtra, boatman's songs of Orissa, love songs/epic poems of Punjab and even in folk tales of Orissa delivered in a song form. The use of rice in love songs is in sharp contrast to other beautiful things/ items as birds, butterflies, rainbow, stars and fairies. See how a lover conveys his feelings to his beloved one:[142]

Chitte chavlo ra reja rekhna
Jeth tera dil lagda
Teth kaat kalya rakhna

"I shall remove the husk and keep only the grains. To fulfill your hearts desire, I am prepared to rip my heart and present to you to prove my faithfulness and immortal love for you."

See! How Oraons of Orissa compare a grown-up daughter with mature ripen rice in field.

When paddy stacks are full of sap
The grains mature and ripen
The pigeon come crowding
I have a grown up daughter
And friends and relatives
Even from distant villages come
Crowding to my house

Rice being base of life is also used as medium to convey paternal love and feelings as is well depicted in folk songs sung at Raja festival in Orissa.

The lake's embankment has given way
Tell my father that he had totally forgotten
Me after sending me away
No dear daughter, I have not forgotten you, look,
I have fried rice for you
Tied in a knot at the clothes end
God alone knows how much I suffer (on your account).

See how a Bangla proverb tells us the reality of life through use of rice,

"Flattened rice, fried rice; but nothing like boiled rice;
Aunt paternal, aunt maternal, but none like one's own mother."

Rice in Various Dialects of Hindi

In addition to various state languages, rice occurs in folklore of various dialects of Hindi such as *Bhojpuri, Awadhi, Kumaoni* and *Garhwali. Bhojpuri* (language of people in western Bihar and eastern Uttar Pradesh), folklore is replete with references to rice as the people are very fond of boiled rice (called *bhat*) and proudly claim to be *bhat*-loving. Rice finds place in different types of *Bhojpuri* folk songs: as *Jaitsar* (songs sung at the time of grinding grains), *ropani* (sung by women at transplanting time), *sohar* (at the birth of child), *Barahmasa* (songs of women separated) and songs sung at time of performance of various rites, rituals and ceremonies of marriage and other rites of passage. Rice used in these songs is depicted as favoured/preferred food item and as offerings to propitiate God for and favor of bestowing a male child, to convey inner feelings of love, express happiness and sadness at times of separation from beloved ones. Its use is sometimes symbolic. Thus, rice equally expresses sadness of a separated lonely woman and desperate feelings of a childless woman in *Jaitsar, Barahmasa* and *Ropani* songs and happiness in *Sohar* songs relating to marriage ceremonies. At the time of ceremonies called *Athghar, Chuamvan* and *Parchan* (paying respect) it is used to express happiness of relatives of brides and bridegroom.

Proverbs and Sayings

The importance of rice in Asian life is best judged from the folk sayings prevalent in various parts of world. Chinese say: "Rice is the stuff of life, so is the rice measure of life". They refer to a steady job as "to have an iron rice bowl" and to be unemployed as "to have a broken rice bowl." To deliberately knock over someone's rice bowl was considered an insult. In Japan, it is said, "he, who wastes rice, will go to hell and be ground to powder". In Japan, rice stands for "we," i.e. whatever social group one belongs to, as in a common expression, "to eat from the same rice-cooking pan," connotes a strong sense of fellowship arising from sharing meals. By contrast, such expressions as "to eat cold rice (rice is usually served hot)" and "to eat someone else's rice" refer to the opposite situation. While in India, *"Adhai chawal ki khichri pakana"* conveys doing work separately from community.

Rice is so pervasive that it has replaced words used for greetings and exchange of pleasantries. Indeed, in Thailand and in China polite way of asking hello in olden times was "Have you taken your rice today." At the beginning of New Year, instead of saying, "Happy New Year", they say, "May your rice never burn".

A Vietnamese proverb highlights the importance of rice farmers cultivating it as "The scholar precedes the peasant but when the rice runs out, it is the peasant who precedes the scholar." A Balinese saying presents truth of life "As the rice bows its head when it is heavy and ready for harvest so a wise man bends his head and is silent."[143]

People of Borneo[144] and Malagasy[145] use rice as a measure of time. Kyan would calculate their children's age by counting number of harvests on different patches of land since each child's birth. Malagasy term *vary* means rice. People use this to denote time, distance and volume. In combination with other elements the word, *vary* is used to denote money. The phrases *manam-bariraventy mewans* and *variifitoventy* denote 'without a penny' and weight of seven grains of rice. The word *vary* is not only used to denote fixed values but also to express proverbs which carefully describe moral attitudes. Malagasy *Mivary lavo* means a man of indifference who indulges in vices and unable to recover. *Atavary fitio tera-bary: tsy vitan'ny efu nijoro fo miondrika indray milo saina;* which literally means "Be like ears of rice, though they have stood upright they decline again for better reflection." The close relationship between rice and water has always been considered as metaphor for friendship: *Tahoka ny vary sy rano: an-tsaha tsy misfanry an-tananbatsy mifandav* (Friendship should be like water and rice: outside they do not part and inside they never separate from each other). Malagasy people do not hesitate to accept that at a certain age man has to take wife: *Ny fitio toy ny ketsa ka raha tsy afrindra tsy mandrobona* (Love is like rice plants; if they are not transplanted, they cannot flourish).

Metaphoric use of rice is not restricted to island of Malagasy but far apart in country of rising sun also. Even today, rice, called "pure rice (junmai)" or "white rice (hakumai)," has an aesthetic quality. The purity of white rice (hakumai) or "pure rice" (junmai) became a powerful metaphor for the purity of the Japanese self.

Folk Sayings

Rice has become a strong medium to convey feelings and varied thoughts. In Orissa, to express that a person behaves in the same way everywhere, they say *"Denki swargaku gale bi dhana kutiba"* i.e. husking machine husks paddy only even if it goes to heaven". Similarly the showiness of a person is expressed in Punjab as *"Andran bhukhiyan be muchh te chawal"* (His stomach is empty but he goes about with grains of rice sticking to his moustache). See, how importance of *kheer* (rice pudding) in raining season is emphasized (*Kyon jamian aparadhian, Je tu sawan kheer na khadi* (Why were you born, if you did not eat *kheer* in a rainy season?). Proverbial use of rice not only expresses the idea of festive food but also conveys peoples' attitude towards good food. *"Bhate khiron di toin raj chhadeo"* (The Brahmin gave up a kingdom for good food (boiled rice and *kheer*) and a man in Karnataka wasted even one years' saving for festive food *"Arasinada kulannu nambi varsada kulanne kaledu konda"* (Devoted to turmeric rice (festival food) he wasted his year's supply).

Rice is used as medium to express even the nature of people and clans *"Gaddi mitter kiskay, bhat khada te khiske"* (Gaddis are nobody's friends. Once they shall have their meals (rice), they disappear). Another saying expresses similar view about another caste: *"Barodari hinai bhat tan needer khadqeen"* (A barad (a low caste) who has rice in his pot cannot sleep).

Rice presents the universal duality of things and thoughts. On the one hand it is considered as the greatest wealth but it denotes petty things too. A Bundelkhand (Madhya Pradesh, India) proverb says: *'Dhan bare, puro ko lekho'* (to worry about petty things in the event of a big loss). Rice cultivation is the greatest example of cooperative and community work but *Dhai chawal ki khichri pakana* denotes to work in isolation. *"Var na viah, cheti laure dhan kute"* means to do something when there is no urgency.

Riddles

A Malagasy riddle asks "Someone does not die when feet are cut but dies when undressed." The answer is rice because it does not die when harvested but it dies when dehusked for eating. Another riddle states: "My father has gone to take away someone's life so that he can live in a better way again." This means that his father has gone to transplant rice.

A Telugu riddle presents the 'grinding stones for rice' as "when Kalinga Kingdom's bull's horns are twisted, it makes noise and cries." While a Manipuri child may put another in a fix by asking *"Yenbi amubi amana marum lising lising kokapa korina? chok hang chaphm"* (What is that black hen that lays thousands of eggs?) The answer is the rice cooking pot. A Filipino riddle elaborates the cooking wonders of rice and presents the brainteaser, too:

> *"What is cooked in the earth that is nice?*
> *Cooked in wood will fetch good price,*
> *Turns into sweet meats if cooked thrice (Rice)"*

Not only riddles about rice and things related to rice but other things of this world are also expressed in terms of rice such as:

"Nili taki chawal baddhe, Dine gawache rati labhe": A handful of rice tied in a piece of blue cloth, lost in the day and found at night" means 'stars'. In Koraput, Orissa (India), *"Addi kutse, addi undke"* (That is a vegetable, that is also rice gruel) and *"Idi gadar bat, nang sapur jol"* (Body full of rice and contain thin curry throughout) both mean snake gourd.

Taboos and Beliefs Associated with Rice

Taboos Associated with Rice

In the rice growing Asian countries, the day begins with 'morning rice' and ends with 'evening rice'. People eat rice, drink rice beer, sleep on rice straw and offer rice as oblation to God.

Through ages, certain taboos got attached with rice especially to avoid wastage of this life saving, sustaining basic cereal.

In China, if a steamer of rice is emptied on ground they feel insulted. A meal without rice is considered like a beautiful girl with one eye. Parents force their children to eat all the servings and for every grain of rice not eaten, they say, there will be marks on the face of their future husbands or wives. To deliberately knock over someone's rice bowl was considered an insult.

In Tibet, upsetting of a basin of rice on the table or elsewhere is looked as unlucky. Similarly, in Japan, it is said, "he who wastes rice will become blind, go to the hell or be ground to powder." It is also believed that any one who did not eat rice at least once in a day was not assured of waking up the following morning.

Certain taboos are associated with various steps of rice cultivation:

In Java, no loud noises are made, lest the rice may miscarry and fail to yield. For the same reason, there is no talk of death or demons in the rice-fields. In Indonesia, from transplanting to grain-filling, eating egg is forbidden in the village. If Nagas encounter a particular type of fern while harvesting paddy, they believe that paddy production will be low. The first sower of rice should be a woman and again the first harvester should also be a woman and she must partake only boiled rice, ginger and fish. During harvest, entering of strangers in *jhum* (slash-and-burn method of rice cultivation) houses is prohibited in Mizoram. There was a taboo for eating bird or rat meat.

Taking rice out of granary in the month of Feb-March is prohibited among Meiteis in India. In eastern Indonesia, men are not allowed to enter granary, women must not enter barn when rice spirit is sleeping. They should be properly dressed and must enter with right foot. Stepping over cooked rice is considered bad.

Similarly, in Malaysia women are not supposed to wear blouse when taking rice out of granary.

- In Tamil Nadu (India), while measuring rice paddy, people avoid the use of number 'one' and substitute it with the word 'first'.
- In Assam, cobras are considered as caretakers of paddy stores. Their presence in pairs is regarded to be a good omen and brings abundance of rice. Paddy is not brought out of stores on certain days.

Beliefs in the Philippines[146]

1. If rice is thrown or wasted, one's hands will become crooked and one will be deprived of abundance and prosperity.

2. One should never eat rice in the dark lest he invite an evil spirit called *momo* (ghost). Another version says that eating rice in the dark might result in swallowing the evil spirit and misfortune.

3. Rice should be given to a beggar on a plate, not in measuring cup, as the latter will result in diminishing the supply of grain and food.

4. The measuring cup should never be placed in the rice bin with its open end down lest it invite a bad harvest.

5. Rice should never be lent or given away at night or early in the morning, as it brings bad luck. One may buy rice at night, but must not be asked to be given some.

6. Showering rice on a newly wed brings prosperity both to the couple and also to one who showers.

Conclusion

In Asia, life starts with rice and ends with rice. Rice to an Iban farmer is life and ancestor; to traditional Japanese, it means cool spirit and, for Javanese, it means more than life and God. His day starts with offering rice in his house temple, household courtyard, at a road crossing, in field shrine, in subak shrine and numerous village temples. Presence of rice can be noticed in all *rites de passage* and other day to day social ceremonies.

Importance of rice in human life, culture and religion cannot be so emphatically described than done by Huggan.[147] To borrow his words "Rice is the mainstay of major population involved in its production. The loss of a rice crop can result in major human catastrophe." Experience must have been the teacher to our ancestors who counseled such eminence and divine reverence to this tiny grain and called it "grain of life." Thus the relationship among rice, religion and social tradition is complementary; rice has permeated into tradition and religion through sustenance and filling the mental appetite of humans.

In addition to being the source of sustenance, rice has become a commercial commodity and a food item with attachment. Change in agricultural practices has led to little time in between the crops.

The advent of fast maturing modern varieties led to widespread changes in traditional annual cycles of labor. Neighboring fields may no longer be worked in synchronized manner, sweeping away sense of organized seasonal activity.

Nowadays concept of Rice mother spirit is rare among young generation especially where modern varieties are grown but is still strong in isolated villages in Asia. Tradition of installing rice goddess image in fields is also disappearing as in Java. New system of transplanting varieties at any time was found to be at a discord with Goddess' erstwhile precepts and has finally made the villagers cease sowing her image in fields.

Accordingly festivals over the past three decades have been altered, redefined to confirm to rapidly changing conditions and are loosing their former function as guide to the actual practice involved in rice cultivation. Most of them have become source of tourist attraction and acquired political touch. Traditional rice festivals have been transformed to extravaganza especially in industrial countries. People in many places still believe and view their traditional rituals valuable as these have provided them with important psychological comfort in their dealings with vagaries of environment. Similarly granaries which were proud of each household are also disappearing as rice has become a commodity and directly taken to market from field.

References

1. Ahuja and Ahuja (2006).
2. Ahuja *et al.* (2000).
3. Williams (1941).
4. Majumdar and Banerji (1960).
5. Ahuja and Ahuja (2006).
6. Ahuja and Ahuja (2006).

7. Walker (1994).
8. Hamilton (2003).
9. Das and Mahapatra (1979).
10. Ismani (1985).
11. Raju (1978).
12. Hamilton (2003).
13. Taryo (1985).
14. Ahuja *et al.* (1997).
15. Kumar (1988).
16. Huke and Huke (1990).
17. Ismani (1985).
18. Ismani (1985).
19. Grist (1986).
20. Taryo (1985).
21. Zefaniasy (1985).
22. Piper (1993).
23. Piper (1993).
24. Terwil (1994).
25. Tuan (1985).
26. Taryo (1985).
27. Huke and Huke (1990).
28. Ahuja *et al.* (2000).
29. Tuan (1985).
30. Piper (1993).
31. Hamilton (2003).
32. Piper (1993).
33. Hamilton (2003).
34. Ismani (1985).
35. Hamilton (2003).
36. Hamilton (2003).
37. See website: http://www.blessingscornucopia.com/Myth
38. See website: http://www.k-mil.gr.jp/kie/en/life/custom.html
39. Piper (1993).
40. Ahuja *et al.* (2001).
41. Panigrahi (1999).
42. Kling (1985).
43. Singh (1992).
44. Bahadur (2003).
45. Yuxiang (1994).
46. Ahuja *et al.* (2001).
47. Kumar (1988).
48. Ahuja *et al.* (2000).
49. Piper (1993).
50. See website: htpp://www.istc.org/sisp/index
51. See website: http://www.istc.org/sisp/index
52. See website: http://www.kansai.gr.jp
53. Wikipedia.

54. Ahuja *et al.* (2001).
55. Das and Mahapatra (1979).
56. Hamilton (2003).
57. Hamilton (2003).
58. Hamilton (2003).
59. Chettiar (1973).
60. Crooke (1926).
61. Crooke (1926).
62. Ahuja *et al.* (2000).
63. Piper (1993).
64. Crooke (1926).
65. Das (1972).
66. Singh (1992).
67. Panigrahi (1999).
68. Panigrahi (1999).
69. Piper (1993).
70. Panikkar (1991).
71. Ray (1993).
72. Bhattacharyya (1978).
73. Raju (1978).
74. Crooke (1926).
75. Crooke (1926).
76. Hamilton (2003).
77. See website: htpp://www.istc.org/sisp/index
78. Hamilton (2003).
79. Hamilton (2003).
80. Hamilton (2003).
81. Ahuja *et al.* (2000).
82. Chettiar (1973).
83. Panigarhi (1999).
84. Ahuja *et al.* (2000).
85. Crooke (1926).
86. Singh (1992).
87. Hamilton (2003).
88. Piper (1993).
89. Hamilton (2003).
90. Hamilton (2003).
91. Ray (1993).
92. Hamilton (2003).
93. Roy (1912).
94. Yuxiang (1994).
95. Hamilton (2003).
96. See website: http://www.sridaladamaligawa.lk/english/rituals_in.html
97. Raju (1978).
98. Crooke (1926).
99. Das (1972).
100. Bhattacharyya (1978).

101. See website: htpp://www.istc.org./index
102. Kumar (1988).
103. See website: http://www.blessingscornucopia.com/Myths
104. Hamilton (2003).
105. Watts (1901).
106. Ahuja *et al.* (2001).
107. Hamilton (2003).
108. Hamilton (2003).
109. Hamilton (2003).
110. Ismani (1985).
111. Hanchett (1988).
112. Gupta (1971).
113. Ahuja *et al.* (2001).
114. Hanchett (1988).
115. Hanchett (1988).
116. Anonymous (1993).
117. Ahuja and Ahuja (2006).
118. Walker (1994).
119. Ahuja and Ahuja (2006).
120. Reid (1984).
121. Ismani (1985).
122. Hamilton (2003).
123. Reid (1984).
124. Hamilton (2003).
125. Hamilton (2003).
126. Panikkar (1991).
127. Hamilton (2003).
128. Crooke (1926).
129. Hamilton (2003).
130. Ponappa (1978).
131. Hanchett (1988).
132. Hamilton (2003).
133. Ahuja and Ahuja (2006).
134. Crooke (1926).
135. Hanchett (1988).
136. Hamilton (2003).
137. Crooke (1926).
138. Ahuja *et al.* (2001).
139. Walker (1999).
140. Bhattacharyya (1978).
141. Piper (1993).
142. Ahuja *et al.* (2000).
143. Piper (1993).
144. Piper (1993).
145. Zefaniasy (1985).
146. Tan (1985).
147. Huggan (1995).

4 | Domestication of Rice in China and Its Cultural Heritage

Shengxiang Tang and *Songnan Xuan*

Domestication and cultivation of graminaceous plants is one of the greatest inventions in human history. Wheat gave birth to the civilization of Asia, North Africa and Europe and maize is the basis of American civilization. The civilization of East Asia, South East Asia and South Asia is based on rice.

Asian cultivated rice (*Oryza sativa* L.) is the staple food crop cultivated widely not only in China but also in Asia. About 60% of Chinese population eats rice as their staple diet. In 2005, the total rice acreage in China was 28.85 million hectares and the total rice production reached 180.6 million tons with an average yield of 6.27 tons per hectare, accounting for 35.1% and 45.3% of all domestic cereal crop acreage and production respectively. Regarding rice ecosystems, irrigated rice occupies 95% of the total rice area, upland rice 2% and rainfed rice 3%. For sub-species, *indica* takes 74.5% of rice land and *japonica* 25.5%. From ecology and bio-environment view-points, rice paddy field is considered as the largest artificial wetland in China with main region in Central and Southern China.

Rice is the most ancient cereal crop of China with long history of its cultivation. The mythological writing *Guan Zi* in Shen-Nong Era (21st century BC, Xia Dynasty) recorded five food crops: *shu* (millet, P. *miliaceum*), *ji* (millet, S. *italica*), *mai* (barley and wheat), *dao* (rice) and *shuu* (soybean). The sub-species *japonica* rice was called as *keng* or *jing*; *indica* rice as *hsien, xia*; and glutinous (waxy) rice as *nuodao*. The concept of elite seed started in Xi-Zhou Dynasty (11th-8th century BC). By the time of Han

Dynasty (2nd century BC-2nd century AD), a technology of panicle selection had been widely practiced. Nowadays, based on many evidences, people believe that China has history of about 10,000 years of rice cultivation.

ORIGIN OF CULTIVATED RICE

Regarding domestication, dispersal and development of cultural practices of rice, a lot of information from the fields of rice genetics, taxonomy, archaeology, methodology, nationality, etc. has been accumulated. Many related papers have been published but some issues still remain to be settled.

However, any affirmation of the original place (ancestral home) of cultivated rice must meet at least four necessary conditions: First, *O. rufipogon* (progenitor of cultivated rice) exists at present or had existed in ancient time in the home area. Second, the conditions of the climate and other environmental factors in the Neolithic time or the concerned period were suitable for growth and domestication of *O. rufipogon*. Third, there should be evidences of human activity for cultivation of the primitive rice in ancient time in the homeland implying great selection pressure on the primitive cultivated races by ancient people as well as by the nature. And fourth, archaeological evidences regarding remains of ancient cultivated rice and related farming tools should be available in the area.

Ancestor of Chinese Cultivated Rice

Multi-disciplinary research has indicated that the immediate wild relative of Chinese cultivated rice (*O. sativa* L.) is the common wild rice— *O. rufipogon* Griff. This wild rice and its spontaneous form were distributed widely in South and Southwest China in ancient time and occur at present. In China, there are three species of wild rices: *O. rufipogon, O. officinalis* and *O. meyeriana*. The common wild rice *O. rufipogon* extends from Taoyuan county (121°15′ E) in Taiwan province in the East to Jinghong county (100°47′ E) in Yunnan province in the Southwest and from Sanya area (18°09′ N) in Hainan province in the South to Dongxiang county (28°14′ N) in Jiangxi province in the North.[1] The common wild rice, *O. rufipogon*, grows in marshlands near river banks, grassy lowland and hill slopes at elevation of 300-600 m in 113 counties of eight provinces, namely, Yunnan, Guangxi, Guangdong, Hainan, Hunan, Fujian, Jiangxi, and Taiwan. *O. rufipogon* found in Dongxiang county of Jiangxi province is considered to be the most northern distribution in China and probably world over. Due to human activities such as construction of houses, factories, roads, irrigation systems, reservoirs, etc. and eco-environmental pollution, it is a pity that about 70% of the recorded habitats of *O. rufipogon* have disappeared in the last three decades in China.

Based on Chinese old writings and the present archaeological findings, it appears that *O. rufipogon* has been growing in South and Central China (the basins of middle and low Yangtze River) since ancient time. The direct evidence is that four grains of common wild rice (*O. rufipogon*) (6950 ± 130 BP) with long awn and other wild characters were discovered along with 146 rice grains excavated at Hemudu site, Zhejiang province in East China (Fig. 4.1). By examining on scanning electronic microscope, they were considered to be grains of wild rice because of their following characters: (a) they have denser bristles on the long awn when compared with the awns of other rice grains excavated at Hemudu as well as with the grains of the present-day cultivated varieties (Fig. 4.2); (b) they show traces of natural shedding at maturity at the base point of the rachis of spikelets; and (c) they have narrow grain shape.[2] The findings reveal that *O. rufipogon* was growing in the area of lower Yangtze River in East China

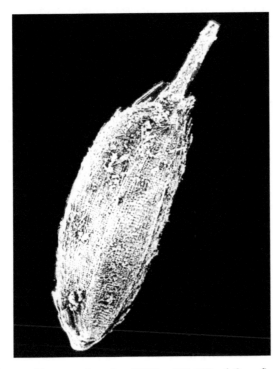

Fig. 4.1. Excavated grains (6950 ± 130 BP) of *O. rufipogon* from Hemudu site, China.

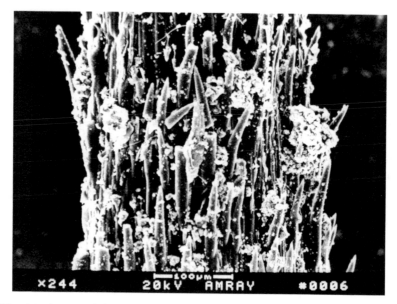

Fig. 4.2. A part of the awn of *O. rufipogon* grain excavated from Hemudu
site (6950 ± 130 BP), there are more dense bristles than on other
primitive cultivated grains from the same Hemudu site.

about 31° N at least 7000 years ago although *O. rufipogon* is not found at
present in this area. Another evidence is that an analysis on rice opal in
the soil and rice grains excavated from Jiahu archaeological site (8285-
7450 BP, 33°37′ N) indicate that Jiahu's ancient rice grains have some
characters of both primitive cultivated rice and *O. rufipogon*.[3]

In historical records, common wild rice (*O. rufipogon* and its weedy
form, *spontanea*) was mentioned as *Ni, Li* and *Lu*. As recently as 1950,
spontanea wild rice of the *keng* type was found in Donghai county, Jiangsu
province about 34.5° N.

Place of Origin

Regarding the origin of cultivated rice, some researchers in 1960s and
1970s proposed that the Chinese rice originated in South China[4] or
Yunnan-Guizhou plateau in Southwest China.[5] The present hypothesis,
however, is that Chinese rice could have originated in Central China,
more specifically in the region of the middle-lower Yangtze River, and the
most primitive rice cultivation can be traced back to about 10,000 years
ago.

Distribution of *O. rufipogon* in Ancient Time

At present, the common wild rice *O. rufipogon* grows in eight provinces in Southern and Southwest China. The most northern habitat called as "Dongxiang wild rice" (28.1° N) is in Jiangxi Province, southern side of Yangtze River. The paleo-meteorological research indicates that the climate of Yangtze River basin during the Neolithic Age (10,000-7,000 years ago) was warmer (about 3-4°C higher) and more humid (more rainfall by 800 mm) than what it is at present and hence was more favorable for the distribution of *O. rufipogon*. The distribution of *O. rufipogon* 10,000-7,000 years ago may have reached northern side of Yangtze River, even up to the southern Huai River.[6] There were 16 records of *"Ni"* and *"Li"* (common wild rice) described in various ancient books mentioning easy shedding and growing naturally in the fields. The northernmost distribution of this self-perpetuating wild rice would have been up to 38° N latitude i.e. up to the Gulf of Qingli Mountain.

Rice Remains Excavated

Till now more than 192 rice remains (grains, hulls, stems, leaves, etc.) spread over a span of 1500 years or more have been excavated in the Central, South and Southwest China. Of these, 144 rice remains belong to Neolithic Age (Fig. 4.3). Among these, there are 16 remains that are the most ancient dated 12,300-7,000 years BP and located in the region of middle-lower basin of Yangtze River (Table 4.1).

In the 70s of the last century, a great quantity of rice remains together with many bone and stone tools used for primary land preparation and harvesting as well as of sub-tropical plants and animals was excavated from Hemudu site (29°58′ N, 121°22′ E) located in the delta of lower Yangtze River, Zhejiang province. Thousands of carbonized rice grains dated 6950 ± 130 BP were found together with rice leaves and straw from these relics (Fig. 4.4). The grains showed yellow-golden colour when they were just excavated from the soil layer but changed immediately to carbonized black. The dimensions of the carbonized grains vary considerably in length and width, being judged as a mixed population of primitive *japonica* rice, primitive *indica* rice and a few of common wild rice (*O. rufipogon*) based on their grain length, width, shape and size as well as the characters of bi-peak-tubercles on lemma and opal.[7] From the Neolithic relics in Jiahu village (33.4° N, 113° E) of Henan province, located near southern region of Huai River, hundreds of carbonized rice caryopses dated 8942-7868 BP were excavated in early 1990s.[8] Morphological study indicated that the major part of Jiahu ancient rice grains belonged to

Table 4.1. Sixteen carbonized rice remains of age 12,300-7,000 years BP, excavated in China

Relics	Location	Era	Rice remain
Hemudu, Yuyao county, Zhejiang province	Lower reach of Yangtze River	6950 ± 130 BP	Thousands of carbonized rice grains, a few of *O. rufipogon* grains
Luojiajiao, Tongxiang county, Zhejiang province	Lower reach of Yangtze River	7040 ± 150 BP	Hundreds of carbonized rice grains, majority of primitive japonica rice
Tongjiaao, Cixi county, Zhejiang province	Lower reach of Yangtze River	About 7000 BP	A few carbonized rice hulls
Shangshan, Pujiang county, Zhejiang province	Lower reach of Yangtze River	11,400-8,600 BP	A great number of carbonized rice hulls in pottery
Kuahujiao, Xiaoshan county, Zhejiang province	Lower reach of Yangtze River	8220-7660 BP	Thousands of carbonized rice grains and hulls
Tianluoshan, Yuyao county, Zhejiang province	Lower reach of Yangtze River	7000-5600 BP	Hundreds of carbonized rice grains
Erjian village, Lianyungang city, Jiangsu province	Middle reach of Huai River	7885 ± 480 BP	A few carbonized rice grains
Lijia village, Xixiang county, Shaanxi province	Middle reach of Yangtze River	About 7600 BP	Vestige of rice hulls on burned soil
Pengtoushan, Li county, Hunan province	Middle reach of Yangtze River	8200-7450 BP	Vestige of carbonized rice grains and hull on roughcast earthenware
Jiahu village, Wuyang county, Henan province	Upper reach of Huai River	8285-7450 BP	Hundreds of carbonized rice grains of primitive *japonica* and *indica* rice
Longqiu village, Gaoyou county, Jiangsu province	Lower reach of Yangtze River	6300-7000 BP	Hundreds of carbonized rice grains
Bashidang, Li county, Hunan province	Middle reach of Yangtze River	8000-9000 BP	Hundreds of carbonized rice grains

Contd.

Table 4.1 continued

Relics	Location	Era	Rice remain
Yuchanyan, Dao county, Hunan province	Middle reach of Yangtze River	12,300 ± 1,200 BP	A few carbonized rice grains and hulls, some with mixed characters of *indica*, *japonica* and *O. rufipogon*
Dulinau, Chalin county, Hunan province	Middle reach of Yangtze River	About 7000 BP	A few of carbonized rice grain
Chengbeixi, Zhicheng county, Hubei province	Middle reach of Yangtze River	8000-7000 BP	A few of carbonized rice hulls in pottery
Zhichengbei, Zhicheng county, Hubei province	Middle reach of Yangtze River	About 7000 BP	A few carbonized rice hulls in pottery

approaching *keng* (*japonica*) type and the minority approaching *hsien* (*indica*) type. In 1980s, many carbonized rice hulls (dated 8200-7450 BP) were discovered in earthenware from Pengtoushan relics, near southern basin of middle Yangtze River, Hunan province with many bone and stone tools roughly found in the same layer. In middle 1990s, the rice hulls and rice opals dated 12,000-10,000 BP were excavated from the soil layer together with plants, animals, stone and bone tools from Yuchanyan relics (25°30′ N, 110°30′ E), Dao county, Hunan province. Yuchanyan rice was considered as the most ancient and primitive cultivated rice of about 10,000 years old.[9] In 1996, hundreds of Bashidong rice grains dated 9000-8000 BP were excavated from Li county, Hunan province. Bashidong rice was identified as small grains of ancient cultivated rice like *keng* (*japonica*) in length and *hsien* (*indica*) in width. In recent years, some of the carbonized ancient rice grain hulls in pottery, tested to be about 11,400-8,600 BP, were found in the relics of Shangshan village, Pujiang county, Zhejiang province. In 2002, thousands of carbonized rice grains and hulls with many primitive farming stone and bone tools, dated 8220~7660 BP, were excavated from Kuahujiao relics, Xiaoshan County, Zhejiang Province located in the low reach of Yangtze Rive, East China.

Centres of Cultivar Diversity

Isozyme analysis of 700 Chinese rice landraces using 9 isozyme loci showed that there were four centres of genetic diversity for the cultivated

92

Fig. 4.3. 144 Neolithic relics (12,300-3,500 BP) with various rice remains excavated in China.

Fig. 4.4. The rice grains excavated from Hemudu relics sites in lower Yangtze River (6950 ± 130 BP).

rice: the Yunnan-Guizhou plateau, upper reaches of Huai River, middle-lower reaches of Yangtze River and South China.[10] Yunnan-Guizhou plateau is considered not to be the original home of Chinese cultivated rice; it is probably the secondary centre of diversity for the cultivated rice and is strongly influenced by the cultivated rice of South Asia. Tang et al.[11] investigated allozyme variation pattern of 6632 Chinese rice cultivars obtained from core collection based on 12 allozyme loci, *Pgi1*, *Pgi2*, *Amp1*, *Amp2*, *Amp3*, *Amp4*, *Sdh1*, *Adh1*, *Est1*, *Est2*, *Est5* and *Est9*. The analysis showed that allele frequencies ranged from 0.001 for *Amp3-1* to 0.976 for *Est5-2*. The gene diversity indices (*He*) varied from 0.017 for *Amp4* to 0.583 for *Est2*. For the 12 allozyme loci, the average gene diversity index (*Ht*), Shannon-Wiener index (S-W index) and degree of polymorphism (*DP*) were 0.271, 0.468 and 21.4% respectively. In terms of allozyme diversity in the six rice eco-regions, namely, South China, Central China, Southwest China, North China, Northeast China and Northwest China, the rice cultivars of the Southwest Plateau and Central China have richer genetic diversity. The results revealed that the region of the middle-lower Yangtze River is one of the centers of genetic diversity of Chinese cultivated rice.

Human Activities Associated with Primitive Rice Cultivation in Neolithic Period

The true domestication process might have first taken place in the region of middle-lower Yangtze River because this region is located in the northern edge of the temperate zone. The cooler weather might have resulted in greater food pressure for survival. The ancient people did some primitive cultivation to supplement their food as not enough meat, fish or fruit could be obtained by hunting-gathering. The excavated rice remains (grains, hulls, straws and leaves) found at relics of Hemudu, Pengtoushan, Yuchanyan, Bashidong, Jiahu, Shangshan, etc. were not only large in quantity but also the oldest in age. Besides, the 16 oldest Neolithic sites shown in Fig. 4.4 have not only large amount of rice remains but also have many bone spades, stone tools and bifacial asymmetrical axes probably being used by ancient people for primitive crop cultivation including primitive rice cultivation.

Remains of Ancient Cultivated Rice Field

The discovery of ancient cultivated rice fields may have special significance for studying the domestication of rice. In the mid-1990s, archaeologists excavated 33 ancient rice paddy fields going back to 6000 BP in Caohaishan relics, Wu County, Jiangsu Province (lower basin of

Yangtze River). The ancient paddy fields vary in size ranging 0.9 m²~12.6 m² with shapes like rectangle, ellipse and irregular and 20 cm~50 cm deep. These rice fields were connected with each other through ditches with small water pools.[12] In Tenghualuo relics (3900-3500 BP) of Lianyungang, Jiangsu Province, not only hundreds of ancient *japonica* rice grains but also an ancient paddy field of about 67 m² size having rough ridges and connected with a water ditch and small water pool have been excavated.

In 1997, a larger ancient rice field with three ridges was found in Pengtoushan relics (6629 ± 896 BP) of Li County, Hunan Province. A longitudinal section of the paddy field clearly showed the carbonized bases and roots of many rice plants suggesting that the seeds were broadcast (direct seeded). Besides, a primitive irrigation system with three small water pools and three ditches was also found. The water pool was at a higher level than the paddy field, had a diameter of 1.2 m~1.6 m and depth of about 1.3 m and was connected to the paddy fields with ditches. It is believed that this ancient paddy field and its irrigation system may be the oldest evidence of rice cultivation in China. This ancient paddy field also indicates that Chinese cultivators exercised great care in water control (irrigation and drainage) from early days of rice cultivation.

So far, all the evidences (distribution of *O. rufipogon* in ancient time, Neolithic primitive cultivated rice grains excavated, old paddy fields and irrigation systems, related stone and bone farming tools excavated, genetic diversity of traditional cultivated rices, human activities in ancient time for primitive rice cultivation, etc.) strongly support the hypothesis that the original homeland of rice could be the middle-low reaches of Yangtze River than anywhere else. Considering that the domestication from common wild rice to primitive cultivated rice must have been a long and gradual process, it is reasonable to believe that the history of primitive rice cultivation in China is about 10,000 years.

Domestication and Expansion of Cultivated Rice

In the long process of evolution and domestication of rice, under strong selection pressure by mankind as well as by nature, the primitive cultivated rice has undergone a series of changes with regard to its agricultural and physiological characteristics. During the multi-directional dissemination of rice to various latitudes and elevations, plains and plateaus, many biological and non-biological factors (such as temperature, sunshine, rainfall, planting season, soil, sowing and trans-planting method, diseases, insect pests, etc.) have affected domestication and development of primitive cultivated rice. For example, the primitive Chinese cultivated rice has been differentiated into *hsien* rice (*indica*) and *keng* rice (*japonica*), paddy rice and upland rice, waxy rice and non-waxy rice, early rice and late rice, etc.

Differentiation of *indica* and *japonica* Rice

Many workers consider *indica* and *japonica* as two subspecies of rice (*O. sativa*) because crosses between them produce partially sterile F_1 hybrids. Besides, there are many characters to differentiate them though all these characters show overlapping variation. Generally speaking, *indica* rice (*hsien*) is mainly planted in South China while *japonica* rice (*keng*) in North China (roughly above 29° N) as well as in hilly areas with higher altitude. However, the Central China grows both the types.

Some early workers thought that upland rice developed from *indica* rice under upland conditions or probably from some plants of *O. rufipogon* in upland condition. Then the upland rice developed into primary *japonica* rice in the conditions of lower temperature, insufficient water supply, shorter growing period and human selection. The early *japonica* cultivars were further differentiated into the glabrous group on the hills and awned group in the valleys.

Nowadays, many workers consider that incipient *indica-japonica* differentiation existed in *O. rufipogon*.[13] In the long process of evolution, the *indica* rice evolved from the *indica*-like wild rice and the *japonica* rice from the *japonica*-like wild rice. Based on his studies on isozyme variation of cultivated rice and *O. rufipogon*, Second[14] opined that *indica* and *japonica* rices were domesticated independently from different progenitors of *O. rufipogon* implying that China could be the centre of origin of *japonica* rice. After isozyme analysis of several allele genes of *Est*, *Cat*, *Acp* and *Amp* of Chinese wild rice *O. rufupogon*, Cai[15] pointed out that the *indica-japonica* differentiation has taken place in the Chinese *O. rufipogon* to some extent. In terms of isozyme variation, *O. rufipogon* seems to resemble *japonica* at sites of higher latitude such as Dongxiang and Chaling in Jiangxi and Hunan provinces whereas it resembles *indica* in lower latitude regions in Guangdong and Guangxi provinces. The isozyme pattern of a traditional *japonica* landrace in Jiangxi province is more like that of *O. rufipogon* of Dongxiang. Chinese *keng* rice (*japonica*) originated from *japonica*-like common wild rice, meaning *hsien* (*indica*) and *keng* rices were domesticated in parallel.[16] The oldest specimens of primitive *japonica* rice in the basin of middle-lower Yangtze River were found in Qiangshanyang (5311 ± 136 BP) and Lujiajiao relics (7040 ± 150 BP), Zhejiang province and in Jiafu relics (8942-7868 BP), Henan province.

Sun[17] studied 76 cultivated varieties and 118 accessions of *O. rufipogon* by RFLP analysis using 7 probes. The result showed that *indica-japonica* differentiation was the main trend in mitochondrial DNA (mtDNA) of cultivated rice. 73% of *O. rufipogon* had *indica*-type mtDNA, 6% *japonica*-type mtDNA and 21% wild rice type. Based on the analysis on chloroplast DNA (cpDNA) of 154 accessions of *O. rufipogon* and 94 cultivars by PCR for the Open Reading Frames (ORF100), there is *indica-japonica* differentiation in cpDNA of cultivars and, to some extent, in *O. rufipogon*.[18] Most plants in Dongxiaiang *O. rufipogon* (higher latitude) are *japonica*-like in their cpDNA but most *O. rufipogon* plants of Guilin and Fusui are *indica*-like suggesting that *indica* and *japonica* cultivated rices originated from *indica*-like and *japonica*-like wild rice respectively.[19]

However, the *indica-japonica* differentiation of *O. rufipogon* was incipient only, showing only a latent tendency and only the potentiality for such differentiation. Similarly, the ancient rice grains could not be clearly identified as *indica* type or *japonica* type according to the present criteria. Actually, ancient cultivated rice was a mixture population undergoing domestication from *O. rufipogon* to cultivated rice thousands of years ago. For example, the shape of rice grains excavated at Bashitan (8000-9000 BP) has tendency towards *japonica*, grain width towards *O. rufipogon* and L/W ratio to *indica*. The L/W ratio of rice excavated at Jiahu (8285-7450 BP) has the tendency towards *indica* but its opal is like *japonica*. Rice excavated at Hemudu (6950 ± 130 BP) has grain shape tendency to *indica* but its bi-peak tubercle to *japonica*. In the plateau of Yunnan province, *indica*, intermediates of *indica* and *japonica* and *japonica* varieties are still found in the rice fields at lower (< 1400 m altitude), middle (1400-1750 m) and high elevations (> 1750 m) respectively.

Based on multidisciplinary studies on ecology, biology, genetics, archaeology, regional distribution of the two types, etc., it can be safely assumed that *indica* rice (*hsien* rice) and *japonica* rice (*keng* rice) both might have originated and been domesticated in parallel from *O. rufipogon* during the same period in the low and middle Yangtze River, respectively.[20]

Direct Seeding to Transplanting

Changing the cultivation technique from direct seeding to transplanting of seedlings must have been a pioneering work in the history of rice cultivation. The technique of transplanting is considered to have first appeared during Han dynasty (25-220 AD) in China. According to the ancient book *Siminyueling* of Han Dynasty, "May could be the season for biedao and basket" in which "biedao" means transplanting rice. The earliest record of how to transplant rice seedlings is in the book *Qiminyaoshu* written by Jiasixie during Beiwei dynasty (386-534 AD) when rice transplanting started becoming popular in the 6th century. In this book, the words used to describe rice transplanting are "replanting rice seedlings of the height of 7-8 inches into soil." In ancient time, the main purpose for transplanting rice seedlings was to avoid weeds. Another purpose of transplanting seedlings was to save time of limited duration of growth in the field. This could have promoted multiple rice cropping in South China during Tang (618-907 AD) and Song (960-1279 AD) dynasties. In the primary stage of rice transplanting during Beiwei dynasty, there was no special rice nursery for sowing. Following the development of transplanting technique, raising seedlings in the nursery and then transplanting in the field became the key steps in rice production during Tang (618-907 AD) and Song dynasties (960-1279 AD).

Unfortunately, the book *Qiminyaoshu* has recorded transplanting rice seedlings in North China only; we have no way of knowing if this technique existed in South China during that period. If it existed, how was it done? Transplanting technique in North China might have started from upland rice cultivation. The ancient book *Iminyaoshu* records that "in long rainy days of May and/or June, pulling out and replanting seedlings" meaning thereby that if there are too many tillers with higher density, expletive seedlings should be removed and replanted at other places during raining days of May and/or June. In that ancient time, farmers just practiced direct seeding; no matter it was paddy (wetland) or upland rice. But when they realized the advantages of transplanting, farmers started to improve transplanting technique step by step.

Expansion of Cultivated Rice

Rice is basically a semi-aquatic plant. The relatively warm climate and swampy habitat south of Yangtze River allowed wild rices to be easily self-perpetuated by shattering of grains. Gathering grains from wild population from low-lying areas must have been practised for several millennia prior to its domestication about 10,000 years ago. Archaeological findings in early sites in Hemudu, Pengtoushan, Yuchanyan, Bashidong, Jiahu, Shangshan, etc. indicate that rice, shellfish, pigs, fruits and legumes constituted the main diet from the dawn of agriculture. The stone and/or bone spades and hoe blades excavated were probably used in preparing the paddy fields. The rice growers lived in houses made of cut timber joined by mortise. Clay pots served as cooking utensils.

The hypothesis that rice spread from China implies that Chinese cultivated rice originated in the basin of middle-low Yangtze River and spread like wave upon wave.[21] Before 7000 years, primitive rice was grown in the reaches of middle-low Yangtze River, especially in the low reach. About 6000-5000 years ago, rice spread to the whole reach of Yangtze River and up to Huai River about 34° N. About 5000-4000 years ago, rice cultivation had reached North and South China. When primitive cultivated rice disseminated to higher elevations, hills, valleys and plateaus, it gradually adapted to the environmental conditions of less water supply and cooler weather. Some changes such as long and thick roots, glabrous (non-hairy) leaves and glumes, heavy grains, long and non-shattering panicles, etc. occurred in the process of domestication under nature as well as human selection. Man made full use of natural variability in rice, spontaneous mutations, natural hybrids and introductions from other areas. Based on the shape of grains excavated from different sites and ages, rice during 8000-6000 BP did not have significant changes in grain length and width. But during 6000-5000 BP, grain length and width as well as grain weight increased significantly, implying the increase in human selection pressure.[22]

Japonica rice spread to Korea and Japan probably by one or/and two or/and three routes: from Shandong peninsula to Korean peninsula at least before 6th century BC and then to Japan around 3rd century BC, or from the delta of low Yangtze River to Korea and/or Japan and/or from Fujian province of Southeast China to Taiwan and then to Japan.

Rice is one of the principal foods of the people of China. It has the longest history, widest distribution and most diversified end-uses in China. Customs of eating rice and its products vary with regions, nationalities, climate, seasons and food habits. Due to wide expansion and multi-ethnic population of China, rice foods show strong local and distinctive features with various nationalities in old as well as present days.

Cooked rice: Cooked rice refers to milled rice that is simply cooked with water. Rice is, however, easier to cook than other cereals and has the highest digestible energy. Generally, in China, cooked rice can be classified into three kinds as steamed rice, braised rice and boiled rice. If meat, fish, vegetable, legume, etc. are added properly while cooking, the cooked rice becomes of distinctive flavor rich in nutrients. Besides, if medicinal ingredients or extracts are added while cooking, rice becomes suitable for certain types of patients.

Rice porridge: Rice porridge is a liquid or semi-liquid food cooked with plenty of water. Easily digested and absorbed, rice porridge with abundant water is especially agreeable to patients and children or serves as summer diet. Cooked together with meat, fish, vegetable or pharmaceuticals, rice can be made to distinctive flavored porridge or pharmacological porridge.

Ground-rice products: Milled rice is grounded into rice flour which is then processed as products of various kinds such as rice cake, rice noodle, New Year cake, stuffed dumpling, sweet dumpling, pudding, bread, cracker, fermented food, etc. Rice cakes can be of many colors if ground rice is red or purple or if colored materials are added to rice flour.

Glutinous rice dumpling: Glutinous rice is first soaked in water and then wrapped with proper stuffing inside special bamboo leaves giving various shapes to it for cooking. It tastes nice for its softy, glutinousness and faint scent and is preferred especially by minority nationalities in the south and southwest regions. In Central China, the famous glutinous rice dumpling is called as *zhongzi* with or without meat and legumes inside.

Rice beverage: Rice is a good material for various alcoholic beverages or types of beer. The famous red rice wines called *Jiafan* and *Nverhong* in Shaoxing area in Zhejiang Province are both made from high-grade *japonica* glutinous rice. The sweet rice wine of red or white color is a traditional homemade drink made from *indica* or *japonica* glutinous rice quite popular in China.

Rice bran: Rice bran is the by-product of rice milling; it contains the embryo, aleurone layer and pericarp of rice grain. Because of the higher content of proteins, oil and vitamins, rice bran is edible or is used as the material for producing oil and feed for fowl and livestock.

Rice hull or husk forms about 20% of the rough (unhusked) rice by weight. It is used as fuel, bedding and incubation material and is sometimes incorporated in livestock feeds. In recent years, rice hull or husk is used as a basic incubating material for mushroom production.

RICE PRODUCTION

Rice Production before 20th Century

Although no information is available concerning the size of the human population and rice productions in prehistory, it can be inferred that the spread of rice cultivation led to larger increase in population than was the case with other food crops.[23] Viewed in historical perspective, rice yield in Qin dynasty (221-220 BC) was much lower, about 1.87 t/ha. With the progress of cultivation technology and development of irrigation system, rice yield gradually increased.[24] Table 4.2 shows the trend of rice yield in the long history of China; this is, however, based on limited information. Because no data are available on area planted with rice during different dynasties, it is difficult to estimate the total rice outputs.

Table 4.2. The trend of rice yield in long history in China

Dynasty	Yield of rough rice (t/ha)	Comparison with Qin dynasty (%)
Qin (221-220 AD)	1.78	100
Dongjin (317-589 AD)	1.97	105.2
Tang (618-907 AD)	2.58	137.6
Song (960-1279 AD)	2.90	154.8
Yuan (1279-1358 AD)	2.90	154.8
Ming (1368-1644 AD)	2.76	147.2
Qing (1644-1840 AD)	2.80	149.6
Former Rep. of China (1931-1947 AD)	2.47	132.0

Based on You, X.L (1992).

According to the old records, during the rule of Song Dynasty, around 1011 AD, the Fujian province of South China suffered from an acute famine caused by severe drought. In this period, a rice variety with short growth duration, called as "Zhancheng Dao (Champa rice)", was introduced from Vietnam. "Zhancheng Dao" was early maturing, drought resistance and insensitive to photo-period. Farmers liked this rice because it matured 20-30 days earlier than other landraces, saving people from hunger to some extent. For this reason, the variety "Zhancheng Dao" spread widely and played an important role in the development of early *indica* rice in the of middle-lower reaches of Yangtze River. During Ming and Qing Dynasty (13th-19th century), some elite rice mutants (early maturing, large grains, long panicles, etc.) were selected from the natural population of rice plants and their grains were kept separately for seeds. Some early-maturing varieties with higher yield such as "Jiugongji" and "Houxiaji" were developed through such method for lean periods. The book on agriculture, *Shoushitongkao* (1742 AD) mentions that there were 3429 landraces of rice distributed over 16 provinces (223 counties) in China as a result of selection over a long period of rice cultivation.

There is an interesting story about the "Emperor Rice" of Qing Dynasty. In 1680s, one day the Emperor Xuan-Ye of Qing Dynasty found a rice plant in the rice field near his palace in Beijing. This rice plant was maturing in a shorter period than other varieties. So he picked it up and planted it in his palace garden for many years. The Emperor Xuan-Ye thought that this rice with shorter growth duration and higher yield could be planted in Central China as early rice or as late rice for double-cropping. He ordered local officers to extend its cultivation in Central China. The officers and farmers named this rice as "Emperor Rice". This "Emperor Rice" became very successful for early rice cultivation and was widely planted in Yangtze River region (Zhejiang, Jiangsu, Jiangxi, Hunan provinces) though it was not so suitable for late cultivation. Indeed, "Emperor Rice" was grown for more than 300 years over large areas. Even in the early part of 20th century, the "Emperor Rice" was being cultivated in many counties in Central China.

Rice Production in the Early Half of 20th Century

At the beginning of the 20th century, rice area in China was about 14 million ha with total production of 1870 million tons. With the increasing population, area under rice cultivation went on increasing. By 1946, it had reached 25 million ha with production of 36.2 million tons, the yield was only 1.3-2.4 t/ha. It was mainly grown as a single crop in a year.

According to available reports, in the early 20th century, the major disease and insect pest of rice were blast (*Pyricularia oryzae*) and striped rice stem borer (*Chilo suppressalis*) and there were no effective ways of controlling these pests though their incidence could be reduced by methods of cultivation.

The important abiotic stresses to rice production in earlier days in China were coldness, drought and soil salinity. Cold damage occurs frequently at seedling stage of early rice and at booting to flowering stage of late rice in Central China. Cold wind from north in middle/late September may also delay or damage panicle initiation and heading of late rice leading to poor flowering, poor seed setting and substantial loss of yield. In North and Northeast China, low temperature in some years causes cold stress, reduced plant growth, poor tillering, small panicle, poor seed setting and sharp reduction in yield.

Drought damage is mainly caused by unbalanced distribution of rainfall or reduced rainfall during growth period. Aridity and less rainfall in Northwest and North China eventually result in inadequate growth, leaf blight, poor pollination and great reduction of yield. The same damages to rice plant appear in the hills, mountains or in poorly irrigated areas in Central, South and Southwest China when rice suffers from high temperature (> 40°C) and extreme drought in summer at booting and heading stages.

In the Southeast costal areas, high salinity of soil can damage proper growth of rice plant. In Northwest and North China, salt content of soil may even reach 0.2-0.6% in some farmlands, creating the same problem. If drought occurs simultaneously, the damages due to soil salinity can still aggravate due to high evaporation from soil. Drought and salinity have been the prominent drawbacks for rice cultivation in rice growing regions of North and Northwest China.

Genetic improvement of rice started in China in the beginning of 20th century with single seed or panicle selection and simple hybridization. The early-maturing *indica* rice "Nantehao" was developed from the landrace "Boyangzao" through pure line selection. "Nantehao" proved to be an outstanding early variety in 1940s and 1950s especially in areas where two crops of rice were raised in a year. It occupied a large area of production and served as a key parent for early *indica* rice breeding. As many as 258 varieties were bred having "Nantehao" as one of their parents. Professor Ding-Ying, the pioneer of rice breeding in the early 20th century in China, started screening elite plants from the progeny of natural hybrids between cultivated *indica* rice and the common wild rice and bred a series of rice varieties like the famous variety Zhongshan 1. However, during 1910s to 1940s, thousands of traditional landraces of rice were being cultivated in China along with a few improved varieties that yielded 2-3 t/ha.

Rice Production since Establishment of PR China

Since the establishment of the People's Republic of China in 1949, rice research and production have achieved great progress. From 1949 to 2007, area planted under rice has increased from 25.71 million ha to 29.23 million ha i.e. an increase by 13.7%; grain yield from 1.89 t/ha to 6.38 t/ha i.e. increase by 237.6% and the total rice production from 48.64 million tons to 186.50 million tons i.e. increase by 283.4% (Fig. 4.5).

With regard to trend of rice production, PR China has gained giant steps four times. The first step was in 1950s when production increased due mainly to the increase of rice acreage. The total rice production increased from 48.64 million tons in 1949 to 80.85 million tons in 1958 due to an additional area of 6.19 million ha (from 25.71 to 31.9 million ha) under rice. Meanwhile, rice yield increased rather slowly by 0.6 t/ha (from 1.9 to 2.5 t/ha). The release and the wide adoption of elite semi-dwarf varieties in 1960s pushed rice production to 125 million tons in 1975 and the yield reached 3.5 t/ha. From the late of 1970s, hybrid rice added a new dimension to rice productivity; it occupied 58% of the national rice area

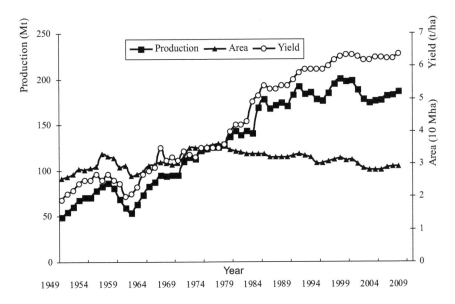

Fig 4.5. Area, yield and total production of rice in China (1949-2007).

by the mid-1990s. The development and extension of hybrid rice was the third step as it led to a breakthrough in yield of rice; it reached a national average of 6.0 t/ha in 1995 and the total production of over 1850 million tons.

Starting from 1996, China has initiated the super rice breeding program at the national level, marking the fourth step. As things now stand, the main targets of super rice breeding program in China are (a) high-yield: the yield potential targets of major inbred varieties and hybrids should be adjusted based on specializations such as different regions (Southern China, Central China and Northern China), subspecies (*indica* and *japonica*) and types (early rice, late rice and single crop rice), (b) good quality to meet market needs: breeding for specialty rice such as Fe, Zn and V_A rich rice and high lysine rice could be conducted, and (c) resistance to major diseases and pests and tolerance to abiotic stresses. In northern *japonica* rice regions, blast resistance and cold tolerance are essential while in *indica* and *japonica* mixed rice regions of Southern and Central China, multiple resistance to blast, bacteria leaf blight as well as to brown plant-hopper and leafhopper will be the main considerations. It is commonly understood that narrow genetic base of restorer lines and hybrids is one of the bottlenecks that limit the development of super rice breeding. The key technologies of super rice breeding are the utilization of elite gene(s) from cultivated varieties and wild rice, the construction of 'ideotype' plants, the production of strong subspecies heterosis by crossing *japonica*, *indica* and *javanica* and the combination of conventional breeding methods and biotechnology tools for developing three-line and two-line hybrid rice and inbred lines. The super rice breeding has been successful in the past 13 years. As a result, though area planted under rice has reduced from 30.7 million ha in 1995 to 29.23 million ha in 2007, its yield has increased from 5.9 t/ha to 6.38 t/ha. It is expected that with the success of super rice breeding, the national rice yield may rise to 6.9 t/ha after 10 years and 7.5 t/ha by 2030.

Ever since its domestication and cultivation about 10,000 years ago, primitive cultivated rice has differentiated into various ecotypes and landraces with significant changes in its characters and adaptations to different environments. Up to 2005, 49,348 traditional rice cultivars (landrace) have been collected in China, including *indica*, *japonica*, upland, lowland, waxy and non-waxy rices.[25]

RICE CULTURE

Posting Chinese Civilization

China developed its civilization based on its own agriculture. In ancient writings, *Sheji* represents the country in Chinese culture, in which *She* and *Ji* mean land god and crop god respectively. The ancestors of Chinese people made eternal linear time into loop cycles using Heavenly-Stem and Earthly-Branch system and designated 24 solar terms. In the ancient carapace-bone script, Chinese character is like someone who has put rice panicles on the head, meaning when grains ripe, one more year passes. In ancient time in China, the emperors used to inaugurate the first plowing or/and the first harvest. The emperors of Nansong Dynasty (1127-1279 AD) used to start the first plowing in a special field called "Eight Diagrams Land". The sites of prayer for good harvest were usually the "Temple of the Heaven", the "Temple of the Land" as well as in the "watching farming tower" in the temple of *Shennong*—the god of the cereals. All these mean the idea of "the first thing for the people to live is the food" and "the country is ruled by its agriculture".

If culture means what we achieve materially and spiritually through our activities for our existence, then rice has definitely provided us with the material and spiritual values through its cultivation: our myths, religion, lifestyle, food, implements of production, manufacturing technique, manufacturing tradition, etc.

The Legend: Rice has come from Heaven

There are a lot of myths and legends relating to the origin of rice, especially, among the minorities who live in the southeastern part of China. In most of these legends, there would be a hero to steal rice seeds from heaven with the risk of his life. And usually, there would be a dog, a sparrow, a mouse, a leech and other animals to help the hero in his act of stealing rice seeds. For this reason, when people perform the ceremony 'first eating of the new rice,' they let these animals eat rice first. In *LiJi* (an ancient book of Han Dynasty), there is the record about the emperor letting the dog taste rice first after the autumn harvest. Some minorities in China (like Miao, Naxi, Lisu, Hani, Yao, Jingpo, Bulang, Zhuang, Shui, Buyi, etc.) have the legend of a dog helping the hero and the custom of letting a dog eat rice first.

Celebrations during Rice Cultivation

There are various customs associated with rice cultivation. However, these rice-related customs are rare in the more developed rural societies. The traditional rice-based festivals are found in the minority areas of south-west China due to their isolation for social or economical reasons or due to lack of communications. For these minority people, rice is the basis of life and grain means fortune. Various sacrifices during rice cultivation are the most important events in a year. Hard efforts during rice cultivation and sincere sacrifices are indispensable for them.

Land-breaking Ceremony

"Miao" minority living in southeast part of Guizhou province has the traditional custom of 'land-breaking ceremony'. The master of every family carries a hoe, a torch made of rice straw and a sign to his rice field for the 'ceremony of land-breaking' in the morning of the first day of the spring. First, he hoes the land three times, then puts the torch and the sign in a mound to express that this rice paddy belongs to him and all animals like rats, birds, insects and plant diseases can't invade and occupy the field and wishes that the weather will be favorable for raising the crop. After this day, people can start farming operations on their rice fields.

Seeding Festival

After the third thunderstorm in the spring, people of "Miao" minority start seeding rice. The first "tiger" day or "rabbit" day (by the Chinese lunar calendar) after thunderstorm is the seeding day because Miao people believe that "tiger" and "rabbit" days are lucky days. The head of the village who has settled first in the village will sow a handful of rice seeds in his own rice field and then others start seeding. It is a mark of respect for the head of the village and also a way of acknowledging his authority in the village. It also means that if all the villagers abide by the regulations, rice crop cannot be damaged by insects or outsiders. After seeding, they put a sign made by a bunch of *cogon* grass because they believe that this sign will help grains germinate vigorously. But some times, the sign could be something different like a 'Dongting' which means a stick to counteract evil forces in Yangpai village, Leishan county, Guizhou province. 'Dongting' is formed by three sticks. One has a wooden chip 5 cm long on its top, the other a red chilli banded and the third a piece of charcoal. It is believed that 'Dongting' could prevent the damage from ominous words by people passing through as well as from diseases and insect pests.

Rice Transplanting Ceremony: Opening Rice Seedling Gate

The ceremony for 'opening rice seedling gate' is called 'Kaiyangmen' in Chinese; it means 'starting a happy rice transplanting for a bumper harvest'. It is also called "Qijiang" in Miao minority language. A legend about how this ceremony came to be is as follows:

Long long ago, the ancestors of "Miao" nationality moved to new place far away from where they lived. They hoped that they could live in the new place rich, peaceful and happy. But they lived still in grotto or hut, wearing leaves and wormwood, eating wild fruits and potherbs. Whenever winter came, the life was even miserable. However, they prayed god constantly to get over their impoverished life. One day, 'Yangong' and 'Yangpo', the two gods (male and female) of rice in the heaven heard their prayer and had compassion for them. They showed them the technique how to reclaim land and transplant rice. After that, people's life became better and better. Therefore, Miao ancestors decided to invite these two gods of heaven for opening seedling gate and have a ceremony before rice transplanting to thank 'Yangong' and 'Yangpo'.

Every year, when the rice seedlings grow up to about 30 cm high, each household ploughs rice land and prepares for transplanting. Before transplanting, farmers in villages will invite rice specialist called 'Huolutou' to organize a ceremony for opening seedling gate. On the day of opening the seedling gate, the Huolutou wears new or clean cloth and burns joss sticks in baldachin of his house. Then he goes to the mountain to cut a gallnut tree, pick three *cogon* grasses and pull out some rice seedlings from rice nursery. After that he returns back to the selected paddy field, inserts a sign made of *cogon* grass into the paddy field and then transplants seven or nine rice seedlings. There are various ways to make the grass sign. The simplest one is just to put a knot in the *cogon* grass. There are also some signs made by *cogon* grass like adding a maple branch or a gallnut branch (Fig. 4.6). Farmers also have different expressions about why they put a grass sign into paddy field. *Cogon* grass means that seedlings will grow strongly without diseases and/or insect pests.

On the day of opening the seedling gate, the Huolutou should avoid meeting anyone on way to his home; he is not allowed to talk with people passing by in order to avoid bringing whammy to the sacred ceremony. After that, he should bring gallnut tree to his home and put it outside the door. Gallnut tree should be bound with the *cogon* grass meaning that the rice seedling gate has been opened. Then, Huolutou and his family put well-cooked glutinous rice, fish and meat on the family alter to pray for a good harvest. They should also burn some incense at the entrance in order to invite gods to visit. They also need recite *shin* ritual player: "This year is a good year, today is a lucky day, this morning we opened the seedling gate, may the grace of gods be with us, may the harvest become abundant and everyone be well fed and well clothed".

After opening the seedling gate, Huolutou will also arrange a feast to thank everyone's service and wish for a good harvest. Then the farmers sing together to thank the Huolutou for his distinguished service. The words of the song are: "You are an able man, diligent and good at plowing. You know how to choose seeds and plant them, leading us to do rice works. Rice granary is all full in autumn, happiness of life lasts for ever."

In Leishan region, the 'opening seedling gate' festival is held on May 15~20. Usually a *"chou"* (cattle) or a *"wei"* (sheep) day in the lunar calendar is chosen for the celebration. On the day, the head of the household gets up early, transplants seven or nine hills of rice seedlings in his rice field and then cuts a branch of a maple tree with its top or three *cogon* grasses which are tied on one side in the middle of rice seedling hills. Then he goes back to his home and cooks fish and flower rice (glutinous rice dyed by saps, like black or yellow) and offers them with liquor on the alter in the memory of the deceased. Then the man starts an incantation, praying god for favorable weather and a good harvest. After the 'seedling gate open', people can either transplant rice the same day or later whenever they like.

On the day of 'opening seedling gate', all the villages of the Miao

Fig. 4.6. A grass sign in the rice field before transplanting
seedlings as a mark of praying for good harvest.

nationality are full of merry-making. There is a common saying in Leishan
district that "children are looking forward to New Year, adults are fond of
opening seedling gate." On the day, every family cooks flower rice and ten
other dishes. Among them, five will be main dishes, meaning a good
harvest. Some villages may also organize Miao traditional dance and
other villages enjoy cockfighting or bullfighting, singing together with
young people. 'Opening seedling gate' is Miao's happiest event and also
the earliest one of Miao's nationality festivals. 50 days after the 'opening
seedling gate' is the festival for 'eating new rice'. And after another 50
days would be the second festival of 'eating new rice'. And 25 days later,
there will be the New Year of 'Miao' nationality.

In the above story about the origin of 'opening seedling gate', the gods of Yangong and Yangpo initiated people to learn the productive skill of growing paddy rice. However, the story does not mention where the seed came from. The villagers believe that Yangong and Yangpo live in heaven and they are the gods of paddy rice for Miao nationality. It is pity that these stories are not rich enough and lack narrative.

'Closing Transplanting' Festival

The day of June 6, according to the lunar calendar, is the day of the 'closing the seedling gate' festival for Miao nationality, meaning closing the transplanting festival. In the early morning, every family cuts a gallnut tree and ties it with *cogon* grass and 3~5 red chilies. There will be a taboo pole for the ceremony, a symbol to close seedling gate. The master of the household will carry this taboo pole to the paddy field where he last transplanted and insert it into the field and then recite "after I have closed, it is your turn; after this event, rice will grow like a *cogon* grass mountain." After inserting the taboo pole, the master goes back home and cooks chicken and ducks. Then, he sets up farming tools, meal and liquor and goes around the fire three times from left to right.

The 'closing the seedling gate' festival is to cerebrate finishing of transplanting and also to pray farming tools for cultivation. It is an agricultural courtesy to pray for better growth of paddy rice.

'Eating New Rice' Festival

Eating new rice festival is the second biggest festival of the year for the Miao nationality of Guizhou province, just next to Miao nationality new year. The significance of eating new rice festival for Miao people is eating new rice together with their ancestors and god. It is usually held around in June or July according to lunar calendar, just when the rice plant is coming to booting stage. People of Miao nationality prefer choosing a *yin* (tiger) or *mao* (rabbit) day in the lunar calendar for the festival. The farmers of Miao nationality living in the area of Qingshuijiang and Leigongshan call eating new rice festival as "Loumao". They celebrate this festival at the first *mao* (rabbit) day according to Lunar June.

On the day of eating new rice festival, every Miao village slaughters a pig and a buffalo, catches fish, prays for ancestors and entertains guests and friends. In the evening before the day of the festival (some on the day of festival), every family picks up rice booting plants in odd number from the whole plants growing best, ties them in a knot and offers it at the family altar for eating new rice festival. On the festival day, rice is cooked together with this rice knot above. After dishes are ready, people put rice and dishes on family alter, burn incense paper, incant blessings, take a pinch of some food and pour liquor as sharing new rice with their ancestors. After the ceremony is the banquet. Hosts eat together with guests and friends. During the festival, young people sing, dance and swing together. In some Miao nationality areas, people wear blue and white skirts. People must bolt the gate during praying to the ancestors to avoid outsiders coming in. In Leishan area, a lot of events like bullfighting, "youfang" and "ganrenao" are held for seven days. Relatives visit each other, celebrate "Loumao" and host guests with food and songs. The just-married girls go back to their parents' home with their husbands with gifts like ducks, fishes, rice cakes, wines, etc.

Miao nationality's 'eating new rice' festival is an agricultural festival to expect a good harvest (Fig. 4.7). Since, by the time of the festival, rice plants are just coming to booting and grains have not yet ripened, the rice they eat on the festival day is last year's rice cooked with new rice plants. Even though the rice plants have reached the booting and flowering stages, there may be many unknown factors that may affect the ripening of the grains. For this reason, people offer sacrifices to gods and ancestors and pray for a good harvest.

Fig. 4.7. People of Sui nationality get together around rice field to celebrate "eating new rice festival".

Traditional Faith Related to Rice

Miao's Life Plant

People of Miao nationality like to plant a flower plant when a member of their family, especially a child, becomes ill. A flower plant symbolizes a Miao's life plant. Miao people think that someone is sick when a minus power invades a body and damages one's energy. They plant a flower plant and pray god and their ancestors to take away the minus power so that the patient gets recharged with energy and recovers. According to them, the flower plant is made up of four stairs representing four generations, namely, great-grandfather, grandfather, father and child's soul i.e. one stair for each generation. There are dolls made of white paper at each edge of the stairs. Before the stairs is a pig's lower jaw bone, meaning that the tribute is a pig. In the middle of second and third stairs is a handful of green bamboo. A small rice bag, a silver bracelet, an oil lantern and the sign of longevity are hanged on the bamboo branches. The rice in the bag is collected from many households; they believe that rice collected from one hundred households can recharge someone's energy. In other words, rice has the supernatural power to restore energy. Some families hang a model of a dustpan when they plant a flower plant to symbolize rice grains, meaning that it restores power to the patient who is the owner of flower plant.

Miao's "Baojia"

Some people of Miao nationality living in Longquan area of Danzhai county, south-eastern part of Guizhou province, ask the Miao shaman to offer a sacrifice on a lucky day of January in lunar calendar. After offering, they hang "Baojia" on the back door lintel to pray for a good harvest and safety of every family member. "Bao Jia" is made of a 50-cm bamboo stick with red and white tickers at both ends, adding 10 thorn sticks, some wild pepper branches, two broad-leaved evergreen branches and a handful of glutinous rice panicles. They tie up these stuffs and put them on the back door lintel and call it "Baojia". According to them, the red and white tickers, thorn sticks and wild pepper, all mean to avoid evil spirit and prevent disaster. Evergreen represents life and rice panicles mean wishing for good harvest in the year. The natives think that sparrows and mice wouldn't eat rice panicles on "Baojia" because the Miao shaman has already incanted and enchanted it. "Dakouzui" is another name of "Baojia" in Leishan County and people living there believe that

"Dakouzui" can keep couples on good terms and make all the family members live peacefully.

Straw Rope to Avoid Evil and Glutinous Rice

As a custom among people of Dong nationality, rice straw, straw rope and rice panicles have special power to ward off evil power and protect people from disasters. They pull up a straw rope, hanging rice straw, glutinous rice panicles and *cogon* grass at the entrance of the village to welcome distinguished guests wishing to protect all guests and villagers from evil power and diseases (Fig. 4.8). Glutinous rice, rice panicles and *cogon* grass are also hanged on the straw rope because, they think, they can prevent invasion by evil power. In Dong nationality villages, when a baby is born, people hang rice panicles of glutinous rice on door lintel to avoid ominous things and diseases affecting the health of the young born. If the baby is a boy, they add a red strap beside these stuffs. It is the same even if a cow has a new calf. In this way, they not only announce the

Fig. 4.8. A custom among Dong nationality for welcoming distinguished guests. A straw rope is pulled up in the entrance of village, hanging rice straw, glutinous rice panicles and *cogon* grass, wishing to protect all guests and villagers from evil power and diseases.

arrival of a new life but also tell others not to visit them. Incidentally, this custom of using straw rope to ward off evil spirits is also common in Japan and Korea.

Minorities living in the south-western part of China, like Miao and Dong nationalities have glorious cultural traditions of glutinous rice as their staple diet and the area is referred to as glutinous rice culture circle. In recent decades, because of the pressure of food supply, most of these minority villages have started planting hybrid rice instead of glutinous rice to ease the pressure. Therefore, glutinous rice has become a kind of special food just used in sacrifices and festivals. The custom of eating glutinous rice as staple diet has been changing generation by generation. But some villages which are rich in land resources still retain the custom of planting glutinous rice and eating it by hand. For instance, Zhanli village in Congjiang County in the south-eastern part of Guizhou Province has a secret recipe inherited from their ancestors for birth control that has resulted in very slow increase in their population since 1949. Due to rather low birth rate, there are enough rice fields for planting glutinous rice. Although most of the villages in southeast of Guizhou Province have no longer the custom of consuming glutinous rice as their staple diet, they still make cakes of glutinous rice for celebrations during festivals or for honoring guests.

References

1. Peng and Ying (1993).
2. Sato, Tang, Yang and Tang (1991), Tang, Sato and Yu (1994).
3. Wang, Zhang, Chen and Zhou (1996).
4. Ding (1957).
5. Liu (1975).
6. You (1987).
7. Tang, Sato and Yu (1994), Tang, Zhang and Liu (1999).
8. Wang, Zhang, Chen and Zhou (1996).
9. Liu (2003).
10. Huang, Sun and Wang (1996).
11. Tang, Wei, Jiang, Brar and Khush (2007).
12. Tang, L.H. (2002).
13. Morishima and Gadrinab (1987), Sano and Morishima (1992).
14. Second (1985).
15. Cai, Wang and Pang (1996).
16. Tang, Min and Sato (1993).
17. Sun, Wang and Yoshimura (1996).
18. Huang, Sun and Wang (1996).

19. Huang, Sun and Wang (1996), Tang, S.X. (1996).
20. Tang, Min and Sato (1993), Tang, S.X. (1996), Tang, S.X. (2007).
21. Yan (1989).
22. Wang, Zhang, Chen and Zhou (1996).
23. Chang (1983).
24. You (1992), You (2004).
25. Han and Cao (2005).

5 | History of Rice Culture in Korea: Origin, Antiquity and Diffusion

Mun-Hue Heu and *Huhn-Pal Moon*

Rice culture in the Korean peninsula is assumed to have settled down as a staple food crop along with the Korean history. Archaeological findings reveal that rice culture in Korea began in the Neolithic era of Korean history around 4000 BP. It appears that rice did not originate in the Korean Peninsula; it was developed as a crop in China from where it was transmitted into the Korean Peninsula. This chapter provides a historical overview of the diffusion of rice culture and the present status of rice production in Korean peninsula.

The first part of this chapter reviews the limited archaeological findings of carbonized rice grains (unhulled and/or hulled) and the remains of rice paddy fields and tries to figure out the transmission route of rice culture into Korea along with the Chinese and Japanese archaeological findings near Korean peninsula. The next part discusses the developing process of rice culture since the beginning of history of Korea based on historical records as available in ancient books on agriculture of Korea and China. Finally, the modern and present state of rice culture in Korea is described.

The Korean nation was formed in the Korean peninsula during the Bronze Age and the Iron Age. Afterwards, the Chinese civilization and culture diffused into the Korean peninsula through interchanges between the two nations. The agricultural system of upland and dryland farming seems to have been replaced by the rice farming with influences from China. Technology innovation of rice culture in the Korean peninsular was initiated with the establishment of the "Gweon-eop mobeom-jang" (Agricultural Demonstration Station), the first Government organization for agriculture in Korea in 1906 and was also subjected to the influence of Japan during the Japanese occupation of 1910~1945. Modern technology of rice production in Korea started with "Tongil type" rice, a new ecotype of rice derived from *indica/japonica* hybridization that brought Green Revolution in Korea.

ARCHAEOLOGICAL REMAINS OF RICE CULTURE

Sites of Archaeological Excavation with Carbonized Rice

The initial reports related to rice came from the archaeological excavation at Kimhae-Paechong (a shell mound) in 1920, in which rice seeds were discovered together with brass coin from a royal tomb of the Xihan, China (AD 9).[1] It suggested that rice culture already existed in the southern sea coast region about 2000 years ago in Korea. During the last 80 years, many archaeological findings with regard to rice have been reported. These include about 10 reports related with rough and/or milled rice, more than five reports on paddy fields, an irrigation pond and a trench, more than 15 reports on the agricultural tools for rice and more than 30 on earthenware with tracing of grains.[2] Here, we would like to present an outline of ancient agriculture in Korean peninsula with archaeological excavates of 10 representative sites in their chronological sequence that include two sites from the Neolithic Age, six sites from the Bronze Age and two sites from the Iron Age (Table 5.1).

Table 5.1. Representative excavation sites and their remains in Korean peninsula

Site No.	Excavated sites	No. of rice grains	Carbonized years BP
1	Ilsan zone-1 (Seongjeo-ri)	22	4330 ± 80
2	Kimpo-Gahyunri	7	4020 ± 25
3	Yeoju-Heunamri	78	2900
4	Pyeongyang-Namkyeongri	150	3000
5	Ilsan-(Gawaji) zone-2	287	2770 ± 60
6	Buyeo-Songgugri	400	2670 ± 60
7	Jinju-Daepyeongri	26	2300
8	Sancheong-Sonamri	10	2200
9	Chungju-Jodongri	27	2065 ± 165
10	Kwangju-Shinchangri	155	2000

Rice Seeds from the Ilsan Zone-1[3]

Some carbonized rice seeds were excavated from the sites of Ilsan zone-1 and zone-2 in 1991, which were located at Songpo-myun, Goyang-gun, Gyeonggi Province before the new Ilsan city was constructed. Twelve rice seeds were unearthed at Ilsan zone-1; these were discovered in the upper part of the brownish peat layer and were relatively well formed and well-preserved in original except two destroyed ones. The grains were well-preserved in their original form presumably because they were located in a peat layer. The grain characteristics of well-formed eight rice seeds are shown in Table 5.2. Seed sample 1 showed the *japonica*-like shape and a typical ancient grain type with narrow bottom and wider top part of the grain with a strong and long apical awn. Even though some of them showed seed size between *japonica* and *indica* in grain length, most of them showed their range of *japonica*-like types in grain characteristics. The archaeological age of the earthenware with rice seeds was determined as 6210 ± 60 BP for earthenware and 4330 ± 80 BP for carbonized rice seeds.[4]

Table 5.2. Grain characteristics of well-formed eight rice seeds from the site of Ilsan zone-1

Grain sample no.	Length (mm)	Width (mm)	Length/Width	Apical awn
Sample 1	6.87	3.23	2.12	Yes
Sample 2	6.45	3.24	1.99	No
Sample 3	6.94	3.14	2.21	Yes
Sample 4	7.18	3.08	2.33	Yes
Sample 5	6.68	3.22	2.07	No
Sample 6	7.04	3.02	2.33	Yes
Sample 7	6.77	3.01	2.25	No
Sample 8	6.69	2.92	2.29	No

Rice Seeds from the Kimpo-Gahyunri Site[5]

Three rice seeds were reported from the site of Gahyunri, Dongtan-myun, Kimpo. The characteristics of rice seeds that were excavated from the site of Kimpo-Gahyunri are shown in Table 5.3. Their historical date was determined as 4010 ± 25 BP. The microscopic measurement of seed size showed similar to *indica*-like with more or less slender grain compared to *japonica* shape. According to the excavator, various agricultural farming tools were also unearthed around areas near the site, which may imply rice cultivation in the areas.

118

Table 5.3. Characteristics of rice seeds excavated
from the site of Kimpo-Gahyunri

Sample	Length (mm)	Width (mm)	Length/Width
Sample 1	7.60	2.80	2.71
Sample 2	6.80	2.40	2.83
Sample 3	7.60	2.80	2.80

Carbonized Rice Grains from the Yeoju-Heunamri Site[6]

Carbonized rice grains were also excavated from the historic dwelling site of Heunamri Jeomdong-myun, Yeojugun, Gyeonggi Province during 1972~1977. These samples showed relatively little variation, even though these had large stress from milling process (Table 5.4). The charcoal around the floor of the 12th dwelling site, in which the carbonized rice grains were unearthed, was measured and found to be of 3210 ± 70 BP and of 2620 ± 100 BP by the Korean Nuclear Energy Institute and of 2980 ± 70 BP and 2920 ± 70 BP by Japan. From these results, the carbonized rice grains are assumed to be roughly 3000 years old (Fig. 5.1). These samples from the site of Yeoju-Heunamri showed the smallest grains in size as compared with those from further southern regions. The grain size appeared to become larger towards the south in the Buyeo in the mid-west

Table 5.4. Characteristics of rice grains excavated
from the site of Yeoju-Heunamri

Sample	Length (mm)	Width (mm)	Length/Width	L/W (mm²)	Remarks
Sample 1	3.60	2.32	1.499	8.763	20 samples
Sample 2	3.67	2.27	1.635	8.315	7 samples
Sample 3	4.45	2.30	1.985	10.005	2 samples
Sample 4	3.80	2.20	1.730	8.360	5 samples
Sample 5	3.48	2.16	1.564	7.784	6~26 samples
Sample 6	4.00	2.34	1.737	9.133	9~6 samples
Sample 7	3.73	2.28	1.635	9.100	6~2 samles
Sample 8	3.30	2.10	-	-	Broken
Sample 9	3.10	1.90	1.630	5.890	1 sample
Sample 10	3.10	1.60	-	-	Broken
Mean	3.72	2.23	1.620	8.595	
δ	± 0.334	± 0.301	± 0.190	± 1.228	

Fig. 5.1. Carbonized rice grains excavated at the site of Yeoju-Heunamri.

and the largest in the Kyeongju in the southeast region of the Korean peninsula. It seems to imply that the rice varieties with larger grains evolved in the southern regions.

Carbonized Rice Grains from the Pyeonyang-Namkyeong Site[7]

Pyeongyang-Namkyeong site is located in Honamri at the outskirts around 10 km towards the northeast from Moranbong, Pyeongyang and further 4 km in the southeast from the Nosanri, Pyeongyang. Five kinds of carbonized grains including rice, sorghum, Chinese millet (*Panicum miliaceum*), Indian millet and soybean were unearthed during 1979~1981 from the middle parts of the 36th habitation site of the Bronze Age in the site of Pyeongyang-Namkyeong. As shown in Table 5.5, the size of rice grains from the site had *japonica*-like shape of short and round grain, which was similar to the excavations from the Paechong (a shell mound) of Kimhae and the site of Buyeo-Songgukri and dated as 1000 BP in the Bronze Age. From these excavations, the authors suggested several facts as follows: (1) rice culture in the Korean peninsula occurred at least 1000 years earlier than the date which has been recorded in the current history books, (2) rice culture in the Korean peninsula might have diffused from the northern parts to the southern, (3) these results might prove that rice culture in Japan might have been transmitted from Korean peninsula to the northern parts of Kyushu, Japan.

However, the excavation of rice seeds from the site of Ilsan zone-1 at the northern outskirt of Seoul was recently reported to be much older than those from the site of Pyeongyang-Namkyeong. Therefore, further archaeological discoveries from North Korea are necessary to support with the facts suggested above.

Table 5.5. Characteristics of carbonized rice grains from the site of Pyeonyang-Namkyeong, Paechong-Kimhae, and Buyeo-Songgukri

Sites	Length (mm)	Width (mm)	Length/Width	Remarks
Pyeongyang-Namkyeong	4.18	2.50	1.80	20 Samples
Paechong-Kimhae	4.18	2.22	1.88	1937
	4.40	2.70	1.63	1969
Buyeo-Songgukri	4.10	2.71	1.53	1969

Rice Seeds from the Gawaji' Site[8]

The Gawaji site (Islan zone-2) which was located in Songpo-myun Goyang-gun, Gyeonggi Province was excavated by Professor Yungjo Lee of the Chungbuk University in 1994, in which many rice seeds were unearthed beneath the blackish peat layer and were assessed as belonging to 2770 ± 60 BP.[9] Although the average size of 300 seeds was slightly slender as compared with the current cultivated rice, these appeared to be typical *japonica* rice judging from the characteristics of rice hull and seemed to be mixed with different grain characteristics such as awned and awnless, easy and hard shattering.

Rice Seeds Unearthed from the Buyeo-Songgukri Site[10]

The ruins of a smooth potter's dwelling site were unearthed at Songgukri Chochon-myun Buyeo-gun during 1975~1978. Archeological remains of Buyeo-Songgukri included about 395 grams of carbonized rice grains at the flat of the 54-1 dwelling. These rice grains also belonged to typical *japonica* type with short and round shape according to the measurement of random 300 grains by the former Prof. Chunyeong Lee of the Seoul National University. As compared with those from the sites of Heunam-Ri, of Gunchangji at Buyeo-Busosan and of Jangchangji at Kyeongju-Namsan, the carbonized rice grains from Buyeo-Songgukri were larger in size than those from Heunam-Ri and Gunchangji but smaller than those of Jangchangji. According to the report of Songgukri archaeological excavation, the Songgukri's remains were measured as of date of 2665 ± 60 BP and 2565 ± 90 BP.

Archaeological Remains from the Jinju-Daepyeongri Site[11]

The Jinju-Daepyeongri remains were unearthed from Eoeun zone of Daepyeongri, Daepyeong-myun Jinju, Gyeongnam Province during 1996 and 1997 by the Museum of the Gyeongsang University. The site constituted of 105,600 m² with 42 field patches. The furrows were unveiled at 4 m layer below the removed surface soil (sand-clay). The furrows constructed were 35 cm wide, 50 cm deep and 10 m long, the longest being 30 m. The archaeological finds were smooth pottery, square-drill fibres, fishing nets, some stoneware, farm applying machine and tools and 10 dwelling sites of size 350 cm × 450 cm. Several kinds of cereal grains were found in field furrows as well as at dwelling sites. The collected grains included 28 grains of rice, 4 each of barley and wheat, 1009 of sorghum, 65 of Chinese millet, 2 of perilla seeds and 3 of beans. They were measured as of 2300 BP.

Carbonized Rice Grains from the Sancheong-Sonamri Site[12]

The site of Sancheong-Sonamri which was located in Sonamri, Danseong-myun Sancheong-gun, Gyeongnam Province was unearthed during 1996~1998 and was measured as of 2600 BP in the Bronze Age. This site was established by an alluvium from the river drift of Namgang (river) in front of the Sonamri in Jinju. A pile of rice grains, which was mixed with soil of 10 m², was discovered in the storage-like lot of the dwelling sites from the site of Sancheong-Sonamri. These rice grains were mixed not only with different sizes of milled and brown rice grains but also with different kinds of cereals including soybean, adzuki bean, mungbean, etc. It was not easy to identify them as *japonica* or *indica* although there were large differences between small and large grains. These excavates were measured as belonging to around 2200 BP.

Paddy and Milled Rice Grains from the Chungju-Jodongri Site[13]

Jodongri excavation was unearthed in Jodongri of Chungju, Chungbuk Province during 1996~1997. It was divided into two different cultural zones: the upper zone of the Bronze Age dated as 2065 ± 130 BP and 2300 ± 250 BP and the lower layer zone dated 2700 ± 165 BP and 2995 ± 135 BP. Several kinds of cereal grains including 27 rice grains and 10 pieces of rice hulls were unearthed in the upper layer zone of the Bronze Age. The size of these rice grains was similar to that of the present cultivated varieties. Out of 10 pieces of rice hulls, one was slender like that of *indica* rice, another had big grain shape of 8.3 mm long and 3.0 mm wide and still another one had long apical awn like that of a *bulu* type. With these results, rice samples from Jodongri belonging to 2000~2300 BP seem to be a mixture of temperate *japonica*, tropical *japonica* and also possibility of *indica* type.

Rice Hulls Excavated from Low, Swampy Ruins of Kwangju-Shinchagdong[14]

This Kwangju-Shinchangdong site, which was unearthed by the National Kwangju Museum in 1997, was located at a low and swampy place in Shinchangdong of Kwangju and revealed a huge complex of a date between the last part of Iron Age and the Three Nations of Korea. Because this excavation site was a low and swampy place, the relics of both artificial and natural organic materials were well preserved as important materials of archaeological research for the 1st century BC. According to the report, this layer contained a heap of rice hulls of 155 cm thick with rice straws and other grasses belonging to the *Gramineae* (Fig. 5.2). However, no normal rice grains were found; the heap contained only rice hulls and straws. DNA analysis of rice hulls, leaves and straws revealed that all of them belonged to *japonica* type with length/width ratio of around 2.0 and were dated as of 2000 BP.

Relic of Paddy Fields and Reservoir and Irrigation Channels

Remains from the Yangsan-Habukjeong Site[15]

The ruins of Yangsan-Habukjeong unearthed by the Museum of Donga University in 1992 were located at the field number 565, 566 and 594, Bukjeongri of Yangsan, Gyeongnam Province. It was surrounded by cultivated fields with tumuli in the mountainous area east of the village and paddy fields in the west. The tumuli seemed to have been constructed in the Yangsan area during late Silla Dynasty, the 6th~7th century. The

Fig. 5.2. Carbonized rice grains excavated at the site of Kwangju-Shinchangri.

paddy fields were investigated with 6 sections of layers in vertical from paddy surface in which rice seeds and job's tears were discovered from the 3rd layer and the pieces of hard stoneware of grey-blue colour and pieces of Buncheongsagi from the 4th and 5th layers. They were dated between the United Silla Dynasty and the early Chosun Dynasty. Sangpyeong-tongbo Dang-ijeon, a kind of brass coin was also discovered from the 2nd layer of the site. Considering that the coin was moulded during the period 1742~1752 in the reign of the King Yeongjo, the Chosun Dynasty, those relics were assumed to have been made prior to the early Chosun Dynasty. According to the excavator, the archaeological excavations showed the perfect structure of paddy fields with a pond, wooden pile, footprint of human and cattle and crop grains although these remains were found to belong to the period of Chosun Dynasty.

Relic from the Nonsan-Majeonri Site[16]

The site of Nonsan-Majeonri which was located at Majeonri of Yeonmu-eub, Nonsan, Chungnam Province, were unearthed by a team of the Burial Culture Institute of the Korea University in 2000. These ruins were spread over a stretch of 3 km in the hills and on slightly low plains and revealed paddy fields, irrigation facilities, water wells, dwelling site and various types of old tombs including stone coffin tomb, earthenware coffin tomb and stoneware-clay tomb at C-zone of three sectioned zones (A, B, C). However, it revealed that they all belong to the Bronze Age.

The lowland regions were divided into two parts that is, irrigation channels at a hillside and plain lands in the lower parts. On the way to the irrigation channels, a pond of around 3 m in square or rectangle had been constructed in the paddy land of the lower plains. There were also the traces of wooden materials to make agricultural implements and an irrigation pond, a rain puddle and irrigation facilities. Two water wells contained wooden plates and large timber crossed at the bottom of wells.

Okhyun Relic from the Ulsan-Mugeodong Site[17]

According to the excavator's report, the Okhyun ruins which were located at 48, Mugeodong of Ulsan, Gyeongnam Province were unearthed by the cooperative teams of the Donga University and the Gyeongnam University in 1997~1999. It was a vestige of human life style from the Bronze Age to the Modern Age. The excavation zones were divided into four sections (A, B, C, D), three (A, B, C) at a hillside and one (D) at lower plains. The C-zone contained 61 items such as dwelling sites, the waste disposal lots and a circular drain of the Bronze Age and dwelling sites of the Chosun Dynasty. The paddy fields which were discovered at D-section during 1998~1999, constituted of several different layers of paddy fields covering different time period from the Bronze Age to the Modern Age (Fig. 5.3). One each from the continuing layers of paddy fields of different ages were selected and investigated three times at different depths on a plane level. Forty-three layers of soil were distinguished along the vertical section of trench from soil surface up to 2.8 m depth. Out of them, those between 34 and 42 layers were determined to be of sedimentary layers and 33rd and upper layers proved to consist of either artificial or natural drift layers. The layers 31~33 revealed paddy fields of the early Bronze Age with accumulation of enough humus in which paddy fields might have started. The layers 25~27 proved to belong to the paddy layers of the Three Kingdom of Korea and the whole area of the valley was levelled and developed with irrigation facilities unto this layers. The layers between 30 and 17 also revealed to be the layers of paddy fields of the Chosun Dynasty and continued to be developed as paddy fields in the whole valley. The layers over the 16th were developed as paddy fields during the Modern Age with repeated artificial soil dressing layers. The excavator[18] described separately the details of different layers that belonged to different Ages as follows:

Fig. 5.3. Ruins of paddy fields excavated at the site of
Ulsan-Mugeodong, The Bronze Age.

1) The layer of Bronze Age: The paddy fields were found at 130~160 cm underground from the current surface of soil. It was confirmed as the surface of paddy fields with the existing ditches and the disturbance of soils and mixing foreign soil with accumulation of oxidized soil and manganese, particularly the extracted plant opal (1500 gr). A patch of paddy field consisted of a small plot with some stepping stairs along the slopes. There were about 100 patches of paddy, each consisting of an irregular form of 3~10 m² in area with a dike of 15~52 cm wide and 14~60 cm high with round top surface.

The bare ground of paddy fields showed the row shape, footprint, pieces of farming tools, stubbles of rice plants, traces of piling holes and cut places of irrigation channels. The large channels were made in the boarder between hill slope and valley with 45 m length, 2~6 m width and 85 cm depth. These irrigation channels seemed to have been constructed in the Bronze Age. Even though these channels were temporarily destroyed, they seemed to have often been repaired and utilized until some time during the Three Nations. The definite ages were determined with the archaeological remains at the surface of paddy fields such as smooth earthenware and dagger-grinding earthenware, net plummet, stoneware, etc. From the discovery of earthenware pieces with hole-line and short slope line out of the smooth earthenware on the surface of paddy field, the paddy field and the circle drain of the dwelling site on the hilly region near to the paddy assumed to be of the same period, early Bronze Age, 6th~7th BC.

2) The layer of Three Nations period: Forty-four patches of paddy fields were discovered at 70~110 cm underground of the surface of the current paddy fields in all levels of valley. A patch of paddy fields was constructed with about 106~145 m², which was much larger than those of the Bronze Age and quite smaller than those of the Chosun Dynasty. The traces of plough and plowing tools, imprint of stalks, footprints and the spoor of cattle were found along with the dikes of paddy fields and the human footprints moved regularly in the same direction. From the separate accumulation of oxidized iron and manganese and the depth of reduced layer at the down-layers of paddy fields, the soil conditions of paddy fields seemed to be semi-dry and/or semi-wet and not of very high productivity. It was dated as 6th~8th century around the period of United Three Nations.

3) The layer of Chosun Dynasty: Paddy fields at layers between 16th and 20th are assumed to belong to Chosun Dynasty and the layers above 16th continued to be established as paddy fields widely in the whole areas; the paddy fields became larger in the scale on the time being to the Modern ages with a long rectangular shape of the terrace. The repeated artificial soil dressing in the paddy field seemed to have accumulated continuously and become thicker from the Chosun Dynasty to the Modern Age.

In summary, Okhyun relic constituted of 33 layers of paddy fields; 33~31 layers of agriculture of the Bronze Age, 27~25 layers of traces of the Three Nations, 20~17 layers for the Chosun Dynasty and above the 16 layers were developed in the present state of paddy fields with continuing accumulation of repeated dressing by artificial soil.

Excavation of Carbonized Rice Seeds in the Territory of China and Japan near to Korean Peninsula

Archaeological Sites of Chinese Territory near to Korean Peninsula

Remains from the Lianyungang-Erkanchun:[19] The site of Lianyungang-Erkanchun is located in the Erkanchun of Lianyungang, Jiangsu Province, China. Rice hulls were unearthed in the site of Lianyungang-Erkanchun, which is located at the point of 34°50′ N latitude and 119°15′ E longitude in the region of 50 km southwest from Lianyungang of Jiangsu Province. These have been dated as belonging to around 4000 BC. According to Yan, the *Shiji-Xiabonji* records that the rice culture in China began with the influence from Vietnam culture.

Remains from Shiha-Yangjiajuan:[20] The site of Shiha-Yangjiajuan is located in the Yangjiajuan of Shiha, Shandong Province, China. Rice seeds were unearthed from the site of Shiha-Yangjiajuan and determined as belonging to 2040 ± 70 BC. The *Shiji-Xiabenje* records that rice culture was already established during the period of the King Yu.

Remains from Darensu-Dachizu:[21] The site of Darensu-Dachizu is located in the Dachizu of Darensu, Liaoning Province, China. Q.Y. Wu reported that the carbonized rice seeds were unearthed at the point of 39°02′ N latitude and 121°43′ E longitude. The earthenware containing the carbonized grains of rice and sorghum were excavated at the dwelling site, which was dated as 2945 ± 75 BP.

Archaeological Remains of Rice Seeds in Japanese Territory near Korean Peninsular

Remains from the site of Karatsusi Nabatake Shoenji:[22] This archaeological site is located in the Shoenji of Nabatake, Karatsusi, Shagahyun, Japan. The remains of agricultural implements and traces of irrigated land surfaces were unearthed from the site of Karatsusi Nabatake Shoenji after

1981, which has been known as the oldest ruins and measured as of 2500~2600 BP.

Remains from the site of Hakada Itatsuke:[23] This site is located in Itatsuke Hakada, Fukuoka, Japan. It was unearthed and investigated during 1951~1969. Carbonized rice seeds, human footprints on the surface of paddy field and some farming tools were excavated with earthen mortar from the Jomon period and measured to a date of 1st~2nd century BC.

HISTORY OF RICE CULTURE AND ITS DISTRIBUTION

Prehistoric Period

Rice Ecology in Korean Peninsula

Korea is located on a peninsula that is bounded to the continent in the north and surrounded by Pacific Ocean on the other three sides in the Far East Asia. The peninsula lies along 33°06′~43°01′ N latitude and 124°11′~131°53′ E longitude. The climate of the peninsula belongs to the temperate monsoon that is characterized by freezing and snowy cold winters and warm and humid summers with a rainy season from June to August. The rest of the year is more or less dry with mild spring and autumn. Annual average temperature ranges from 5 to 12°C with some regional differences from south to north and plains to mountains. Annual precipitation is 1000~1300 mm on the average with about 70% of total rainfall in summer season of June through August. The day length ranges from 12 to 14 hours with the longest in the late June and the shortest in the late December.

Rice season occurs from April to October; only one crop of rice is grown in a year. During the rice season, the weather is usually dry at seedling and maturing stages of rice growth and often some cold injuries occur at early and late stages of rice growth particularly in the mountainous and coastal regions. In the past, there were rain-fed paddy fields; naturally the crop was very uncertain.

From the Prehistory to the Iron Age

Classification of the prehistoric periods of Korea:[24] The timeline of Korean history from the prehistoric to the Three Nations is shown in Table 5.6.

The Neolithic Age: Archaeologically, the Neolithic Age in Korea was around 7000~3000 BP and is characterized as a period of grinding devices both of stoneware and bone-ware. It has been known that the climate was

Table 5.6. Classification of historic times of Korea

Archeological Age	Chronology	Example archeological site
Paleolithic Age	10,000 + BP	Goolpo-ri, stonewares
Mesolithic Age	10,000~7000 BP	Sangrodae-chon, millet
Neolithic Age	7000~3000 BP	Daechun-ri, cereal grains
Bronze Age (Earth pottery)	3000~2300 BP	Songkoo-ri, rice grains
Early Iron Age	2300~2100 BP	Ilsan-eup, rice grains
Proto-kingdom	2100~1700 BP	Suyanggae, rice grains
The Three Nations:		
Koguryeo	37 BC~658 BC	North-east, rice farming
Baekje	18 BC~660 BC	Mid-west, rice farming
Silla	57 BC~935 AD	South-east, rice farming
Koryeo	918 AD~1392 AD	
Chosun Dynasty	1392 AD~1910 AD	

warmer with the warmest period around 6000 BP with the sea level being 2 m higher and the coolest and wettest period around of 2000 BP. Afterwards, it has gradually recovered to the state of present times.

Many archaeologists have assumed that millets and rice were the earliest crop plants domesticated in the prehistoric period in Korean peninsula and rice has developed as the historical staple food crop of Korean people. Some archaeological finds of rice grains from several Neolithic sites such as Seongjeori (4070 BP) and Gawaji (4330 BP) of Goyang, Gyeonggi Province[25] and Gahyunri (4020 BP) of Kimpo, Gyeonggi Province,[26] reveal that rice culture was practiced in the Neolithic Age 4000 years ago in Korea. However, it is assumed that rice was just utilized without any cultivation and production or with a limited production before the Bronze Age because very few rice grains have been found from archaeological excavations at the limited Neolithic sites and no typical traces of rice cultivation have been found. Most of the farming tools unearthed from the Neolithic era were stoneware with few wood-wares probably due to their decay and disappearance. Therefore, major farming tools in the Neolithic Age are assumed to be made of woods and stones with a few of bones, horns and tooth of animal and/or of shells. The stoneware was identified by the usages such as digging, harvesting, threshing, processing and storage.

In an historical perspective, the rice farming followed was upland and/or dry-land farming. Therefore, there might not be farming machines and tools specifically for rice farming but ones developed originally for dryland farming might have been used for rice farming as well. According to the specialists, these seemed to be iron-hoe, stone-shovel, ploughshare and bone-hoe.[27] The farming implements also seemed to be utilized simply for land preparation and seeding with stone shovel, harvesting with stone sickle, pounding rice grains with mortar or pestle, milling flour with grinder and storage with earthenware (Fig. 5.4).

The Bronze Age: The development of the bronze wares and the use of a

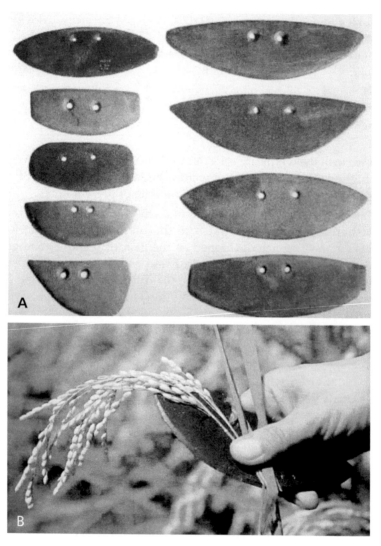

Fig. 5.4. Stone knife (A) and how to use it (B) for harvesting rice panicles.

sharp ploughshare assumed to improve and diffuse rapidly the farming methods and techniques in Korean peninsula during the Bronze Age. Plenty of rice grains have been excavated from several archaeological sites that belong to the Bronze Age. These are about 300 rice grains dating to 2770 ± 60 BP in Gawaji,[28] 80 grains dating to 2920 ± 70 BP in Heunamri, Gyeonggi Province,[29] 20 grains dating to 2700 ± 165 BP in Jodongri of Chungju, Chungbuk Province[30] and about 400 grains dating to 2670 ± 60 BP in Songgukri of Buyeo, Chungnam Province.[31] The development of stoneware in the Bronze Age also seems to bring the development of storage and processing techniques of rice grains.

According to the archaeological findings, rice farming might have followed the dry land and/or upland crops and diffused to the southward river streams along with the western coast of the Han River and Geumgang in Korean peninsula. Rice culture in the Bronze Age is assumed to have diffused from Goyang, the middle parts of Hanriver, to Ilsan, a branch of Han River in the Iron Age around 2270 BP[32] and from Chungju to Suyanggae of Danyang around 100 AD.[33] Rice culture of Songgukri of Buyeo in the Bronze Age diffused from Geumgang to Shinchangdong of Kwangju which is located in the Yeongsangang (river), the southwest region in the Iron Age dated 2100 BP[34] and Gangruri of Sancheong in the region of the Namgang (river) through the Seomjingang (river) around 2200 BP.[35]

The archaeological remains from several sites including Habukjeong,[36] Majeonri of Nonsan[37] and Mugeodong of Ulsan[38] reveal that rice cultivation was likely extended to the larger areas during the Bronze Age.[39] Rice paddies from the ruins of Majeonri reveal to be in patches of 3 m² of rectangular shape with artificial facilities for irrigation and drainage and those from Mugeodong in Ulsan seem to be established on sedimentary soil on the plain surfaces of paddy fields for all the periods from the Bronze Age to the present times through the Iron Age. In the Bronze Age, people manufactured necessary agricultural implements for the land development, water control, plowing, harvesting and threshing and the implements and facilities for transportation and storage and used them for rice farming. Such technical improvement of agricultural implements seems to have brought the change in productivity and social life style of the community.[40]

The Iron Age: The development of the Iron culture is assumed to have advanced agricultural production one more step in Korean peninsula. Korea seems to have developed her own unique types of rice culture with an influence of the Iron culture from China during 4~3 century BC. With the archaeological remains of a big hoe which was called as the Cheolfak from the ruins of a cast near Buyeo,[41] a set of iron implements for agriculture seems to have been introduced from the Yan State of China into Korean peninsula around 2nd century BC. Many of the farming tools such as a weed-plough, hoes and iron sickles appeared from the remains of 1st century BC. providing evidence that the implements for ploughing and harvesting were replaced with iron implements (Fig. 5.5). The flat hoes are assumed to have been used for different purposes with different types of handles attached; as a chopper with the finest iron part on the upper parts of wooden handle and as an adze and digging hoe with L-shape of handle. The semi-lunar shape of iron sickle seems to have been replaced by new types of sickles around 3~4 century AD. The wooden implements such as hoes, handles, rakes and sickles with handles also appear to have been utilized for agriculture during this period.[42] The wooden hoes and wooden rakes are assumed to have been used for leveling land with wooden weed-plough and iron tool as major farming tools for plowing and harvesting. The wooden hoes and rakes were unearthed in the lower parts of ponds from the site of Shinchangdong, Kwangju.

Fig. 5.5. Iron rake (left), iron shovel (middle) and iron hoe (right).

The early rice culture using iron-wooden farming implements seems to have developed in the riverside, delta areas or swamp areas without any facilities of irrigation. The vestiges of reservoir and banks were discovered from the excavation of 3rd century AD, a long handled spade from the 4th century AD and a U-shape weed-plough and iron rakes from 5th century AD. The rice cultivation probably extended rapidly during the period of

5th century AD. The *Samguksaki*, a famous book of the Three Nations period in Korea, records that the cattle plowing was encouraged by King Jijeung of the Silla Dynasty in 502 AD and gave importance to reservoirs in 536 AD.[43] Similarly, the *Suhseo*, a famous Chinese book that was written in 619 AD, records that the lands of the Silla were highly fertile both for dry land and lowland farming, implying that double cropping of rice-barley was carried out during the Silla Dynasty.[44]

Transmission of rice culture into Korean peninsula: As mentioned above, the first archaeological remains related to rice appeared around 4300 years ago in Korean peninsula. The archaeological remains of rice grains and related items from the Bronze Age around 3000 years ago particularly reveal the practical development of agricultural implements. Rice cultivation is assumed to have widely diffused and became common throughout the Korean peninsula in the Iron Age. The first archaeological remains of rice were unearthed in the region of Han River and belong to *ca.* 4000 years ago.[45]

Rice cultivation had already developed on the lower valley of Yangtze River and Huaihe River and had become established in the region of 30~35° N in China around 7000 years ago. Whereas agriculture in the lower Yangtze valley was based on rice culture as evidenced at Hemudu,[46] the agriculture in the northeast region of China was developed with upland crops such as sorghum and millet in the regions of the Yellow river for several thousand years. This civilization diffused toward the northeast, in which upland farming became widespread in the region of 40° N in 4000 BP.

The dry land and/or upland agriculture that seems to have developed with grains like sorghum, millet and soybean diffused northeast-ward with wide dissemination in the region around 40° N by 4000 BP. It is assumed that the agriculture in the dry land region of northeast China extended and diffused spontaneously into the Korean peninsula. It is also assumed that the crop domestication was not so easy to switch from dry land farming system to rice farming in the dry land region of northeast China around 4000 BP.

As mentioned already, the results of archaeological excavation reveal that rice culture reached to Shandong region of China around 4440 ± 130 BP[47] and extended to Liaohe region of northeast China around 3000 years ago, as the archaeological remains of rice unearthed at Dachizu of Darensu, Liaodong peninsula, China[48] belonging to a date of 2945 ± 75 BP. Dry land farming developed in the period of 6000~4000 BC in China; it is assumed to have been transmitted from Liaodong region to the northern parts of Korean peninsula around 4000~2000 BC and into Japan around 2000-1000 BC through the southern regions of Korean peninsula. Diffusion and development of *japonica* rice have been proved by the imprints of rice hulls from the Liangzhu site, implying that rice farming diffused to Shandong and later on to the Liaodong region through a chain of the Miaodao Islands and diffused into Korean peninsula together with the dry land farming. Thus, the rice farming during this period seems to have been positioned as a supplement of the dry land farming. The Liaodong region with the only archaeological remains of the dry land farming around 4000 years ago, would be a possible connecting region as a pathway for the transmission of civilization between Hebei region in China and Korean peninsula. However, it was reported that no archaeological remains of rice culture was found but only of dry land agriculture

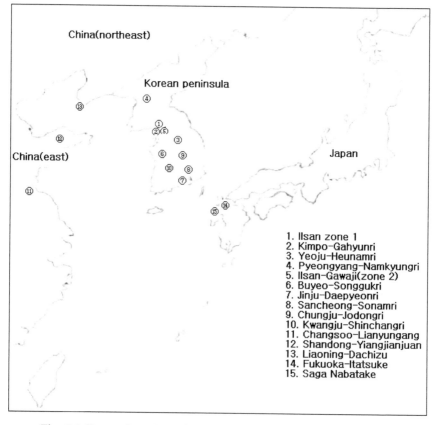

1. Ilsan zone 1
2. Kimpo–Gahyunri
3. Yeoju–Heunamri
4. Pyeongyang–Namkyungri
5. Ilsan–Gawaji(zone 2)
6. Buyeo–Songgukri
7. Jinju–Daepyeonri
8. Sancheong–Sonamri
9. Chungju–Jodongri
10. Kwangju–Shinchangri
11. Changsoo–Lianyungang
12. Shandong–Yiangjianjuan
13. Liaoning–Dachizu
14. Fukuoka–Itatsuke
15. Saga Nabatake

Fig. 5.6. Excavation sites where carbonized rice has been found in the Korean peninsula and in the Chinese and Japanese territories near to the Korean peninsula.

at the point around 41°10′ N, the upper stream course of the Liaohe River which was a representative region for the dry land farming. So, the rice culture in the Liaodong region, in which rice culture was diffused from the Shandong region, would have no alternative course but to transmit through a chain of islands of the Miaodao.

According to the Japanese archaeologist Takakura,[49] the most reasonable pathway for introduction of rice cultivation into Japan seems to be a route through Manchuria and Korean peninsula. According to his point of view, rice culture began in the region of the Hangzhou bay which was the lower course of Zhangjiang River from where rice culture gradually diffused to Shandong region and then into the riverside of the Han River through the Miaodao Islands. Afterwards, rice culture extended toward the southern parts of Korea where techniques of rice cultivation developed and matured. From there, the developed technology of rice cultivation was transmitted into Japan (Fig. 5.6).

Rice Culture in History

The Three Nations to the End of Koryeo

The rice culture reached Shandong region around 4000 BP[50] and to the Liaodong region, China around 3000 BP.[51] It seems that rice culture took over 3000 years to diffuse into the regions of dry land farming in the northeast of China. Afterward, it was introduced into the Korean peninsula and gradually merged with the dry land farming which already existed in the Korean peninsula. The results of archaeological excavations indicate that the rice culture in the Korean peninsula began in the prehistoric period along the riversides of the Han-river in the middle parts of peninsula and then diffused to the Geumgang River and the Seomjingang River in the southwestern parts and the Nakdonggang River in the south-eastern part.

After the conquest of the Gochosun, a prehistoric state of Korea by the Han state of China in 109 BC, the Chinese immigrants from China settled down in Korean Chinese characters) and religion.[52] The history of Korea is divided into the prehistoric and historic periods based on the system of writing and recorded culture. The Chinese civilization together with its dry land farming which was practised in the regions of Shandong and Liaodong was introduced into the Korean peninsula around 2000 BP. This introduction of the Chinese civilization in the Korean peninsula dramatically changed the culture and social system of Korea and provided a system of writing and history began in the Korean peninsula.[53]

The Three Nations, namely, the Koryeo, the Baekje and the Silla, in the Korean peninsula and some territory of the northeastern region of China in the neighborhood of the Korean peninsula developed between 1st century BC and 9th century,[54] so called the Three Nations period which belongs to the post-historic period. In this period, the Three Nations commonly used the Chinese characters for recording history. The Three Nations were unified into the Silla around 7th century AD and united into Koryeo (Korea) in the early 10th century AD. The rice culture[55] existed in the prehistoric era and might have been practiced from the early stage in the Baekje Dynasty, which was situated in the southwestern region of the peninsula, near to the northeastern coast of China. According to the *Samguksagi*, a famous historical literature written during the period of the Three Nations, the reclamation of paddy fields and rice culture were encouraged by the States, in the Baekje from 33 AD and in the Silla from 186 AD.

The Byeokgolje, a famous old reservoir during the Baekje Dynasty, was constructed at Kimje in 330 AD and more reservoirs were constructed afterwards. *Samguksagi*, the historical book of the Three Nations period in Korea, records that King Jijeung of the Silla Kingdom encouraged plowing by cattle in 502 AD by and importance of reservoir in 536 AD. As mentioned already, the *Suhseo*, a famous Chinese historical book that was written in 619 AD, records that the lands of Silla were highly fertile with both upland and lowland farming, implying that double cropping of rice-barley was carried out around 500 AD during the Silla Dynasty.[56]

The archaeological findings indicate the use of various agricultural implements such as a ploughshare, a mould board, a long handled spade, a weed-plough, a rake, a sickle and a short handled hoe, the rice culti-vation seemed to have spread to all the corners of the state during the unified Silla period. The farming characteristics during 4~5 century AD was the use of the farming tools and irrigation facilities, a large scale and deep plowing by cattle and development of more precision farming tools using the sickle and a short handed hoes.

During the period of the Three Nations, many literary works related to agricultural practices were introduced into the Korean peninsula such as *Fanshengzhishu, Saminweolryeong, Zhaiminyaoshu, Nongsangjibyo*, etc. The *Fanshengzhishu* written around the 1st century BC records the cultural method of rice and rice varieties and mentions non-glutinous and glutinous rice separately.[57] *Saminweolryeong* written in 220 AD also records rice varieties, their usages and different methods of cultural practices for different seasons. *Jiminyaosu*, a kind of encyclopedia for agriculture written during 532~544 AD, records the characteristics of 10 different rice varieties and their usage.[58] These agricultural literatures seem to have widely influenced the agricultural policy in China.

Considering the increased trade relationships between China and Korean peninsula from the early periods of the Three Nations, it was not so difficult that these agricultural literatures were transmitted into the Korean peninsula. The Chinese civilization including rice varieties and its cultivation methods were probably introduced into the Peninsula, particularly during the period of the Koryeo Dynasty which was located in the border of the two countries.[59] The brownish paddy rice and polished rice were discovered from an image of Buddha in 1302 AD.[60] The grain shape and amylose content of these rice grains showed that there were both *japonica* and *indica* rice and glutinous and non-glutinous as well during that period.[61] Another Chinese literature on agriculture, *Nongsangjibyo* written in 1273 AD was probably introduced immediately after its publication and was widely used towards the end of the Koryeo Dynasty.

Period of the Chosun Dynasty

The *Nongsajikseol* which was written by Cho-Jeong in 1429 AD and was used as a technical manual for farming in the early period of the Chosun Dynasty, indicates that the Dynasty emphasized agriculture from the beginning of the Dynasty and the agricultural pattern was changed from the dry land farming to a new pattern with emphasis on rice cultivation.[62]

The Dynasty also published many monographic and general books related with the agricultural farming, particularly 33 general volumes including 7 volumes with the description of rice varieties.[63] Seventy-one rice varieties were listed with their names and some characteristics such as their maturity time, awns, panicle colour, threshing ability, wind tolerance and soil adaptability. From these records, it is assumed that by this period rice culture was already established as the present system of rice farming. The *Sanlimgyeongje* published in 1682 AD, also records 36 rice varieties and classifies them into three different groups such as Shen (*indica*), Gaeng (*non-glutinous*) and Na (*glutinous*) from the rice culture of the southern region in China.[64] The characteristics of three groups of rice varieties are described as follows; Shen (*indica*) with the early maturity and long slender shape of grain, Gaeng (non-glutinous) with the late maturity, aroma and smoothness and Na (glutinous) with the intermediate maturity, waxy and white rice.[65] From their characteristics, it is inferred that seven varieties belonged to Shen (*indica*). Another volume, the *Haedongnongseo* published in 1798 AD, also indicates that the Gaeng was similar to the previous descriptions and the Shen (*indica*) had smaller size of grains with early maturity.[66]

There is also a record of sando (upland rice) with three different variety groups, namely, *wudeugsando, jeomsando* and mosando. The sando variety, Jeomseonggukdo that belongs to mosando group was introduced for its tolerance to flood and drought. Furthermore, *indica* rice was also introduced from China during the dynasty.[67] The agricultural basis of this dynasty definitely seems to be established on rice farming and rice was utilized as a means of taxation such as the payment for irrigation and other revenues during this dynasty.[68]

According to *Nongsajikseol* (1492 AD), the methods of rice culture were classified into three types based on different methods of planting: musami (wet direct seeding), geonsami (dry direct seeding) and transplanting. Wet direct seeding was generally practiced in lowland having supply of water from reservoirs or in marshy lands and direct seeding in dry land. In dry direct seeding, rice seeds were directly sown in dry condition of the soil; they received water from the rains. According to the Hanjeongrok (1610~1616 AD), seedlings were raised and transplanting of rice was carried out; this allowed double cropping with wheat or barley. The transplanting method of rice culture was widely expanded with large scale of rice cultivation.[69]

Rice Varieties in the Modern Age

Rice Groups Cultivated in Modern Age

Old records of rice varieties: From the rice culture practiced since the era of the Three Nations in the Korean peninsula, it is assumed that there were many rice varieties. However, the first description of rice varieties is found in the *Nongsajikseol*, which just classified rice varieties as Jodo (early maturing rice) and Mando (late maturing rice) by maturity. The *Geumyang-jabrok*, an assay on agriculture written by Hi-Myeong Kang in 1449 AD, records 27 rice varieties by their names and characteristics such as awn, color and adaptability to ecological and soil conditions;[70] seven varieties were classified as awnless and 20 as awned.

With increase in the number of rice varieties since then, *Limweong-yeonjeji* written by Yoo-Guh Suh around 1760 AD recorded 68 rice varieties, implying that the selection of rice variety was practiced unconsciously by rice growers. By collecting information on characteristics of different rice varieties from historical records, it now appears that a large number of rice varieties had existed. Seventy-one rice varieties were recorded by the end period of the Chosun Dynasty. The number of rice varieties increased presumably through introduction from China and by pure-line selections by the rice growers. From the list of about 10 rice varieties missing in the *Geumyangjabrok* (1492 AD) compared with those of *Zhaiminyaoshu* (530 AD);[71] one may assume that some varieties might either had disappeared due to lack of their adaptability or could have been available in different ecological regions.

Classification of cultivated rice group: The cultivated rice in the Korean peninsula was generally classified into three groups such as the food grains, animal feeds and weedy rice. There were two different varietal groups for food grains: *japonica* rice and Tongil type, which were also subdivided into glutinous and non-glutinous by the characteristics of endosperm and as low-land and upland rice by their difference in cultivation ecology. Even though the cultivated rice varieties in the world have diverged, the rice varieties of the Korean peninsula are quite simple and differ from those of the south Asia and Africa. The African rice species, *Oryza glabberima*, was never cultivated in the Korean peninsula; only the Asian species, *Oryza sativa* has been cultivated. There are mainly three ecotypes of rice varieties, *indica*, temperate *japonica* and tropical *japonica* within the Asian rice, *Oryza sativa*. Among them, major ecotype cultivated in the Peninsula was temperate *japonica* with additional ecotype of Tongil type, which was derived from the hybridization between *japonica* and *indica* in Korea. However, archaeological evidences reveal that some *indica* rice was cultivated in very few cases. There were some *indica* rices among the native varieties. Out of 27 rice varieties recorded in the *Keumyangjabrok* (1492 AD) two varieties Sando and Udeuksando appear to be *indica*. These might have been introduced from China.

Weedy Rice

Origin of weedy rice: The off-types of rice plant that are often observed in the farmers' fields should be removed from the population of the cultivated plants. They usually have the characteristics of easy shattering and brownish color of pericarp; they are called 'aengmi' (red rice) and often deteriorate the rice quality as it is difficult to remove the red pericarp color completely during the milling process. These weedy rices observed in the field of the native varieties generally show easy shattering in the early stage of ripening and survive by germinating several years after they fall in the paddy field. Investigations reveal that weedy rices were commonly observed in the fields of both *japonica* and *indica* groups and the occasional hybrids between *indica* and *japonica* rice often segregate and develop into the weedy rice. These phenomena continue to occur along with the riverside of Nakdonggang and the Seomjingang (rivers), even though it was a rare situation. The rice seeds of the different ecotypes that shatter in the previous years germinate naturally and hybridize with each other to produce the distantly hybridized plants. Besides, the seeds which drop naturally in the direct-seeded field germinate the next year, get mixed with plants of the new genotype grown and produce weedy rice plants.

Utilization of weedy rice as genetic resources:[72] Some weedy rices that are the product of natural hybridization between different ecotypes could be utilized as useful genetic resources. Professor Suh made a collection and evaluation of weedy rices that occur naturally in the cultivated rice fields all over the country and made a database of 1114 genotypes with a view to use them as genetic resources in rice breeding.[73] He made a catalogue of their characteristics as follows: (1) they have the surviving ability as semi-wild in the field, (2) they shatter easily, (3) their pericarp colour is generally red, (4) they are generally long-awned, (5) they are taller in height than the cultivated rice, (6) many of them have red hull color, (7) they are low-yielding, (8) some of them are wild forms of *jeomsando* from China, (9) some of them have long seed dormancy, (10) some of them have tolerance to some adverse environments including cold and drought, (11) some have resistance to major diseases and insect pests such as rice blast, stripe virus and brown plant-hopper, and (12) the range of their amylose content is 15~27%. The similarity test showed that with long red grains genotypes were similar to *indica* and short red grain genotypes to *japonica* rice.

PRESENT STATUS OF RICE PRODUCTION IN KOREA

As discussed above, rice culture in the Korean peninsula is assumed to have diffused from China in the Neolithic Age and has become the staple food crop along with the Korean history for the last 5000 years. The present technology of rice cultivation had its beginning in the early 20th century when the "Gweon-eop mobeom-jang" (the Agricultural Demonstration Station), the first public research organization, was founded by the Korean Empire in 1906. During the first half of the 20th century, the total areas of rice cultivation in Korean peninsula ranged from 1.402 to 1.656 million ha of which around 70% area was in the southern part (presently South Korea) and around 30% in the northern part (presently North Korea) of the peninsula.

During the second half of the 20th century, the total rice area of South Korea ranged between 1.10 and 1.26 million ha and is currently reduced to around 1.0 million ha in 2006. Average rice area per farmer has changed from less than 1.0 ha in 1990 to 1.06 ha in 2006. Rice is mostly grown as lowland paddy by mechanical implements; about 10% by direct seeding and 1~2% as upland rice. It is estimated that 78% of rice area is irrigated with water supplies from reservoirs, multi-purpose dams, rivers and wells equipped with pumping facilities. Rest of the area is rain-fed, especially in the areas that are at the bottom of the valleys and hilly terrace with little control.

The rice growing areas are distributed in three major regions, the northern, the central and the southern based on the climatic conditions and into six sub-regions with respect to ecosystems and distribution of rice varieties. These sub-regions are (1) the south-western sea coast with very late maturing varieties, (2) the southern plains with late maturing varieties, (3) the central plains including mid-western and south-eastern sea coast with intermediate maturing varieties, (4) the northern plains and the mid-mountainous areas with early maturing varieties, (5) the alpine and high elevation areas with very early maturing varieties, and (6) double cropping areas where short duration rice varieties are grown with cash crops.

In the following sections, the authors have discussed the present state of rice cultivation with a brief description of changes in the rice varieties, cultural practices, productivity and utilization of rice in Korea in different time periods.

Present State of Rice Cultivation

Chronological Changes of Cultivated Rice Varieties

The comprehensive work of rice variety improvement was initiated with a systematic collection of native rice varieties and their pure-line selections followed by exotic introductions and their adaptability tests during the early period of 20th century. These were replaced by hybrid varieties bred through hybridization within the traditional temperate *japonica* varieties since the late 1930s. During 1970s, the *indica* and *japonica* types were hybridized and Tongil types were developed with a view to achieve Green Revolution in Korea. These Tongil type varieties were completely replaced by the improved modern *japonica* varieties with improved plant architecture, high yield potential and premium grain quality in the early 1990s.

Native and/or landrace rice varieties: Prior to 20th century, a large number of native rice varieties were being cultivated by the farmers throughout the Korean peninsula. However, it is not clear how they were formed or developed; they were probably the outcome of unconscious selections by growers. The Agricultural Demonstration Station (Gweonupmobeomjang) collected 3331 accessions of native varieties and/or land races which were growing in the farmers' fields all over the country in 1910~1912 and identified 1451 distinct genotypes (excluding duplicates due to similar names and characteristics) among them. These included 876 non-glutinous and 383 glutinous lowland paddy, and 117 non-glutinous and 75 glutinous upland rice. It also revealed that 124 varieties out of them occupied 20% of total rice acreage at that time.[74] An additional collection of 448 varieties was listed in 1923 and 581 varieties including 388 non-glutinous and 193 glutinous in 1931.[75] These native collections were characterized for awn (presence or absence), maturity duration, number of tillers, height, number of grains per panicle, grain size and shape, period for ripening, tolerance to drought, susceptibility to lodging and blast disease and germinability under moisture stress and soil conditions.[76]

Major rice varieties widely cultivated were as follows:[77]

Non-glutinous

Nation-wide: Jodongji, Noinjo, Dadajo, Maekjo
Mainly in the southern region: Namjo, Woaejo, Mijo, Guwangjo
Mainly in the northern region: Yongcheonjo, Seoncheonjo, Naengjo, Daegujo

Glutinous

Noinna, Ilweolna, Jado, Geumnado Jeoknado.

Varieties introduced from Japan: During Japanese occupation from 1910~1945, some of the Japanese varieties were introduced into the Korean peninsula as the farmers were compelled to produce food grains mainly for the Japanese people. Their yield potential was 16~21% higher. Rice varieties introduced from Japan replaced the native varieties and reached 57.5% of total rice acreage by 1920 and expanded to around 85% by 1937. About 47 varieties were introduced for general cultivation by 1939. Major rice varieties introduced were Ilchul, Joshinryeok, Gokryangdo, Goomi, Eunbangju, Ryukwoo 132, Norin 6, etc.[78]

Cultivation of domestic bred traditional japonica varieties: The first hybridization programme to develop domestic-bred rice varieties in the Korean peninsula began in 1915 and the first homebred rice varieties, Namseon 13 and Poongok, were released to the farmers in 1932.[79] Since then, a total of 36 home-bred rice varieties including Iljin, Palgwaeng, Paldal, Palkeum, Jinheung, Nongbaek, Gwanok, etc were developed; these covered 80% of the total rice area by 1970. During this period, most of the genetic resources utilized for rice breeding were purely the traditional temperate *japonica* rice including the native and foreign introductions mainly from Japan. Major breeding method employed was the pedigree method and partly mutation technique targeting high yield potential, tolerance to heavy doses of nitrogen application and lodging and resistance to major diseases and insect pests including rice blast and stripe virus.

4) Cultivation of Tongil type rice varieties: Tongil type is designated from "Tongil", the first commercial variety developed from a selection from hybridization between *indica* and *japonica* in 1971. It was selected from the progeny of IR 667, which was derived by 3-way cross (IR8//Yukara/ Taichung Native 1) and released to the farmers in 1972. Yield potential of 'Tongil" was 5.13 MT/ha (metric tonne per hectare) of milled rice that was 28% higher than "Jinheung" a representative *japonica* variety of that period. Afterwards, other Tongil type varieties like Milyang 23, Geumgang, Raekyeong, etc. were released that rapidly increased the area under their cultivation. The acreage reached 76.2% of the total rice area in 1977, resulting in Green Revolution in Korea. Forty varieties of Tongil type rice were developed and cultivated during 1970s and 1980s.[80]

Successful cultivation of Tongil type rice varieties opened not only a new milestone for future improvement of rice varieties but also offered a practical opportunity to utilize *indica* germplasm in temperate environments. Major characteristics of Tongil type rice varieties were short stature in their plant architecture with erect leaves, high yield potential, tolerant to heavy doses of nitrogen and lodging and resistance to major diseases and insect pests, particularly more or less neutral responses to photoperiod and longer period of basic vegetative stage of rice growth. However, there were some shortcomings too such as susceptibility to low temperature, easy shattering of grains and unacceptability of grain quality and palatability to Korean consumers. The area under Tongil type varieties gradually declined to 14.5% in 1989 and was completely replaced with the improved *japonica* varieties by 1992, mainly due to the drawbacks of Tongil types mentioned already and also due to increased consumers' preference for better grain quality and palatability as a result of socioeconomic development and self-sufficiency in rice production.[81]

Cultivation of improved modern japonica rice varieties: The world free trade situations have been the driving force for the changed agricultural system i.e. the ecological adaptation of rice cultivation to the mechanical transplanting, increased acreage under direct-seeding and extending double cropping with cash crops in the paddy fields during the last two decades. The improved socio-economic life style of the consumers has also demanded better and diversified quality of food stuff. As a result, rice varieties having diverse qualities of grain and processing have gradually increased the acreage of rice cultivation.

To meet such situations, the rice breeding programs have concentrated on developing better grain quality with short stature and high yield potential by improving the traditional *japonica* varieties. During the period 1990~2006, 157 rice varieties have been developed and released for general cultivation. They include 145 improved *japonicas* with high yield and premium quality and 12 Tongil types for super yield and processing. *Japonica* varieties include 117 varieties for cooking purpose and 28 for processing. Twenty-eight varieties for processing include 6 glutinous, 7 aromatic, 8 for pericarp colour and 5 high nutrition types such as high hemi-cellulose content, large embryo, etc. Fourteen rice varieties including Hwaan, Pyeongan, Hwagang, Dongjin No. 2 and Juan were well adapted for direct seeding with adaptability for dense planting and 12 rice varieties including Manweolbyeo, Manchbyeo, Manhobyeo, Joanbyeo, Gemho No. 3 and Mannabyeo were adapted to late planting for double cropping with short growth duration of less than 120 days.[82]

Technical Changes in Rice Cultural Practices

According to old records, the transplanting method of rice culture began in the early stage of the Chosun Dynasty in the 15th century and expanded rapidly to the whole peninsula during the 16th century.[83] The modern techniques of rice culture were systematically established with hand transplanting after the foundation of the "Gweonubmobeomjang" in the early 20th century (1906), the first public Agricultural Research Station in Korea and have developed into the present scientific system after the Liberation from Japan in 1945 along with intensive research works. Particularly, the technology of rice cultural practices has been greatly advanced by the cultivation of Tongil type rice varieties since the early 1970s because of different ecological responses of *japonica* varieties. Tongil type rice varieties needed relatively longer basic vegetative growth with more or less neutral response to photoperiod as compared with traditional temperate *japonica* and relatively higher temperature during the growth stage of grain ripening in the Korean climatic situations.

Seeding and transplanting: Until 1970s when *japonica* types were being grown, rice was usually sown in late April to early May and transplanted in early to late June and hand-transplanting was practiced. However, with the cultivation of Tongil types since 1970s, sowing was advanced by 10~14 days to early April to mid April and, to protect the seedlings from low temperature, the seedlings were covered with polythene sheets. The earlier production of rice seedlings provided the critical growing period to the rice plants. However, since 1980s, rice seedlings are grown in indoor seed-box system and transplanted with transplanting machines. This has shortened the period for raising seedlings. Earlier when hand-transplanting was practiced, 40~50 days old seedling were used; when seedlings were raised in polythene covered seed beds for Tongil types, 20~35 days old seedlings were used and now for machine transplanting of seed-box raised infant seedlings, infant seedlings of 8~10 days are used.[84] The regular distance planting method of rice transplanting was settled down since the 1950s. The planting density was adjusted to 72 hills/3.3 m² with 3~5 seedlings in the plain areas and 80~90 hills with 4~5 seedlings in the rice-barley double cropping areas. Furthermore, various methods of direct seeding of rice have also been practiced since 1993 which occupy about 10% of total rice area in these days.

Fertilizer application: The majority of paddy soil in Korea consists of Inceptisols, which is generally shallow in plough depth and has very low infiltration rate. It is very compact and has high hardness, bulk density and low clay content. Thus, paddy soils are usually infertile with lower contents of nitrogen, potassium and phosphorous. The resources of nitrogen and potassium fertilizers were mainly supplied with the homemade manure such as barnyard manure, soybean cake and oilseed dregs; and the phosphorous fertilizer with sulfurous phosphate by 1920s. The chemical fertilizers were not applied until late 1930s; ammonium-sulfate was domestically produced by the construction of Hamheung fertilizer plant in 1936. The chemical fertilizer of nitrogen was increasingly applied together with the homemade manure by the 1960s. In the late 1960s, ammonium sulfate was replaced by urea as the source of nitrogen fertilizer to reform the soil that was acidified due to continuous application of ammonium sulfate. Amount of fertilizer applied was 60-40-40 kg/ha of N-P-K before 1945 and increased to 80-50-60 kg/ha by 1970. It was enhanced to 150-90-110 kg/ha with Tongil type varieties but reduced to 110-70-80 kg/ha with improved *japonica* varieties after 1970. However, it was readjusted to 90-45-75 kg/ha since the 1990s. Application of nitrogenous fertilizer is generally split to 40~50% as basal, 20~30% during tillering stage and 20~30% at the time of panicle initiation. Phosphorous fertilizer was applied 100% as basal and potassium as 70% basal and 30% at the time of panicle initiation.

Weed and pest management: Prior to 1960s, hand weeding was the most common method of weed control and usage of a short-handed hoe for weed control provided simultaneously the effect of inter-tillage. The first herbicide, used for controlling weeds in the paddy field in Korea in 1956 was 2,4-D followed by other herbicides such as PCP, TOK and Pamucone and was rapidly increased to include Swep, MO and Machete during the 1970s.

Rice crop used to suffer from damages due to various diseases and insect pests that varied in different years and regions in Korea. The annual yield losses of rice due to diseases and insect pests have been estimated at around 4%. According to literature, 43 species of rice diseases and about 140 insect pests attack rice in Korea. Major diseases are rice blast, bacterial blight, sheath blight, viruses including stripe and dwarf viruses, scald, false smut and panicle blight. Major insect pests are brown plant-hopper, white-backed plant-hopper, stem-borer, rice leaf folder, small brown plant-hopper, rice green leafhopper, etc. Major strategy to control them has been the cultivation of resistant or tolerant varieties against the key diseases and insect pests coupled with necessary application of agricultural chemicals. However, the development of *japonica* varieties resistant to brown plant-hopper in particular has not been so successful.

Harvest and post-harvest management: Rice harvest was commonly practiced by hand-labour with sickle prior to 1970s. Rice threshing methods evolved from threshing panicles on a log threshing-stand into striping on a threshing hackle followed by employing tread-thresher in the 1940s, a threshing machine in the 1960s and large scale motor thresher in the 1970s.

The combine harvester was introduced in 1975 and became popular in the 1980s. The introduction of combine thresher greatly improved the rice harvesting process including threshing. The combine harvester also improved soil fertility greatly as it simultaneously returned the chopped rice straw into the paddy soil. Drying methods of harvested rice also progressed from manual drying of un-threshed rice plants such as drying on paddy racks, standing small bundles and spreading on paddy field to the current machine drying methods.

Chronological Changes of Rice Productivity

Yield and Productivity of Native and Introduced Rice Varieties[85]

Not much information on rice productivity prior to the 20th century is available in the historical records. However, according to available historical records and farmers' diaries, it is assumed to be less than one ton per ha. It was not until the early 20th century that the official collection of data on rice productivity was attempted. From the evaluation of the native collections and adaptability test of foreign introductions, the improvement of rice was targeted simply at increasing rice productivity. Milled rice productivity from the evaluation plot was around 2.37 MT/ha of Jodongji, the best native variety, in the 1920s and 3.10~3.60 MT/ha of Joshinryeok and Eunbangju, the best introduced varieties, in the 1930s. The national average of milled rice yield from the farmers' field was 0.9~1.1 MT/ha with native varieties and 1.5~1.8 MT/ha with the introduced varieties and about 2.0 million MT for the total rice production.

Yield and Productivity of Domestic Bred Modern Rice Varieties

The milled rice yield potential of rice varieties that were domestic-bred by hybridization and were first cultivated in the late 1930s, progressed to 4.06~4.57 MT/ha for the leading varieties such as Palkweng, Jinheung and Palgeum by 1970 and greatly increased to 3.3 MT/ha as the national average on the farmers' field due to improvement of both varieties and cultural practices and to 3.7 million MT/ha for total production. Although productivity of rice varieties increased significantly by the 1960s, it was still behind in meeting self-sufficiency in rice production as a staple food crop in Korea.[86]

The development of "Tongil" rice variety, the first variety derived from *indica-japonica* hybridization in 1971, lifted up milled rice productivity to 5.13 MT/ha which was 28% higher than that of the best *japonica* rice variety at that time. The productivity of subsequent Tongil type varieties steadily increased to 5.76 MT/ha for Milyang 23 in 1976 and 6.05 MT/ha for Yongmubyeo in 1985. The national average of milled rice yield in the farmer's field dramatically increased to 4.93 MT/ha ranging 4.37~5.53 MT/ha in 1977 as compared with *japonica* varieties yielding 3.37~4.69 MT/ha of milled rice. Total rice production reached 4.67 million MT in 1976 which was the first self-sufficiency of rice in Korea, and 5.21 million MT in 1977 and 6.01 million MT in 1978 which was the first time to exceed 5 million MT of rice production in the history, resulting in a significant achievement for green revolution in Korea.[87]

The modern *japonica* rice is greatly improved, yielding 5.34 MT/ha in the case of several leading varieties such as Ilpumbyeo and Unbongbyeo with semi-dwarf plant type, high yielding and premium quality in the early 1990s. The national average of milled rice yield reached 4.60 MT/ha compared with 3.30 MT/ha of traditional *japonica* varieties during the 1970s. The milled rice productivity of modern *japonica* varieties steadily increased to 5.80 MT/ha. The national average in the farmers' field also increased significantly to 5.0 MT/ha during the last decade and total rice production has been sustained of 5.0 million MT, even though rice acreage has gradually declined to below one million ha. The productivity of Tongil rice varieties also greatly increased to 7.19 MT/ha in the case of Keunseom and 7.53 MT/ha for Hanareum.[88]

Rice Processing and Its Utilization

Rice is not merely a food; it is also an ecological force. It has determined the appearance of the landscape of Korean peninsula, the distinctive features of which have developed because of the need to grow as much rice as possible. Rice has been commonly utilized in the form of whole milled grains for staple food, of whole grains and/or rice flour for the processed foods. Rice straw and rice bran are used as animal feeds. It has been consumed mostly as boiled white rice (*scalar* in Korean) for staple food and partly as the stuff for the processing foods such as rice gruel (*juke*), alcohol (*sulk*), rice cakes (*took*), rice snacks (*hangwa*), rice noodles (*ssalguksu*), fermented foods (*sikhye*), instant boiled rice (*gimbap*), etc.

Rice used as Staple Food

About 95% of total rice produced in Korea is consumed as the staple food mainly in the form of *bap* (a boiled white rice) and 4~5% as processed rice. The methods of preparing rice cuisine have changed over the centuries. According to historical records, the first form of rice cuisine as staple food was not the boiled rice but in the form of roasted rice with probably matured paddy rice and as *jjinbap* (a steamed parboiled rice) probably by steaming immature paddy rice during the period between the Bronze Age and the early Three Nations. It was probably followed by a form of *juk* and *tteok* and developed into the present form of *bap* (boiled white rice). The method of preparing boiled rice (*bab*) seemed to have developed along with the discovery of cauldron in the 4~5th century and progressed to form a unique food culture of current daily life style. It appears that *juk, jjinbap* and *tteok* were treated as minor foods for special occasions such as national festive and/or formal ceremony after rice culture had settled down as a form of agriculture in the period of the Three Nations.[89]

There are various kinds of *bap* (a boiled white rice): *maebap* (boiled white rice) with non-glutinous milled rice, *hyeonmibap* (boiled brown rice) with brown rice and *chalbap* (boiled waxy rice) with glutinous rice depending upon the kind of rice used for cooking and *ibap* or *ssalbap* rice cooked alone, *jabgogbap* with coarse grains, *kongbap* with beans and *ogogbap* with special five grains including glutinous rice, depending upon different kinds of the supplemental grains and stuffs mixed with rice when cooked. According to historical records, there were more than 90 different kinds of *bap*. The *ibap* (*ssalbap*), which is common boiled white rice, is made of non-glutinous milled rice alone and the most familiar form of *bap* nowadays. It was common only among the rich or noble class in the Chosun Dynasty. The *jabgogbap* is made of rice mixed with other kinds of grains such as barley, foxtail millet, common millet, sorghum and pulses, particularly *kongbap* that is mixed with soybean. The poor people used to mix rice with other kinds of grains and pulses in antiquity and even during food shortages prior to 1970s. Even now some people like to mix some barley and foxtail millet and occasionally beans with their rice. *Ogogbap* which is prepared at Daeboreum (a traditional festive holiday on the first full moon day) is a unique type of boiled rice which is made of glutinous rice mixed with four other staple cereal grains including waxy millet, waxy sorghum, black soybean and red bean.

Before the advent of the electric rice cooker that is now a stock appliance in every Korean home, rice was boiled in some kind of a cauldron. Only since 1980s did stainless steel or tin pots became common. Nowadays families often keep their electric cookers on for the entire day so that there is always some hot rice available. Boiled rice has settled down as the main dish of any menu with several side dishes such as soybean sauce, *kimchi*, meat, fish and various kinds of seasonal vegetables. From the traditional unique characteristic of principal menu, *bap* has differentiated into various types of secondary processed dishes of such mix-up types together with other ingredients on boiled rice as *bibimbap*, *deopbap* and some instant *bap* in these days.

The quality rice for Korean consumers is usually *japonica*-like grain such as short, round, translucent in appearance, sticky with low amylose content and low gelatinizing temperature. Thus, most of the commercial varieties grown in Korea meet the consumers' preferences with short and round shape ranging 4.6~5.8 mm in length, 1.64~1.99 in length/width ratio and 17.3~25.1 g as thousand grain weight. Most of the high quality *japonica* varieties showed low amylose content (below 20%) and gelatinizing temperature below 70°C.[90]

Rice Used as Processed Food[91]

About 4~5% rice in Korea is consumed as processed rice and utilized mainly in the form of flour of glutinous and non-glutinous rice. Major products of processed rice are grouped as *juk* (rice gruel), *sul* (alcohol), *tteok* (rice cake) and *hangwa* (rice noodles and cookies), *sikhye* (fermented rice drinks), etc. depending upon types and methods of processing rice. Glutinous rice flour is sometimes added to make them sticky. Traditional processed foods using rice have been usually produced by home manufactory.

Tteok is the most common Korean rice cake which is differentiated by the types and methods of making them, such as *jjintteok* (steamed rice cake made of rice flour), *chintteok* (pounded rice cake with steamed whole grains of glutinous rice) and *bizeuntteok* (stuffed and boiled rice cake). *Tteok* is served as a snack and also as a substitute meal for *bap*. According to historical records, more than 200 kinds of *tteok* have been handed down with more than 100 different kinds of stuffed material. *Juk* (rice gruel), which is the next common category of processed rice, is made of whole grains of rice mixed with other ingredients such as ground meat and fish, abalone, mungbean, adzuki bean and various types of vegetables. *Juk* has been recorded for the first time in the Chosun Dynasty; it was used for convalescence and aged people and has differentiated into more than 170 kinds, mainly depending upon the kinds of mixed stuff (*Imweongyeonjeji*).

Rice alcohol is another famous traditional item of processed rice in Korea; it has differentiated into more than 200 kinds of traditional alcohol since the ancient time. *Maggeolli, Cheongju, Soju* and *Beopju* are the most common and popular rice alcohol nowadays. *Maggeolli* is made from the fermented rice. *Cheongju* is obtained from the filtration of *Maggeolli* and *Soju* is also obtained by distillation of *Cheongju*. *Beopju* is brewed from glutinous rice by fermentation. Rice alcohol is assumed to have come down from the early days of the Three Nations; the brewing technique was modified by replacing the brewing material from non-glutinous to glutinous rice during the Koryeo Dynasty. Several rice varieties adaptable for brewing have been developed and cultivated since 1990. They are Yangjobyeo with large white belly and white centre, Seolgaeng with opaque traits, Daeribbyeo No. 1 with large and soft grains, some coloured rice including dark purple and red color, several aroma rice and some glutinous rice.

Another traditional drink is *sikhye*, a fermented sweet drink of rice. Rice is first fermented and then added to liquor; it has a transparent and gold colour. It has a honey like, malt taste. Another similar type called *gamju* is prepared in a different way and is usually opaque. Rice taffy is also obtained by boiling *gamju*. *Sikhye* is a familiar fermented drink of rice since the ancient period: it appeared for the first time in Samgugyusa and seems to have become common when boiled rice became popular during the Three Nations. *Hangwa* is a type of traditional cookie-like snack of processed rice, both glutinous and non-glutinous. There are different kinds of it such as Yugwa, Gangjeong and Dasik. Yugwa is made with flour of glutinous rice and liquid rice taffy (malt syrup). Gangjeong is made with popped or toasted rice grains mixed with liquid rice taffy (malt syrup) and allowed to harden. Dasik is made of rice flour kneaded together with liquid rice taffy (malt syrup) in moulds.

References

1. Hamada and Umehara (1920).
2. National Museum of Korea (2000).
3. Son (1992).
4. Lee *et al.* (1994).
5. Yim (1978).
6. Yim (1978).
7. Kim and Suk (1984).
8. Lee *et al.* (1994).
9. Lee *et al.* (1994).
10. Lee and Park (1979).
11. Lee (1997).
12. Ahn (1999).
13. Heu *et al.* (1997).
14. Kwangju Museum (1997).
15. Shim (1992).
16. Lee, H.J. (1999).
17. Lee, S.K. (1999).
18. Lee, S.K. (1999).
19. Yan (1995).
20. Wang and Wang (1995).
21. Wu (1995).
22. Nakashima (1995).
23. Yamazaki (1995).
24. Kim (1986).
25. Son (1992).
26. Yim (1990).
27. National Museum of Korea (2000).
28. Lee *et al.* (1994).
29. Yim (1978).
30. Heu (1997).
31. Lee and Park (1979).
32. Lee *et al.* (1994).
33. Heu (1998).
34. Kwangju Museum (1997).
35. Ahn (1999).
36. Shim (1992).

37. Lee, H.J. (1999).
38. Lee, S.K. (1999).
39. Lee, H.J. (1999).
40. National Museum of Korea (2000).
41. National Museum of Korea (2000).
42. National Museum of Korea (2000).
43. National Museum of Korea (2000).
44. Lee *et al.* (1992).
45. Son (1992).
46. Yan (1995).
47. Wang and Wang (1995).
48. Wu (1995).
49. Takakura (1995).
50. Yan (1995).
51. Takakura (1995).
52. National Museum of Korea (2000).
53. Yan (1995).
54. Kim (1986).
55. Heu (1996).
56. National Museum of Korea (2000).
57. Okajima and Shida (1986).
58. Nishiyama and Kumashiro (1957).
59. Heu *et al.* (1991).
60. Heu (1991).
61. Heu (1999).
62. Lee, S.K. *et al.* (1992).
63. Lee, S.K. (1992).
64. Heu (1996).
65. Heu (1996).
66. Lee, S.K. *et al.* (1992).
67. Heu (2000).
68. Lee, S.K. *et al.* (1992).
69. Lee, S.K. *et al.* (1992).
70. Lee, S.K. *et al.* (1992).
71. Nishiyama and Kumashiro (1957).
72. Suh (2003).
73. Suh (2003).
74. Gweonupmobeomjang (1913).
75. Anonymous (1976).
76. Choi, H.O. (1978), Lee, J.H. (1965).
77. Lim *et al.* (1983).
78. Choi, H.O. (1978), Lim *et al.* (1983).
79. Choi, H.O. (1978), Lim *et al.* (1983).
80. Crop Expt Sta (1990).
81. Moon (2005).
82. RDA (2006).

83. Yeom (2007).
84. Moon (2005).
85. Lim *et al.* (1983).
86. Crop Exp Sta (1990).
87. Crop Exp Sta (1990).
88. Moon (2005), RDA (2006).
89. Choi, H.C. (2005).
90. RDA (2006).
91. Choi, H.C. (2005).

6 | History of Rice in Japan

Chukichi Kaneda

MYTHS AND LEGENDS ABOUT THE ORIGIN OF RICE

Kojiki (Record of Ancient Matters)[1] the first book ever compiled in Japan in 712 AD, describes as follows: the god Susano-o, the younger brother of Amaterasu, the supreme goddess of Japan, killed the goddess Ohgetsu-hime in a fit of anger, then various crops originated from her corpse. Silkworms appeared from her head, rice seeds from both eyes, millet from both ears, red (adzuki) beans from her nose, wheat from her genitals, and soybeans from her buttocks. The second, but far voluminous book compiled in 720 AD, *Nihon-Shoki* (Chronicles of Japan)[2], described that, when the goddess Izanami gave birth to the god of fire, Kagutsuchi, she almost died from the heat, but further gave birth to the goddess of soil, Haniyama-hime and the goddess of water, Mitsuhanome. From the couple of fire and soil, the goddess Wakumusuhi was born; and on her head silkworm and mulberry appeared; rice, millet, wheat and beans on her belly. This couple might be considered to symbolize the origin of primitive, slash-and-burn agriculture.

However, so far as the written documents narrate, we come to know that rice farming in Japan was conducted in irrigated lowlands from the beginning. Both *Kojiki* and *Nihon-Shoki* describe the violent behavior by the god Susano-o, who destructed ditches and levees of paddy fields owned by his elder sister and the supreme goddess Amaterasu. In *Kojiki*, where to develop paddy fields is described in the story on brothers Yamasachi-hiko and Umisachi-hiko, too. No stories can be found in relation to upland rice cultivation so far as described in the oldest books compiled in Japan in early 8th century.

A legend in Iwate Prefecture (north-eastern Japan) tells us about the source of rice introduced to Japan. The god named Inari-sama took the rice seeds in 'Kara' (ancient China) and hid the seeds in the hollow stem of a large reed. Plugging the open end of the reed with paper, and using it as a walking stick, Inari-sama returned to Japan. The two letters denoting 'Inari' in Chinese scripts express 'luggage of rice', and Inari-sama has long been enshrined everywhere in Japan. Scientific works on the route of introducing rice to Japan will be discussed later, but this legend seems to indicate one of possible routes of introduction of rice very simply.[3]

ARCHAEOLOGICAL RECORDS ON RICE CULTIVATION

Rice cultivation in China dates back to at least around 7000 BC of Peng-toushan, and carbonized rice remains of more than 10,000 years ago have been found from several sites.

In Japan, after the World War II, archaeological studies started to become quite active and excavation works were conducted in many places. The first great achievement was excavation of the Toro Ruins in Shizuoka City, which started in 1947. The Toro Ruins was the agricultural village that flourished on the basis of well developed rice cultivation around AD 100 to AD 300, with 12 houses, 2 storehouses with elevated floors, 30 rice paddy fields, 370 m of waterways, wooden farming tools such as hoes, spades, spatulas and stone knives, weaving tools, musical instruments, etc. The academic research related to this excavation led to the establishment of the Japanese Archaeological Association, and from many other excavation sites, it became evident that Japan had Paleolithic Period in 30,000 to 10,000 years ago. During the period, people lived by hunting and gathering, used fire, and except a wide variety of Paleolithic tools, no bone or horn artifacts have been found, neither pottery.

This era was followed by the Jomon period (7500 BC-250 BC, Mesolithic), and the Yayoi period (*ca.* 250 BC-*ca.* AD 250, Neolithic). During these periods, rice was introduced to Japan in various forms and routes, the facts being clarified by pollens, plant opals, and DNA studies of carbonized rice grains excavated from ruins and unearthed sites. Though the written document is not available before AD 712 (the year of publication of *Kojiki*), history of rice cultivation became clearer, though not conclusive, owing to archaeological and molecular genetic studies.

Division of Archaeological Periods

Dividing the Jomon and Yayoi Periods is still disputable, for new discoveries of ruins may lead to further modification of the division of the periods, and would be different by regions within Japan. This is remarkable especially for subdividing the Yayoi period.

The Jomon period is subdivided into several stages as follows: Initial stage (11,000-8,000 BC), Very Early (8000-4000 BC), Early (4000-3000 BC), Middle (3000-2000 BC), Late (2000-1000 BC), and Very Late (1000-200 BC).[4]

Dividing the Yayoi period can be different, but the National Museum of Japanese History proposes that the Yayoi period in Kyushu started from 1000 BC, more than 500 years earlier than considered before, and overlapping the Very Late Jomon stage. According to the proposal, Very Early stage lasted for about 200 years, and Early stage lasted up to 400 BC, Middle stage up to some decades in AD.[5]

Traces of Rice Cultivation in Archaeological Sites

The Kyushu region was most advanced in introducing rice cultivation, for its location facing East China Sea and had more opportunities of accepting immigrants from China and southern islands. Another was the Chugoku region facing to the Tsushima Straits and Korea.

The earliest record of rice in Japan was found in 2005 from Hikosaki Kaizuka (shell mound) in Okayama city.[6] From the soil layer of around 6000 years ago, and 2.5 m underground, a great amount of plant opals of rice (2000-3000 per gram of soil) were found. The size of plant opals found in Hikosaki was 30-60 micrometer and the shape indicated that the rice grown there was of *japonica* type.

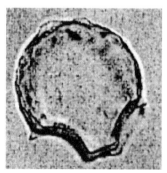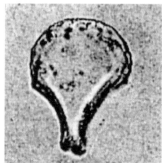

Fig. 6.1. Type of plant opal: *indica* (left) and *japonica* (right) rice.[7]

Plant opals are microscopic plant-derived silica body (SiO_2) deposited within motor cells, and are found when plant tissue breaks down in the soil, remaining undecayed almost forever. Silica bodies were classified into the α-type and β-type. Survey of 96 Asian rice cultivars clarified that typical indica had α type silica bodies and typical japonica had β-type.[8]

Plant opals may not necessarily indicate the existence of rice cultivation at the place, because when the amount of plant opal is limited, that might come from other places. The amount found in the Hikosaki Kaizuka is good enough to estimate existence of rice cultivation, but no other clue to show rice cultivation. Besides, the era was the early Jomon, far ahead of the era when rice was considered to be first grown in Kyushu.

Remains of agricultural tools and paddy fields were excavated from ruins of much later than the era of plant opals, carbonized rice grains, or figures of rice grains impressed on earthen wares that are the late Jomon to the early Yayoi. The earliest finding of plant opal in Kyushu was in mid-Jomon, carbonized grains in later half of mid-Jomon, and ruins of paddy fields in later half of the very late Jomon to early Yayoi, while in Chugoku region, plant opals in early Jomon, and ruins of paddy fields in early Yayoi.[9]

Kinds of Rice Introduced to Japan

Rice widely grown in present Japan is temperate *japonica*, cultivated under irrigated lowland conditions. The story in *Kojiki* also suggests that lowland rice cultivation was prevalent in those ages. However, accumulated information from excavations of many sites in Japan tells us that rice introduced in Jomon periods was tropical japonica probably introduced via southern part of Korea, or by the route of southern islands of Japan, and grown under upland conditions. The type of rice plant opals found from ruins of the Jomon stages are known to be mostly of tropical *japonica*.

In the late Jomon period, a new rice technology, lowland rice cultivation, began to be introduced into north-western part of Kyushu from China, and then disseminated gradually to eastern Japan. The type of rice was temperate japonica. According to Sato,[10] the fact that new rice varieties came to Japan directly from China may be proved by difference in the type of specific region of SSR (single sequence repeat) of DNA. The region named RM1 is polymorphic, and out of 8 types (RM1-a to RM1-h) found in China, RM1-b occupies nearly 2/3 of 90 varieties tested, while in Korea 55 varieties contained 7 types, excluding RM1-b. In Japan, majority of varieties had either RM1-a or RM1-b, and RM1-c was a single minor type, without other types.

Matsuo[11] wrote in the summary of his famous research report on classification of 666 rice varieties collected worldwide: "the Japanese rice is probably the A type in middle China brought direct to northern Kyushu and differentiated," based on his findings that the group of rice varieties, most similar to lowland rice varieties of Japan, was found in japonica rice in central China.

Agricultural Tools in Different Archaeological Stages

The oldest remains of rice cultivation so far found in Japan, where lowland rice fields and several kinds of farming tools were excavated, are 'Itazuke' remains in Fukuoka Prefecture, and 'Nabatake' remains in Saga Prefecture, both in northern Kyushu. The survey of the Itazuke ruins in 1951-1954 identified carbonized rice grains, and arc-shaped ditches, and in 1978, paddy fields with many footprints, irrigation canals, gates for irrigation and drainage were found from the soil layers of the Jomon period. The village was surrounded by deep ditches (4 m wide and 2 m deep) probably for protecting against hostile outsiders. In the Jomon period, food was stocked in holes dug in the ground. The Nabatake ruins was excavated in 1980-81, and the features of lowland rice cultivation operated in late Jomon period to mid Yayoi period could be realized: carbonized rice grains, stone axes and sickles, gates and ditches for irrigation, wooden 'eburi' for leveling field after puddling, and other farming tools were collected.[12]

As not much space is available to look in detail into developments of farming tools in ancient periods in Japan, one of the laborious works made in different regions is briefly excerpted from Yamaguchi.[13] He compiled many drawings of wooden farming tools excavated from various sites, not only in northern Kyushu but also many other regions in Japan, and viewed regional difference of developmental stages of rice cultivation in Japan. Fig. 6.2 shows many tools, especially tools for cultivation, in northern Kyushu.

Very Late Jomon Stage

About 80% of wooden farming tools, excavated in northern Kyushu, eastern Kyushu and the Seto Inland Sea area, are for cultivating lands, and wood material was found to be from *Quercus acutissima* Carr. The most common type of hoe is shown as (1, 2).

Early Yayoi Stage

In the older stage, the type of hoe was similar to the previous stage, but the material was obtained from *Cychrobalanopsis,* and a new type of hoe (3), 'eburi' for leveling land, pestle (4), and bar for beating rice panicles appeared. In the newer stage, new types of hoes appeared (5, 6, and 7). At this stage, wooden farming tools can be found in sites in San-in, Kinki and Tokai regions. Yamaguchi notices regional differentiation of types of hoes after this stage, and considers that lowland rice cultivation was fully established in northern and eastern Kyushu and San-in regions.[14] He considers that in Kinki region lowland rice cultivation started during this stage.

Middle Part of Mid Yayoi

Long hoe (8), trident hoe (9) and trident spade with a part to put foot on (10), and mortar, etc. appeared. Also, cramps to fit hoe more tightly to the handle were used. The G-type hoes (11) introduced from Korea became the principal tool in Kyushu, and denotes that lowland cultivation in the plain area became common, as well as in San-in, Seto Inland, and Kinki regions. In the Tohoku region, rice cultivation was introduced during this stage with technologies of manufacturing and using farming tools.

Last Part of Mid Yayoi to Initial Part of Late Yayoi

Shape and functions of hoes (12, 13, and 14), spades (15) and 'eburi' (16, 17) became more diversified. The specific patterns of combination of tools could be found in topographically different sites, low alluvial area with or without frequent flooding, higher alluvial or sandy places. In Kyushu and Kinki regions, existence of dams indicates that starting from this stage, development of fields and water management was led by local chiefs. In southern Kanto, technologies of lowland rice cultivation and manufacturing farming tools were introduced in this stage. He also noticed that development of paddy fields proceeded to higher alluvial places in northern Kyushu during this stage. Iron tools for harvesting rice panicles started to be used from this stage.

Late Yayoi to Tumulus Period

Hoes of various types (18, 19), spade (20) and pestles (21) were excavated. 'Ta-geta' (a pair of special footwear, (22), probably a specific tool only to Japan, appeared during this stage to be used for very deep muddy

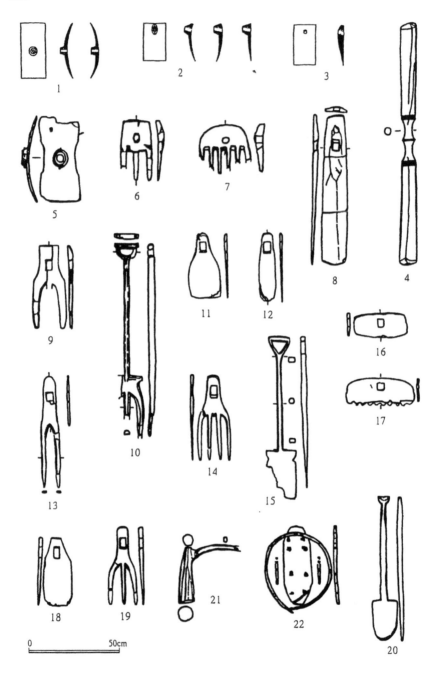

Fig. 6.2. Examples of wooden farming tools excavated, mostly in northern Kyushu.[15]

161

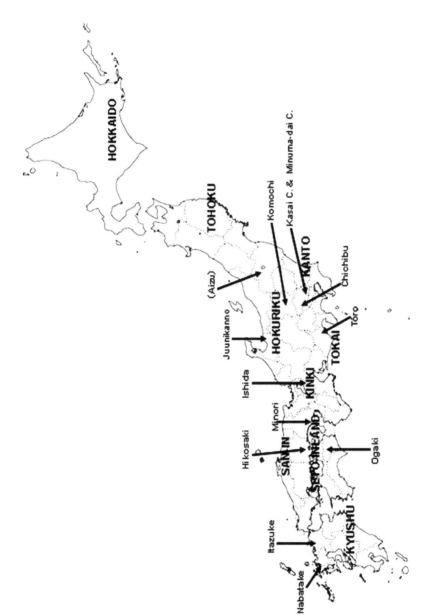

Fig. 6.3. Map of selected excavation sites where rice has been found in Jomon and Yayoi periods.

fields, and this suggests that development of lowland rice fields progressed even to marshy areas. The area of rice cultivation expanded to northern Kanto, Hokuriku and central highland regions during this stage.

It is noticed that tools for plowing and harrowing using animal power did not exist in Japan during these periods. A wooden harrow was excavated in May, 2000 from Ishida ruins in Shiga Prefecture. This site was buried in the period of late Yayoi to early Tumulus, and the wooden harrow of 70 cm with 12 teeth, the longest one 53 cm, is considered to be the oldest so far found in Japan, and was used in 4th century.[16]

RICE CULTIVATION IN ANCIENT TO MEDIEVAL PERIODS

Both of 'Kojiki' and 'Nihon Shoki' describe events mostly centered on Imperial courts, though mythological and legendary especially for the first nine generations of Emperors. This period corresponds to the late Yayoi and the Tumulus period (mid 3rd-mid 7th century). The name of Tumulus period came from huge tombs which many of powerful local governors constructed during the periods. Some of these governors are kings of 'Wa' countries recorded in the history of China, and the Yamato imperial court, located in Nara and the vicinity area, is one of these countries.

Kojiki and *Nihon Shoki* describe that the tenth Emperor 'Sujin', in his 62nd year on the throne, ordered to construct water reservoirs to encourage farmers, and three dams, 'Yosami', 'Karisaka' and 'Sakaori', were constructed. The 15th Emperor 'Oujin' constructed water reservoirs 'Tsurugi' and 'Kudara'. The next Emperor 'Nintoku' did various civil engineering works, construction of canals, banks, water reservoirs, and development of wide areas for fields. Such records seem to show that political leaders noticed that disseminating lowland rice cultivation was necessary to keep social order and their power by supplying more food to sustain the population growth.

Historical Processes of Paddy Field Development

Initial development of paddy fields is considered to take place in skirts of hills/mountains with water-flow from brooks or springs, and then expanded to construction of banks to reserve water at the outlet area from a valley, and develop paddy fields in the flat area downwards. Then came the more positive activities such as construction of water reservoirs.

The remain in Chichibu, Saitama Prefecture shows that regularly arranged paddy fields of about 40 ha were developed in 7th to 8th century, with roads, levees, irrigation and drainage canals.

In AD 701, a great legislation system, 'Taiho Ritsuryo', was established. The Ritsuryo included laws and regulations relating to administration, taxation, military system, criminal, etc. In the Ritsuryo, unit area of paddy fields was defined as: 1 'tan' = 30 'bu' (54.5 m) × 12 'bu' (21.8 m) = 1188 m², 10 'tan' = 1 'cho'. Presently one 'cho' is almost equal to 1 hectare. The units, 'bu', 'tan' and 'cho' had been used until mid 20th century.

The productivity of the field was defined by bundles of rice harvested per 1 'tan' as: upper class field 50, middle class 40, lower class 30 and very low class 15. Tax was 20% each of the harvest. The worst class field was prohibited to plant every year, but be planted every other year due to poor soil fertility. Based on statistical data in documents of *Shosoin* (The Shosoin Repository, Imperial Household Agency), it is estimated that the yield (of brown rice) as 1.27 t/ha for the upper class fields and 0.76 t/ha for the lower class.[17] This taxation system informs us that the soil fertility was an important matter even in the ancient period.

In the Ritsuryo system, which lasted up to mid 10th century, each farmer was given definite area of lands (both lowland and upland fields) to grow rice, barley, wheat, millet, or buck-wheat. In some places, sericulture was also conducted. However, regulations on upland were quite limited and documents were few, denoting that the more importance was given to rice by rulers to collect benefits.

Technologies of Rice Cultivation

Site excavations have clarified various phases of rice cultivation in different ages. However, after Taiho Ritsuryo, records describing about rice cultivation are rather few until 17th century due to social instabilities.

Animal plowing seems to be commonly practiced in 7th century, as Taiho Ritsuryo described that each farm household should keep an ox for plowing. Method of fertilizing soil was application of grasses/weeds until medieval period. Planting rice was practiced by direct seeding initially. In 1985, a ruin considered to be seedbeds was found in Komochi village, northern Kanto, which seemed to be buried with volcanic pumice stones from Mt. Haruna in mid 6th century. Upland seedbeds were set near living area perhaps to protect from birds. Transplanting became common perhaps in 10th century.

Harvesting was done by cutting panicles with stone knives, and it was early 10th century that sickles were used to cut rice plants at the base, then two bamboo sticks were used to hold panicles in between sticks and thresh off grains. This style of threshing lasted until the end of 17th century. Dehulling and milling with a mortar and pestle was the style since the Yayoi period.

Varieties cultivated in the Nara Period (8th century) were very simple, and differentiated by maturity, waxy/non-waxy, or red/white grains, as described in *Manyo-shu*,[18] or in records of *Shosoin* collections, etc.

In July 1999, records on wooden plates excavated from 7 ruins of northern Honshu to Kyushu were deciphered to be names of rice varieties of nearly 1000 years ago. By this finding, it was known that rice varieties were much differentiated than what has been considered, and perhaps were used as a measure of reigning rural leaders by lending superior varieties. In the Medieval period (until 16th century), varieties with names such as Meguro, Houshi, etc. were grown, and red kernel varieties were relatively common due to better adaptability to low fertility soil, cold irrigation water, drought, or insect pests, though usually low yields. A total of about 20 varieties were identified from wooden plates, some of which have been grown even in 17th century.[19]

Referring to documents in the 'Nara' period, taxes were collected in forms of various farm products, and rice was the most common and important to be used not only for food but also in place of currency. The form of rice for storage was mostly (95%) paddy (rough rice), followed by panicles (1.5%), then cooked and sun-dried rice perhaps convenient for provision for the army troops.

Double cropping of rice with barley, wheat, or other upland crops, started around 12th century. From 'Minori' ruins of late 11th to early 12th century in Kakogawa city, carbonized rice and barley were found in the ancient cultivated land. The ridges of paddy field soil were as high as 1.5 m, almost twice of normal height,[20] and this must indicate that winter cropping was conducted after rice.

The oldest rice terrace in Japan was excavated in 'Oogaki' ruins of Tokushima Pref. in 1998. The terrace, with five stairs of long-shaped field about 2 m wide, and height difference of about 20-50 cm, was estimated to be constructed in the last part of early Yayoi period. The area seemed to have often suffered from floods by Yoshino river, and people moved to higher locations.[21]

RICE CULTIVATION IN THE 'EDO' PERIOD (17-19th C)

The turbulent ages finally ceased in early 17th century when the Tokugawa shogunate was established, and this period of 260 years without war is called 'Edo' period. During this period, rice was the core of national economy: the rank of all 'daimyo' (feudal lords) was expressed by the amount of rice (areas of territory producing the amount) given by the shogunate, and the same to their retainers. Therefore, feudal lords made their best efforts to increase rice production within their territory by developing fields, irrigation systems and strict protection of forests as the important source of irrigation water. Thus, until the mid 'Edo' period, cultivated land area increased 3.5 times, and rice production 3 times more, compared to the period 13th to 16th century. During the period, the population increased from 10 millions to 30 millions. It is estimated that 'Edo' (presently central Tokyo) had the biggest population in the world in those days. Tomiyama estimates that upland farming in Europe in 19th century supported 1 person per 1.5 ha, while rice farming in Japan in 'Edo' period could support 10 persons per 1 ha.[22]

Hydraulic Engineering and Large Scale Paddy Field Development

Efforts of the Tokugawa Shogunate and most 'daimyo's to increase their income (tax in the form of rice) were directed to improvement of environments for cultivating lowland rice, through civil engineering and various technologies.

(1) Development of Irrigation Canals

The Central 'Kanto' Plain was upland area until 16th century, but the shogunate succeeded in development of a great canal system of 'Kasai' (completed in 1667) and 'Minuma-dai' (completed in 1727). The lake 'Minuma' was reclaimed, and lowland rice fields were developed. Both canals got irrigation water from 'Tone' river. To compensate ('dai' means 'instead') the water resource of the lake Minuma for farmers, the shogunate constructed the trunk canal of 96 km using underpass (siphon) and overpass waterway in half a year, mostly for irrigating the central Kanto plateau area. This canal also had the system like that of the Panama Canal.[23] There are many similar stories during this period throughout Japan relating to developing canals to open lowland fields.

Not a few canals were developed using tunnels through mountainside. Such kind of civil engineering was supported by technologies from gold/silver mining industry, which was vitally important for 'daimyos' to their economy, and the know-how of excavating a tunnel in mining was applied to civil engineering for developing irrigation canals. Shiina Michizou developed the irrigation canal, 'Junikanno', of 30 km with 16 tunnels to open paddy fields in the highland in Toyama, in early 19th century.

(2) Land Reclamation

In Kyushu, Saga 'han' (feudal clan) started reclamation of the Ariake Sea from early 17th century, and could increase rice production by 40% in 100 years. Reclamation has been continued to the present and up to now 266 km² is diverted to very fertile field. Prefectural average yield of Saga was the highest in Japan until the Tohoku region became top yielder by success in improving low temperature tolerance (cf. p. 14).

(3) Efforts of Planting and Protecting Forests

In order to secure abundant irrigation water, it was very important to have and maintain forests in good conditions. Many episodes are available on this topic, and therefore, it is said, in other words, that forest in Japan was brought up by rice. Under the shogunate, unlawful cutting-down of one tree resulted in cutting of his head.

(4) Role of 'Wasan' (Japanese Mathematics)

On the background of many achievements in civil engineering, 'wasan' is recognized to have played a significant role. Yoshida Koyu (1598-1672) wrote a mathematical book 'Jin-Gou-Ki' (1627-1643), which contains: decimals, divisions, calculation methods used for sale of rice and/or gold/silver, money exchange, interest calculation, taxing, civil engineering of rivers, etc. and also various interesting quizzes, such as: "How do you divide 10 liter of oil into 5 liter each, using two measuring square boxes of 3 and 7 liter?" The series of the book became best sellers for many decades, and even in countryside, leader farmers liked to read and learn in arithmetic schools. Another famous scholar is Seki Takakazu (1632?-1708), author of differential-integral calculus, determinant, etc.

Differentiation of Rice Varieties

During the Edo period, it was not easy for common people to travel to other 'daimyo's territories except on the occasion of pilgrimage to the 'Ise' Shrine, or famous temple such as 'Zenkoji'. Introduction of rice varieties from other localities was seen on the occasion of such travels, and new rice varieties were good souvenir for home villages. Those varieties had commemorative names, such as Ise-nishiki, Zenkoji, or Izumo.

Aizu Nousho (Book on Agriculture in the Aizu District, Fukushima prefecture) published in 1684 by a leader farmer, Sase Yojiemon, listed rice varieties according to adaptability to 13 different soil types of 2 ecological conditions (plain or mountainous area) of paddy field. In the list, aromatic rice, red rice, and a variety seemingly by name introduced from Indo-China, are included.

Koka-shunju (Descriptions of Annual Farming Works), published in 1707, describes rice varieties in Ishikawa county near Kanazawa city: by maturity, 28 early, 20 intermediate, 32 late; and 16 waxy, 5 *indica* were included. In late 19th century, leader farmers in different localities conducted experiments for comparison of rice varieties.

Methods of Rice Cultivation

The various features of farming (rice farming as the main activity) in the district near Kanazawa city was well illustrated in the book, *Nogyo Zue* (Agriculture Pictorial), and described in detail in the book *Koka-Shunju*, both by Tsuchiya Matasaburo (1644-1719).

The first important activity of the year for farmers was courtesy call to town households carrying products from gardens as small gifts. The purpose was to reserve the right of collecting excreta throughout the year. This system also well fitted from the political point of view to maintain good sanitary conditions in the urban area, and by order of the ruler, farmers stocked manure in tanks or holes dug in the fields to be used as organic fertilizers.

The steps of rice cultivation are briefly excerpted from the pictorial book as follows:

In March, seeds are soaked in cold water to let them absorb enough water until the time for germination is ready. This treatment is important especially in cooler regions. Then seeds are sown to seedbeds on a fine day. In this age, there was not yet practice of applying charred rice husk for absorbing more heat from the sun to warm the seedbed soil. After plowing wet paddy fields and breaking soil lumps into small pieces, water is let in to the field and levees are scraped off for weed control, and prepare for coating the levees with wet mud, so that holes made by rats are filled in and loss of water is reduced. Usually, soybean was grown on levees.

Tillage was done by hoe of awfully various types depending on locality, soil property, or purpose. Okura Nagatsune (1822) illustrated many types from different regions of Japan in his book *Nogu Benri-ron* (Introduction to Useful Agricultural Tools), for example, standard type for cultivating upland fields, one good for very stony soil, one very good for plowing and leveling paddy fields, one good for heavy soils, and others good for weeding of upland fields, etc. Not only hoes, he described almost every kinds of agricultural tools and machinery: for example, the wooden structure for irrigating as much as 63 tons of water in 24 hours up to the height of 4.5 m with the power by a bull, or a fire extinguisher which could also be used for draining water from deep holes, etc. As sizes of parts are shown to illustrations, this book seems interesting and useful for developing countries still now.

For puddling, horse or ox was used due to limited period of farming, and usually 3 puddlings (rough, intermediate and finishing) were conducted, with the last one as important for preventing leak of water and reducing percolation.

As for transplanting, *Nogyo-zensho* (Encyclopedia of Agriculture, 10 volumes, first published in late 1697 by Miyazaki Antei) recommends 30 hills per square meter, with 3-4 seedlings, usually planted by village farmers in co-operation with each other. *Koka-shunju* describes that a woman can plant 1100 m² in a day. Many farmers planted barley, wheat or rape-seed as winter crops, thus transplanting rice might be delayed in some year due to cooler climate conditions.

After transplanting, hand weeding was conducted three times, and then fertilizers (diluted liquid manure: fully dissolved human manure) were applied in August. Tread-wheel for drawing up water from the lower level stream to irrigate fields was invented in 1660s and prevailed by 1750s. Indica type varieties, called 'Daito-ine' or Champa, which were introduced from Indo-China through China, were planted along levees. They were adapted to lower soil fertility and matured earlier than traditional japonica varieties. Therefore, when farmers had eaten up all the stock of rice, indica varieties were just good for harvesting, and easy to reach from the levees. Yield was not high, and tended to lodge by wind due to tall height. Such kind of varieties is threshed on the day of harvesting, due to easy shattering.

For harvesting rice, sickles of normal type or with saw-like blades were used, and reaping near the base of hills, making a bundle for 5 to 6 hills, then dry bundles by putting their base on the ground, or hang bundles to ropes or sticks high in the air. After long history of primitive method of threshing, i.e. the two bamboo sticks method, a revolutionary implement called 'Senba' (meaning 'thousand teeth') was invented in early 18th century (Fig. 6.4A). This implement was so efficient that many women, especially widows, lost their job, and the 'semba' was called as 'widow killer'. Semba was continuously improved for almost two centuries. Dehulling was done using earthen- or stone-mill (Fig. 6.4B) driven by two men. The containers of rice were 'tawara', made of rice straw by farmers themselves as night works.

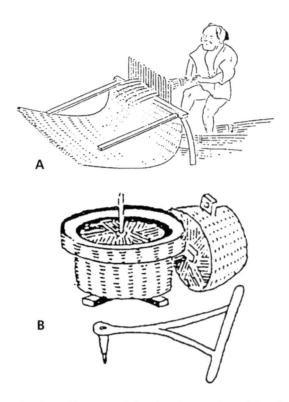

Fig. 6.4. 'Semba', efficient tool for threshing (A), and hand-mill (B).

The source of fertilizer was, other than human manure, mostly weeds and grasses collected for making stable manure and compost, or fresh sprouts from trees plucked and applied directly to fields before puddling as green manure. The way of management of mountains/hills owned in common by the community was very important in order to maintain forests orderly.

Insect pest control has been a great problem; especially brown plant-hopper was known since 9th century. A book published in 807 described about this pest in a mythological tale.[24] In 1732, nearly 900 thousand people died from famine caused by outbreak of brown planthopper, which ravaged from Kyushu to central Tohoku regions. In the book *Jokouroku* (Manual for Controlling Insect Pests) published by Okura Nagatsune in 1826, he described on kinds of pest insects, climatic effects, controlling methods, insecticidal plants such as morning glory, *Brassica juncea*, *Dioscorea tokoro*, *Pieris japonica*, etc. In this book, he illustrated the method of preventing damage by the brown planthopper: kerosene was dripped onto the surface of shallow flooded field, and then lower parts of rice hills were hit or swept to let insects fall and be drowned. After that, by draining the water, insects are largely removed out of the field. Light traps using kerosene lamps were also used for collecting adults of insect pests after late 19th century.

Units for Measuring Rice and Other Cereals

Rice has been so important since ancient ages, that it was placed in the center of national economy in the 'Edo' period, and was used in place of money for payment of tax, or land rent, etc. The system of measuring rice was determined in 1669; to be handled by units of volume. The standard of measure box has not been changed until now, though the metric system is officially used at present. The units of volume are: 10 'gou' = 1 'shou', 10 'shou' = 1 'to', 10 'to' = 1 'koku'. The standard of 1 'shou' box is 14.857 cm × 14.857 cm × 8.18 cm = 1804 cm^3 (1.8 liter). One 'tawara' contains 4 'to', and weighs 60 kg. One 'koku' weighs 150 kg.

When farmers finished harvesting rice, they brought rice packed in 'tawara' to the warehouses of their landlords, and then rice would be turned over to brokerage and wholesalers.

RICE CULTIVATION IN LATE 19th TO MID 20th CENTURY

In 1867, Tokugawa shogunate collapsed and the political power was restored to the imperial court. This great historical turning is called 'Meiji Restoration', and Japan started to open doors to overseas to introduce modern technologies. Rice cultivation in Japan also experienced substantial changes in technologies, especially in fertilizers and soil management, agricultural machinery and tools. Such changes affected important characteristics of rice varieties, resulting in an increase of yields by 50% and in reduction of working days also by 50% during this period.

There were prominent leader farmers in different regions in the late 19th century, who contributed in various ways to the promotion of rice cultivation. Among them Nara Senji in Kagawa Pref., Funatsu Denjihei in Gumma Pref. and Nakamura Naozo in Nara Pref. were called 'the Three Great Farmers'.

Rice Varieties

In the end of 'Edo' period, several progressive farmers started rice observational trials. For example, Nakamura Naozo initiated a series of varietal trials in 1863, and in 3 years, the number of his test entries reached 100 including those from fairly remote areas within Japan. In 1881, varieties from western countries, such as Carolina (USA), Java, Italy, etc. were tested, but were found much taller and of inferior quality than traditional Japanese varieties. It is not known nowadays that Japan had once deep-water rice, but they were planted in inundated areas such as in Gifu Pref., where three rivers, Kiso, Ibi and Nagara, confluent. The varieties were: Ikesoko, Minazoko (both meaning 'bottom of pond or water'), Mizu-kuguri (living in water), etc.

In 1901, Ishikawa Rikinosuke, a colleague of Nakamura Naozo at the botanical station of Akita Prefecture, published the book, 'Ine-shu Toku-shitsu no Ben' (descriptions on merit/demerit of rice varieties), describing characteristics of a total of 103 varieties, dividing them by color of awns: awnless 18, white awn 32, black 4, red 35, and yellow 14. Firstly, general description was given on classification of varieties by maturity, and quality of various soil types, then for each variety: name, origin, adapted region and soil, plant characteristics, quality, tolerance/susceptibility, effect from continuous planting, etc.

Many devoted farmers searched for better looking plants from their fields, and some of those varieties selected by farmers were widely grown, and became Japan's leading varieties in late 19th to early 20th century. The national breeding system started its work only in 1893 and the first arti-ficial hybridization was made in 1897. For this national program, the most important genetic resources were these farmers' varieties. The followings are only a few examples of varieties selected by farmers.

Omachi: selected around 1870 by Kishimoto, J. in Okayama Pref., and planted to 110,000 ha (max.) in 1907

Kameji: selected around 1870-75 by Hirota, K. in Shimane Pref., planted to about 60,000 ha

Shinriki: selected in 1877 by Maruo, J. in Hyogo Pref., planted to 510,400 ha in 1907

Kameno-o: selected in 1893 by Abe, K. in Yamagata Pref. planted to 190,000 ha

Tougou: selected in 1901 by Sato, M. in Yamagata Pref.

Aikoku: found in 1889 from a population given to Honda, S. in Miyagi Pref.

In accordance with increased use of fertilizers, rice varieties with improved plant type and resistance/tolerance to biotic and abiotic problem were selected. Application of dried sardine, soybean cakes, rape oil cakes and fish meal in late 19th century, and ammonium sulphate in early 20th century, necessitated varieties more tolerant against lodging, low temperatures, diseases and insect pests such as blast and stemborers. Shinriki and Aikoku were widely grown with such a background. When the amount of nitrogen applied further increased, these varieties were replaced by Asahi etc.

Plowing and harrowing with horses and bulls became popular only in the end of 19th century, and diffusion of this practice stimulated drainage of so far ill-drained fields. The 'dry field effect', namely release of soil retained nitrogen when the field became well drained, resulted in higher incidence of blast, and naturally more disease tolerant varieties were needed. Kameno-o was welcome by farmers under such conditions.

The first widely grown variety, bred through hybridization, was Rikuu 132, released in 1921 for Tohoku regions, and was grown to 2,00,000 ha in 1935 replacing Kameno-o which was not tolerant to lodging. After this, Norin number varieties (bred by the national breeding system) dominated, such as Norin 1 (Rikuu 132 is the male parent), 6, 8, 22 (offspring of Norin 8/Norin 6), and many of the present varieties with good eating quality were derived from these ancestry.

Frequent attacks of cool weather in north-eastern Japan in 1920's and 1930's caused serious famines and social problems. Various aspects of research were tackled to overcome cold damages, starting in 1935 by establishing experiment stations in each of the 6 prefectures of the region. The average rice yields in the region was lower than the national average at the beginning, but breeding of highly cold resistant varieties as well as new cultural technologies brought about the effects after 1950s, and now the region became the highest yielding, and the significant rice bowl of Japan.

Cultural Technologies

In cooler rice growing regions like in Japan, it was said that the quality of seedlings determines almost half of the success in rice cultivation. Nara Senji (1881) published the book 'Shinsen Beisaku Kairyoho' (Newly-edited Methods of Improving Rice Cultivation), explaining methods of seed selection and pre-germination, seedbed management, transplanting and field management, harvesting, and soil management comparing different kinds of fertilizers. It is noticed that he had already considered about conditions of exporting rice to Europe and other countries.

Yokoi Tokiyoshi (1882) published a paper describing the method of salt water selection of seeds of different crops with the specific gravity. For rice, non-waxy rice is treated with the gravity of 1.13 and waxy rice with 1.08.

Presently, ammonium sulphate is used instead of salt; 2.64 g and 1.50 g, respectively, in 10 litres of water. Effect of this method of selecting good seed was said to be about 10% of yield increase. NARA (1881), discussing on the water selection of seeds, said that salt water method might have demerit of cost and labor, and also possible damage by salt.

Preparing seedbeds of long rectangular shape (width about 1.5 m) was recommended around 1892, which made it efficient to look for and remove stem borer egg masses and other management. Protecting seedbeds with oiled paper (later plastic sheet) made it possible to extend the growth period of rice more than 3 weeks forward, thus making it easy to adopt high yielding, later maturing varieties.

In Hokkaido, transplanting was considered to give negative effect to the limited short growth period due to low temperatures, and 'octopus' rice seeder (Fig. 6.5) developed in 1905 was widely adopted, and accelerated by adoption of the new awnless variety 'Bozu'. This seeder could seed 16 hills per one move, 0.5 ha per day, and in 1936, 82% of planted area was sown by this method. The problem was, however, slow initial seedling growth in the cold water. After 1950s, direct seeding by this method was displaced by transplanting which was enabled by the new technology i.e. protected nursery.

Laborious weeding of rice fields has long been done by hand with/ without a simple tool. The prototype of rotary weeder is said to have existed even in 'Edo' era in some districts, but a leader farmer Nakai Taichiro in Tottori Pref. produced his own type, 'Taichi-kuruma' in 1892. He traveled throughout the country to alleviate hard work of weeding.[26]

Fig. 6.5. 'Octopus' seeder, product by Kuroda Umetaro.[25]

The other side of effect of this weeder was rapid diffusion of straight line transplanting among farmers.

Plow which existed in the 'Edo' era in some areas of western Japan for horse plowing, was long-soled and therefore stable in handling, but inefficient in performance. No-soled type plow in northern Kyushu was fitted for deep plowing due to its low soil resistance, but unstable and difficult for holding. In 1900, Matsuyama Harazo succeeded in modifying the short-soled plow which had been used in a limited area in Kumamoto Pref. by the name 'Higo-suki'. This plow Matsuyama-suki could turn the soil on both sides by switching the reversible blade, and greatly promoted dissemination of horse plowing in the early 1900s.

In 1910, the pedal thresher was invented by a local factory worker, who was commuting by bicycle and happened to observe rice grains were threshed from panicles on the roadside by spokes of his bicycle. In 1918, motorization of the thresher, coupled with winnowing apparatus and various improvements of the system thereafter, accelerated adoption of varieties with harder- or non-shattering habits of different degrees, depending on warmer or cooler regions. Especially in cooler regions, easy shattering varieties are avoided due to yield losses caused by easier formation of abscission layer under lower temperatures at the maturing stage. As the threshing machines were useful for handling wheat and barley, too, the area for winter cropping of these two crops was much increased. Earthen hand mill was much improved by adjusting the distance between upper and lower rollers, resulting in remarkable reduction of both labour for rotating and the breakage of rice, and also enabled extension of rice storage period. In 1921, rubber roll mills were invented, and motorization together with attaching steel-wire separators in the late 1920s greatly alleviated hard work in the season of sowing wheat and barley, contributing to the increase of the acreage for winter cropping in paddy fields.

RICE SITUATION AFTER WORLD WAR II (1945-2007)

In August, 1945, Japan was defeated in the World War II, and GHQ (General Headquarters) of the Allied Forces started to control the government. The most important agricultural policy which changed the traditional system was the Agrarian Reform. Non-resident landlords in the village had to dispose of all the land, and resident landlords were allowed to hold only one hectare within the mainland Japan, and 4 ha in Hokkaido. As of 1951, the total farmland area, collected as such, of 1,987,000 ha was sold to tenants with low prices, and the number of independent farmers increased from 2.84 million to 5.41 million. The farm rent so far collected in the form of rice was prohibited and was ordered to be paid in cash only. This reform stimulated volition of farmers and encouraged increasing application of fertilizers and chemicals, or usage of various machineries, but the ways of farmland management had to be changed from large scale ones to minor, inefficient operations.

During these 60 years, rice situation in Japan has greatly changed. Up to early 1960s, rice consumption per capita was nearly 120 kg and every effort was paid to increase the production. In 1949, the Rice Yield Competition Program was initiated by the Asahi Shimbun Company, and the government joined the program in 1952. Rice farmers poured all of their knowledge and technologies into their fields, and Mr. Kudo, Akita Pref., achieved the record of 10.522 t/ha of brown rice in 1960, which has not been surpassed by anybody until now.

The total rice area in 1960 was at the peak of 3.3 million ha, of which upland rice occupied around 5%. After that, rice consumption decreased to lower than 100 kg in late 1960s, continuously decreasing to 61 kg in 2006 due to the change of food habit and increased consumption of bread, animal food products and other foods.

Under the circumstances during these 60 years, drastic changes had occurred to varieties, methods of cultivation including plant protection and land management.[27]

Rice Varieties

In the 1940s when the supply of fertilizers was not enough, varieties adapted to low level of fertilizer application, with tall culms and long, heavy panicles, were prevalent. In the following decade, when domestic production of fertilizers became sufficient, and demand for rice was very pressing, varieties with high fertilizer response, high yielding ability and tolerant to diseases and low temperatures were required. Several years earlier than Asian Green Revolution, Japan succeeded in breeding the first semi-dwarf rice variety 'Hoyoku' in 1961 in Kyushu Agricultural Experiment Station. The regional average yield per unit area in Kyushu jumped up rapidly in accordance with the area planted to semi-dwarf varieties. This type of varieties was essential also for mechanized cultivation, especially for binder- or combine-harvesting which was adopted by most farmers in the 1960s. Reimei, another semi-dwarf variety for northern Japan, successfully bred by gamma-ray irradiation, was released and widely grown after 1966.

As soon as the balance of production and consumption is achieved in late 1960s, consumers began to call for better eating quality rice. High yielding varieties so far predominant were generally poorer in table quality, and when overproduction became apparent in 1970s, good palatability was the most important trait of rice varieties. 'Koshihikari', first released in 1956 and once avoided by farmers due to its susceptibility to lodging and blast disease, revived after mid 1970s, and was applauded by consumers for its good palatability in most prefectures. It is noticed that the other significant reason of the wide adoption of this variety is its unusual wide adaptability to day length and temperature. Because of these important characteristics, Koshihikari is now cultivated in many places even in Southeast Asian countries, Australia, and the USA.

In 1972, the new research project 'Super High-yielding Rice' was initiated, considering the necessity of maintaining sufficient areas of irrigated fields in view of environmental and political reasons. In this project, *indica* rice varieties were widely introduced to produce hybrids for heterosis in productivity on the one hand and, to cope with the various demands from consumers, quite new table quality such as hard and fluffy rice or aromatic rice on the other hand. From this project, 'Hoshi-yutaka' with high amylose content, and 'Sari-queen' with Basmati type rice, and various varieties with new characteristics were obtained. Even after the completion of the project, diversification of varieties continued, producing varieties with direct seeding adaptability, with very low amylose content, with colored kernels (dark purple, reddish, etc.), with medicinal effects such as lowering blood pressure, blood sugar, or good for diabetics, or adapted for animal feed as silage.

Rice breeding has been so far conducted only by national experiment stations and a few designated prefectural stations. When rice supply became sufficient, and table quality became the most important to appeal to consumers, many prefectures initiated breeding of their own varieties to secure customers to sell the produce. The issue of intellectual property rights accelerated this movement. Lowering amylose content was effective to get good eating quality rice, and now not a few varieties from cooler areas can compete with those from other regions.

Plant Protection

In rice cultivation under cooler climates, blast would be the most serious disease. Spraying Bordeau mixture (solution of $CuSO_4$ and $Ca(OH)_2$) was the common measure to prevent expansion of the disease in late 19th to mid 20th century.

Development of resistant varieties took several steps, starting from combining resistance of Japanese varieties, then introducing genes from Taiwanese upland rice, and then expanding sources of resistance to varieties of China, Philippines, Vietnam, India, USA and others. However, in early 1960s, all the newly released varieties with 'true' (qualitative or vertical) resistance genes showed breakdown after a few years of cultivation, and the strategy was switched to the breeding for 'field' (quantitative or horizontal) resistance. At present, multi-line varieties which are made up of several isogenic lines concerning blast resistance genes are available, with an example of new generation 'Koshihikari'. Recently no severe incidence of blast is reported partly due to the level of fertilizer application which is much lower than before and aims at producing better table quality rice.

Introducing resistance genes from indica varieties was conducted in other diseases and insect pests. Stripe virus became the serious problem in late 1960s to mid 1970s in several regions. When no control measures were undertaken, yield loss was more than 90% in fields of the Central Agricultural Experiment Station at Konosu, Saitama Pref. Screening of "blast" resistant lines, bred from Modan/6*Norin8 at Chugoku Agric. Expt. Station, succeeded in identifying a highly stripe virus resistant line St. No. 1, and varieties bred from this selection effectively controlled the virus disease in a few years. Development of resistant lines against insect pests, such as green leafhopper, *Nephotettix cincticeps*, and brown planthopper (BPH), *Nilaparvata lugens*, was also tried in 1970s to 1980s. However, when the promising resistant selections were obtained, the level of table quality was not sufficient for dissemination, due to indica-oriented gene fragments related to inferior eating quality, and so far no commercial varieties are available for BPH resistance.

As for chemical control, synthetic agricultural chemicals became a main countermeasure of disease and pest control after 1950. Aerial application, which was first introduced in 1958, rapidly expanded the area and in 1990 the total area of aerial application reached 1.67 million ha. At present, due to strict regulations to secure safety for food and environment, only limited number of registered chemicals are used avoiding drift as much as possible, such as application to seedling boxes for mechanized transplanting, or using granules, or spraying/dusting at the minimum amount needed. For controlling weeds, the first commonly used chemical was 2,4-D, which appeared in mid 1940s. Since then various weedicides have appeared, but the significant aspects is regulation to the amount of residual effects to the environment. In recent years, organic agriculture farmers use solution of fermented rice bran soon after transplanting. It is said that the solution activates algae on the soil surface and prevent germination of weed seeds.

Mechanization

Motorization of agricultural machinery started already in early 20th century especially in threshing and husking. Air-cooled high speed diesel engines were widely used for tillers/cultivators and sprayers/dusters after mid 1950s, and the area cultivated by such machinery reached 80% of the total rice fields in 1960s.

The rapid growth of national economy in 1960s caused the outflow of labor force from rural to urban areas. The situation enhanced further mechanization of farming activities especially for transplanting and harvesting. Until mid 1960s, it was common to use sickles to harvest rice, but in 1966 power cutter-binders were introduced, and in a few years small type combines took place. The area harvested with powered machinery reached 80% in 1975. Mechanized transplanters were developed around 1970, and were rapidly adopted by farmers. Large scale nursery service centers were set up in each village to supply box-raised seedlings. This system changed the ways of pest control; by applying systemic pesticides and fungicides onto the seedling boxes. In 1980, the area planted with machines exceeded 90%. It is noticed that the mechanized transplanting contributed to the increase of yield by increasing tillers from lower nodes, which might have been lost by deeper hand transplanting. The mechanization proceeded further, and initial walking-type machines were replaced by riding-types. Many transplanters were equipped with row fertilizer applicators.

Mechanization of farming practices resulted in the reduction of the ratio of labour in the production cost, but high input to machinery became a great burden to rice farmers

Shifting of Rice Policy and the Present Situation

The pricing and distribution of rice had been strictly controlled by the government since 1942 under the Foodstuff Control Law. Under this law, the government fully controlled the rice-related marketing system: buying all the amount of rice produced, and strict regulation of marketing. The rice market had long been closed to importation except a few cases, such as for brewing special rice wine ('awamori') in Okinawa, or some amount of waxy rice for confectionery, etc.

However, after late 1960s, increasing unbalance of rice production and consumption, and financial burden forced the government to loosen marketing regulation and to reduce the cultivated area after 1970. Ministry of Agriculture, Forestry and Fisheries (MAFF) took all the responsibility to adjust the production by deciding the planted area in each of prefectures paying compensations to the areas reduced or converted to upland crops such as wheat, soybean, etc. The area planted to lowland rice in 2007 was 1,669,000 ha, only 52.6% of that in 1969 (Fig. 6.6).

The government also introduced the Law for Stabilization of Supply/Demand and Prices of Staple Food in 1995 to reduce buying and selling of rice by limiting to the two categories only: storage for emergencies, and importation of rice. In 2004, government completely abandoned control of rice marketing, and in 2007, responsibility of area adjustment considering the balance of supply and demand was fully transferred to farmers' organizations. MAFF's support to the organizations was to encourage their own ideas to stabilize the balance of supply and demand in each locality.

As for importation of rice, Japan suffered severe shortage of rice supply in 1993 due to cold climates, and had to import 2.3 million tons of rice. Also, based on GATT Uruguay Round Agreement, Japan started importation of Minimum Access Rice in 1995. Since then, a total of 8.32 million tons of brown rice had been imported by 2006. Of this amount, 2.81

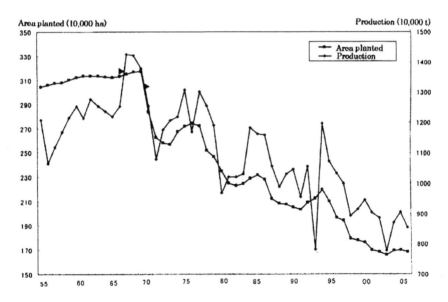

Fig. 6.6. Changes of planted area and total production of rice, 1955-2006 (MAFF Statistics).

million tons was used for food processing, 2.12 million for overseas assistance, and 0.4 million for feed, and as of March 2007, 1.75 million tons is in stock.

Rice farming in Japan is now at its critical stage because of aging of farmers with few successors, and low farm management efficiency due to the very small cultivated land area. MAFF once stimulated farmers to assemble land and form a larger scale farming, but recent drawback in prices of land as well as rice, large scale (only a few scores of ha in Japan) rice farmers are facing difficulties of managements. The number of bearers (carriers) of rice farming is rapidly decreasing within communities. Under such circumstances, MAFF is now encouraging formation of community-managed farming with the minimum size of 20-30 ha per unit, operated by a leader and a few cooperators. This movement, however, should not scare large scale farmers, who have so far assembled small pieces of lands from aged farmers by renting their idle lands.

CEREMONIES AND FESTIVALS RELATED TO RICE CULTIVATION

As rice has been the most important, so to speak 'supreme' crop in Japan, three ceremonies have been performed at the Imperial Palace every year: 'Kinen-sai' on 17 February to pray for the bumper crop of rice in the year, 'Kanname-sai' on 17 October and 'Niiname-sai' on 23 November to report and thank the god for the bumper crop. Also various types of ceremonial events can be seen in different regions[28] of which the followings are some of the examples generally observed; though some of these might not be practiced in recent years. Traditionally, every 'maiko' (dancing) girl in Kyoto adorns her head on New Year's day with a panicle of rice as a part of hairpins.

Preparatory Ceremony

On a new year's day of lunar calendar, the head of family makes gestures of digging land and transplanting rice seedlings, or practically insert rice straws or pine leaves onto the snow-covered field (in snowing regions). On the 15 January (lunar calendar), many small glutinous rice balls are put onto ends of tiny branches of willow or a specific tree species, and the bough is bound to the main pillar of the house, to pray for bumper harvest of rice. The ceremony is especially common in northern Japan.

Nursery Bed Ceremony

When seeds are sown to the nursery bed, or upland rice is sown to the fields, a sacred cut-paper is placed to a stick and set up at the inlet of irrigation water or a corner of the field. This stick is believed to be the place where the god comes down from the mountain areas and stays to guard the field.

Transplanting Festivals

In the end of the transplanting season, finely dressed maiden planters and cattle are collected to a field, usually owned by the shrine or landlord of villages, and perform cultivating, leveling and transplanting with a big band of musicians of drums, gongs and flutes. This festival is widely seen especially in western Japan.

Incantations

During the growth period of rice, various small festivities are conducted depending on localities. To pray for avoiding damages by insect pests, a group of children hit the ground by bamboo sticks, or carry burning sticks and shout loudly to drive away insect pests. In case of drought, villagers pray for rain in various ways.

Harvesting Festivals

Various types of festivals are observed in different regions. At the start of harvest, 'ho-kake' is practiced in many regions, which is a small bundle of rice panicles offered to the family altar to thank the god. The 'tawara' containing paddy rice is considered sacred and nobody is allowed to sit or stand on it, based on the belief in 'spirit of rice'.

The shadow 'sumo' wrestling performed twice a year at the 'Oyamatsumi' Shrine in Omishima Island, Ehime Pref. after the transplanting ceremony and the harvesting ceremony, is a good example of various site-specific ceremonies observed in different regions of Japan to praise the spirit of rice. A 'sumo' wrestler fights with the shadow (no visible counterpart) several innings, and is always defeated. At the Ise-Jingu, Mie Pref. that enshrines the Sun Goddess 'Amaterasu' holds important rituals related to rice especially on the occasion of 'Kanname-sai'.

References

1. *Kojiki:* The first written document that describes legends about the first to 33rd Emperor so far orally conveyed from ancient time using letters introduced from China.
2. *Nihon-shoki:* Similar to *Kojiki*, describes up to 41st emperor.
3. Ohbayashi Taryo, UNESCO Courier, http://findarticles.com/p/articles/mi_m1310/ is_1984_ Dec/ai_3540665
4. See website: <http://www.komenet.jp/database/culture/culture01/culture01-3.html>
5. See website: <http://www.rekihaku.ac.jp>
6. See website: <http://osaka.yomiuri.co.jp/inishie/news/is50219a.htm>
7. Fujiwara (1998).
8. Sato *et al.* (1990).
9. See website: <www.coara.or.jp/~sasakiak/ SiteList.html>
10. Sato (2002).
11. Matsuo (1952).
12. See website: <http://inoues.net/ruins/itazuke.html, http://inoues.net/ruins/matura.html>
13. Yamaguchi (2000).
14. Yamaguchi (2000).
15. Extracted from Yamaguchi (2000).
16. See website: <http://www.bcap.co.jp/s-hochi/bno/2002/02-11/n021110.html>
17. Report by SAWADA, See website: <http://www.nouminren.ne.jp/dat/200208/2002080514.htm>
18. The great collections of Japanese short poems made in mid to late 8th century by various people, from emperors to farmers and soldiers. Characteristics of rice in those poems are very simple, mostly 'early maturing'.
19. See website: <http://www.asahi.com /culture/news_culture/.>
20. See website: <http://www.nouminren.ne.jp/dat/200200/2002110422.htm>
21. See website: <http://www.awakouko.info/modules/pukiwiki/273.html>
22. Tomiyama (1993).
23. Tomiyama (1993).
24. The book *Kogo Shui* (collection of old stories) mentions that the god of field feasted people with bull meat on the festival and the god of the year got angry at the report, released insect pests to the field, resulting in total withering of rice.
25. See website: http://www.hokkaido-jin.jp/zukan/picturebook/itemview.php?iid=1300100057
26. See website: http://www.yashio.net/modules/tinyd1/index.php?id=2
27. Kaneda (1995).
28. The website such as http://ja.wikipedia.org/wiki describes in detail about the god of rice field and the events.

7 | History of Rice in Southeast Asia and Australia

Lindsay Falvey

Southeast Asia is the home of rice. Although its origins may be elsewhere, it is in this region that technologies have been refined in the last millennium and where rice is synonymous with food in most languages. Beginning with selection of grains that did not naturally fall to the ground prior to ripening as an unconscious action of our forebears, rice was quickly domesticated[1] as a rather narrow gene group was associated with the shattering of grain. Hence selecting plants that did not shatter led to exclusion of most of the other types—yet by coincidence an abscission layer that facilitated harvesting was retained. From such recent genetic work, we know now work tells us that rice, *Oryza sativa*, was domesticated from either of two wild types distributed from India to South East Asia. Previously we had postulated various sites for the origin of rice, as no doubt is mentioned elsewhere in this book, the most common being the southwest Himalayas.[2]

The prehistory of Southeast Asia includes evidence of human settlement dated to 40,000 BCE and what may be tools dated to 75,000 BCE years, both discovered in Malaysia. By the Mesolithic period, early Southeast Asian agricultural societies had domesticated the chicken and pig as part of their food production systems that were apparently easier to maintain in Southeast Asia than in some other regions. It is said that this ease produced the oldest customs of Southeast Asia that centre on food ceremonies. But we know almost nothing of these early people.

The eleven countries of Southeast Asia today are an arbitrary construct that, without care, can lead to underestimation of the older empires of

Srivijaya, Malacca, Pagan, Khmer, Ayutthaya and so on. It was such kingdoms that attracted Indian, Chinese and European trade and accepted aspects of the new cultures to form today's countries. The peoples of Southeast Asia themselves possibly came from southern China around 2500 BCE, initially to the Philippines. The ancient Greeks knew the Malay Peninsula as *Aurea Chersonesus* (golden peninsula) and the ancient Indians knew mainland southeast, possibly the river deltas of what is now Thailand as *Suvarnabhumi* (land of gold), which today is interpreted to refer to agricultural potential. The Chinese also knew the region well and the mixture of cultures that ensued makes it impossible to determine exact stories of the origins of Southeast Asian rice, although cultural traits do provide hints as introduced later in this chapter.

Rice cultivation in China seems to predate that of India and is the most likely source of rice in the northern mainland parts of Southeast Asia, but this is more as a result of geographical proximity and migration in antiquity. In fact, the earlier introduction of rice around coastal Southeast Asia probably came from India for the western coasts of Southeast Asia and from China for those eastern, with Singapore in the middle. Even though the earliest writings, which of course are from India, do not indicate the presence of rice, this may be taken only as evidence of its earlier domestication in China not of the source of the spread of rice and its technologies to Southeast Asia. So, having perhaps settled some of the parochial claims to having introduced rice to Southeast Asia, let's see what we can be a little surer about.

Rice cultivation in coastal Southeast Asia probably preceded that of the inland[3] and it is supposed that the rise in sea level between 8000-4000 BCE submerged most evidence of coastal rice production that would otherwise have complemented such findings as those of Khok Phanom Di, Thailand that date from 6000-4000 BCE. Some claim that cultivation of root crops predated rice[4] in Southeast Asia, but the relatively recent domestication (*ca.* 2000 BCE) of yams in the region[5] suggests that only in the later settled hills did yams preceded rice cultivation, which in these regions would have been dryland rice. Rice was the storable foodstuff between seasons, and its origin in Asia is yet to be fully revealed to us.

If rice cultivation in India and China is placed at nearly 10,000 years ago, and it is known that rice was cultivated in central and east China at 6000-5000 BCE, we may suppose that it would have required only a thousand or two years to disperse to Southeast Asia. Linguistic similarities for the word 'rice' hint at early associations around the food; etymologically, we may link the northern Chinese word for rice, *tao* or *dao* or *dau*, to words of southern China, Thailand and Indo-China, *k'au* (for grain), *hao, ho, heu, deu,* and *khaw,* while the Dravidian Indian root *arisi* may be linked to *ris, riz, arroz,* rice, *oruza,* and *arrazz.* Similarly, in peninsular Southeast Asia, the terms *padi* and *paray* for rice and *bras* or *beras* for milled rice are of Austronesian origin, and the Chinese words *ni* or *ne* (for wild rice) seems to be related to the word *nu* (for glutinous rice) in Southeast Asia. But all this provides no clear path for the dissemination of rice cultivation.

Historical records suggest a line by which rice spread from South Asia, Southeast Asia, and China to other regions or countries, though exact dates may be lacking. In one direction, the mainland Southeast Asia may have been the source of Sri Lankan rice before 543 BCE as it is similar to that of the Malay Archipelago and Indonesia from between 2000 and 1400 BCE, which has mutated from its original Yangtze River type of around 5000 BCE.[6] *Javanica* rice originated in mainland Asia and then differentiated into a dryland ecotype related to the hill rices of Southeast Asia, and the wetland ecotypes of *bulu* and *gundil* in Indonesia, which in turn spread to the Philippines in about 1000 BCE.

What is now the north of Vietnam became prosperous with rice cultivation about 250 BCE in association with harbors for trade along the coast, and thereby attracted the attention of China, which in about 207 BCE sent an imperial delegate to the Red River region to rule an annexed kingdom called Nam-Viet that became a Chinese province in 111 BCE. Within a century, Indian influence entered what is now the south of Vietnam in the Champa kingdom.

The ancient kingdoms were either rice-based as in mainland Southeast Asia, or maritime states dependent on sea trade, such as Malacca and Srivijaya. Trade between China and India influenced agricultural development with goods carried across the Isthmus of Kra in Thailand, at least until the 6th century when the Srivijaya kingdom on Sumatra developed ships that could sail around the southern coast. The Srivijaya kingdom was the first major kingdom in the region with its capital in Palembang where Indian and Chinese influences blended. The great Palembang rice fields are mentioned in various ancient reports and seem to have used Indian technologies.

By the 10th century, shipping technologies allowed Srivijaya to be bypassed and it was duly plundered by the Indian Chola state. Meanwhile, a series of rice-based kingdoms competed on the extremely fertile Java, while Muslim traders increasingly influenced Southeast Asia from the 12th century, although Malacca displaced Srivijaya with the assistance of Chinese patronage. By the 16th century, Europeans arrived as traders, first from Portugal, then the Netherlands and Spain and eventually

France and Britain, to carve up the region as colonial powers with only Thailand escaping direct colonization. All of this history had its particular impact on rice in Southeast Asia.

But let us go back a step to review the detail of what may have happened. The first cultivation of rice, as distinct from its natural wild distribution, may have been in southern China bordering on Southeast Asia, but could have been in Southeast Asia itself, or in India; what is more certain although still with an element of doubt is that rice was probably first cultivated some 9000 years ago. Today rice is assumed to be an irrigated crop, yet for most of its agricultural history it has been planted in naturally flooding areas. We think its first irrigation was in 780 BCE in China.[7] Rice seed broadcasted into receding flood water areas was probably the earliest form of wet rice domestication[8] as indicated from early prehistoric archaeological sites.[9] Opportunistic harvesting had long given way to agriculture before any written records of South East Asia and certainly before the major rice kingdoms of the Khmer and Tai. Early Mon and Khmer influence[10] probably began the rice revolution of Southeast Asia[11] by building on the agro-cities of the shallow and gentle floodplains.

Expansion of the Rice Countries

Rice is synonymous with lowland Southeast Asian agriculture. Shifting cultivation, possibly including some dry rice within a range of vegetable crops are thought to have predated the use of wild wet, and certainly, domesticated, wet rice production in inland areas. One theory suggests that dry rice sown with a digging stick into the ashes of a cleared and burned forest predates the cultivation of wet rice. The theory which is based on traditional stories in Vietnam, that rice originated in the mountains and moved to the plains, may be challenged in terms of river valley based migration patterns, and the apparent absence of relevant historic sites in the mountains. Rice seed broadcasted into receding flood water areas was labor efficient, and probably became the earliest form of wet rice domestication as a simple modification of primitive husbanding of useful plants in their natural environments.

Shifting cultivation in upland and mountainous regions of Southeast Asia long predates that of today's hill tribe groups. The extensive use of fire in forest farming provided a labor cost-effective means of introducing root and tree crops, particularly along the water courses of lowland and contiguous rising regions before the creation of irrigation fields.[12] From the Neolithic seeking out of natural swamps with slowly receding water regimes co-existing with hunters and gathers for millennia there was a slow transition to integrating the techniques of migrants[13] from China who arrived by sea as well as down the river valleys of mainland Southeast Asia. The Khok Phanom Di site dated at 2000-1400 BCE is now land-locked but was once an estuary with mangroves and fresh ponds suited to rice production with the benefits of alluvial deposition to maintain fertility. The Ban Kao culture of Kanchanaburi dated at 2000-500 BCE further supports the likelihood of agricultural technology at least co-originating from sea migration.

With such new technologies in agriculture, seasonal variations in rice yields could be reduced, albeit with increased labor inputs. However, with larger population densities supportable through these systems, division of labor, and increased efficiency for its use would soon develop through the Iron Age allowing further increases in settlement size. Prior to the iron age, three hectares seems to have been a maximum area for an independent site compared to more than twenty hectares, possibly in association with reservoirs or moats, once iron was introduced. This more managed rice production allowed the development of politics, social ranking systems, and military organization.

Once introduced, rice encouraged foreign contact and technological development. Sea trade widened technological awareness and food supply which allowed more free time for development of a society. Technical innovations of puddling, plowing and even contrived annual replenishment of alluvium, led to a reliable form of low intensity rice-agriculture by the 8th century across the mainland. The greater potential of the wet rice cultivation system to sustain the development of a civilization was now clear.[14] The alternatives, hunting and gathering or reliance on another staple, could not have produced this situation. Hunting and gathering relied on small groups and low population densities. The best available alternative cereal was the widely adaptable species, millet, which had predated rice as a staple throughout the region; however, its shifting cultivation prohibited large population concentrations with the labor economies of wet rice.

In what is now Thailand and Cambodia, Indian scripts record the regularity of the rice surpluses.[15] Fragments of Funan (Chinese) records also describe the inhabitants of these areas in a manner suggestive of their being Austronesians, and also describe their honest nature and devotion to agriculture. Noting that . . . . *they sow one year and harvest for three*, ... records also indicate the people's involvement in ornamental engraving, silver utensil production, and trade in gold, silver, pearls, and perfumes. Later documents suggest Mon, Khmer and Tai residents, although the influence of Funan beyond coastal areas appears to have been minimal and their understanding of changes inland was probably limited.[16] Other Chinese records nevertheless do confirm the existence of significant cities in the Chaophraya Basin from the 7th century CE, particularly around Nakhon Pathom and U-Thong.[17]

Early settlement of U-Thong, probably from the 1st century BCE, suggests the emergence of irrigation canal engineering skills in Thailand.[18] A thirteen kilometer straight geological formation running east from U-Thong to, what would have been at that time, the head of the Gulf of Siam suggests separate development from coastal trading settlements. The ability to control water links directly to the subsequent Khmer Empire and suggests that the intervening Dvaravati cultural period of the region probably focused more on trade than political domination.[19]

The Dvaravati culture appears to have arisen in what is now Burma and Thailand and beyond between the 6th and 9th centuries and seems to have been based on Buddhism, the Mon language, and overland trade between the Gulf of Martaban and the Gulf of Siam via the Three Pagoda Pass between Burma and Thailand. More a civilization than an Empire, no capital is known to have existed although archaeological sites appear to be densest around the fringes of the central plain. Sites fan out from those around the Gulf along trade routes to Burma, Cambodia, Chiang Mai, towards northern Laos, and northeast towards the Khorat Plateau. Frequent finds of foreign objects provide further evidence of the trade orientation of the civilization. Foreign ideas, tools and innovations flowed speedily along trading routes and demand for rice stimulated the testing of new techniques for producing food surpluses along trading routes.[20] Lasting until the 11th or 12th century CE, Dvaravati influence is otherwise poorly understood. Ethnically it is suggested that it was controlled by peoples of Mon[21] or Mon-Khmer origin although there appears little supporting or contrary evidence.

While the Dvaravati Empire is difficult to define, the production of the centre of U Thong contains evidence of its Mon origin, Indian influence, and ability to absorb diverse pre-existing cultures, migrants, and seafarers, such as from the Funan trading sites. Its culture appears to have extended beyond its governed realm, interfacing easily with the expanding Khmer rice culture. It was around this time that migration from the southeast China and Vietnam introduced the water buffalo which displaced draught cattle and ultimately assisted expansion of rice production within the Chaophraya Delta.[22]

Meanwhile, coastal areas showed different development patterns. By the 6th century a widespread network of agricultural communities existed in peninsular Southeast Asia as much as they did in the inland areas and the great river deltas. The cultural differences of the Peninsula and Indonesia and Malaysia today reflect these different origins, and histories, even in some agricultural practices such as raceme rather than whole-stalk harvesting techniques. However, the deltas and plains have long been a focus of the region, both because of their subsequent history and their potential, which was clearly apparent to Indian missionaries of the 3rd to 2nd century BCE. Upland river valleys in the west and southwest of what is now Thailand leading into areas of northern Laos and southern Yunnan remained sparsely populated by the aboriginal Austronesian or Austro-Asiatic speaking groups, possibly ancestors of some of today's hill tribes. These peoples were poorly equipped to deal with the technologically superior wet rice growers.

Wet rice irrigation probably evolved to river off-takes to augment natural pondages. Ponding and canalling of water to maintain a stable rice growing environment would have been an easy development with rice terraces evolving as an adjunct of nature's own micro-environments. In contrast to this hydraulic domination, populations closer to the sea where water was abundant, or in the delta where water remained mainly uncontrollable, adapted their lives to the flux of water and its control.[23] In all cases, life in the mainland of Southeast Asia was increasingly dominated by water; the ancient name *Sayam* or 'Siam' may have even contained the meaning of 'people of the river'[24] or 'water people'.

Within the first millennium CE, inland communities had discovered means of reliably producing rice surpluses and within centuries, organizational skills to continually increase surpluses would allow the emergence of the Khmer Empire centered at Angkor. Technologies developed through Khmer agriculture provided a revolutionary fillip for rice agriculture.

Khmer Agriculture

The agricultural settlements which gradually displaced hunters and gatherers grew to agricultural cities, some of which were subsumed into the emerging State-religious Empire of the Khmer. Such agro-cities required an assured rice production base, which in the case of the Khmer, relied on supplemental water management, and appropriate rice varieties. Judged by today's standards, such systems might be considered sustainable within the parameters of the technical applications, and they did last for centuries before finally failing. In the event, Khmer wet rice culture proved less sustainable than the pre-Green Revolution river-basin wet rice systems of the ethnic group known as Tai.

Agro-cities in the shallow and gentle floodplain areas have been found in the central plains of the Chaophraya and around the delta of the Mae Klong River, an area now dominated by sugarcane on the higher ground and rice in the floodplains. They extended through the Mekong River delta in what is now Cambodia and Vietnam. The transition from agricultural settlements to agro-cities[25] arose from the apparent abandonment of prehistoric villages and concentration into larger settlements, often surrounded by more than one moat with radiating canals. The agro-cities were overwhelmingly associated with the gentle flooding regimes around the boundaries of large flood plains. This significant change in settlement patterns was associated with the adoption of monocultural flooded rice production which reduced labor inputs and risk compared to that of the smaller agricultural settlements. Nevertheless, the significant individual earthworks undertaken appear to have been related to governance within each agro-city without coordination across a wide area. Many of these large settlements contained no religious edifices and hence the term 'agro-city' has been adopted to indicate this stage of agricultural development prior to the emergence of religious States.

In the 9th century, the civilization of the north western shore of the large natural overflow reservoir of the Mekong River, the Tonlesap in modern Cambodia, grew to dominate the areas including much of what is today Thailand and Laos. Rice fields around the edge of the Tonlesap, down into the lower reaches of the Mekong River, and into the Mun and Chaophraya basins, allowed the Khmer Empire to establish itself with rice as the primary source of growth and wealth.

Various Khmer attempts to consolidate power were constrained until they understood the central significance of a secure rice supply. This allowed, with foreign influence, the development of a State-religious Empire in which temples owned land and agrarian workers contributed labor motivated by both coercion and an afterlife reward. The Empire eventually crumbled from within, as a result of, among other factors, alternating strong kings and interregnal disorder which disrupted maintenance of domestic water systems which incidentally served rice production. Policies to ensure rice surpluses were also negated. Around the same time an increase in trade and commerce may have encouraged the disillusioned Khmer to abandon the high cost and increasingly difficult to manage site of Angkor in the 1430s in favor of better sites for trading along the coast. The Khmer Kingdom supported a population in excess of one million in its Angkor capital at a time when the Norman army marched on the city of London (1066) and its population of 35,000. The agricultural system to support this major world centre required skilled engineering and rice agronomy.

The Angkor agricultural system was based on the natural rise of flood waters and their rapid recession in the Tonlesap, and their supplementation by a network of dams and bunds to divert or retain receding waters. No large dam technology is evident. Phnom Kulen, approximately fifty kilometers northwest of Angkor was the centre of the water management network, which as the civilization evolved was increasingly dedicated to religious and domestic water supply purposes. Control of land as well as water was essential to the development of the Empire. Landed elites donated their land and its farmers to the temple and registered these transactions for possible spiritual and probable commercial gains. Control of labor and production, including management responsibilities, seems to have been handed to temples while the donor continued to retain a percentage of the harvest. Donations also included domestic stock such as cattle, buffalo and goats, tree crops such as coconuts, fruit, areca nuts, and other agriculturally related items such as threshing floors and clothing. Concentration of economic power in the temple consolidated political development, which in turn was reflected in agricultural legislation, for example, a 10th century edict concerning negligent grazing of buffalo in proximity to rice fields. The King, difficult to separate from the temple, retained the right of ownership of all unused and unallocated land and could also influence ownership rights in all areas. The modern Thai system was to retain these elements centuries later.

Thus the temple was the central agricultural institution. As a source of investment it was the agricultural bank. It had the capital and land, and increasingly became the repository of technical information for agriculture itself, albeit with a cosmological emphasis. The temple managed agricultural labour, including war captives, through promises of spiritual rewards as they opened unpopulated lands donated to the temple. The water management system required large infrastructure to control receding floodwaters, and for canals to supplement irrigation, and thereby proscribed small private agricultural producers, who would have in any case been inconsistent with the evolving political system.

By the 12th century, the Empire was producing around 38,000 tonnes of hulled rice each year[26] for the Pra Khan temple complex from a system with no formalized bureaucracy but simply a temple-King assignment of land rights balanced with spiritual and subsistence rewards to the poor. The King, as the largest land owner and the temple as the owner of labor led easily to accommodate the God-King system compatible with the adopted Indian religions of Angkor.[27] Inscriptions from the 9th to 13th centuries proclaimed the King as both creator and director of public works which irrigated some five million hectares (31 million rai),[28] incidentally with providing water for domestic and religious purposes. The water system which has been termed 'theocratic hydraulics' as many water sources were, latterly at least, of symbolic or religious importance rather than having been designed for a central irrigation purpose. Through this period, Angkor was known through the region for its 55 million rice fields—'millions of rice fields' being a measure of a kingdom's power.

The Khmer selected sites for high labor efficiency in the simple rice water management system. The sites themselves suggest use of rice varieties with relatively low water requirements and probably modest yields. Ranking reliability of production over maximizing of yields reflect the limitation of the water management systems, and the State's emphasis on stability of production. As the Khmer Empire waned, Sukhothai, one of its outposts, was progressively dominated by Tai whose own irrigation technologies (discussed below) had been integrated with those of the Khmer. However, infrastructure developed for rice is now difficult to discern from that developed for other purposes.

The Sukhothai and Sisatchanalai sites in what is now northern Thailand include a 100 kilometer long earthwork extending as far as Kamphaengphet which was probably a flood-controlling barrage. The two fifty-five and sixty-eight kilometer constructions are not considered to have been a canal even though Sukhothai hydraulic engineers are known to have gained considerable experience in canal construction by this time. Nevertheless, they avoided attempts to manage the major rivers and areas subjected to deep inundation, preferring to concentrate on diversion of flood waters. It was the Tai who mastered the management of water directly from medium sized rivers such as the Ping at the Kamphaengphet site.[29] A barrage construction also serving as a road would have assisted Khmer management of regions away from the Sukhothai and Sisatchanalai complexes; we may assume that such developments reflect a mode of extending Khmer political influence and incidentally a new approach to rice culture. The significance in barrage construction to Thai agriculture lies in its blending with Tai irrigation systems for eventual control of the waters of the Central Plain of the Chaophraya River.

The construction of urban dams and dykes by the Khmer appears to have been based on gravity tanks feeding fields via canals with water control managed through wooden sluice gates.[30] That these are considered by most observers to have been oriented to religious purposes actually hide an earlier agricultural purpose overtaken by the religious State. Nevertheless, rice culture in the Khmer period seems to have been simple

Table 7.1. Water related construction sites of the Khmer

Name	Size	Comment
Indratataka		Reservoir built 877.
Yashodharatataka (eastern Baray)	1.8 × 7 km; 30,000,000 cu m reservoir	Also linked to modification of the course of the Siem Reap River.
Rahal	360 × 1200 meters	Built on tributary of the Siem Reap River south east of Prasat Thom.
	2.2 × 8 km; 40,000,000 cu m capacity	Largest of all Khmer reservoirs; eastern section silted.
	Up to 14 × 100 m; brick ponds and fountains	Fed by Siem Reap River and rain water; drinking, fish ponds and bathing.
Jaytataka (north Baray)	900 × 3700 meters	Designated the holiest of the waters.

and reliable, and was probably only marginally dependant on the major water works presented in Table 7.1.

Much of the information concerning Khmer agriculture and life is derived from Zhou Daguan, a Chinese adventurer who wrote of his visit in 1296-1297. From his descriptions and other evidence we know that rice was hulled through bruising with mortar and pestle rather than by grinding stones, and that women were a dominant part of agriculture, and in particular trading. Small trading transactions at the time were effected through barter of rice, cereals, and objects from China, medium sized transactions included fabrics, and large transactions included gold or silver. Such a civilization required a sound land use and rice production system.

The Khmer land use system was an evolution of India's as an adjunct to the religions. Initially and for some 400 to 500 years, rice production was based on the use of naturally flooding areas. Forested areas were lightly used until the later large Empire converted forests to bunded rice fields. Resulting square rice fields and bunds suited an overall auspicious shape for city layout, possibly planned to reflect Khmer cosmology. The clearing of forests for rice and city development limited water run-off through the millions of paddy fields, retained wet season silt in these fields and in canals and reservoirs, and changed soil chemical and physical characteristics in paddy fields on a scale hitherto unknown. The wetting and drying of soils allowed reduction and oxidation of silica among other soil components, increasing crystallization and hence the sand component of the soil profile. Agriculturally generated environmental change, in some cases irreversible, appears to start at this stage of Southeast Asian rice agriculture around 1200 CE. Nevertheless, wet rice cultivation under the different Tai traditional conditions of the *müang fai* (discussed below) continued to yield satisfactorily on such impoverished soils for centuries because of the essential benefits of the modified aquatic environment for the rice plant.[31]

Khmer influence on agriculture extended beyond techniques adopted by the Tai as it extended deep into the psyche of the persons that would assume the Khmer cities of Lopburi, Ratburi, and Muang Singh, among others.[32] The slow immigration of Tai from the north down the river valleys led to a significant number of Tai persons in the Khmer Empire. This force may have developed influence and seized an opportunity at a time of weakness of the Khmer Empire in outlying Sukhothai as the Empire began to decline after 1150 when massive investment in construction and deification of kings caused neglect of water management and food production. Canals accumulated silt, and rice production plummeted, forcing large scale emigration to other flood plains in the Mekong delta system.[33] The imposts of malaria and Tai attacks possibly hastened the final rapid fall of the Empire.[34]

So we see that in early Cambodia, the Khmer annual rice production of 38,000 ton of hulled rice[35] from some five million hectare[36] substantiated the region's huge potential. Rice in Vietnam and Burma had also allowed the development of agricultural kingdoms and in the case of what is now Thailand, Tai wet rice cultivators[37] migrated south from China and blended their unique *müang fai* irrigation with Khmer technology to control increasingly larger rivers until eventually the flooding of the delta was controlled.

The Khmer Empire provided a pervasive Indian influence in religion and culture which, matched with influence from Burma, continues to flow through rice culture toward the more Chinese influence of Vietnam. Through much of the Khmer period, a parallel although technologically different form of rice culture evolved in what is now Burma centering on Pagan.

Pagan Agriculture

Khmer influence from the east met Mon influence from the west and with Tai infusions, produced the modern rice-agriculture of the region. While the Khmer Empire developed large State-religious edifices, the ancient Mon culture in the west of mainland Southeast Asia was less well represented architecturally and as a consequence, is less well understood. Each developed rice technologies from external contact and their interaction formed part of the emerging rice revolutions of the region.

Mon-Pyu authority across large areas of Burma was interrupted by immigrants displaced from Nanchao in southern China who possibly assumed the power of the Pyu from about the 9th century. As a consequence, the Mon centre on the coast at Thaton fell and the Mon migrated predominantly to Pagan. The Pagan Empire of the Mon, who united with the Pyu and Burmans from about 1200 CE to repel invasions from both the mountains and the seas, was based on a rice-sufficient empire located in the dry zone of Burma on the banks of the Irrawaddy River. The initial headquarters was between the two rice production areas of Minbu and Kyaukse, both of which had extensive irrigation systems. Pagan's success relied on its ability to produce rice and, in common with the Khmer, they develop an inland rice-based culture which overshadowed coastal trading cultures. Pagan also developed monumental religious sites, at least in its immediate area of control. By the 13th century, various power struggles, including land disputes within the monkhood, led to the establishment of a new Mon Kingdom at Pegu.[38]

The period known as Dvaravati was linked to the rise of Mon influence from the west. Intensification of iron production and probably of copper, lead, and silver, indicate an advanced culture. The attraction of the region appears to have been the consistent ability to produce rice surplus from relatively low labor inputs. This stability and wealth stimulated trading in rice and forest products through Thailand and brought new ideas and technology. Named from a coin found at the site, the Dvaravati is named after an inscription on a coin found in Nakhon Pathom, which appears to have been a major centre as it was then on the coast and offered trading access to the protected Gulf. Other Dvaravati sites include U Thong in modern Thailand, which was probably a sub-center of Nakhon Pathom, Kubua southwest of Nakhon Pathom with access across the Tenasserim mountain range to Mons in the west, Khao Ngu caves in Ratchburi province, and overland routes through Petchaburi and other centers in peninsula Southeast Asia.

By the 7th century, the three important cities of Nakhon Pathom, U Thong and Kubua provided the western interface[39] with the rising Khmer culture which produced the Mon-Khmer period of Southeast Asia. Throughout this period the sticky rice-growing Tai ethnic group was increasing in number through continual southward migration from China. Insignificant at first, this tribe was to become important as an integrator of agricultural and other technologies across the region. The first indication of a rising political ambition of the Tai appears in this disjointed history of Burma and is integrally associated with control over rice production.

Neither the Mon nor the Burmans appear to have been interested in the upland valleys of the Shan States which were being populated by Tai with their specific rice growing technology. The Mon culture was largely absorbed into other cultures including the Pyu, Burman and Khmer as a result of its inferior military force in the 9th and 10th century when the Kingdoms of Burma were smaller than that of the Khmer. A strong military pressure from Nanchao from the mid-8th century until the mid-9th century accelerated the demise of the Pyu State at Prome and Shwebo allowing Burmans to move into the extensive irrigated rice lands of the Mandalay region. The relatively smaller new State at Pagan developed from the mid-9th century coincided with the new Mon Kingdom which was developing at Pegu.

The influence of the Tai in the Burmese centre of Pagan rose around the 13th century when Tai Shan from northern river valleys assisted the then weak Pagan to repel the Mongols. In helping the Pagan kings, the Shan gained sufficient influence to assume power. The subsequent establishment of the centre at Ava adjacent to the Kyaukse rice fields and the Mon centre at Pegu was one of the first mixed Tai States. However, their power was balanced against the other close by independent Kingdoms at Arakan and Prome.[40] Further information about Tai peoples and these Kingdoms is limited; early Shan contact with Pagan was probably as slaves and soldiers, which accounts for Tai presence down to the Isthmus of Kra as part of the 12th century campaigns of the Burmans against the Malays.

The irrigated agriculture of the Mon and Burman cultures complemented that of the Khmer. They included canal irrigation associated with major rivers across ancient alluvial flood plains. Tai with their small *Müang fai* river valley irrigation[41] systems were to learn from this for their eventual domination of the Chaophraya Delta. With the rise of the inland rice Kingdoms, coastal regions remained exposed to foreign trade and ideas.

Peninsula Southeast Asia

The rice culture of peninsula Southeast Asia has been historically determined by the sedimentation of clay and mud in this relatively young geological area. As soils determined the patterns of agricultural settlement, the geographical location of trading centers and subsequent Indianization follows the development of agriculture. Variations in rice cultivation methods across the peninsula reflect its many micro-environments, as well as variations in cultural influences associated with trading and migration.[42]

Sea trade routes and the narrow land connection across the Isthmus of Kra shaped further development of the South. With new nautical technologies in the 4th century, trade via the Straits of Melaka led to Palembang and Sumatra becoming a major trading centre, incidentally attracting Buddhism and Chinese culture. The Malay-controlled trading system was managed on a cooperative basis which attracted avaricious invaders including Javanese and Tai who sought to vassalize the Malay rulers of the Straits region.[43] The rich archaeology of the ceremonial centre Palembang derives from its regularity of rice production from extensive rice fields. Gaining further influence through international trade, it dominated the Srivijaya Kingdom of Java. The Majapahit State of Java relied on a decentralized agrarian culture which was unprepared for the dealings of wealthy commercial centers which its own wealth had helped to create. Such transition from rice security through rice-based Kingdoms to domination by trading powers flows through Southeast Asian political history.[44]

The agriculture of peninsula Southeast Asia therefore combined technologies from Java, Malaysia, India, and China from extensive trading connections. Technologies emanating from Java and Sumatra, which differ from mainland Southeast Asia are still evident today, such as rice harvesting techniques. Prior to these developments, a rapid rise of Tai power in Nakhon Si Thammarat occurred in the 13th century to subjugate Khmer, Malay, Burmese, Mon, and south Indian rulers in what was probably the major centre of the region. We therefore turn next to the peculiar traits of the Tai rice farming system that has been central to rice expansion.

Tai Rice Culture

The emergence of the Tai tribe coincident with the decline of Mon-Khmer domination reformed rice culture in the region. While known in China before the 11th century as wet rice cultivators, Tai-specific technologies may better be deduced from cultural associations with similar technologies across the Tai diaspora. With the gradual southward movement of the Tai ethnic group prior to the 12th century, and in particular in the 13th century and after, Tai technologies in irrigation and rice culture mixed with those of the Khmer and Mon. As valley dwellers, they had developed and refined technologies through experience and contact across valleys from Assam to Vietnam. Their water management systems complemented those developed by Mon-Khmer. In particular, the *müang fai* irrigation system represented a technologically and socially sophisticated system which proved sustainable through at least eight centuries.

The Tai, of whom the Chinese wrote, universally lived in lowlands and valleys, having developed an economy based on wet rice cultivation.[45] Linguistic and cultural associations suggest contact between the Tai culture and for example, Hua Xia culture of southern China more than one thousand years ago.[46] The progressive southward migration of the Tai introduced their *müang fai* irrigation system to the narrow river valleys of northern Thailand, as indicated by the associated innovation, the *luk*, a huge bamboo water lifting wheel used in Thailand since before the Sukhothai period. The *luk*, powered by the river current, used short sections of bamboo attached to the outer rim of the paddle-wheel to collect water and lift it above the level of the riverbank and empty the water from each bamboo cylinder into a drain leading to a field. Dismantled or abandoned prior to the river rising each wet season,[47] *luk* seems to have been used by Tai for more than ten centuries with *müang fai* irrigation systems and glutinous rice culture in what is now southern China and northern Thailand.[48]

Müang Fai

The *müang fai* irrigation system was used on fast flowing streams up to twenty meters in width, across which weirs elevated water by up to two or more meters.[49] The *fai* held back water which was directed to major and minor canals known as *muang* in which gates, *tang*, controlled flow rates. Where a *muang* could be constructed by diverting water from a river, no *fai* was needed. Constructed from bamboo and wooden stakes driven into the river bed against which rocks, poles and sand were placed, the *fai* allowed water to pass through and over the barrier while restricting the rate of flow and thus raising the water level. Annual maintenance necessitated by peak wet season water flows and siltation formed the basis of the community ownership of these resources and the development of a democratic Tai administrative system. The system allowed the development of States with a ruler over several *müang fai* in a river valley, although independent systems appear to have existed in parallel with consolidated arrangements through to the 19th century in the larger northern rivers.[50] The porous weirs with water brimming over the top enabled successive *fai* to

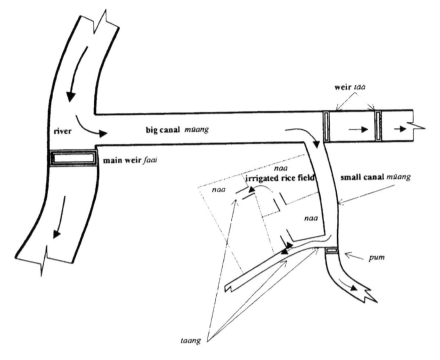

Fig. 7.1. The *Müang Fai* Irrigation System[52]

be built on a river. The system as depicted in Fig. 7.1 required sound social organization[51] and appeared as early as 757 CE to have been managed through the local rulers as a means of coordinating irrigation or rice fields belonging to a significant proportion of the populace. The well documented northern reign of Mengrai in the 13th century indicates a widespread and well managed irrigation system in the northern river valleys.

The social organisation allowing the management of the rice system evolved to rely on officials, such as the *Khun Nai Fai* and the *Hua Na Fai*, as managers of rice irrigation systems on behalf of the ruler. These offices became local leaders and were elected by those participating in the irrigation system. The irrigation manager's responsibility was to:

- calculate the amount of water and its allocation to individual farmers
- coordinate the initial construction of weirs and canals
- coordinate annual repairs required after each wet season
- manage propitiatory and other rituals associated with rice culture
- collect fees for irrigation system maintenance and associated rituals.

Requirements of farmers to provide labor on the basis of their area of paddy fields formed part of an early user-pay system within a community-based activity which was uniquely Tai. This sustainable social system was critical to its technical sustainability. Elaborate rules evolved to maintain systems and to avoid damage, with policing power vested in the irrigation manager. Serving rice agriculture until the 20th century, the *müang fai* system was eventually incorporated into the modern Irrigation Department until it was superseded by developments in pumping and piping technology.

The *müang fai* system well suited the valleys draining northern highland of Southeast Asia although it was universally popular with other ethnic groups. Even in this century, some hill tribe groups have adopted these techniques of irrigation, weir construction, and canal orientation, with an incidental outcome of adopting the rituals and prayers directly from remnant Tai language as part of the 'development package'.[53] The *müang fai* system was less suited to delta areas with their heavy river sediment loads and extensive flood plains. In these areas, the opportunistic use of minor earthworks to delay receding flood waters continued as the basis for rice cultivation until later merging of Mon-Khmer and Tai approaches to water control enabled settlement of the hitherto lightly populated deltas. The talent of the Tai people with the *müang fai* rice system was a critical input to the development of Sukhothai, and subsequently, Ayutthaya. While the rising influence of the Tai at Sukhothai is difficult to separate from their assumption of Khmer ideology, culture, and technologies, the later Sukhothai period when Tai control was well established suggests that sediment settling ponds used as ritual architectural artefacts were of less interest to the Tai than the Khmer. Tai inscriptions from Sukhothai refer to the use of the *müang fai* system to irrigate crops adjacent to smaller streams in conjunction with more opportunistic system of bunding to retain receding water in areas protected by flood barrages. Thus Sukhothai, represents a blending of the smaller scale community *müang fai* irrigation system of the Tai with the extensive system of the Khmer and their agriculturally less significant pond and gravity feed system.

Integrating Technologies

The interface between Tai and Khmer rice irrigation technology led to the diversion of major river waters into canals built at the river's natural height. This allowed swamps and old river bows to fill when flows were high, and for that water to be trapped and subsequently drained quickly towards the end of the rainy season. By this means, the simple earthworks which delayed receding flood waters as practiced from the time of the agro-cities through to the Khmer period could be used to greater effect. As the Tai had not been associated with large rivers and broad flat plains for centuries,[54] it is probable that these technologies were developed elsewhere and transmitted through the extensive network of Tai people extending from what is now southern China through Vietnam, Laos, Thailand, Burma, Bangladesh, India, and Bhutan, and along trade routes which passed Angkor and Sukhothai. Nevertheless, this technology, while widely used on rivers with a significant gradient, was inappropriate for the meandering delta distributaries such as those of the Chaophraya River. The relatively scant rice based-agriculture of the deltas supported a small population based on simple water engineering. Abundant flood water enabled areas which flooded naturally for four to five months a year to be used with minor earthworks to enhance the depth of water, and retention rate at the margins. Through this period, the deposition of sediment and a gradual fall in sea level, led to the creation of natural distributary canals in the alluvial mud, thereby creating potential rice fields in adjacent areas. These changes in the natural environment provided a significant advantage for the rising population in the river deltas. The annual replenishment of fertility through sedimentation and the abundance of water to irrigate rice provided a basis for further expansion which modified irrigation technologies derived from the *müang fai* and Mon-Khmer systems.The bountiful irrigated rice production systems allowed the development of crafts as part of their evolving culture. Even at the village level today these crafts can still be seen in the sensitive and time consuming skills of designing such utilitarian items as fish baskets, for fish are an adjunct of rice throughout the region.

Dissemination of Rice Technologies

Continuing for the moment with this unique contribution to rice culture in Southeast Asia, we can see that the southward migration of the glutinous or sticky rice growing Tai took place over centuries, rising to a critical peak at the time of Kublai Khan. Migration is confirmed by Tai presence in Angkor records around the beginning of the 12th century. Around 1300 CE, Chinese annals note that the Tai who migrated to Southeast Asia benefited from the fertile soils of the region, while those who migrated to Laos inherited soils less suited to wet rice agriculture. The Tai found themselves as the occupants of upland river valleys surrounded by the Vietnamese State centered in the Red River Valley and delta regions, the Kingdom of Champa on the coast of central Vietnam, the Khmer Empire centered at Angkor and the Kingdoms of the Mon and Pyu of Burma. These States were oriented to either the coast or their own State-religious Empires and rice was the basis of their security after commerce.

Independent Tai States of southern Yunan were recognized in Chinese chronicles as having entitlements to revenues from rice and other economic endeavor from what is now northern Thailand, northern Vietnam, Laos, and Sipsongpanna. By the end of the 13th century, Tai chieftains had established their right to the manpower needed to build many rice fields and they began to assimilate with other cultures or be absorbed as they assumed leadership roles in a new culture, as occurred in the case of the Ahom Tai of Assam.[55] In any case, it appears that rice was a unifying theme of the Tai-influenced groups. The numerous Tai settlements of the era emerged as four groupings according to the river basin of their location:

- The Mekong River group which is found in Sipsongpanna (the '12,000 rice fields' kingdom) region of Yunan Province of China and extending down through the northern areas of Myanmar, Laos, and Thailand, and including the northeastern region of Thailand.
- The Salween River group which is concentrated predominantly in northern Burma, although part of this group subsequently moved to the Phrommabutr River Basin in India.
- The Red and Black Rivers groups which include the Tai speaking groups of present day Vietnam.
- The Chaophraya River group which includes the Tai speaking groups of present day Thailand.

The small cultural differences between the different groups are overshadowed by their overwhelming similarities in language, rice cultivation systems, consumption of glutinous rice, and the distinctive form of raised-floor house construction. In particular, the association of Tai people with glutinous rice appears to have been almost absolute. While during the modern era, groups with whom the Tai mixed in delta regions have forsaken glutinous rice production and consumption, their rituals continue to reflect Tai origins; sticky rice is used in spirit and ancestor offerings, marriage ceremonies, as well as a component of specialized dishes, now regarded as delicacies. Glutinous rice remains the preferred diet in the northern part of the Mekong area[56] and in the northern river valleys across to Burma. The association with wet rice production saved many of these groups from the population pressure which had outstripped agricultural production in their Chinese homeland leading to the 1793 Malthusian predictions of Hung Ling-Chi.[57] Extreme food shortages along the Yangtze fuelled the civil war known as the Taiping Rebellion in the mid 19th century. Wet rice agriculture which had in part fuelled their southward migration had yielded its harvest for Southeast Asia in a manner reminiscent of 2nd century BCE observations that the proto-Tai people of the Yangtze River were then blessed by a bounty of food in an area not subjected to floods or droughts.[58] Wet sticky rice was the hallmark of the Tai-associated groups.

Expansion of Rice from 1200 CE

Merging of Tai and Mon-Khmer systems allowed expansion into new areas of Southeast Asia but the the *müang fai* irrigation system, with its reliance on strong community organization[59] including water rights and pricing systems, and the recognition of the water manager as a leader, served as a basis for emerging governance. The King became the ultimate water manager and societal representative to propitiate water controlling spirits is as much a logical extension of the social system of the Tai *müang fai* as it is the Mon-Khmer systems. The Burmese Kyauskse[60] river-tributary based irrigation system, the Khmer dam and canal system, and Tai weir and community-based irrigation technologies were precursors of irrigation systems that eventually tamed the Chaophraya and remaining deltas of the region. With the fundamental ethic of secure rice production, society's stability relied on sound rice management—the alternative may have occurred in the frenetic monument building in the last decades of the Angkor Kingdom when siltation of irrigation schemes seems to have been uncontrolled, or uncontrollable.[61]

The centrality of water control in rice development in the region led to the evolution of central governance to manage reliable food surpluses and to then develop trade. The Javanese design of a U-shaped dyke across a river constructed in 11th century Angkor territory is indicative of trading contact.[62]

Such contact presumably showed that the Khmer rice management system was doomed, or perhaps the management of the rainfall and flood recording systems, surveying contours, operating sluice gates and siphons, and the constant maintenance of reservoirs and canals was just too much work compared to the simple weir diversion system.

Tai assumption of the Khmer mantle may have tempered rice production culture. The intricate association with cosmology in the architecture of the Angkor civilization has oriented modern interpretations of the water diversion systems to be similarly associated with the religion. The thousand *lingas* through which the Siem Riep River flowed before the waters were to be used by the city at Angkor is interpreted as a religious hydrological system.[63] However, the societies which created this culture relied on a continuous supply of rice and such diversion systems would more likely have originated as small interventions to serve agriculture and the city, and have been attributed a religious significance to protect their integrity. As the society became more refined, mundane reasons for religious associations may have been subjugated to the expansive religious architectural work that increasingly deified the King. And rice production became a secondary pursuit and eventually faltered.

Larger and larger rivers were tapped by the *müang fai* system until, by the Ayutthaya era, a large kingdom was founded on the rice security of modified main river flooding.[64] The combination of the technologies is evidenced in these systems and the construction of the first storage irrigation system in 1633 at Ayutthaya,[65] an echo of the Khmer storage *barai*, and the comparatively tiny *müang fai* systems of the traditional intermontane Tai.

The rice system of mountain Southeast Asia, based on *müang fai* irrigation, remained viable long after extensive northern Lanna ('million rice fields') Kingdom's fall. By this time, Ayutthaya had codified water control works to reduce floods, supplement rainfall to create a known environment for rice cultivation, and even to grow off-season rice crops.[66] Not to be confused with today's systems, these early delta irrigation systems delivered the requisite water from an inundation *khlong* with simple control systems that could fail completely if a season was overly wet or dry. However, the extensive delta and its relative under-population ensured that there was always an area from which sufficient rice could be harvested.

The evolution of water management and rice production techniques reflect cultural evolution.[67] Traditional Tai agriculture used broadcasting of seed to plant rice in flooded swamps and river overflows, whereas the Khmer rice production on elevated sites required transplanting to maximize use of scarce water. Transplanting was later adopted widely, even for the Tai staple of glutinous rice which remained important for local consumption. As irrigation systems became more complex, organizational systems that could manage rice and water were developed and emerging over more than fourteen centuries as modified forms of the ancient administrative units of *ban, muang,* and *nakhorn*.

The *ban, muang,* and *nakhorn* of what is now Thailand are mainly located facing rivers, reflecting original transportation systems for rice production and all communication. Expansion from rivers was along canals dug to improve irrigation, or to open new areas to irrigation. Thus *muang* developed where river tributaries joined the main river, and *nakhorn* where navigable rivers and land or sea routes intersected. By contrast, the *ban* and *muang* reflect the expansion of a community to a neighboring naturally-raised area surrounded by swamps in which rice was culti-vated. On the coast, rice and trading settlements developed along sand dune ridges and at river mouths, with the administrative units reflecting trading importance or population density. In the mountainous north, the interconnection of separate *ban* on a river where the *müang fai* irrigation system was practiced, created an affinity which continues today;[68] *muang* consolidated a group of *ban* in a valley or geographical contiguous area, with *nakhorn* being the major trading centres in larger valleys.

Under the influence of commercial success in trading, the coast[69] relied on rice as elsewhere, and learned from influential trading groups inclu-ding Indians, Persians and Chinese. As distinct as the Khmer stream and reservoir system to supplement rainfed rice on elevated terraces was from the Tai *müang fai* system, so is the coastal technique of only harvesting the rice raceme distinct from the usual harvesting of the whole plant in inland Southeast Asia.

The security of rice production provided security of kingdoms that then required skilled political management to grow. Reliance on the balance between manpower and rice production for political security fuelled military expansionism and development of democratic rice production systems. Since before the time of Ayutthaya, an early democratic approach to ownership of the productive capability had been institutionalized as *sakdi na* or 'field power', whereby an area of rice fields was allocated according to a persons social level, with use of this right in the settlement of legal disputes.

Development of large delta areas was not achieved until after 20th century at what is now seen to be significant environmental cost. The floods common to large parts of the deltas required new approaches to water control, which ultimately benefited from globalizing forces.

Rice in Southeast Asia has thus defined lifestyle, law, war, and religion. It was the chief cause of civil litigation in Khmer and succeeding Tai Kingdoms. Wars of the era, which frequently seem to have been indecisive to analysts today, are made intelligible when it is recognized that they were conducted between rice growing seasons. Returns to the battlefield depended on successful rice harvests, themselves increasingly dependent on the valued spoil of war, labor.[70]

Rice Agribusiness to the 20th Century

Development to this time across mainland Southeast Asia had relied on clearing of rainforests and to an extent this had restricted population expansion to alluvial riverbanks and volcanic loam soils. Alluvial areas benefited from silt deposition to maintain a level of fertility along the Mekong, Chaophraya, Irrawady, and Red River plains.[71] With the emergence of rice surpluses, significant States arose and international trade developed from around the 14th century. Economic development relied on expertise in wet rice production. The Ayutthaya Kingdom, for example, was established on modified water recession from flooded areas on the delta; minor earth-works slowed natural drainage according to the stage of rice maturity. Supplementary water was drawn from some canals that served agriculture after their primary transportation purpose and symbolic functions that recalled something of the Mon-Khmer traditions.

The global cooling of the 17th century, which created famines in Europe, was only experienced in Southeast Asia in the form of reduced rainfall; teak tree ring analyses from 1514 to 1929 in Java[72] indicate that the years 1645 to 1672 each received less rain than the average of the four centuries. With such longer dry seasons, crop failures were common and populations declined in many Asian communities. In delta areas, the ease of moving closer to the river in dry years and of exploiting gradients rising away from the river to ensure at least some production, provided advantages over the swamp and lake wet rice systems and some rice exports continued throughout the cold period.

By the 1840s, the rice-based settlements were small core communities or States with small outlying provinces, all in the lowlands where rice culture required minimal effort and provided greatest reliability. Even delta kingdoms may not have extended complete influence more than ten kilometers away from rivers until the early 19th century, when the population was estimated to be three million for the whole of what is now Thailand.

Western contact brought a view of technological dominance of the environment[73] that contrasted with views of small-holder farmers[74] more than with the Palace classes involved with trading. In any case, teak and other forests attracted the interests of colonial groups to the extent that by the 19th century, western political ends were subjugated to trading benefits gained from guaranteed access to valuable forests. This expanded rice production from the river and coastal basins in association with the gathering of valuable items from the hinterland.

Export Rice

Rice gradually changed from a solely domestic to an export crop. Under colonial influences, trade stimulated widespread city expansion in Asia fed from the granaries of Thailand, Burma, and Indo-China where some 80% of the population may have been engaged in some aspect of rice agribusiness.[75]

From the north with its durable *müang fai* irrigation system, to the Indian-influenced rice production systems of the coast, to the rainfed and receding swamp systems of the Khmer and the rivers and flood plains of the delta, Southeast Asian rice agriculture came of age in the 19th century. Adequate production for domestic requirements in all but exceptional years is echoed in the numerous reports which emanated from the exploring, proselytizing and trading Europeans.[76]

Through the period 1870 to 1934, rice exports increased 20 fold, the population doubled, and the area planted to rice rose several fold, for Thailand.[77] Such increases indicate much more than an economy seeking to export occasional surpluses of its staple. Indeed the influence of foreign traders in freer trade through the 1855 Bowring Treaty encouraged the production of a surplus as a means for Crown revenue raising through taxation to purchase more readily available foreign goods. However, the decision to expand production was ultimately made by the individual rice farmers themselves.

Rural expansion proceeded along the rivers and swamps lands, and eventually into the uplands where earth tanks could hold supplemental water for rice. Areas missed previously were developed through small canals and contour barriers to facilitate drainage on a small scale.[78] At first these developments were primitive and temporary in the manner of frontier agriculturists, but successive generations and migrants improved these systems to their full capacity under a human and draft animal powered rice agriculture.

Foreign Influence

Europeans followed the Persians,[79] Indians and Chinese who had established influential roles in the growing kingdoms. The development of rice production relied on the inputs of these foreigners for new irrigation technologies.

The first European awareness of rice export potential seems to be in 1502,[80] and the first European written reference is a letter from the Portuguese Governor of India in 1510, after which the Portuguese monopolized European trade until the rise of Manila in 1565.[81] Drawing on such information as Marco Polo,[82] Nicolo Conti[83] and Vasco de Gama, the Portuguese preceded the English, Dutch, and Danes by a century. Arabic references to the region also contain records of traders from Southeast Asia frequenting a port in the Persian Gulf[84] where contact with Portuguese and Spanish was likely. However, Portuguese contact effectively began with their conquest of Malacca in 1511 when Albuquerque sent ambassadors to important Southeast Asian kingdoms.

Portuguese, French, Dutch, Greek, and other nationalities were involved in the active foreign life that developed around the main city of Ayutthaya,[85] reflecting a preference for this Kingdom as a trading centre above other local possibilities including Vietnam.[86] The arrival of Europeans is associated with a marked increase in the documentation about the region, in both local and European languages, in the western style of recording trade, exploits, and accomplishments. With major concentrations in Batavia, Manila, and Ayutthaya at the time, Chinese maintained access to markets, trade routes, and supplies.

Experimental farms for rice and other crops began in the 19th century under colonial terms and began the modern era of rice breeding and development. Small-holder rice production systems had usually yielded poorly as yields had not been a preoccupation before this time. Entering a rice trading marketplace with more defined land rights meant that small-holders and landlords sought to maximize returns to land rather to labor. Yields and in particular output increased from about 1860 to 1900 as a result of improvements in traditional stalky rice varieties, and perhaps the use of fertilizer.[87]

Large-scale Rice Irrigation

Floods had rendered rice production tenuous and rice shortages occurred following major floods—a flood in Thailand in 1917 destroyed more than 450,000 hectare of rice, equivalent to 21% of the crop.[88] What proceeded in various ways in colonized countries was a different process than in the more independent Thailand where the Rangsit Project of the 1890s was designed to expand rice production in the depression areas near Bangkok. The first comprehensive irrigation scheme, it included 1600 kilometers of waterways and large mechanized dam gates to control water flows. A monopoly was granted to a Thai and Italian group registered as the Siam Land, Canals, and Irrigation Company, to develop, use and sell the agricultural land created by the scheme.[89] By 1900, the government rescinded the monopoly that was in fact determining settlement patterns around the capital, and formed the Royal Irrigation Department in 1903 under the name of the Department of *Khlong*, reflecting the primary function of canals.

To this department came the Dutch expert, van der Heide who conceived the comprehensive barrage system for the Chaophraya delta. Van der Heide presented three plans over the more than six years that he worked on irrigation designs for Thailand, the latter two being scaling down of his initial comprehensive plan. None were accepted at any stage, with the most consistent reason given being the lack of population to utilize the new agricultural lands that the scheme would create.[90] Following floods in 1912 and 1913, the Minister of Agriculture appointed a British adviser who accommodated criticisms and strengths of the van der Heide scheme to produce an acceptable plan that was partially implemented, and industrial rice mills were introduced to the region.

The first steam-powered rice mill in mainland Southeast Asia was an American built machine erected in Bangkok in 1858. It was followed by British versions installed in Rangoon and Bangkok in 1869 and 1870,[91] and by 1900 there were some 50 rice mills and 20 sawmills in Bangkok, mainly associated with the port and Chinese labor.[92] Thus Southeast Asian rice production entered the industrial age, in which the Green Revolution was to be one event.

The world's largest exporter in recent times has been Thailand, as it had been in some earlier periods. That Thailand has been eclipsed by Vietnam in some recent years is a reflection of the recovery of Vietnam at the time that Thailand's economy is moving into less basic commodities for at least some of the areas nearer Bangkok, just as in the colonial era, Burma was the world's largest exporter. That all these countries are in lowland Southeast Asia is no coincidence, for rice is not only the staple of these people and it is well suited to the environment that has persisted there with the global cooling of past centuries. But we should be careful not to equate success in exports with high levels of production—and extreme example being the minor producer and consumer of Australia. Some 90% of the world's rice is produced and consumed within Asia, mainly in the country of production. As shown in Table 7.2 below, the large producers are the populous countries of China and India; in South East Asia, the ranking of production has been and probably still is Indonesia, Vietnam, Thailand, Myanmar (Burma) and the Philippines.[93]

Table 7.2. Rice production (yield) for selected countries (1991-1995)[a]

Country	Production ('000 ton)				
	1991	1992	1993	1994	1995
World Total	515,431	526,161	523,743	536,432	549,291
Asia	**474,720**	**481,106**	**480,587**	**489,748**	**501,980**
China	186,086	188,255	179,977	178,031	187,192
India	110,591	109,001	118,464	121,997	122,372
Indonesia	44,688	48,240	48,181	46,641	49,860
Bangladesh	27,377	27,510	27,062	25,248	24,659
Vietnam	19,622	21,590	22,837	23,528	24,000
Thailand	10,400	19,917	18,447	21,111	22,016
Myanmar	13,199	14,837	16,760	18,195	20,109
Japan	12,005	13,216	9,793	14,976	12,625
Philippines	9,673	9,129	9,434	10,538	11,002
Korea, South	7,293	7,303	6,507	6,932	6,519
Pakistan	4,865	4,676	5,992	5,170	5,714
Nepal	3,223	2,585	3,493	2,928	2,906
Sri Lanka	2,389	2,340	2,570	2,684	2,685
Korea, North	4,420	2,439	2,300	2,500	2,580
Others	8,889	10,070	8,770	9,269	7,742

[a] Office of Agricultural Economics (1998).

Rice Production Systems

Rice production systems today vary by types of rice and region, just as they have been for a millennium and more. Four types of *Oryza sativa* widely used in South East Asia are:[95]

• **Wetland rice** (*Oryza sativa* var. *dura*) is often referred to as paddy, being produced in controlled flooded fields during the entire growing period and hence is restricted to the delta and valleys with developed irrigation systems. It supplies most export rice in countries such as Thailand and domestic rice in countries such as the Philippines.

• **Glutinous rice** (*Oryza sativa* var. *glutinousa*), also known as sticky rice, varies in grain colour and cooking characteristics from paddy. Its translucent color in the uncooked state and sticky characteristics upon cooking make readily distinguish it among informed consumers. Some 10% of the major producer, Thailand's production is exported to neighboring Tai-related groups, particularly in Lao-PDR. It is grown mainly in the Northeast and North production areas of Thailand and is strongly cultural specific, as we will discuss later. Production systems for glutinous rice are essentially similar to those for paddy.[96]

• **Upland rice** (*Oryza sativa* var. *montana)* is grown under shifting and permanent cultivation in the mountainous and poorer upland regions of the Southeast Asia. Entirely dependent on rainfall, it is a subsistence crop which has not been considered of statistical importance in past official surveys, because such environments are felt to be aberrant for rice—but, as we shall see, flooded paddy rice will not be necessarily the future production system.[97]

• **Floating rice** (*Oryza sativa* var. *fluitans*) has been long grown in the region, being used earlier than the Mon-Khmer period. It is a type of wetland rice which can rapidly accelerate internodal growth in response to rising floodwaters up to two meters in depth by nutrient-uptake from water more than soil based roots. Low grain and very high stem yields, and modern water control works, have reduced its area—for example, in Thailand, it is now restricted to less than half the estimated five million rai (800,000 hectare) of the 1960s, and used only in flood-prone areas of the Northeast. It remains important in the Tonlesap area of Cambodia.

Regional variations define much of South East Asian rice production. The river plains form national rice bowls in all major countries especially, Burma, Thailand and Vietnam. Rainfed rice produces lower yields yet occupies large areas and is associated with relative poverty—again in the case of Thailand, upland rice in the Northeast covers some 43% of total acreage and produces around 30% of rice. Such differences can reflect inferior water and soil regimes, but may also reflect dietary preferences for the lower yielding glutinous varieties.[98]

Glutinous Rice

The long cultural and regional association with wet rice[99] includes effects from trading, migration, and market demand. Technologies and varieties of continental South East Asia reflect the southern China migration route of the Tai and other rice growing groups, while the larger rivers reflect Indian influence through the Mon-Khmer and other cultures of the Mekong delta region in its Bengal rice cultivation approaches, as also found in Peninsula and Island South East Asia.[100] The trend is perhaps best illustrated by the type of rice least affected by modern breeding and cross-introduction—glutinous rice.

Glutinous rice is a peculiar phenomenon of the Tai ethnic group. Today it is synonymous with subsistence production with some cross-Tai group

212

trading, and an embryonic boutique market. Its different cooking, eating, and taste qualities separate sticky rice eaters from others in terms of kitchen, meal, and snacking behavior. Cohescing at temperatures as low as 72°C which produce no change in non-glutinous varieties,[101] sticky rices[102] exhibit different amylose: amylopectin ratios, and contain four to 5% of dextrose in their endosperm compared to very low levels in non-glutinous varieties.

Glutinous rice growing is today a cultural preference, not an indication of environmental variation. It covers an estimated one-half million square km[103] across several ethnic groups, the majority of which are of Tai origin and all of which have some Tai association (Fig. 7.2). Originating as a short growing season variety suited to low rainfall regimes and light soils with minimal water control and being a recessive mutant that can only be differentiated at harvest, suggests that glutinous varieties were originally selected from non-glutinous varieties.

With the migration of the Tai, large scale adoption of non-glutinous varieties for other markets since the 1600s was consolidated through controlling of water environments to favor the original, longer growing

Fig. 7.2. The glutinous rice zone of Southeast Asia.[104]

season non-glutinous varieties.[105] The recent nature of the change to non-glutinous varieties with incidental changes in diets is indicated from the now export-oriented Chiang Mai valley, which as recently as 1974 grew glutinous varieties on more than 80% of the rice area.[106] Genetic modification of non-glutinous varieties led to their yield capacities exceeding those of glutinous varieties, thereby ensuring their commercial dominance.

Variations in photoperiodicity,[107] photosynthetic and temperature responses, water regime requirements, grain production characteristics, suitability for harvest, and a range of other factors were considered in breeding new non-glutinous varieties. These produced the high-yielding varieties of the Green Revolution emanating from programs of the Rockefeller and Ford Foundations and the International Rice Research Institute.[108] High yielding varieties caused the apparent cyclical movement between broadcasting and transplanting techniques in delta regions and with the advent of rice mono-culture in these areas, the preferred transplanting of rice seedlings shifted to broadcasting, to cover the larger areas available.[109] Progressive intensification of rice production with the advent of high yielding varieties then again favored seedling trans-plantation to allow double cropping and efficient water management. The less labor intensive broadcast sowing required growing periods of up to nine months.[110] By the 1970s, 80% of the rice area was planted by trans-planted seedlings.[111]

Rice Breeding

Rice in South East Asia today is a narrow shadow of its once rich genetic diversity. Thousands of rice varieties chosen over millennia were once cultivated from India across southern China and through Southeast Asia when communities remained relatively separated. With migration, knowledge of varietal suitabilities to sites led to sharing of genetic material which initially expanded local diversity such that as recently as the 1900s some 300 varieties were in use in the Red River delta region of Vietnam.[112] By 1991, traditional varieties represent only about 2% of the planted area in Thailand and projected reductions in subsistence farmer numbers will exacerbate the trend of genetic uniformity. Modern rice breeding began in the early 1900s with the production of longer grains to suit domestic markets and Thai rice achieved notoriety at the 1933 World Grain Exhibition Conference in Canada. However, until the 1950s, rice yields remained low averaging 1.6 ton per hectare—thereafter the production of IR8 and other high yielding varieties changed rice, perhaps forever.

The Green Revolution is documented in Chapter 16 (A Century of Rice Breeding, Its Impact and Challenges Ahead) in this book, but it is worth noting a special South East Asian trait, at least in some cultures of this region—that is the insistence that taste is of value even if not reflected in commodity oriented markets. Discerning Thai palates deemed the Green Revolution IR8 rice variety as poor, causing its slow acceptance until it was crossed with a native variety, *Luang Thong*, to produce two palatable photoperiod-insensitive varieties. Further, glutinous and non-glutinous varieties were developed through the Thai Rice Research Institute for major rice production environments and dry season cropping.

As for elsewhere in South East Asia, a range of lowland rice varieties produced through hybridization, mutation breeding or pure line selection, as well as (in Tai-influenced areas of Thailand, Lao-PDR, Burma and Vietnam) non-glutinous varieties. Five floating rice and upland rice varieties were also produced and by the 1990s, high demand rices such as Basmati were also imported from India and Pakistan.[113]

Rice Husbandry

To understand changes in rice husbandry techniques, we must examine labour inputs. Table 7.3 indicates that higher labor inputs of transplanting produced lower outputs than broadcasting in the 1960s, thereby confirming that transplanting was used for reasons other than saving of labor. Broadcasting had been more common in many areas from 1890 to 1935, after which transplanting assumed importance. Shifting cultivation, where a field is used for a few years and then abandoned when another is cleared for use, remained important in some areas throughout until the 1980s, although in major rice production areas it was of significance only up to the mid-19th century.[114]

Vietnam, probably a very early centre of rice cultivation, is often overlooked in rice culture yet today has the capacity to be the world's largest exporter, as it has been in some recent cyclone damage-free years. Rice

Table 7.3. Labour: harvest output comparisons
for rice production systems[115]

System	Labour days per crop season	Output (ton)	Output: Input
Shifting	245	2.5	9
Broadcast	301	6.2	20
Transplanting	430	5.4	12

surpluses began after the vast Mekong Delta was able to be partly controlled about 300 years ago and developed with joint local and colonial inputs in a manner akin to the other deltas, with the local embellishment of using tidal rises to back fill higher canals with fresh water. Improved rice varieties dominate today although some local favorites such as *Nep Mot, Tam Thom and Nang Huong* are still grown on significant areas. Commercial rice is produced predominantly under irrigated conditions in the northern and central areas. Among the top five rice consumers in the world, the Vietnamese have integrated rice cultures from coastal China, the Tai, the Khmer and India; a glutinous rice cake associated with the New Year festival (*Tet*) provides an indication of one link that legends relate back some three millennia.

Rice cultivation in Cambodia covers some 90% of the agricultural area, producing an estimated 4.5 million tonnes in 1996, which is said by government to represent national rice self-sufficiency. As in neighboring Thailand, rice is produced in various ecosystems including rainfed lowlands, rainfed uplands, floating rice areas and dry season irrigated areas—with nearly 60% of the harvest from rainfed lowland areas and most dry season rice being high yielding varieties. Risk-reduction strategies often lead to several rice varieties being planted in a field.

The Philippines has a distinguished history of rice cultivation from about 3200 BCE, as indicated from an archeological site on the fertile plain around Andarayan that has yielded samples of both wild and cultivated rice. The Philippines also shows the early use of rice in non-lowland areas through early use of terraces and rice varieties suited to the lower temperatures of higher elevations. While such formations as the majestic Banawe rice terraces are dated from antiquity, Spanish records do not mention them in the 1600s, and it is supposed that the terraces at this time were limited to rainfed rice production. The remnants of a traditional calendar based on the annual rice cycle and the scheduling of planting and other events from natural changes may be seen in the survival of rice gods in upland cultures and the setting of harvest dates by village elders' observations of natural signs. However, in the lowlands, rituals are largely absent as a result of foreign contact and perhaps the later adoption of rice in these areas.[116] Today, rice is produced mainly in Luzon, Western Visayas, Southern Mindanao, and Central Mindanao with an annual harvest of around 10 thousand tonnes. Yields do not meet those of other Southeast Asian countries but production is sufficient for domestic use in most years. As the host to the International Rice Research Institute since 1961, the Phillipines was an early beneficiary of the Green Revolution through the variety IR8 with its higher demands for water, fertilizer and management although enthusiasts for higher yields claim it has never reached its potential.

Last, we come to the populous country of Indonesia, Southeast Asia's leading and one of the world's largest rice producers with annual production of around 50 million tonnes from some 12 million hectares with the highest yields of the region. Often overshadowed by Thailand (which produces about 30 million tons but could produce much more) because of the latter's dominance in rice exports including to Indonesia and its longer history, Indonesia grows about three-quarters of its rice in irrigated areas and on the island of Java. This remarkable density of high-yielding rice is a phenomenon of the Green Revolution—and it indicates the operation of that revolution as building on known rice potential on fertile sites such as Java. Output is now in decline as land is assigned to other uses and this places further demands on the skilled Indonesian researchers trained at IRRI over the decades.

To understand rice in Southeast Asia, one must look at both Indonesia the largest producer and Thailand the largest exporter. Exporting mainly to Indonesia, Nigeria, Iran, the USA and Singapore, Thailand's success derives from its high quality, long-grain white rice, which enjoys premium over other rices. The focus on quality limits use of very high-yielding varieties and explains the reason for yields appearing to be less than their potential. Grown on some 10 million hectares in all provinces, only about 25% of rice is irrigated, and this is the source of most of the exported rice—yet most rice farmers in the country are semi-subsistence farmers, selling only their excess production. By contrast, Indonesia's millions consume their subsistence and commercial production of rice, as well as that of other countries. Hence the focus of Indonesian production is on yield more than quality. Between Thailand and Indonesia, other Southeast Asian countries exhibit these tendencies to greater and less degrees.

Another rice producer in the region though culturally and geologically separated from Southeast Asia is Australia, which has developed its own rice industry in a manner similar to its production of sugar. Also considered a 'third world crop' like sugar, rice is produced under the Australian subsidy-free system of its modern agricultural policies and has been able to compete through rigorous application of technologies, some of which were developed by its regional competitors.

A Short History of Rice in Australia

Rice was probably first introduced to Australia by Asian migrants as they visited Australia for fishing and trade with the aborigines of northern

Australia. Possible Chinese settlements in Australia long before any European presence may also have introduced rice. But to date, no evidence for these suggestions exists and it is commonly suggested that it was Chinese gold prospectors who introduced rice to southeastern Australia in the 1850s. As the gold rush declined and diggers scattered, no continuity of rice growing is known and it is assumed by many that the different Australian environment hampered rice production. However, the Australian environment is not so different from many areas in Asia that has grown rice for a millennium.

Various European attempts to grow rice in Australia failed until the first record of rice cultivation in the 1850s in Queensland where a small rice industry slowly developed, reaching a total annual production of 30,000 tonnes before being displaced by sugar cane production in 1893.[117] Through the late 19th century, rice was also grown in the Northern Territory, later with significant technical successes in terms of breeding and at that time extremely high yields from experimental sites. Rice was also grown in Western Australia after World War II, but its European beginnings were those in Queensland until the early 20th century when successful trials for commercial rice production began in southeastern Australia. This led to the Victorian government allocating 80 hectares to a former Japanese parliamentarian, Isaburo (Jo) Takasuka who had moved to Melbourne in 1905 from Matsuyama in Japan to demonstrate commercial rice growing at Swan Hill. Allocated 200 acres of flooding land on the Murray River, Takasuka brought 15 bags of *Japonica* rice from Japan and stayed until in 1914 producing rice for commercial sale.

At the same time, irrigation development along the Murrumbidgee River in New South Wales led to rice plantings and the creation of the Yanco Experiment Farm, from which the State Government supported the idea of a large irrigation scheme. The damming of the Murrumbidgee River for irrigation first led to fruit, vegetables and dairy production until 1914 when the same Japanese entrepreneur Takasuka introduced rice seed from his successful rice harvest at Swan Hill. By 1915, the first government rice growing experiments had begun at Yanco and these were soon supplemented by new ideas from California. By 1922 rice seed could be offered to farmers who in turn produced 222 tonnes. By 1924, commercial crops were grown and from 1928 rice breeding commenced to develop varieties better suited to this environment. By 1930, rice production was sufficient for the very small Australian domestic market and export became the objective of the emerging rice industry.

World War II stimulated expansion of rice production to other areas such as Wakool in the Murray Valley and later Deniliquin was developed for rice—all were government irrigation districts until 1988 when a free-market production was allowed and rice production expanded to Hay, Carathool, Hillston, Forbes and Echuca. It seems that World War II also stimulated renewed interest in the Northern Territory where rice had been tried many times and in fact grew well. Folklore and politics combined to cease the successful Northern experiments with stories of the loss of high yields before harvest due to the endemic magpie geese; those involved in the industry recalled that their stories of fitting in with tropical nature required planting enough to share with the limited numbers of geese were dismissed by southern advocates who favored their own rice industry. In fact, the extremely low population density of the north of Australia and its distance from domestic markets and voters were probably the main reasons that tropical Australia, despite having more suitable habitats for rice than the temperate and subtropical industries areas, contains the rice industry of today.

The short history contains a lovely anecdote of rice production in Australia being provided in modern times by a Japanese, who was eventually honored by the industry a century after his pioneering work when Australia began not only rice exports to Japan but entered into joint ventures with the Japanese rice industry. Today production is nearly 2 million tonnes of which 85% is exported through a growers' cooperative to more than 75 countries, producing some AUD $500 million per annum from the only unsubsidized rice industry in the world.

With annual average output of less than two million tons, Australia is not a major rice producer. But it is an important rice exporter and up to 40 million people across the globe eat Australian rice every day. Australians themselves consume just 10 kg per head a year. Rice is grown on some 145,000 ha of land, mainly in the irrigated areas of south-eastern Australia. Eighty percent of rice produced in Australia is of medium-grain Japonica varieties, which are well suited to high summer temperatures without the humidity of tropical climates.

Direct seeding is the main method of crop establishment. Production is highly advanced and mechanized, and rice is often planted in rotation with pasture crops. The main planting season in eastern Australia is October, with harvesting in March-April. Thanks to the use of improved varieties and better farm-level management, rice yields have increased from 5-7 tonnes per ha in the early 1970s to more than 10.2 tonnes per hain 2003, with maximum yields of 14 tonnes. Australian rice growers are considered among the most efficient and productive in the world. Asian immigration has introduced many new rice dishes to the national cuisine,

Detailed Historic Timeline of the Australian Rice Industry[118]

CE-1426	Possible rice introduction by Southeast Asians, Chinese and Indian ships, undocumented.
1850	Possible rice introduction into Southern Australian Gold fields by Chinese prospectors.
1860	Small 'upland' Chinese rice fields in Northern Queensland for the local gold fields.
1882	Irrigation on the Murrumbidgee River recommended to the Lyne Royal Commission.
1891	NSW Department of Agriculture begins trials of 'upland' rice varieties in Northern NSW.
1897	Retired Indian Civil Service officer Home recommends dams (Murray and Murrumbidgee).
1906	Isaburo (Jo) Takasuka plants Japanese (Japonica) rice near Swan Hill, Victoria.
1913	Irrigation dams deemed a failure for conventional crops and techniques.
1914-1915	Takasuka sells seed to the NSW Department of Agriculture where they are tested.
1920	Japonica rice varieties imported from California.
1924	Seed harvested from a successful 1922 trial of Japonica varieties is offered to settlers.
1925-1926	Policy changes; MIA Rice Growers' Co-operative Society formed; yield 1570 tonnes.
1927-1928	5114 tonnes from 2110 hectare; duty to imported rice; Australia self sufficient in rice.
1930-1931	34,405 tonnes from 8093 hectares; 250,000 recipe books and schools education program.
1935	36,553 tonnes from 8847 hectares (4.13 tonnes per hectare).
1943-45	1662 new hectares for wartime rice production; 32,809 tonnes from 9953 hectares.
1950	Co-operative mills agreed to solve bickering between small mills; 17,780 tonnes.
1955-56	'Sunwhite' brand of rice launched; aerated bulk storage technique developed.
1961	Coleambally Irrigation Area leads to a substantial increase in rice acreage.
1969	Planted area increased by 25%; largest rice mill built in Deniliquin.

Contd.

1970-1980	Green Revolution successes reduce export markets; incomes fall; protection reduced.
1980-1985	Farmers rationalize overproduction and restructure rice industry.
1986	Rice Cakes, Rice Bran and other products developed to broaden market; world oversupply.
1987-1989	Australian growers emerged from trade war as the last 'free trader' of rice in the world.
1990	SunRice 'Sculptures' advertising campaign increases domestic consumption by 17%.
1990-1993	Swan Hill erects a memorial to Takasuka's pioneering rice growing experiments.
1994	180,000 tonnes of Australian rice accepted by Japan under emergency conditions.
1995	'SunRice' joint venture with Japan; new variety 'Millin' developed for Japanese market.
1998	SunRice Arborio, Doongara, Jasmine, Koshihikari and Wild Blend rice launched.
2001	1,744,066 tonnes; Flour Mill built and old mills upgraded; industry poised to grow further.

but a traditional favorite remains the classic Australian rice pudding, which is the way most European Australian were introduced to rice—as a dessert.

Trends for the Future

Variations in rice yields with rainfall and temperature have been made manageable with irrigation across Southeast Asia and now Australia, although irrigation water availability, competitive water regimes for other crops, and drought in rainfed and partially irrigated rice areas continue to affect production. A critical period in Southeast Asia remains that of the early wet season when rice is planted before monsoon rains replace intermittent thunderstorms. Reliable access to water will determine the future of regional rice production and allow application of known technologies which can increase yields by two to four times the current levels of many Southeast Asian situations. This may well occur regardless of government or cultural intent as tradition continues to give way to commerce.

Notwithstanding the substitution of other crops for rice whether as a result of farmer initiatives, government diversification policies, declines in domestic rice consumption, or depressed global prices, rice will remain fundamental to South East Asian agriculture. One cannot be so sure about Australian rice production in the face of domestic pressure for realistic water pricing. Nevertheless, existing large infrastructural investments, including irrigation systems, are more suited to rice than other crops and have decades of service left at marginal cost in most countries—and global demand for cereals remains high. Many South East Asian environments and cultures continue to favor rice above other crops, and current production systems allow scope for large yield increases or possibly sustainable innovations as a benefit of relatively low current yields. Unrealized potential to patent research outcomes in rice production and high value processing could consolidate South East Asia's historical leadership in rice export, although ownership of such genetic material is already under threat.[119] Whatever the future, rice will continue to define Southeast Asia.

Further Reading

Falvey, Lindsay (2000) Thai Agriculture: Golden Cradle of Millennia. Kasetsart University Press (international distributor, White Lotus), Bangkok. 490pp (2000).

References

1. Changbao Li, Ailing Zhou and Sang Tao (2006).
2. Vavilov, N.I. (1930).
3. Chang, T.T. (1988).
4. For example: Sauer, C.O. (1952).
5. Alexander, J. and Coursey, D.G. (1969).
6. Lu, J.J. and Chang, T.T. (1980).
7. Grist, D.H. (1959).
8. Wyatt, D.K. (1989).
9. Yen, D.E. (1977).
10. Rogers, P. (1996).
11. Van Beek, S. (1995).
12. Pelzer, K.J (1978).
13. Shoocongdea, Rasmi (1996).
14. Hall, K.R. (1992).
15. Hall, K.R. (1992).
16. Rogers, P. (1996).
17. Van Beek, S. (1995).
18. Van Beek, S. (1995).
19. Rogers, P. (1996).
20. Wyatt, D.K. (1998).

21. Guillon, E. (1999).
22. Suchitta, Pornchai (1989).
23. Jumsai, S. (1997).
24. Van Beek, S. (1995).
25. Van Liere, W.J. (1989).
26. Hall, K.R. (1992).
27. Hall, K.R. (1992).
28. Van Liere, W.J. (1980).
29. Van Beek, S. (1995).
30. Murray, S.O. (1996).
31. Van Liere, W.J. (1989).
32. Rogers, P. (1996).
33. Rogers, P. (1996).
34. Groslier, B.P. (1962).
35. Hall, K.R. (1992).
36. Van Liere, W.J. (1980).
37. Kato, K. (1998).
38. Taylor, K.W. (1992).
39. Saraya, D. (1989).
40. Taylor, K.W. (1992).
41. Surareks, Vanpen (1998).
42. Trebuil, G. (1984).
43. Taylor, K.W. (1992).
44. Hall, K.R. (1992).
45. Kato, K. (1998).
46. Yamchong, Cheah (1996).
47. Penth, H. (1994).
48. Van Beek, S. (1995).
49. Surareks, Vanpen (1998).
50. Cohen, P.T. (1980).
51. Attwater, R. (1998).
52. Surareks, Vanpen (1998).
53. Kunstadter, P. *et al.* (1978).
54. Van Liere, W.J. (1989).
55. Wyatt, D.K. (1984).
56. Wongthes, E. and Wongthes, S. (1989).
57. Hung, Ling-Chi (1793).
58. Gutkind, E.A. (1946).
59. Attwater, R. (1998).
60. Taylor, K.W. (1992).
61. Rogers, P. (1996).
62. Dumarcay, J. and Smithies, M. (1995).
63. Siribhadra, S. (1999).
64. van Beek, S. (1995).
65. Arbhabhirama, A. *et al.* (1987).
66. Donner, W. (1978).

67. Ishii, Y. (1978).
68. Vallibhotama, S. (1989).
69. Taylor, K.V. (1992).
70. Thompson, V. (1967).
71. Steinberg, D.J. (1987).
72. Lamb, H.H. (1977).
73. Croll, E. and Parkin, D. (1992).
74. Tanabe, S. (1994).
75. Phongpaichit, Pasuk and Baker, C. (1998).
76. Ingram, J.C. (1971).
77. Sukwong, Somsak (1989).
78. Phongpaichit, Pasuk and Baker, C. (1998).
79. Ibrahim, M. (1972).
80. Mouhet, M.H. (1864).
81. Lourido, R.A. (1996).
82. Yule and Cordier (1903).
83. Major, R.H. (1957).
84. de Campos, J. (1940).
85. Hutchinson, E.W. (1940).
86. Anon (1884).
87. Phongpaichit, Pasuk and Baker, C. (1998).
88. van Beek, S. (1995).
89. van Beek, S. (1995).
90. Brown, I. (1988).
91. Owen, N.G. (1971).
92. Phongpaichit, Pasuk and Baker, C. (1998).
93. Haanant, Juanjai *et al.* (1987).
94. Office of Agricultural Economics (1998).
95. Donner, W. (1978).
96. Judd, L. (1964).
97. De Datta, S.K. (1975).
98. Chomcalow, N. (1993).
99. Kato, K. (1998).
100. Watabe, T. (1978).
101. Chandraratna, M.F. (1964).
102. Chang, T.T. and Bardenas, E. (1965).
103. Watabe, T. (1967).
104. Watabe, T. (1967).
105. Golomb, L. (1972).
106. Tanabe, S. (1994).
107. Tanaka, A. *et al.* (1966).
108. IRRI, (1995).
109. Tanabe, S. (1978).
110. Kulthong Kham, S. *et al.* (1964).
111. Donner, W. (1978).
112. Steinberg, T.J. (1987).

113. Setboonsarng and Evenson (1991).
114. Hanks, L.M. (1972).
115. Hanks, L.M. (1972).
116. Centro Escolar University (1999) Beyond Rice.
117. Grist (1975).
118. Adapted from the Rice Growers 'Detailed Historic Timeline of the Australian Rice Industry'.
119. Rerkasem, Benjawan (1999).

8 | History of Rice in South Asia (Up to 1947)

Mofarahus Sattar, S.D. Sharma and *Anil K. Pokharia*

Up to 1947, India, Pakistan and Bangladesh formed a single country that was known as India. The history of these three countries is, therefore, inseparable up to 1947. In 1947, India was partitioned into two countries, namely, India (with its present boundaries) and Pakistan. In 1971, Pakistan was further split into two countries, namely, Pakistan (with its present-day limits) and Bangladesh. Sri Lanka, however, was a separate country all along its history except for a short period when a part of this island was under the rulers of South India. It became an independent country in 1948. This chapter discusses the antiquity and history of rice in India (including Pakistan and Bangladesh) and Sri Lanka up to 1947 only.

In the north of the sub-continent, the Himalaya mountain ranges run from east to west. In the west, Karakoram and Hindukush mountain ranges run from northeast to southwest but there are passes in between through which people have come into this sub-continent in the past.

The northern part of this sub-continent has three river valleys: Indus River valley in the west (Pakistan), Ganga River valley in the centre and Brahmaputra River in the east. Bangladesh lies in the delta region of Ganga and Brahmaputra rivers. The Vindhyan mountain ranges and their forests in the central India were a barrier for social and cultural exchanges and for political integration in the pre-historic times. South of Vindhyas is a plateau known as Deccan plateau except the coastal belts. Rivers in the Deccan plateau flow from east to west. The coastal belts, the river valleys and their deltas provide the ideal settings for rice cultivation.

The languages of the four southern states of the present-day India belong to a group (Dravidian) that is different from the group (Indo-Aryan) of the north Indian languages. It is generally believed that the people of south India came to this sub-continent much before the Aryans and settled in this sub-continent. The Aryans came about four thousand years ago from Central Asia. Greeks, Sakas, Arabs, Turks and Persians have come in subsequent periods and have settled here. They all came from the central Asia from Afghanistan side. The last to come were the British. They came by the sea route from southern side with no intention to settle; their purpose was to exploit the sub-continent economically.

The northwest region of the sub-continent has been traditionally a barley and wheat growing area only. Rice has been grown in this region only after irrigation was developed. The eastern part of the country, the coastal belts and the river valleys in the south have been traditionally rice growing areas. India has been considered as one of the centers of origin and genetic diversity of the cultivated rice. Rice has been grown in this region since the pre-historic times as the archaeological findings indicate.

ARCHAEOLOGY

The region of Middle Ganga Plain has been considered as a part of early hearths of rice domestication and cultivation.[1] However, the antiquity of rice domestication in the subcontinent has long been a subject of debate in archaeology due to problems in chronology and lack of thorough archaeo-botanical sampling and analysis.[2]

Lahuradewa, an early lake-side Neolithic settlement of architecturally well expressed village-farming community in Sant Kabir Nagar district of Uttar Pradesh (India) has provided new evidences of rice domestication in the Middle Ganga Plain.[3] In the very beginning of Neolithic phase at Lahuradewa, the presence of domesticated rice (*Oryza sativa*) has been recorded in the two deepest layers of two different trenches (Fig. 8.1), in the mixture of the remains of wild or weedy form of rice (*Oryza rufipogon*) and broken pieces of the foxtail millet (*Setaria* sp.). The powdery charcoal associated with these remains in the bottom layers of both the trenches dates to 6th-5th millennia BC. The AMS date (6409 BC) of husk-piece of domesticated rice carried out at Physikalisches Institüt der Universität Erlangen, Nürnberg, Germany has been quite surprising.

Further, the presence of wild-rice phytoliths since 10,000 yr BP and cultivated rice phytoliths from 8300 yr BP onwards in Lahuradewa lake deposits supports beginning of agriculture during later half of 7th millennium BC.[4] In their preliminary investigation of soil samples from lake

1 mm

Fig. 8.1. Grains of domesticated rice excavated from the
Neolithic phase of Lahuradewa.

sediments, Chauhan and his associates[5] have reported Cerealia and other
culture pollen taxa from the level dated to 7500 yr BP and inferred early
activities of man associated with some sort of cereal-based agriculture.

The extensive excavations at Chopani Mando, Koldihwa and Mahagara
(all in the Middle Ganga valley) revolutionized our understanding regarding the sequence of transition from the stage of food-gathering and
selective hunting (in Epi-palaeolithic period) to settled farming in the
Neolithic times.[6] Based on radiocarbon dates[7], the evidence of Neolithic
culture at Koldihwa and Mahagara was regarded to date back during 7th-
6th millennia BC. Pottery from Koldihwa and Mahagara was reported to
contain copious impressions of cultivated rice (*O. sativa*), along with those
of annual and perennial forms of wild rice, viz. *O. nivara* and *O. rufipogon*.
Grains of *O. sativa* were also recovered from Neolithic Mahagara.[8] Rice of
wild variety was used in this region even in the advanced Mesolithic or
Proto-Neolithic cultural phase as evident from Mahagara.

Saraswat recorded the grains of *O. sativa* along with wild perennial *O.*

rufipogon and *Setaria* cf. *glauca* from a pit sealed by a layer of Iron Age deposit.[9] The carbonized material dates around 4620 BC i.e. about 3000 yr earlier to the beginning phase of the Iron-Age occupation at the site. The single early date from the deposit in the Iron-Age pit merely suggests some place near the settlement in the past time to which the range of some early human activity is expected to lie. Some earlier deposit containing the organic material, obviously, would have been in the close catchments area and accidentally dumped in the pit by the settlers, during the Iron Age. This is the only possibility, which would have moved earlier mud with organic refuse into the much late habitation, to have given such an early date. The probability of some early farming culture and the rice cultivation somewhere in the region seems quite high and indisputable.

Fujiwara discovered phytolith derived from leaf-blade of rice plant in the excavated soil samples at Pirak site, Pakistan (*ca.* 3800 yr BP) suggesting that people in the Indus region had practiced rice cultivation at that time.[10]

Equally important is the cultivation of rice during third-second millennium BC in the wheat and barley growing zones of northwestern and peninsular India during Harappan Culture.[11] It may only have been possible when its potential as a cultigen in Ganga Plains would have been realized at much earlier dates. Copious husk impressions in mudclods and pot-sherds from ancient sites in this region could be the result of tempering and indicate that rice could have been established as a staple crop at these settlements and, in all likelihood, in the surrounding regions as a whole.

The published data indicates negligible presence of rice on the Indian peninsula during second millennium BC but widespread cultivation during first millennium BC in at least some areas.[12]

PREHISTORIC PERIOD

Indus Valley Civilization

The Indus Valley Civilization (also known as Harappan civilization) was spread over a wide area in Pakistan and in the border areas of India. The mature period of this civilization lasted from *ca.* 2600 BC to *ca.* 1900 BC. The Harappan civilization was mainly city-based and it had its trade links with Mesopotamia. The main excavated cities of this civilization are Harappa, Mohenjo Daro (on Pakistan side), Dholavira, Rojdi, Rangpura, Surkotada, Kanmer and Lothal (on Indian side). Each of them would have controlled a large hinterland with regard to their agricultural produce, natural resources and crafts. The main crop seems to be barley and wheat; they might have had rice but this was not their staple food. According to C. Gates[13]

> The well-being of Harappan cities depended on successful agriculture . . . . Agriculture prospered through the natural flooding of the river. Evidence is lacking since alluvial deposits have covered any traces of irrigation systems. Among the crops grown wheat, barley and sorghum were staple cereals. Rice may have been grown in the southeast sector on the Indian coast. Mohenjo Daro and Lothal have yielded cotton cloth.

As the archaeological findings indicate, rice cultivation was already being practised in the middle Ganga valley from the very early stages of Neolithic culture when there is no evidence of their contact with the Harappan civilization. As Peter Bellwood puts it,

> In my opinion, the best explanation for the spread of agriculture through the Ganga Basin is that a cultural assemblage with cord-marked pottery and rice cultivation was moving upstream from the east at about the same time as a separate complex with southwest Asian crops and OCP pottery was moving downstream from the west that is at about 3000 BC. The resulting fusion appears to have been rapid and without undue social upheaval.[14]

In the early second millennium BC, the Indus Valley Civilization disappeared without any trace of its continuity.

Vedic Period

The next people, who called themselves Aryans, came from the central Asia through Afghanistan. They were basically pastoral people and their main wealth was cattle. These Aryans neither inherited the culture of the Harappan people nor carried it forward. The earliest literary works available in India are the Vedas (Books of Knowledge) composed by the Aryans. These were composed during 1700 BC to 900 BC and contain prayers to gods, rituals prescribed for the betterment of life, some philosophical discussions, etc. There are four *Vedas: Rig, Yajur, Sama* and *Atharva*. Of these, the earliest one is *Rig Veda*. The geographical limit of the area described in the *Rig Veda* is *sapta-sindhu* i.e. the land of the seven rivers meaning the Indus and its tributaries. Cattle were the most valuable property for the Aryans and as John Keay has put it,

No doubt they then also planted their grain crop which, watered by the rains and fertilized by the manure from their cattle pens, would have been harvested during the winter months. The grain was probably barley. Rice, although apparently cultivated by the Harappans, does not feature in the earliest of Vedas. Nor is the word used to designate it Sanskritic. It, too, was probably acquired from one of India's aboriginal peoples. Later, however, after the arya had adopted a settled life, rice receives its first mention, and later still, following their colonization of the middle Ganga in the early centuries of the first millennium BC, the cultivation of irrigated padi would become crucial to their pattern of settlement.[15]

By the later Vedic period, the Aryans had moved to the Indo-Gangetic divide and the land between the rivers Ganga and Jamuna. In the later Vedic literature, rice (vrihi) finds frequent mention as an item used in their rituals.[16] For example, Taittiriya Samhita mentions five types of rice (Ashudhanya, Hayana, Krishnavrihi, Mahavrihi, Shuklavrihi) and wild rice (nivara). Rice was eaten either by simply boiling it with water (odana) or by boiling it with milk (krishara). Sometimes, it was cooked with sesame or mungbean or meat. It was also mixed with curd or ghee (butter oil) before eating. It was also popped (laja) or parched (parivapa).

Aryans moved to the divide between Indus basin and Ganga Basin around 1100 BC. Iron was used for the first time around 1000 BC and its use, especially in weaponry, was known around 800 BC.[17] Around 950 BC, there was a major battle between two groups of them as described in the book Mahabharata. Even during this period, cattle were valuable property for them and fire was still used to clear the forests as Mahabharata narrates that Krishna's childhood was passed as a cowherd boy and the Kauravas raided the capital of King Virat to take away his cattle. Similarly, to build their new capital, the Pandavas set fire to the Khandava forest. Stone seems to have been used for making arrowheads for quite some time as the use of word shilimukh (stone-headed) for arrows indicates.

The period between 800 BC and 500 BC is the sutra period because the books of this period were composed in the form of aphorism (sutra); these were supposed to be memorized and recited instead of being written and read. Many such books dealt with rituals and household duties. Rice has been prescribed in many of these rituals. For example:

The Srauta Sutras refer to several sacrifices for fulfillment of specific purposes, viz. (i) for success in learning and recognition, rice cooked in milk was offered to Soma-Rudra (Katyayana Srauta Sutra 15.3.22, Manav Srauta Sutra 5.1.10) or nivara, wild rice grains were offered to Marutvant Indra; (ii) to regain one's kingdom from the conqueror, white grained rice cooked in clarified butter was offered to Adityas (*Manav Srauta Sutra*) but black grained rice was offered to Varuna (*Baudhayan Srauta Sutra*); . . . .[18]

PROTOHISTORIC PERIOD

By 6th century BC, Aryan settlements had moved up to the middle Ganga valley and had formed large settlements called *mahajanapadas*. This period saw the urbanization of India, first use of money and organized trade that can happen only if there is agricultural surplus. With the shift of the centre of Indian civilization to the middle Gangetic valley that had warmer climate, higher rainfall and more fertile land; rice was now a more productive cereal than barley that had been the staple food of Aryans when they were in the Indus basin or in the upper Gangetic valley. Iron was no more a new metal and was available in this region (Bihar) for making axes (to clear the forests and acquire land for agriculture) or to prepare plough shares (to cultivate the land).

The oldest *mahajanapadas* that came into prominence were Kosala, Kasi and Videha. Regarding Kosala, Burton Stein writes:

> The region around its capital at Ayodhya was able to sustain a rich agriculture owing to the plenitude of its irrigation. Reliable crops of barley and rice were cultivated with ploughs using iron shares to increase the depth of tillage. Conflicts are recorded among local groups and settlements over the control of watercourses and the labor required for the more intensive cultivation of wet rice and for the construction and maintenance of dams needed to regulate water use and frequent floods.[19]

Kapilavastu was another *janapada* in the foothills of the Himalayas where Siddhartha Gautama (the Buddha) was born. His father's name was Shuddhodana which means "pure rice." When Buddha was fasting in the forest and had become too weak, two merchants of Kalinga (modern Orissa) traveling through that forest offered him rice cakes. When he was sitting under the Bodhi tree where he got enlightenment, a thirteen year old girl named Sujata offered him rice cooked in milk.[20]

The *Jatakas* (stories regarding the Former Births of Lord Buddha) were written around 300 BC) and contain references to hill rice, red husked rice (*ratta salivam*), dehusked rice (*tandula*), scented rices, liquor from rice and rice having medicinal properties and tanks for irrigation.[21]

The use of iron for tools is widely attested in the early Buddhist literature. We hear of *ayokapala* (iron pot), *ayokuta* (iron hammer), *ayokhila* (iron stake), *ayoghana* or *ayomuggara* (iron club), *ayosanku* (iron spike) and *ayonangala* (iron plough). Axe or *pharasu* is mentioned but apparently not specifically iron axe. The iron ploughshare (*phala*) is vividly described at several places as heated during the day and then sizzling and steaming when plunged in water. Perhaps the word *phala* not merely meant the ploughshare but any piercing or cutting tip.[22]

Charaka Samhita (Charaka's *Compendium*), was written in around 7th century BC. It is the earliest text regarding medical science and lists a number of rice varieties with their medicinal properties. Another treatise on medical science, *Sushruta Samhita* (Sushruta's *Compendium*), composed around 400 BC lists many more rice varieties and provides their medicinal properties. Both Charaka and Sushruta treated rice in detail, giving names of varieties and their effects on human physiology.

Charaka classified rice into three main groups: Shali, Shashtika and Vrihi. Some of the rice varieties mentioned by Charaka in his Shali group are Raktashali, Mahashali, Kalama, Shakunahrita, Turnaka, Deerghashuka, Panduka, Langula, Sugandhika, Lohawala, Shariva, Pramodaka, Patanga, Yavaka, Hayana, Vapya and Naishadhaka. The Shashtika varieties were so-named because they matured in 60 days. Vrihi varieties matured in four months, they were either red or white grained. Sushruta mentions some more Shali varieties of rice such as Lohitaka, Kardamaka, Pundarika, Pushpandaka, Shitabhiruka, Rodhrapushpaka, Kanchanaka, Mahishamastaka, Hayanaka, Dushika and Mahadushika. He lists Pramodaka, Kedaraka, Pitaka and Mahashashtika as varieties in his Shashtika group and Kishnavrihi, Jatumukha, Nandimukha, Lavakshaka, Tvaritaka, Kukkutandaka, Parvataka and Patala in Vrihi group. He mentions Naivaram (wild rices) as another group of rice.[23]

Panini was a grammarian who lived in northwest India (now Pakistan) in the 4th century BC and composed the book on Sanskrit grammar called *Ashtadhyayi* (a book in eight chapters). The book mentions rice frequently in its aphorisms (*sutras*).

The cultivation of rice required water-immersed fields which were called *kedara* and Panini has a special *sutra* to signify the collection of *kedaras*. Several varieties of rice are mentioned including the famous variety ripening in sixty days. Although Panini belonged to north-west, it is obvious that the expansion of the settlements towards the east in the later Vedic age brought in general changes in the cultivation of crops. The cultivation of sugar-cane was important from a commercial point of view.[24]

Panini mentions cakes of rice (*anupa*), boiled rice (*bhakta*), cakes made of powdered rice (*pistaka*), etc. He names the style of joining two words when it means a third thing as *bahuvrihi* style. The word *bahuvrihi* means neither *bahu* (a lot of) nor *vrihi* (rice) but a field which produces a lot of rice.[25]

Between 400-100 BC *Krishi Parashara* (a treatise on agriculture) was written by Parashara. His treatise mentions (i) the process of selecting phenotypically similar seed, which in turn must have led to establishing better varieties of rice; (ii) emphasis on seed purity; (iii) classification of varieties into two groups, viz., direct sowing types and transplanted types; (iv) ensuring availability of water during the growing season; and (v) identification of pests.[26]

ANCIENT PERIOD

North India

In 327 BC Alexander the Great invaded India (now Pakistan) and India came into full light of the recorded history. The Greeks came to know about rice and rice cultivation and introduced it in West Asia and Egypt. The Mauryan Empire was established by Chandragupta Maurya in 322 BC. Kautilya, his political mentor, has written *Arthashastra* (a treatise on financial management) that offers detailed advice on agriculture in general and rice cultivation in particular. For example, he advises that *shali* (late maturing, transplanted?) and *vrihi* (early maturing, broadcast?) rice should be sown with the commencement of rain. According to Kautilya, the standard food of a gentleman (*arya*) was rice, broth, some clarified butter or oil and some salt. Megasthenes, the Greek ambassador of Seleucus Nicator was in Pataliputra (the capital of Chandragupta Maurya) and, according to him, the food of Indians was principally a rice pottage.[27]

Chandragupta's grandson, Ashoka, became a great Emperor. At that time, Kalinga (modern Orissa) was a rich country due to rice cultivation in the delta of River Mahanadi and its trade with Sri Lanka and Southeast Asian countries. Kalinga was, however, not under Ashoka's empire when he took over the reign. Ashoka invaded Kalinga and occupied Kalinga after a bloody war. Ashoka was pained due to this bloodshed and converted himself to Buddhism. He propagated the religion of non-violence and compassion and sent his own son and daughter to Sri Lanka to preach this religion. Ashoka had sent 160 loads of hill paddy to the king of Sri Lanka as the local production there was not sufficient.

The Gupta period (4th and 5th century) is considered the golden age in Indian history. When Guptas established an empire, peace and prosperity returned. With increasing security of life and property and introduction of a more extensive and intensive cultivation, population multiplied. This meant greater dependence on land. Providing irrigation water was mostly the responsibility of the State. Irrigation wells were also dug. Though it is likely that the villages were left to themselves as self-sufficient economic units, the prosperity of the country depended on industrial progress. According to the Chinese pilgrim Fa-hsien [Fa Hian] who visited India *ca.* 400-410, 'the people were very well-off.' After the death of Skanda Gupta in *ca.* 467, power of Gupta dynasty declined due to several reasons: the Huns attacked India from the northwestern side; they blocked the trade routes of China and demolished the Roman Empire. The industry and international trade of India was greatly affected. The south Indian trade with Southeast Asian countries however was not much affected.[28]

In the 7th century AD, Harsha became a powerful ruler of north India. During his reign, Xuanzang [Huen Tsang], a Chinese Buddhist monk visited India for pilgrimage and in search of sacred texts. He traveled widely in India and has left a vivid description of what he saw in India. He was an honoured guest at the Nalanda University between 631 and 635 AD. According to Mishi Saran,

> Each day he received 120 *jambiras*, a kind of fruit, twenty areca nuts, twenty nutmegs, a *tael* of camphor, butter and other necessities, and a peck of *Mahasali* rice, a special fat grain that cooked into a shining aromatic heap and was only available in Magadha and only offered to kings or distinguished priests. Every month he was given three measures of oil. He was also gifted an elephant to ride on and two attendants were assigned to take care of his every need.[29]

Harsha died in 648 AD. There was no great ruler in north India during the next five hundred years. India was divided into many kingdoms but on the whole there was peace and prosperity in the country as reflected in the construction of great temples throughout the country and flourishing art and literature. The kings used to donate lands and even the entire villages to the temples, to their priests and learned scholars without charging any revenue. It implied that the revenue of these villages would go to those temples for their maintenance, or to priests or scholars to carry out their professions unhindered by pecuniary worries. These grants were generally engraved in copper plates with royal seal and were held by the donees and their descendents until they have been acquired in modern times by the archaeological department. Many of the copper plates, particularly those that belong to eastern and southern parts of the country, mention rice fields or the amount of rice the fields produced. Some inscriptions were engraved on the stones of the walls of the temples.

Kashyapa circa 800 AD wrote *Kashyapiya Krishisukti* (recommendations on agriculture by Kashyapa) wherein he gives a detailed account of agriculture. It was written for areas which had good rainfall and hence has detailed recommendations for rice cultivation. He divides the land into two types, one that is low-lying and hence suitable for irrigation and suitable for rice cultivation and the other that is upland and hence suitable for growing other crops. He classifies the rice varieties into four groups and mentions wild rice as the fifth group; he lists 4 varieties under *shali* group, 5 under *kalama* group, 4 under *sambaka* group, 11 under *vrihi* group, and 2 within *nivara* group.

It appears that wild rice was available in abundance in its natural habitats and was often harvested and consumed. For example, Ashvaghosha (2nd century) mentions in his epic *Saundarananda* that wild rice (*nivara*) was consumed by the ascetics. Kalidasa (4th century) again makes such statements in his drama *Abhigyan-shakuntalam*. Banabhatta (7th century) mentions in his novel *Kadambari* that birds used to eat so much wild rice that its husk when dropped on the ground under a tree acted as a cushion for a young bird that fell from its nest.

South India

In central India, the Vindhyan mountain ranges run east to west. During pre-historic period these were covered with dense forest, hence movement across this forest was difficult. Only when the population had grown to a sizeable number and iron tools (axes for clearing the forest, ploughshares to cultivate the land, sickles to harvest the grains) became common, Vindhyas became accessible and north-south movement was possible.

Starting somewhere around 1000 BC, the Aryans started moving south of the Vindhyas 'more or less steadily and peacefully' and the process reached its southern extreme before the extension of the Mauryan Empire to that region.[30]

There are stray references to south India in literature prior to Kautilya (4th century BC). Kautilya, the political advisor to the Mauryan Emperor, knew south India and referred to the region as Dakshinapath (the Southern Route). According to him, south India was economically of greater value than the side that was in the north of his country. This may explain the extension of the Chandragupta Maurya's empire up to the present Karnataka in the south in the 4th century BC. The fact that Ashokan edicts are in Brahmi script and Prakrit language has been used in as south as Mysore (Karnataka State) implies that India was culturally unified as back as 3rd century BC. Ashoka mentions Cholas, Pandyas Cheras and Keralaputras in his edict as kingdoms in the south of his Empire and he professes to maintain friendly relations with them.

After the death of Ashoka in *ca.* 232 BC, his empire gradually disintegrated; the first ones to become independent were the distant provinces. The Kalingas in the Mahanadi delta became independent under the rule of Kharavela. Satavahanas became independent rulers in the Godavari and Krishna delta in the 1st century AD; their rule extended from the east coast to west coast in the northern part of the peninsular India. They ruled the area up to 3rd century AD. The valleys and deltas of these rivers had been and still are renowned for their rice production. The capital of Satavahanas was in the Krishna valley at Dhanyakataka (*dhanya* = rice, *kataka* = fort, a central place).[31] This place (later called Dharanikonda) became an important centre for trade by land and sea. The Satavahanas were having trade relations with the Romans and also with the Southeast Asia as attested by archaeological findings and also Roman records. On the west coast of India, their trading post was Sopara near Mumbai.

In far south (now known as Tamil Nadu and Kerala states), it appears that the Cholas had occupied the Kaveri Delta since pre-historic times. In the 2nd century BC, there was the demographic shift from the upland settlements to the alluvial soils of the deltas with construction of irrigation works and wet rice cultivation. Rice was a productive crop in the Kaveri delta.[32] In the 1st century AD, Egypt came under Roman occupation and, from Egypt the Romans had direct access to India as well as Rome by sea route. The Romans, who had by then become familiar with the monsoon trade winds, could sail direct from the Red Sea ports to the west coast of India. Romans were visiting the south Indian ports in the 1st century AD for Indian pearls, spices and ivory and paid for these items in Roman gold coins. Around the same time, south India had trade links with Southeast Asia as well. One may assume that they also carried rice from India to Egypt and Rome. The trade between India and Rome continued during this period for about a hundred years.[33]

The Sangam literature is an anthology of ancient Tamil poems composed by an academy of poets during the 1st century BC to 2nd century AD.

The poems reflect the transition from tribal society to settled agriculture and early state formation. It mentions five kinds of land (*tinai*): mountains, forests and pastures, dry barren lands, the valley of the great rivers and the coast. The river valleys (*marutham*) were considered the best lands for rice cultivation with natural or artificial irrigation. Rice was already a well-established crop in the Kaveri basin by the 2nd century AD. Murugan who was essentially the Tamil god for war and fertility was worshipped with the offerings of rice and blood. Later, his identity was merged with that of Kartikeya, a god of Aryan mythology and blood was substituted with red rice as offering. During this period,

> The land was fertile and there was plenty of grain, meat and fish; the Chera country [*Kerala*] was noted for its buffaloes, jackfruit, pepper and turmeric. In the Chola country [*Tamil Nadu*] watered by the Kaveri, it was said that the space in which an elephant can lie down produced enough to feed seven and a veli of land yielded around thousand *kalams* of paddy. The little principality of Pari abounded in forest produce like 'bamboo-rice', jackfruit, valli-root and honey. Many rural activities like the cultivation of *ragi* [*Eleusine coracana*] and sugarcane, the making of sugar from the cane and the harvesting and drying of grain are described in the Sangam poems in vivid and realistic manner.[34]

The *Shilappadikaram*, a romantic novel written in the ancient Tamil language either at the end of the 2nd century or in the beginning of the 3rd century AD mentions rice frequently in various contexts. From the book, one gets the impression that rice cultivation in the Kaveri river valley at that time was not very much different from what it is now:

> Passing through a fertile country, they came to a pillared hall surrounded by trees and gardens. In all directions, they could see rich rice fields scattered between shining sheets of water. And all around them tender crops were undulating in the breeze.[35]

The *Manimekhalai* written in the 2nd/3rd century AD continues the story of *Shilappadikaram* further. It mentions that rice and wild rice were sold in the bazaar,

> In the alleys of the bazaar, eight varieties of grains were sold: rice, grass seed (wheat?), four kinds of millet and two sorts of wild rice.[36]

With the decline of Roman power since 170 AD, the trade between India and Rome declined and the Indians focused their attention on Southeast Asian countries as is evident from the archaeological finds in Myanmar, Thailand, Vietnam and Indonesia.

From middle of 6th century, for the next 300 years, south India was occupied by three contenders: Pallavas of Kanchi, Pandyas of Madurai and Chalukyas of Badami, but the life of the peasants was not very much affected.

MEDIEVAL PERIOD

The medieval period of Indian history starts with the arrival of Muslims into India in 1192 when Muhammad Ghori from Ghazni, Afghanistan conquered Delhi and ends with British colonization in late 18th century. Compared to earlier periods, more information is available for the medieval period of India but the available records pertain mostly to the royal dynasties, their political adventures, their life-styles, travelogues of the foreign visitors and religious writings. These records do not provide much information about the condition of agriculture or that of the farmers of the period. Information on rice cultivation continues to be scanty. We can, however, discuss the factors that affected or influenced the farming community and agricultural production of the period. We will also glean whatever little information is available on rice cultivation, production and marketing.

North India

The Muslim rule quickly expanded its hegemony over most of northern India under the Ghorids extending its sway up to Bengal in the east by 1206 while much of south India remained under the then existing local rulers. Successive Muslim dynasties, known as the Delhi Sultanate (Mamluks: 1206-90, Khiljis: 1290-1320, Tughlaqs: 1320-1430, Sayyids: 1414-51, and Lodhis: 1451-1526) consolidated their power in the north and gradually extended their hegemony to the south. These rulers not only brought a new religion into the sub-continent but also caused movement of people and goods from the west. Warriors, administrators, intellectuals, artisans and *sufis* came from the west of Indus into the subcontinent and received patronage under the new rulers. These new rulers replaced the old guards in the administrative and military hierarchy with people who followed their own faith and were loyal to them. The agrarian community, however, continued as before and followed the traditional agricultural practices as usual.

Delhi Sultanate (1206-1526)

Vast tracts of land that were under forest cover in the Gangetic plains were brought under cultivation in 14th and 15th centuries. Ziauddin Barani[37] (14th century) mentions about forests in the middle Doab, the tract between Jamuna and Ganga. By 16th century, the Doab was almost fully under cultivation.[38] Ibn Battuta (1304-1378), the famous Moroccan traveler, and Barani have mentioned that Muhammad Tughlaq (1325-51) gave loans to dig wells for expanding agriculture. Raychaudhury and Habib[39] state that Ghiyasuddin Tughlaq (1320-25) was the first ruler to dig canals for promoting agriculture and Firuz Tughlaq (1351-86) dug many canals that remained the biggest network until 19th century. This greatly enhanced cultivation of spring (*rabi*) crops, especially wheat, in addition to the autumn (*kharif*) crops, a major one being rice. Raychaudhury and Habib also cite Thakkura Pheru of Delhi who wrote in 1290 about cultivation of 25 different crops which in addition to rice and wheat included barley, gram, lentil, mung, *jowar* (sorghum), *bajra* (pearl millet), cotton, sugarcane, mash, moth, mustard, linseed, sesame, etc.

Ibn Battuta also mentions that rice cooked with pulses and *ghee* was eaten by Muhammad Tughlaq as his breakfast food. Rich people of Egypt, who could afford, imported rice from India. He was greeted by the sultan of Mogadishu with betel nut and areca nut, an Indian tradition. Rice was cooked with *ghee* (clarified butter) in Mogadishu and was eaten in Indian style. Aden was an Indian emporium where Indian and Arab ships anchored and made voyages on the Arabian sea and visited ports on the west coast of India, east coast of Africa and those of the Red Sea. Rice was also exported to Southeast Asia. He also mentions that India suffered from a famine during 1334 to 41.[40]

The agrarian administration and land tax system under the Muslim rulers of Delhi experienced several shifts and turns, although essentially the idea remained central to derive income for the empire from growth of agriculture and its surpluses. Alauddin Khilji (1296-1316) decreed three taxes, viz., *kharaj* or tax on cultivation, *charai*—tax on milch cattle, and *ghari*—tax on house. *Kharaj* was imposed at half the production in a very strict and harsh manner. Muhammad Tughlaq's overvaluation of *kharaj* for yield and price of the produce affected the peasantry, which caused revolt or abandonment of cultivation and ultimately started the famine in 1334-5 that lasted for seven years.[41] Although in later years the Sultan took measures to promote agriculture and advanced between seven and twenty million *tankas* as *sondhar* (loans), the cultivation did not extend even a hundredth of the pledged coverage as opportunists mostly availed of those loans, and taking advantage of the Sultan's long absence away from the capital at the battlefront, did almost nothing. His successor Firuz Tughlaq (1351-86) eased the tax burden, forbade *ghari* and *charai*, dug many canals as mentioned earlier and consequently brought about new levels of production and prosperity. The system of tax collection was either through *khots* or *muqaddams* (headmen) or appointed *shiqdars* and *faujdars* (revenue collectors and commanders). There were also other feudatory systems of revenue collection through *jagirdars, zamindars* or *chaudhuris* as hereditary rural landlords and aristocrats in possession of certain tracts of land who exercised their power over the peasants or *ryots*.

Mughal Period (1526-1764)

The year 1526 AD is a landmark in the history of India as the Mughals established their rule when Babur defeated Ibrahim Lodhi. However, between 1540 and 1555 the Afghans briefly displaced the Mughals when Sher Shah Suri drove Humayun out of India.

Sher Shah Suri ruled only for five years, during 1540-45, but much of the civil and administrative reforms, public service and institutions which are credited with Akbar or the Mughals actually began under him. Sher Shah Suri reorganized the administration, the army, system of tax collections and built roads (the most famous one being the Grand Trunk Road from Bengal to Punjab) and travelers' inns (*sarais*), rest houses and wells, improved the judiciary, founded refuges and hospitals, introduced free and/or gruel kitchens, and organized a mail services and the police. Of special mention for our purpose are the land revenue reforms, which were wisely devised following humane principles and served as the model for future agrarian systems. After a careful and proper survey of lands, he settled the land revenue directly for the cultivators, and fixed at one-fourth or one-third of the average produce, payable either in kind or in cash, the latter method being preferred. [42] As Mughals returned, they took the advantage to build on his measures.

Humayun returned with the help of the Safavid king Shah Abbas of Persia and a new era started in India as Persian culture and language slowly displaced that of hitherto practiced Turkish from central Asia. The connections enhanced overland trade routes through Mughal, Safavid and Ottoman empires in the sixteenth and 17th centuries as never before, and significantly influenced agrarian situation in India. These empires depended for their wealth on networks of trade that linked them to one another by land and sea. Across southern Eurasia, the net flow of manufactured goods and spices moved from east to west and the net flow of precious metals moved in the opposite direction. Inland trade with the west justified great Mughal expanse and to keep the mountain passes to Kabul open for safe travel.

The most famous treatise of the medieval period *Ain-i-Akbari* was written by Abul Fazl Allami around 1590 AD during Akbar's rule (1554-1605). It is an excellent document about Akbar's administration, particularly regarding Akbar's revenue collection. The following paragraph gives an idea of how rice cultivars were obtained for the "imperial" kitchen during Akbar's period. ". . . . at the beginning of every quarter, the Diwan-i-buyutat and the Mir Bakawal collect whatever they think will be necessary e.g. Sukhdas rice from Bharaij (present-day Bahraich in Uttar Pradesh), Dewzira rice from Gwaliar (present day Gwalior in Madhya Pradesh), Jinjin rice from Rajori (present day Rajauri in Kashmir?) and Nimlah . . . ." He provides the prices of these varieties in dam (one-fortieth of a rupee) for a maund (37 kg) of unhusked rice: Mushkin (180), Sada (160), Sukhdas (100), Dunaparsad (90), Samjira (90), Shakarchini (90), Dewzira (90), Jinjin (80), Dakah (50), Zirhi (40) and Sathi (20). It appears that the strains like Devzira, Mushkin, Samjira Shakarchini, Sukhdas, etc. were three times costlier than the common varieties like Zirhi, Sathi and Sada. Some of these are still grown in different parts of India. For example, Mushkin is described as a small-grained white rice with fragrance and "pleasant to taste", perhaps with musky scent. Sada, Samjira, and Shakarchini are still grown in West Bengal, India. Sukhdas was cultivated in Oudh (Kannauj to Gorakhpur in Uttar Pradesh, India). This is a white-grained, delicate, and scented rice variety "scarcely to be matched". Dewzira could be Dahijira of West Bengal.[43]

Following the footsteps of Sher Shah Suri, Akbar's land revenue system evolved under his revenue and finance minister Raja Todarmal who with Diwan Shah Mansur divided the Mughal Empire into 12 provinces, each administered by a governor and a Diwan. In 1571, they introduced a rational revenue assessment based on intimately surveyed land holdings. Basically, the length of the period of cultivation was considered for land classification. The Mughals extended special concessions to the cultivators for bringing new areas under cultivation and were vigilant to expand the cultivated land; in no case the cultivated land was permitted to be degraded into barren (*banjar*) lands.

The land policy of Akbar was peasant-friendly. He introduced the quality assessment to ensure enhanced land revenue for his expanding empire. But he did not overburden the cultivators. The officers were given order (*farman*) to protect the crops and interest of the peasants. The peasant society in pre-British India had more autonomy and was adored with respect and admiration.[44] The Canal Act of Akbar (1568) detailed the Emperor's desire to "supply the wants of the poor" and to "establish the permanent marks of greatness" of his rule.[45]

Though India was politically unified by the Mughal dynasty in the 16th and 17th centuries, it could not create an integrated grain market for the country. Bullock carts were the mode of transport but there were not enough roads to transport grains over long distances. The Mughal army used to employ thousands of pack animals for transport of grains during its wars. Grains were, however, transported by boats on rivers and by ships overseas. So the grain markets over long distances were limited to areas where transport by river or sea was feasible. For example, Bengal rice was sent to north Indian markets by river (only Ganga and its major tributaries were navigable by boats) or to Sri Lanka by sea. From Gujarat, rice and other commodities were exported to ports of Persian Gulf and East Africa. Inside the country, grain rarely ever reached the next regional market even during famines. Thus the inland grain markets were localized.[46]

One of the earliest European travelers who mentioned rice was, perhaps, Jean Baptiste Tavernier, who described a rice field seen by him on a march south from Surat (Gujarat, India) in 1654. He specially described the musk-scented rices of Surat: "All the rice which grows in this country possesses a particular quality, causing it to be much esteemed. Its grain is half as small again as that of common rice, and when it is cooked, snow is not whiter than it is, besides which, it smells like musk and all the nobles of India eat no other. When you wish to make an acceptable present to anyone in Persia, you take him a sack of this rice." It is interesting to note that this popular rice cultivar was small-grained, white, and with a musky scent.[47]

During Mughal rule, monetization of land revenue was essential for public financing of the administration and army of the vast empire. Having no silver mines in India, the Mughal rulers had to depend on imported silver brought by European traders for minting coins. With Indian textiles gaining great demand in Europe, more and more silver came to India in its exchange, excess of which led to inflation. Coupled with the cost of their lavish expenditure (Shah Jahan built Taj Mahal, Aurangzeb's long-drawn wars in Deccan), the Mughals stepped up their land revenue demand.[48] Aurangzeb, the last great Mughal, died in 1707 and soon after his death the central authority of Delhi declined.

Once the central authority of great Mughals weakened, there was scramble for power among the regional rulers. In the process, there was no repair of whatever roads Mughals had built and maintained; the bridges, *caravanserais* and wells on the roads went neglected, there was no more a single currency for trade in the country and robbers made the roads unsafe for travel. The only transport possible was on pack animals which was economical only for high value merchandise. Transport of grains like rice over long distance was not feasible. As a result, the market of rice was highly localized even during famines. Transport was confined to navigable rivers only such as Ganga and its tributaries or near the seacoasts.

South India and Deccan

While the Muslim rulers of Delhi were quick to establish their rule over northern India in the 13th century, their southward expansion took few more centuries. During this inter-regnum, various regional powers ruled over different regions—Cholas in Tamil country (851-1279), Hoysalas in Mysore (950-1343), Paramaras in Malwa (800-1305), Kakatiyas of Warangal (Andhra Pradesh) (1083-1323), Devagiri Yadavas in Maharashtra (1175-1317) and Gangas and Suryavamsis in Orissa (1038-1568). During this period, agrarian territories were expanding all over Europe and Asia. In India, territories of agricultural expansion developed continuously in Punjab, Rajasthan, Malwa, Orissa and the Ganga plains. Networks of trade connected territories by land across the Silk Road and by sea across the Indian Ocean.

Cholas

In South India, Cholas' resurgence in circa 850 started one of the most illustrious empires of India. At the height of their power during the 10th and 11th centuries, Cholas expanded their empire in Sri Lanka and across the Indian Ocean into as far as Java (*Srivijaya*), probably the only Indian empire to expand overseas. The power of these rulers was partly based on the control of the fertile rice basin of the Kaveri delta but also on their intimate links with prosperous and influential merchant guilds which controlled long-distance trade. The Cholas had trade relations with China, Burma (Myanmar), Srivijaya and other Southeast Asian countries. King Rajaraja's embassy, which reached China in 1015 AD, advertised his realm as an important member of the new emporia network. The issue at stake was undoubtedly the access to the Chinese market which developed very fast under the Song dynasty.[49] Much of the systematic organisation of agrarian and other administration of the Tamil country were put in place by the Cholas, upon which succeeding rulers adopted modifications to bring in their advantage.

Peasants during Chola rule had customary rights to land but did not have exclusive rights and the lords were *Nattar*, who claimed the agricultural surplus as revenue. *Nattar* from village clusters organised themselves in collective bodies, known as *Nadus*, which acted as local level lordly organization for administrative, economic and political affairs.[50] Importantly, much of the revenue was spent by the *Nadus* locally for maintaining irrigation channels, tanks, new investments in cultivation and other repairs. They paid a part of their income as tax to the imperial Chola power. The Cholas expanded rice cultivation to new areas as the rich soil only needed irrigation, access of which was ensured through donation of new tracts as villages to brahmans (as *brahmadeyas*) and temples, who arranged for the necessary infrastructure. The creation of temple villages and *brahmadeyas* on one hand increased cultivation, on the other hand created demands for manufactured goods, such as textiles, luxury items, etc. This also necessitated concentration of artisans, masons, traders, builders and manufacturers to give rise to a number of *nagaram* or urban centres.

Hoysalas

Hoysalas who hailed from western hills, established their dynasty in 950 AD and controlled much of present day Karnataka at their height. As is the case with others, the Hoysala administration derived their revenues mainly from an agrarian economy. Paddy and corn were staple crops in the tropical plains (*bailnad*), while the highlands (*malnad* regions) with their temperate climate were suitable for raising cattle and the planting of orchards and spices. Irrigation tanks such as *Vishnusagara, Shantisagara* and *Ballalarayasagara* were built with expenses of the state while taxes were collected on irrigation systems including tanks, reservoirs with sluices, canals and wells, including the ones built and maintained at the expense of local villagers. Their rule ended in 1343 by the conquest of the Vijayanagara empire from the north.

Gangas of Orissa

In Orissa, the Gangas consolidated their power as sovereigns in 1038 AD. The most important economic aspect of the Gangas of Orissa was their agrarian structure. From the time of Chodaganga, the Ganga kings adopted significantly the policy of developing agrarian tracts and continued the process which contributed to stabilize their rule.[51] Rice was the principal produce from the land grants and various kinds of taxes, such as, *kshetra kar* (land tax), *utpanna kar* (product tax), *mamvatsy* (house tax) and some other occasional taxes were realized from the peasantry. From mid-14th century, Gangas gradually weakened due to invasions from Vijayanagara Empire in the south and Firuz Tughlaq from the north. Ultimately, Suryavamsis replaced them in 1440 who finally collapsed at the hands of Afghans from Bengal in 1568.

The Vijayanagara Empire

The Vijayanagara Empire was the most powerful and the largest empire in South India who conquered the whole area south of Tungabhadra River including the prized Krishna-Tungabhadra Doab. Successive kings extended their domination further north and east. As northern India was crumbling under the onslaught of Muslims, the Vijayanagara Empire stood strong as a Hindu kingdom and attained the highest level of prosperity in the contemporary period. The success of the empire was based on a well-organized land administration which favored farmers and tenants, great improvements in irrigation through ponds, canals and aqueducts, a standard justice system, promotion of overland inter-regional and overseas trade and ensuring harmonious and peaceful living for all communities and faiths. They established their capital at Vijayanagara near present day Hampi which was so beautiful that according to the travelogue of Abdul Razzaq from Central Asia *'The city is such that eye has not seen nor ear heard of any place resembling it upon earth'*. Similarly, Nicolo Conti from Italy visiting in 1420 and later, Domingo Paes, a visitor from Portugal left identical accounts comparing it to be larger than Rome and more prosperous.

The empire's economy was basically driven by agriculture and productive lands grew rice, wheat and sugarcane while drier areas grew corn, pulses and cotton. During their rule, water conservation gained importance and many large tanks were constructed. In the empire, majority of the people were involved in agriculture growing rice, ragi (*Eleusine coracana*), cotton, sugarcane, pulses, etc. and plantations growing coconut, areca nut and betel. Later, Portuguese introduced new crops such as groundnut, potatoes, chillies, tomato and tobacco into the sub-continent. Major exports were rice, jaggery, sugar, spices and textiles. Vijayanagara conducted overseas trade with China in the east to the Red Sea ports of Aden and to Mecca selling goods up to Venice in the west. Exports included both agricultural products including rice and manufactured items.

Bengal

It is now more than common knowledge that Bengal's prosperity in the medieval period was legendary. Most of the prosperity was gained through expansion of rice culture. Bengal produced so much of rice that during Aurangzeb's time eight maunds (296 kg) of rice were sold for a rupee. As his Governor Shaista Khan was leaving the capital Dhaka after completing his tenure, he famously closed its outer gate forbidding its opening unless rice could be sold again this cheap. From early medieval times, Bengal exported rice to Malayan states, Java, Maldives and Sri Lanka. Barbosa Duarte, a Portuguese trader mentioned that cowries came from the Maldive Islands in exchange of Bengal rice.[52]

Economic Condition during Medieval Period

Famines

Most of the famines during the medieval period broke out as a result of crop failure caused by lack of rainfall or flood. However, some of the famines were also due to endemic warfare and resultant displacement of people.

The famine of 1334-42 affected the northern India mostly when Muhammad Tughlaq imposed excessive taxes and many people fled the country. The Durga Devi famine lasted for twelve years between 1396 and 1407 and ruined the country between the Godavari and the Krishna rivers. The 1630-31 Deccan famine was caused due to crop failures for three consecutive years when about two million people died. Mention has been made of 1661 famine in India when not a drop of rain fell for two years—no rain since 1659. The extent of area and loss of lives, however, could not be gathered. 1702-1704 famine in Deccan killed two million people but the cause of the famine could not be ascertained. It is highly probable that failure of crops due to drought could have caused this famine.

Estimate of Rice Production during Medieval Period

There have been attempts by various scholars to assess the population of India during the medieval period. For the year around 1600, Raychaudhury and Habib put the Indian population estimate to be about 142 million while various other scholars arrived at their own estimates: Moreland put it at 100 million, Kingsly Davis 125 million and Shireen Moosvi 144.3 million. Based on the arguments and methods presented by the authors, it is highly probable that India's population stood somewhere around 140 million. After considering the manufacturing, overseas trade, export descriptions, volumes of export-import and overall economic prosperity of 17th century India, the present authors are of the opinion that rice could have been grown in about 12 to 15 million hectares producing between 10 and 13 million tons of rice.

Growth of Indian Trade and Role of Rice

Around 1000 AD there was a remarkable change in the structure of Asian maritime trade also. The previous pattern of pre-emporia trade changed into the new pattern of emporia trade. Whereas in the phase of pre-emporia trade, goods were shipped directly from the place of origin to that of final consumption, the rise of emporia, particularly along the Indian coasts, implied new practices of re-export, breaking bulk, assorting shipments according to the demands of various ports of call, etc. This major change in the pattern of Asian maritime trade was related to the simultaneous rise of powerful empires in several parts of Asia. The Chola Empire of South India, the Khmer empire of Cambodia, the empire of Champa in Vietnam and China under the Song dynasty emerged in the 11th century AD, which witnessed a rapid extension of rice cultivation and a large-scale increase both in local and long-distance trade. The goods traded were no longer only a few luxury items but a wide variety of commodities such as processed iron, spices, sandalwood, camphor, pearls, textiles as well as horses and elephants.[53]

Indian economy between the 16th and 18th centuries was at its prime when the goods it produced were in heavy demand in Asia, Africa, Europe and the Americas.[54] Most in demand were cotton textiles, silk, jewellery, certain metal products, food and spice products, etc.

Before the British traders subdued the local rulers and became rulers instead, the various regions of India enjoyed high prosperity from industry, commerce and above all a remarkable state of excellent agricultural production. In their own accounts, the British writers bear testimony to the fact. Luke Scrafton, a member of Robert Clive's Council, in his book *Reflections of the Government of Hindustan* (1770) mentions,

> In passing through the Rampore [present day Rampur, Uttar Pradesh, India] territory, we could not fail to notice the high state of cultivation which it had attained . . . . If the comparison for the same territory be made between the management of the Rohillas and that of our own Government, it is painful to think that the balance of advantage is clearly in favour of the former.

Sir John Malcolm visited the Maratha country around 1803 and according to his records,

> 'It has not happened to me ever to see countries better cultivated, and more abounding in all produce of the soil, as well as in commercial wealth, than the Southern Maharashtra districts . . . . Poona, the capital of the Peshwah, was very wealthy and a thriving commercial town, and there was as much cultivation in the Deccan as it was possible an arid and unfruitful country would admit.[55]

The situation however, became different when the British occupied the sub-continent.

BRITISH PERIOD

Stories from Marco Polo's famous journey to China and India in the 13th century describing the riches of these Asian countries might have prompted the European traders to get an access to these nations. The Ottoman and Persian traders, however, controlled the land route and hence the European traders had to look for a sea route. Columbus started his voyage to find a sea route to India and in the process landed in America. Soon thereafter, the Portuguese sailor Vasco da Gama circumnavigated Africa and discovered the actual sea route to India in 1498. The western European nations (especially the Portuguese, Spanish, British and French) occupied the new world, exploited its gold and silver and in exchange purchased cotton textiles from India, spices from Southeast Asia and silk, tea, porcelain and other goods from China.

When European traders entered Indian Ocean in the beginning of the 16th century, the maritime trade in the Indian Ocean was dominated by the Arab traders. It took almost a century for the European traders to gain dominance in Indian Ocean and another century to gain a sound foothold on the South Asian and Southeast Asian ports. Among the European traders, first the Portuguese and then the Dutch, French and British were the main contestants in the Indian sub-continent but ultimately the British succeeded in establishing their supremacy in India. By 1750, the Indian coastline was becoming more and more exposed to the Europeans and their culture but the Mughal heartland of India still remained integrated with Iran, Afghanistan and Uzbekistan.

British Supremacy

The British East India Company defeated the Nawab of Bengal at the Battle of Buxar in 1764 despite the latter's support from the Mughal Emperor and Nawab of Awadh. After the battle, the Company concluded a treaty with the Mughal Emperor and obtained *diwani* (authority to collect land revenue) in Bengal (present day Bangladesh and the states of West Bengal, Bihar and Jharkhand of India). In effect, the East India Company became the ruler of this large territory without of course its crown.

After gaining *diwani* (authority to collect land revenue) for Bengal from the Mughal Emperor, the British continued with the same way of plunder and extortion. Revenue was exacted from the *zamindars* (landlords) on land and its produce at a very high and arbitrary rate unprecedented before. Inland trading was made duty-free for the British while salt and certain other items were monopolized, ruining business of local traders completely. Leaving business, many of the local traders turned to land for a living but adverse weather decreased production and by 1769 the infamous famine broke out, which is discussed in a separate section. Reports of the famine and consequent death of one-third of the population of Bengal forced the British Parliament to issue the Regulating Act in 1773 to improve the affairs of British administration in India. The British parliament arranged for a Governing Council in India with a Governor General to oversee all the affairs and possession of the East India Company.

However, little did change in revenue administration and ruination of the country continued by both official and private trading of the Company. Ultimately in 1773, Permanent Settlement of land revenue (the so-called *zamindari* system) by Governor General Cornwallis provided for definite tax and assessment system for the land administration in Bengal. This and the other land revenue systems introduced in course of the colonial rule are described under agrarian system later.

The British introduced or undertook several measures in their colonial and administrative interests, many of which benefited Indians in short and long terms. Since India did not exist as a single political unit when British took it over, they had to introduce certain uniform measures for effective control and administration. For example, they introduced a single currency as the legal tender, a common system of weights and measurements, telegraph and postal services and railways and abolished all inland and town duties. Banditry was controlled; especially the *thugies* (mafia of cheats) and thus inland travel became safer. Introduction of modern education, record keeping, research systems and delivery of justice became the hallmark of British rule. Most of the British scholars and a section of Indian educated class cite these as boon for the Indians and conclude that these would not have happened at all without the British rule. But a careful examination of the historical facts indicates a terrible decline of the Indian economy and throwing majority of its people into distress under abject poverty.

It has become increasingly clear from recent researches into economic history that the British colonizers transformed India, the richest and the most industrialized region of the world, into a pauper nation. Around 1750 India alone produced 25% of world's total industrial manufacturing output;[56] by the end of their 190-year colonial rule, the Indian economy stagnated and experienced zero growth.[57]

The policies and actions of East India Company had annoyed the Indian rulers who had come under the suzerainty of the Company and so were the Indian soldiers serving under the Company. Also repeated famines and lack of action by the ruling Company angered common people. As a result, there was a rebellion mainly in north and central India in 1857, which the British termed as Sepoy Mutiny. The Company had a hard time and ultimately was able to suppress the mutiny at a huge cost of life and property. As a result, the British Government took direct control of Indian Administration in 1858 from the Company. A Secretary of State for India was appointed at London and a Viceroy ruled India as a representative of the Crown.

Over the next ninety years, the British administration in India introduced several acts, regulations and laws, mostly to bring more control of the British Parliament in the affairs of administering India, partly to reform social system and, at the later stage, to accommodate Indian representation in the political affairs. However, the traditional bond of co-ordination between Indian agriculture and its rural and cottage industry were being destroyed and a new bond of Indian agriculture and British industry was being established. "Destruction of rural and cottage industries, allied trades and commerce ruined all supportive vocations of the bulk of population and the people thus thrown out from their vocations crowded in the agriculture sector."[58]

For example, Nagpur, in the year 1800, had a very flourishing textile trade. Before the downfall of Peshwa in 1818, the export of cloth woven in Nagpur city and around Pune was of the value between Rs. 1,200,000 to Rs. 1,400,000. The agricultural prosperity of Pune people in turn helped the artisans and tradesmen in Nagpur area to prosper. Within eight years of British Raj i.e., in 1826, it was hardly Rs. 300,000.[59]

While Europe was going through industrialization in the late 18th and early 19th century, India was going through a phase of de-industriali-zation. In due course, this discussion will reveal the factors which either paralyzed or promoted Indian agriculture during the British adminis-tration in India.

Famines in British India

The British East India Company arrived in India for trade with a motive to maximize its profit. After taking over the *diwani* in Bengal, the profit motive extended to revenue as well as to trade tariffs. These coupled with maladministration by the Company and later by the British Indian govern-ment caused a series of famines unprecedented in Indian history. The famines that occurred in the rice growing areas only are mentioned below:

The Bengal Famine of 1770

Excessive revenue collection in Bengal caused serious hardships for the local peasantry as well as gentry. It was further exacerbated by mono-polization of internal trade by the Company and private trading by Company persons including even the Governor Warren Hastings himself. The displaced local traders had to seek refuge in the land for a living. To make matters worse, rice crop failed partly in 1768 while a drought next year affected its production severely and led to the infamous famine in 1769. Again, excessive rainfall in 1770 caused overflow of rivers, standing rice crop was damaged and the situation worsened further. The famine lasted for about three years and at least 10 million people starved to death. But the British remained indifferent to the plight of the people and continued their plunder. Warren Hastings, the Governor of the Company wrote on 3rd November 1772,

> Notwithstanding the loss of at least one-third of the inhabi-tants of the province, and the consequent decrease of the culti-vation, the net collections of the year 1771 exceeded even those of 1768 . . . . It was naturally to be expected that the diminution of the revenue should have kept an equal pace with the other consequences of so great a calamity. That it did not was owing to its being violently kept up to its former standard.[60]

Orissa, a state adjacent to Bengal, was ruled at that time by Bhonsla of Nagpur. According to a modern historian of Orissa,

> Maratha rulers constructed many dams on rivers to save the crops. During famines, they used to grant farmers relief from

land revenue. In the year 1775 AD, the crop of Cuttack district was damaged. Therefore, they had forgone revenue worth seven hundred thousand rupees. At the time of distress, they used to grant taccavi loans. They made efforts to improve agriculture.[61]

The famine affected the present day Bangladesh and the states of West Bengal, Bihar, Jharkhand and Orissa. Many areas were completely depopulated and agricultural production, particularly rice, declined substantially. However, revenue collection though lower, did not decline at that rate. Following this famine, "successive British governments were anxious not to add to the burden of taxation."[62]

Famine of Uttar Pradesh, 1838

In 1838, there was a serious famine in North-West Provinces (present-day Uttar Pradesh, India), when about 800,000 people perished. It was caused by drought of 1837 in the *doab* i.e. the region lying between Delhi and Allahabad and both *kharif* (summer) and *rabi* (winter) crops failed. Governor General Auckland toured the affected areas and ordered relief measures but it was conducted in exchange for work of able-bodied persons.[63]

The Great Orissa Famine of 1866-68

The first major famine after the takeover of the Indian administration from the Company by the British Government was in 1866-68 in Bengal and its adjacent areas. The total area affected was estimated at 180,000 square miles with a population of 47.5 million but the distress was greatest in Orissa which was at that time practically isolated from the rest of India. Scanty rainfall and short monsoon depleted food-stocks especially rice and caused the onset of famine. The gravity of the situation was not realized till the end of May 1866 and then the monsoon had set in. Carriage by sea was extremely difficult and even when grains reached the coast, it could not be conveyed inland. At great cost, some 10,000 tons of rice were imported but this did not reach the people until September. Meanwhile the mortality had been very great. It was estimated that at least a million people or one-third of the population died in Orissa alone. Later, Secretary of State for India, Lord Salisbury regretted,

I did nothing for two months. Before that time the monsoon had closed the ports of Orissa—help was impossible—and—it is said—a million people died. The Governments of India and

Bengal had taken in effect no precautions whatever . . . . I never could feel that I was free from all blame for the result.[64]

The troubles of Orissa did not cease in 1866 as heavy rains caused floods and destroyed rice in low-lying lands. In the following year, relief measures were again undertaken. Then, as an apparent result of the reaction following the want of foresight and activity in the preceding year, the relief operations were marked by a profusion and absence of check hitherto unexampled. Altogether about 40,000 tons of rice was imported, even half of that could not be disposed of. While it cost four times the usual price to procure, the residual stock had to be sold for almost nothing. The monsoon of 1867 resulted in a fine harvest and put an end to the famine in 1868. In the two years, about 35,000,000 units were relieved at a cost of Rs. 95 lakh, two-thirds of which were debitable to the expense of importing grain. Adding loss of revenue in all departments, the famine in Orissa is said to have cost the state about Rs. 1-2 crore. The Famine Commission presided over by Sir George Campbell laid down for the first time the guidelines upon which subsequent famine-relief should be organized in British India.

The Great Deccan Famine of 1876-78

This was perhaps the most severe famine of the 19th century in India. It was widely spread in about 650 thousand square kilometer area in southern India, western India, Punjab and Uttar Pradesh. It is estimated that 55 million people were affected and more than five million people died due to this famine.

The drought of 1876-78 affected all of south India. From October 1875 to October 1877, three successive monsoons failed to bring full supply of rain. The harvest of 1875 was generally below the average and remissions of revenue were found necessary but it was not till the close of 1876 that actual famine occurred. During the whole of 1877, people suffered from famine.[65]

Lytton, the then Governor General of India, discouraged relief efforts in the pre-text that 'free and abundant trade cannot co-exist with Government importation' and that 'more food will reach the famine stricken districts if private enterprise is left to itself (beyond receiving every possible facility and information from the government) than if it were paralysed by Government competition.'[66] Reacting against calls for relief, Lytton instructed the district officers to "discourage relief works in every possible way . . . . Mere distress is not a sufficient reason for opening a relief work."[67] Lytton's era saw approximately one million tons of rice exported to Europe from India while millions of Indians starved to death. Besides, to checkmate Russian advances into central India and to expand the British Empire to Afghanistan, he led an Afghan war the cost of which was borne by the Indian Government. Besides, "Lytton had been absorbed in organizing the immense Imperial Assemblage in Delhi to proclaim Victoria Empress of India (Kaiser-i-Hind) . . . . Its "climacteric ceremonial" included a weeklong feast for 68,000 officials, satraps and maharajas: the most colossal and expensive meal in world history. An English journalist later estimated that 100,000 of the Queen-Empress's subjects starved to death in Madras and Mysore in the course of Lytton's spectacular durbar."[68]

According to Mike Davis, the famines killed between 12 and 29 million Indians; he demonstrates that these people were 'murdered' by British state policy. He cites that when an El Niño drought made the Deccan farmers destitute in 1876, there was a net surplus of rice and wheat in India. But the viceroy, Lord Lytton's Anti-Charitable Contributions Act of 1877 prohibited "at the pain of imprisonment private relief donations that potentially interfered with the market fixing of grain prices". The only relief permitted in most districts was hard labour, from which anyone in an advanced state of starvation was turned away. In the labour camps, the workers were given less food than inmates of Buchenwald. In 1877, monthly mortality in the camps equated to an annual death rate of 94%. In the northwestern provinces, Oudh and Punjab, which had brought in record harvests in the preceding three years, at least 1.25 million died.

Between 1880 and 1896, there were two famines and five scarcities, all of them of local characters. The famine of 1896-97 was, however, severe and affected almost every province of the country though the intensity varied from province to province. About 35 million people were affected. The casualty was most severe in Central India.[69]

The Famine of 1899-1900

A year after Lord Curzon became the Viceroy and Governor-General of India, he had to deal with the famine of 1899-1900, another severe and widespread famine of India. It affected an area of six hundred thousand square kilometers and a population of 28 million. Lord Curzon spent Rs. 68 million (about £10,000,000) to reduce the effects of the famine and, at its peak, 4.5 million people were on famine relief. It is estimated by census commissioners that, in the famine of 1901, four million people died—three million people in the native states and one million in British territory.[70]

Lord Curzon appointed a Commission under the presidency of Sir Anthony MacDonnel to go into its causes and suggest remedy for future. Some of the measures taken by Lord Curzon were long lasting and are discussed later in this chapter.

The Great Bengal Famine 1943

The last famine during the British Empire in India was the Bengal Famine of 1943 when World War II was in full swing. The Japanese defeated British forces in Singapore and advanced to Burma (now called Myanmar) while *Azad Hind Fauj* (Indian National Army) of the Indian nationalist leader, Subhas Chandra Bose with Japanese support liberated the Andaman and Nicobar Islands from British control and advanced up to Imphal and Kohima in the northeast of India. The Indian Administration feared that Japanese forces may invade India any moment and therefore became too much preoccupied in defense preparations. The Indian railways and roads got busy shipping army, arms and rations for the soldiers to the frontline in northeast India where British forces took positions. Around that time, Mahatma Gandhi, the nationalist leader, also waged the famous "Quit India" movement against the British. The British Government in India was very hard-pressed facing the external aggression as well as internal agitation. It had also to export food to its soldiers in the Middle East and, in the process, neglected its civic responsibility. Rumors of grain shortage further added to the scarcity as traders and public who could afford hoarded grains while others starved.

Amartya Sen, the Nobel Laureate economist, witnessed the famine as a kid[71] when beggars lined up endlessly on their and other peoples' doors. Later he dwelt at length on the event and showed that there was no production shortage. British scorched-earth policy at the border near Chittagong fearing Japanese invasion and other factors mentioned above fuelled rice prices to shoot-up beyond ordinary peoples' reach. This was further compounded by incompetent handling by the Government and their inability to keep rice prices down. As a result, an estimated 2.5 million people died of hunger and resultant epidemics.

It is strange that the British administration in India started with a famine and ended with a famine, the first one was partly due to natural causes and mostly due to lack of administrative responsibility and the second one purely for the latter reason. A number of researchers attested to the fact that British policies and administration were largely to blame for most of the famines when millions of people died.[72] It is to be noted that during the pre-British periods, famines also broke-out in various parts of India but the rulers of the day took measures to ease the hardships by arranging food supply and distribution, reduced or exempted taxes of the affected people and took immediate measures to boost food production by providing credit, arranging irrigation, supplying seeds, etc. In a few cases when the British tried to provide relief and took measures to reduce suffering of starving masses, famines were avoided.

Famine Commissions were constituted after every major famine and actions were taken on some of these recommendations of the Commissions. These Commissions recommended the construction of canals (to irrigate the crops during the drought periods) and railways (to transport grains to the famine-affected areas) and the district officers were made responsible to save people from all preventable deaths.

Introducing Modern Systems in Agriculture

Agrarian Systems

It is interesting to note that the techniques used in land surveying in many parts of India (and even the terminology used for land records) even today remain substantially unchanged since their introduction by Raja Todarmal during the reign of Mughal Emperor Akbar. The British superimposed a system over the existing pattern in tune with British customs and laws relating to land. Broadly three principal types of land revenue system were introduced in British India. The basic characteristic of each system was the attempt to incorporate elements of the preceding agrarian structure of the area. The interaction of colonial policy and existing systems produced widely different local results and hybrid forms.

Zamindari System: The Permanent Settlement Act was passed in 1793 and initially introduced in Bengal. The system was also introduced in large parts of Bihar and Orissa. The system was introduced to ensure the revenue receipt of the British colonial power where a *zamindar* was declared the proprietor of land on condition of fixed revenue payments to the British regime. The peasants were turned into tenant farmers and deprived of the land title including other rights and privileges enjoyed during the Mughal period. This revenue system accounted for 57% of cultivated area in the country.

Ryotwari System: Ryot means subject, tenant or, in the wider sense, citizen. This system was introduced in Madras Presidency in 1792 and in Bombay Presidency in 1817-18. This system recognized the proprietary right of the peasant on land and resembled the revenue system of the

Mughal to a great extent. The system covered nearly all the southern states and many western states of India including the erstwhile Central Province (present-day Chhattisgarh and parts of Madhya Pradesh). Even the princely states of Jaipur and Jodhpur had this category of revenue system, as it existed during the Mughal period. However, the pockets of *zamindari* prevailed within the *ryotwari* regions, particularly in the princely states. The *ryotwari* system covered nearly 38% of the cultivated area in India.

Mahalwari System: This system was introduced between 1840 and 1850. In this system the entire village constituted a collective unit for revenue settlement. The peasants paid the revenue share of the whole village in proportion according to their individual holdings. The system covered the present day Punjab, parts of Orissa and Madhya Pradesh and Uttar Pradesh. The system included only 5% of the cultivated land in India.

Departments of Agriculture

Lord Mayo (Viceroy of India, 1869-1872), who had been a practicing farmer, first proposed to establish 'something like an agricultural department' which 'is almost entirely neglected by the Government'.[73] At the same time, the textile industry of Manchester faced crisis because of the stoppage of supply of cotton from the United States of America. The lobby of the textile industry which was very strong in England made Lord Mayo propose for a separate department of agriculture in India. His proposal was not fully accepted as only a secretariat of Revenue and Agriculture could be set up under the Home Department of the Government of India in 1871.

The Famine Commission appointed by Lord Lytton under the presidency of Richard Strachey stressed the urgent need for establishing agriculture departments at the Centre as well as in provinces. In pursuance of the recommendations of this Commission, the Revenue and Agriculture Department of the Government of India was set up and similar departments were established in provinces. Directors of Agriculture were appointed for United Provinces in 1875, for the Punjab in 1880, for Bombay in 1884 and for Madras in 1882. In 1889, the Secretary of State for India sent J. Voelcker, then the Consulting Chemist to the Royal Agricultural Society, to India to undertake a survey of Indian agriculture. Voelcker spent a little more than a year in India and travelled widely in this country. Voelcker's report received official endorsement and is often cited as a landmark document on Indian agriculture. Voelcker advocated adding manure to the soil to maintain soil fertility, social forestry along the canal banks and railway tracts to solve the fuel problem and schools for agricultural education. No action, however, was taken on Voelcker's report until another famine struck India in 1899. In that year, George Nathaniel Curzon came as Viceroy of India.

Irrigation

Before the East India Company rule, tanks were very common in central and south India and they had respect in society. It was a part of social culture.

> "Mysore State had over 39,000 tanks in 1800. It was said that the system of tanks here was so remarkable that if a drop of water falls anywhere in the state, there is some tank somewhere to harvest it. The state used to generously support maintenance of the tanks. When British took over the state, one of the first steps they took was to reduce this allocation by half. Then in 1863 when Public Works Department (PWD) was opened, the ownership of all the tanks was handed over to it. Thus, the state took away the resources for, and then ownership of the tanks away from the society. Thus, communities were no longer able to maintain the tanks."[74]

Sir William Willcock, a distinguished hydraulic engineer while investigating the conditions in Bengal, discovered that innumerable small destructive rivers of the delta region, constantly changing their course, were originally canals which under the English regime were allowed to escape from their channels and run wild. Formerly these canals distributed the floodwaters of the Ganges and provided for proper drainage of the land, undoubtedly accounting for that prosperity of Bengal which lured the rapacious East India Company merchants there in the 18th century.[75]

In the second half of the 19th century, the British Government of India promoted irrigation works in the river deltas of India all of which produced rice and sugarcane. The renovation of the dam across Kaveri and the construction of Coleroon anicut, dam across the river Godavari and the Ganga canal assured irrigation in the drought hit areas. Arthur Thomas Cotton was the architect of the irrigation works in the southern region and Proby Thomas Cautley was the architect of the Ganga canal in the north. Thus the rice production and productivity increased in the river deltas of India during the British period.[76] The reason for the interest of the British in irrigation is said to be the collection of land revenue which was not possible to collect when crops failed during the drought years. At least that was the justification provided while pleading for investment in irrigation.[77] Stockpiling of rice for the army outside India was another reason cited.

By about 1880, the old settled regions of the Punjab had been provided with irrigation facilities either by constructing new canals or by remodeling and restoring the old canals. An Irrigation Commission set up to examine the then situation and suggest future plans submitted an exhaustive report in 1903. It proposed many improvements in the existing system. According to the Report of the Royal Commission on Agriculture (1928),

> On an average, for the five years from 1921-22 to 1925-26, nearly 50 million acres were irrigated by government and private irrigation works, the percentage of irrigated area to area sown being 19.4. Practically half the total area irrigated is irrigated by canals, the remainder being irrigated by tanks, wells and other sources.

In 1932, the Sakkar Barrage was constructed on Indus River in Sindh.[78] *Warabandi* (rotational) system of water distribution amongst users is the hallmark of Indus basin that has worked well for centuries.[79] In the next twenty years of the 19th century, major canal schemes were instituted in the Indus basin that more than doubled the area under irrigation.

Agricultural Research and Education

The Famine Commission of 1901 made many recommendations. The Commission stressed the desirability of improvement of irrigation facilities, transport facilities, improved methods of agriculture, etc. and to set up an Agricultural Research Institute for the country. The Government accepted most of the recommendations of the Commission. Based on its recommendation, an Imperial Agricultural Research Institute was established at Pusa now in Bihar State of India. Agricultural farms were established in provinces. These farms were also to function as depots for seeds, manures and implements.

In 1876, an agricultural college had been opened at the Saidapet Experiment Station in Madras. A college near Poona had begun teaching agriculture in 1879 and had begun issuing diplomas in agriculture since 1890. Following the recommendations of the Famine Commission, Agricultural Colleges were established at Coimbatore (Tamil Nadu), Pune, Nagpur (Maharashtra), Kanpur (Uttar Pradesh), Sabour (Bihar) and Lyallpur (now Faislabad, Pakistan). These colleges combined teaching and research. In 1905, the Government of India decided to set apart annually two million rupees to assist the development of agricultural research, demonstration and education in the provinces.

Henry Phipps, a philanthropist of Pennsylvania (USA), who was a friend of Curzons and came to India probably to attend the ceremony of King Edward's accession to the throne, gave Curzon an unrestricted grant of $100,000. With it, Curzon built an agricultural research station named as Imperial Agricultural Research Institute (IARI) and the campus was named as PUSA after 'Phipps of USA'.

Albert Howard who worked earlier in Barbados on arrowroot and at Kent on hops was appointed as its first Director. In India, Howard turned to wheat and by the 1920s some two million acres of India were sown with the so-called 'Pusa' wheats. After an earthquake in Bihar in 1935, the Institute was shifted to New Delhi. After independence, its name was changed to Indian Agricultural Research Institute (IARI).

Since the IARI did not represent any particular rice zone, the Government soon set up two experimental farms for rice research, one in East India at Dhaka (now in Bangladesh) and the other at Coimbatore (now in Tamil Nadu State of India) in South India. Besides, some experiment stations/demonstration farms were opened at Maruteru (now in Andhra Pradesh), Raipur (now in Chhattisgarh State of India), Pattambi (now in Kerala State of India), Kanpur (now in Uttar Pradesh State of India).

Royal Commission on Agriculture

To examine the condition of agriculture and rural economy in India and recommend steps for its improvement, the Government of India appointed a Royal Commission on Agriculture in 1926. It was headed by Lord Linlithgow who later became Viceroy of India. The commission worked for about a year and a half, traveled 18,000 miles, spent about 100,000 pounds sterling and produced a voluminous report.

The Commission suggested creation of an Imperial Council of Agricultural Research, the primary function of which should be to promote, guide and co-ordinate agricultural research throughout India. The Commission stated that their object in proposing the constitution of such a body was to provide provincial governments with an organization embracing all the research activities of the country in the field of agricultural sciences including animal husbandry and veterinary sciences. The Commission further observed that one of the most important functions of the Council would be to impart training to the research workers and hence part of its fund should be utilized in the provision of research, scholarships and intensive training in scientific research in agriculture. It would also take over the publication work done by the Agricultural

Advisor to the Government of India and would arrange for sectional meetings of experts in particular branches of agricultural science. On the basis of this recommendation, the Government of India set up the Imperial Council of Agricultural Research (ICAR) in 1929. With a non-lapsing fund of Rupees five million, the ICAR was expected to supplement research activities of provinces and train scientific manpower. Concomitantly, a number of central commodity committees were constituted, mainly for commercial crops. The funding of these committees from cesses was the first attempt to link research funding with the beneficiaries.

With the recommendation and the financial support from the ICAR, rice research stations were opened up in almost all the agro-climatic zones of India in 1930s. These research stations did pioneering work on rice research, on manuring (there were no chemical fertilizers in those days), insect pests of rice and their control (there were no insecticides in those days), varietal trials and pure line selections from the best of the local rice varieties. In fact, by 1946, 82 rice research stations were established throughout the country (India, Pakistan and Bangladesh) and 445 rice varieties were bred mainly by pure line selections from the traditional rice varieties. The characters selected for were mainly early maturity, tolerance to flood and deep water and flood, resistance to diseases and resistance to lodging and drought. Non-shattering of grains, dormancy of seed, control of wild rice and higher response to increased manuring were also emphasized. Table 8.1 provides a list of important varieties bred during this period.

Table 8.1. Some popular improved rice varieties bred in India during 1921 to 1950[80]

Wide adaptability	GEB 24, PTB 10, MTU 15, BAM 9, N 136, T141
Fine grain	R 11 (Dubraj)
Aromatic	Basmati-370, R 11 (Dubraj), Ambemohar, Badshabhog
Earliness	MTU 9, MTU 15, TKM 6, T 136
Tolerance to deep water	PTB 15, PTB 16, MTU 16, CO 14, HBJ 1, HBJ 2
Tolerance to drought	MTU 17, PTB 18, PTB 30, Lal Nakanda
Tolerance to flood	FR 13A, FR 43B
Tolerance to salinity	SR 26B, Nizersail, Kalarata, Bhurarata, Jhona 349, Chinsurah 13, Sathra 278
Resistance to blast	CO 25, CO 26, S. 67
Chinese introductions	CH 45, CH 988, CH 1039, CH 1007

Department of Statistics

The Department of Statistics was founded in 1847. In 1848, the first census relating to the area and revenue of North-West Provinces (the present-day Uttar Pradesh) was released. In 1853, the Department released the first series of statistical papers on India. W.W. Hunter became Director-General of Statistics of India in 1869 and, under him, *The Statistical Account of Bengal* was published in 20 volumes which included data for each district and, among other information, details of agricultural situation.[81]

The need for timely and accurate collection of agricultural data was felt by the Indian Famine Commission and agricultural departments were organized in various provinces which resulted in the publication of *Agricultural Statistics of British India* in 1886.[82] The publication of *Dictionary of Economic Products of India* by George Watt in 1891 was another valuable documentation of the period.[83]

Agricultural statistics from British Government of India reveal that in 1894-95 rice was grown in about 69 million acres producing about 23 million tons. By 1920, about 78 million acres produced 32 million tons of rice. Bengal had the highest acreage and production followed by Bihar and Orissa and Madras Presidency.

Indian Agriculture under British Rule

British ruled India for almost the whole of 19th century and during the first half of the 20th century. During the first half of the 19th century, it was the East India Company which ruled India. The Company's time and effort were spent mainly in fighting battles to acquire new territories, consolidating power in India and to maximize profit as, basically, it was a trading Company. In the second half of the 19th century, India came under the British Crown but all efforts of the Government were dissipated in tackling successive famines. In the first half of the 20th century, the British Government had to face two World Wars and an economic Depression. Under such situations, it was hard to expect any development of agriculture from British rule that was wedded to the welfare of an alien people and answerable to an alien Government.

The prolonged neglect of agriculture in India meant that there was almost no growth in the agricultural sector. From 1891 to 1946, output of all crops grew at 0.4% a year; the rate for food grains was only 0.1% per year. The land tenure system led to exploitative agrarian relations and stagnation. Farmers had little incentive to invest and, despite great strides in foreign agricultural technology, Indian agricultural technology stagnated. Specifically, there were few improvements in seeds, agricultural implements, machines or chemical fertilizers.[84]

The subcontinent became independent in 1947 when the British partitioned it into India and Pakistan and left. The overall achievement of the sub-continent during the entire period of British rule may be summarized in one sentence: "There was no increase in India's per capita income from 1757 to 1947."[85] Or, as M.S. Swaminathan, a renowned agricultural scientist of India puts it "The growth rate in food production during the 1900-1947 period was hardly 0.1%."[86]

After partition, agricultural production, particularly rice production increased dramatically in each of the three countries (India, Pakistan and Bangladesh) as discussed in the next chapter. Takashi Kurasaki's analysis shows that total output growth rates turned from zero or little to significantly positive levels, which were sustained throughout the post-independence period.[87] He also shows that a substantial increase in aggregate land productivity occurred *before* the introduction of high-yielding varieties of rice, the fact that the sustained growth in the total output began just after the partition, well before the introduction of the Green Revolution technology. Over and above, independence from foreign rule proved its essentiality and ushered in new growth in agriculture and for rice in particular.

SRI LANKA

Geography

Sri Lanka is an island situated at the southern tip of the Indian subcontinent. The 50-km wide Palk Strait separates it from India. The island is just 350 km long and 180 km at it's widest. The southern half of the island is dominated by rugged hills. The entire northern half comprises of a large plain. The highest mountain Mt Pidurutalagala (2524 m) is near Nuwara Eliya and the longest river is Mahaweli which courses from the centre and empties into the Indian Ocean at Trincomalee. Sri Lanka is a typically tropical country. Temperatures in the low-lying coastal regions are high all the year round but fall rapidly with the rise in the altitude. The south, southwest and central highlands get heavy rainfall while the northern and north-central regions are arid.

Sri Lanka has an agro-based economy with rice, tea, rubber and coconut as its chief agricultural crops. Rice is cultivated by farmers on small-scales. The principal cultivation season, known as "maha", is from October to March. During this season, there is usually enough water to sustain the cultivation of all rice fields. The subsidiary cultivation season, known as "yala", is from April to September.

Rice is the single most important crop occupying 34% (0.77 million ha) of the total cultivated area in Sri Lanka. On average 560,000 ha are cultivated during *maha* and 310,000 ha during *yala* making the average annual extent sown with rice to about 870,000 ha. About 1.8 million farm families are engaged in paddy cultivation island-wide. Sri Lanka currently produces 2.7 million ton of rough rice annually and satisfies around 95% of its domestic requirement. Rice provides 45% total calorie and 40% total protein requirement of an average Sri Lankan.[88]

Early History

The main source of information about the ancient period of Sri Lanka is *Mahavansa* (The Great Genealogy), a chronicle compiled in Pali by Buddhist monks in the 5th century AD. Later, another chronicle named *Chulavansa* (The Lesser Genealogy) was compiled by another Buddhist monk in the 13th century. This latter chronicle was expanded by a second monk in the following century and was concluded by a third monk in the late 18th century. Information on rice cultivation is also available in *Saddharmaratnavaliya, Pujavaliya, Saddharmalankara* (13th century) and *Butsarana* (possibly 12th or 13th century). More detailed information on rice cultivation in the Island becomes available from various sources as we advance chronologically towards the modern times.

There is hardly any information about rice cultivation in Sri Lanka before the 6th century BC. According to *Mahavansa*, Vijaya was the founder of the Sinhalese race in the 6th century BC. He presumably went from Kalinga (modern Orissa of India) and distributed rice among his followers there. Buddhism was introduced in Sri Lanka in 247 BC by Mahinda, the son of Emperor Ashoka of India; it was the most important event in Sri Lankan history as it set the country on the road to cultural greatness. Local production of rice was so insufficient that Emperor Ashoka sent 160 loads of hill paddy from India among the presents sent to King Devanampiya Tissa of Ceylon.[89]

The first extensive Sinhalese settlements were along rivers in the dry northern zone of the island. Because early agricultural activity—primarily the cultivation of wet rice—was dependent on unreliable monsoon rains, the Sinhalese constructed canals, channels, water-storage tanks and reservoirs to provide an elaborate irrigation system to counter the risks posed by periodic drought. Such early attempts at engineering reveal the brilliant understanding of these ancient people about hydraulic principles.

The discovery of the principle of the valve tower, or valve pit, for regulating the escape of water is credited to Sinhalese ingenuity more than 2000 years ago. By the 1st century AD, several large-scale irrigation works had been completed. The mastery of hydraulic engineering and irrigated agriculture facilitated the concentration of large number of people in the northern dry zone where early settlements appeared to be under the control of semi-independent rulers. In time, the mechanism for political control became more refined, and the city-state of Anuradhapura emerged and attempted to gain sovereignty over the entire island.

Scripts and pillar inscription of the 10th century prove the point that farmers had to adhere to certain laws laid down by the king or regional chieftains in relation to the repair, maintenance and management of small irrigation systems. The adherence to these laws over many generations resulted in the birth of customs and traditions, which gave the irrigation systems a discipline which continued up to British times. The ancient *rajakariya* system was a compulsory personal labor obligation that helped to guarantee the maintenance of these small irrigation systems over several centuries.[90]

Towards the end of the 10th century, the capital was shifted to Polonnaruwa as Anuradhapura became vulnerable to attacks from Chola kings of south India. However, the fall of the ancient hydraulic civilization of Anuradhapura in around 1220 AD was, according to A. Denis N. Fernando, due to sudden natural cataclysmic change of the river course of the Mahaweli Ganga. This sudden geological cataclysm changed the river course that used to sustain the ancient hydraulic civilization. This resulted in the major part of the population to abandon these areas and move to the wet and intermediate zone.

The period between 1200 and 1500 was the period of instability in Sri Lankan history due to foreign invasions, political turmoil, repeated shifting of the Singhalese capital southwest-ward and weakening of the central administration. The breakdown of the administrative and social order led to decline in the management of this complex irrigation system. The new rulers of the dry zone lacked the expertise and the experience to manage the elaborate hydraulic system.[91] As a result, many of the larger reservoirs were breached and smaller tanks that were fed by excess waters from them lost their supply. The amount of water available for cultivation was reduced which in turn reduced the area of cultivable land.[92] Rice cultivation became dependent on rainfall, which was not sufficient. Some of the destruction was deliberately caused by rival armies as logistics of their warfare. The abandoned tanks and channels became an ideal breeding ground for mosquitoes causing spread of malaria in the 10th century.[93]

In the wet Zone, which received the benefit of two monsoons, there was no scarcity of water but the terrain was difficult, particularly in the Kandyan hills. As a result, the farmers were forced to construct their fields in terraces like so many stairs up the hills. Terraced paddy fields are still a beautiful feature of this area of Sri Lanka. The water is allowed to trickle from the top terraces to those immediately below them and so on, until it escapes into the depths of the valley. Farming had to be of subsistence nature in the absence of any royal support.

With the advent of the new millennia, the maritime trade in the Indian Ocean had gained prominence. With the spread of Islam, the Arab traders started playing a major role in spice trade and Sri Lanka was a good source of cinnamon. The custom duties on spices were a better and easier income for the kings than the cumbersome system of collecting land revenues or tax on agricultural production. Consequently, the kings paid less attention on agriculture and its improvement.

The Portuguese and the Dutch

Portuguese arrived in Sri Lanka in 1505 and established their head-quarters at Kotte, in the western lowlands. By 1619, they had occupied most of the areas of Sri Lanka. Their major interest was in the cinnamon of the island. By 1640, the Dutch out-maneuvered the Portuguese and gradually occupied the country. The Dutch ruled the country for 140 years and left their mark in irrigation, plantation, transport, jurisprudence and even in the language. The Dutch attended to irrigation for rice cultivation and to coconut plantation.

In order to encourage and increase the cultivation of rice and other crops and to develop the agricultural resources of the country, the Dutch initiated and carried out many important works of irrigation. Two of the best known of these were the Urubokka and Kirima dams, which were a monument to the skill and energy of Captain Poenander, the Dutch Engineer, who successfully completed the work. The object of the Urubokka dam was to turn the abundant water which periodically inundated and ruined some of the richest tracts of land in the Matara district into that of Tangalla in the Southern Province, whose extensive tracts were previously abandoned owing to the scarcity of water. This magnificent work improved the cultivation of 8000 acres of paddy fields.[94] The Mulhiriyawa tank, perhaps the largest fresh-water reservoir made by the Dutch for irrigation purposes in the Western Province too served 2000 acres. The Giant's Tank in the Mannar district was constructed to serve the needs of that district. Unfortunately, these great works of utility constructed by the Dutch fell into neglect after they left the Island. The Government is now devoting much time and attention to renewing these useful projects for the benefit of agriculture.[95]

The portion of the Island from Colombo southwards which was consi-dered as waste land, was surveyed by the Dutch and divided among the people to be planted with coconuts. When the British took over, the whole of the south-western coast presented the scene of unbroken groves of coconut which we find to this day."[96]

Paddy cultivation by the farmers of Sri Lanka in the 17th century has been vividly described by Robert Knox[97] in his *An Historical Relation of the Island Ceylon in the East-Indies* (1681),

> The whole town [village] joins together in tilling [and] so in their harvest also. For all fall in together in reaping one man's field, and so the next, until every man's corn be down. And the custom is that every man, during the reaping of his corn finds victuals for the rest. The women's work is to gather up the corn after the reaper and carry it all together . . . . Then the women, whose proper work it is, brings each their burden of reaped corn upon their heads and go round in the Pit three times and then fling it down. And after this without any more ado, bringeth the rest of the corn as fast as they can. For this labour and that of weeding, the women have a fee due to them which they call *warapol* that is as much corn as shall cover the stone and the Conjuration Instruments at the bottom of the Pit.

He also mentions that every village had a small or big tank and the end of its bank was cut to irrigate rice fields.

British Period

In 1796, the British took over the Dutch possession of Sri Lanka and in 1815 the kingdom of Kandy, the last remnant of independent Sri Lanka, was also ceded to the British who now controlled the whole island.

British commercial interests saw the opportunities for cultivation of cash crops. Cinnamon was important for the British too just as it was for the Portuguese and the Dutch but after 1815 coffee plantation was spread to the Kandyan Hills. Despite ups and downs, production increased dramatically until 1875, when a catastrophic attack of a fungus disease (*Hemileia vastratrix*) wiped out almost the entire crop. It was replaced, particularly in the higher regions by tea. Rubber plantation was started in 1895 and became successful in the next century. The major interest of the

British administration was on cash crops; improvement of rice cultivation was overlooked.

Between 1815 and 1872, hardly any government assistance was given to the people to restore or repair tanks. Besides, the practice of getting these repaired and maintained by voluntary services of the people (*rajakriya*) was abolished by the British Government on grounds that it was a form of slavery. No alternative system was introduced for the repair and maintenance of tanks by the community and many minor irrigation works fell into neglect and general decay. In 1873, the Government took initiative to repair and restore these tanks and by the turn of the century, almost all small village tanks had been supplied with durable sluices which helped to conserve tank water supply; the wasteful practice of 'cutting a gap in one end of the bank to draw water little-by-little for watering their corn' (as described by Robert Knox) was given up.[98]

Food production received serious attention of the Government only when the Island was engulfed with a famine in 1912-13. To avoid recurrence of such famines in future, the Government opened large-scale private commercial farms. When these farms failed, the Government persuaded the rich farmers of the wet zone to take up irrigated paddy farming in commercial scale. But this measure only helped some rich farmers to grab more lands. The failures of these efforts prompted the Government to appoint a commission to suggest measures for increasing food production. In 1925, the Land Commission recommended the creation of a Ministry of Agriculture and Lands and to adopt the policy of expanding irrigation for increasing rice production. This policy has been continuing ever since then.[99]

In the 1930s and 1940s, the policy of the Government was to develop the dry zone for rice production. To meet the shortage of rice during the Second World War period, the Government started the universal rice rationing scheme in 1942. The British transferred the power to the people of the island in 1948. They left the country united under a single administration with British tradition of parliamentary democracy and administrative service.[100]

References

1. Saraswat (2005).
2. Fuller (2002), Harvey *et al.* (2005).
3. Saraswat and Pokharia (2004).
4. Saxena *et al.* (2006), Tewari *et al.* (2006).

5. Chauhan *et al.* (2004).
6. Sharma (1985).
7. Possehl and Rissman (1992).
8. Sharma (1985).
9. Saraswat (2004).
10. Fujiwara (1996).
11. Pokharia (2006), Kharakwal *et al.* (2007).
12. Fuller (2002).
13. Gates (2003).
14. Bellwood (2004:95).
15. Keay (2001:26).
16. Pande (1990:76).
17. Thapar (2003).
18. Mehra (2005).
19. Stein (1998:60).
20. Now archaeologists have been able to locate the house of Sujata. See Chakrabarti (2001).
21. Kumar (1988).
22. Pande (1990:85).
23. Nene (2006).
24. Pande (1990:85).
25. This style of forming words is followed even in the modern Indian languages and is called in grammar by the same name i.e. *bahuvrihi samasa* as designated by Panini in the 4th century BC.
26. Nene (2005).
27. Pande (1990:25).
28. Chopra (2003).
29. Saran (2005:278).
30. Nilakanta Sastri (2004).
31. In Sanskrit, the verb *dha* means 'to hold' and the word *dhani* means 'any container or place that holds something'. The word *dhanya* literally means 'anything that can be stored in a *dhani*'. In course of time, the word *dhanya* was used for 'grains of any kind' and, later in eastern India, where rice was (and is) the most common grain; it came to mean exclusively rice. In modern Indian languages, therefore, the word *dhan* has come to mean rice only.
32. Keay (2001:118).
33. Keay (2001).
34. Nilakanta Sastri (2004:118).
35. Danielou (1965:69).
36. Danielou and Kopalayyar (1989:143).
37. Ziauddin Barani, *Tarikh-i-Firuz Shahi* (1357).
38. Moreland (1922), Habib (1964).
39. Tapan Raychaudhury and Irfan Habib (1982).
40. See website: http://www.sfusd.k12.ca.us/schwww/sch618/Ibn_Battuta/Ibn_Battuta_Rihla.html, and also website: www.saudiaramcoworld.com
41. Barani (1357).

42. Haroon Mohsini, See website: www.afghan-network.net/Culture/shershah.html
43. Nene (2005).
44. Saha (n.d.).
45. Barker and Molle (2002).
46. Rothermund (1993).
47. Nene (2005).
48. Rothermund (1993).
49. Rothermund (1993).
50. Vivek Chibber (1998).
51. Dash (1997).
52. Barbosa, Duarte (1918) as mentioned by Tarafdar (1950).
53. Rothermund (1993).
54. Washbrook (2007).
55. Bedekar (2002).
56. Washbrook (2007).
57. Davis (2001).
58. Bedekar (2002).
59. Harnetty (1991).
60. Romesh Dutt (1902).
61. Das (1993), translation by the author (SDS).
62. Banglapedia.
63. See website: http.www://en.wikipedia.org
64. Quoted in Davis (2001:32).
65. Encyclopaedia Britannica, 1902 Edition.
66. Belfour (1899).
67. Quoted in Davis (2001:52).
68. Davis (2001).
69. Grover and Grover (2004).
70. Encyclopaedia Britannica, 11th edition, 1911.
71. David Barsamian (2001).
72. Amartya Sen (1981), Bhatia (1985), Romesh Dutt (1906).
73. Randhwa (1979).
74. Thakkar (1999), Mishra (1993).
75. See website: http://india_resource.tripod.com/colonial.html
76. Bret Wallach (2004).
77. Bret Wallach (2004).
78. Jaffrelot and Beaumont (2004:154).
79. Schultz et al. (2004).
80. Based on Ramiah and Rao (1953), Richharia and Govindaswami (1990).
81. Mohan (2007).
82. Mohan (2007).
83. Watt (1891).
84. See website: www.indianchild.com/economic_development_in_india.htm
85. Davis (2001).
86. Swaminathan (2007).
87. Kurasaki (2006).

88. See website: http://www.agridept.gov.lk
89. Harischandra (1998), Page 4.
90. Convocation address by Dr. C.R. Panabokke at the Rajarata University. See website: http://www.lankalibrary.com/geo/ancient/rajarata2.htm
91. See website: http://damindu.tripod.com/lanka/sl-11.htm
92. See website: http://damindu.tripod.com/lanka/sl-11.htm
92. See website: http://members.tripod.com/hettiarachchi/history.htm
94. See website: www.mediquipment.com/history.htm
95. Mottau (1980).
96. Mottau (1980).
97. Robert Knox (1641-1720) was an English sea captain in the service of the British East India Company. He was driven ashore on Ceylon in a storm in 1659 while on his way home. He was captured in the name of the King of Kandy along with 17 other members of the crew. The sailors were free to travel in the Kingdom of Kandy, marry and have their own houses and businesses but were not allowed to leave the Kingdom. Robert Knox survived by knitting caps, selling goods and dealing in rice and 'corn'. Knox eventually escaped with one companion after nineteen years of captivity. In 1681 Knox wrote an account of his experiences, accompanied by engravings showing the inhabitants, their customs and agricultural techniques. The book is one of the earliest and most detailed European account of life on Ceylon and is today seen as an invaluable record of the island in the 17th century. *Source:* Wikipaedia.
98. Convocation address by Dr. C.R. Panabokke at the Rajarata University. See website: http://www.lankalibrary.com/geo/ancient/rajarata2.htm
99. See website: http://www.etc-lanka.org/spotlight/spotlight_sep_2005.pdf
100. See website: http://www.mediquipment.com/history.htm

9 | History of Rice in South Asia (1947-2007)

Swarna S. Vepa, Mofarahus Sattar and *S.D. Sharma*

The countries that are now known as India, Pakistan and Bangladesh were, historically, a single country known as India and were under the British rule from the 19th century up to the end of the Second World War. In 1947, India was partitioned into two countries, namely, India and Pakistan. In 1971, Pakistan was further split into two countries, namely, Pakistan (earlier known as West Pakistan) and Bangladesh (earlier called East Pakistan). Sri Lanka was also under the British rule since the 19th century up to the end of the Second World War and known as Ceylon. It became independent in 1948. Nepal and Bhutan, the two sub-Himalayan countries were all along independent. This chapter deals with the history of rice in India, Pakistan, Bangladesh and Sri Lanka that are the major rice growing countries of the region since their independence.

Before the Second World War when India, Pakistan and Bangladesh formed a single country, their agriculture and industries were developed as for a single integrated country. After their independence, they had to re-orient their agricultural and industrial production as independent nations. After about a quarter century of formation of Pakistan, it had to undergo another partition. This upset the economy, especially of Bangladesh, once again. For the last 25 years, Sri Lanka went through political turmoil that disturbed its agricultural production. The history of rice in these countries should be viewed in this socio-political background.

INDIA
(Contributed by Swarna S. Vepa)

India is one of the world's largest producers and consumers of rice, next only to China. It accounts for about 21% of the world's rice production.[1] It is a traditional staple food and is grown in most parts of India. As late as 1950, more than 50,000 traditional rice varieties were being grown by the farmers of India. Aromatic and non-aromatic, long grain and short grain, fine as well as coarse varieties are grown in India. There are sticky varieties and non-Basmati aromatic rice varieties also native to India. Most Indians prefer rice varieties that are not sticky, unlike the rice varieties of many Southeast Asian countries that are often gluey. Many varieties are specific to a region and suit the local tastes. Hence to meet the demands and tastes of the population, it is important for India to produce location specific varieties of rice for location specific consumption. Though the rice production in India is just enough to meet the needs of the people without imports, the rate of growth of production is decelerating and not keeping pace with the population growth. The growth prospects are not very encouraging due to multiple reasons; lower profitability, competition from other crops, wide fluctuations in yields and prices to cite a few. In addition there are research constraints, soil degradation woes, climate change concerns, input inefficiencies and so on.

Both domestic and international trade policies bring anomalies and distortions in price signals to which the production responds.

Production and Productivity Growth of Rice in India—Ups and Downs

Rice in pre-Independent India

In pre-independent India, rice was produced in much larger quantity than wheat. India was almost self-sufficient in rice until 1938 when it started importing small quantities of rice from Burma (now called Myanmar) which was a part of the then British Empire. The rice production in India was estimated to be about 8.474 million tons in 1938. However, the production was fluctuating between a low of 6.7 million in 1941 to 9.3 million tons in 1942 (Table 9.1). In the famine year of 1943, the production was lower than the bumper year of 1942 at 7.628 million but not as low as that of 1941. Pointing to this, Prof. Amartya Sen has argued that failure of rice production was not the major factor responsible for the famine.[2]

Table 9.1. Rice production in India during 1938-1943

Year	Rice production (million tons)
1938	8.474
1939	7.922
1940	8.223
1941	6.768
1942	9.296
1943	7.628

Memory of the famine in pre-independent India influenced the priorities of independent India's first five year plan and the resolve of the subsequent governments to achieve food self-sufficiency. The example of 1943 Bengal famine causing death of millions of people was used for dire predictions in the fifties that severe shortages in food grain production would lead to mass deaths of people in India.[3] Subsequent to the famine of 1943 and the end of the Second World War, rice production improved in India. From an estimated peak of 9.2 million tons in the early forties, independent India was producing about 20 million tonnes of rice by 1950-51.[4] Over the decade the rice production doubled.

Rice in Independent India

In 1950-51, the rice production of independent India was of the order of 20 million tonnes. India has come a long way since then defying the predictions of doom. In 2007-08 India recorded an all time high production of 94.08 million tonnes of rice till date. However the target production that would meet the requirements of the country was projected for 2006-2007 as 100 million tonnes and that for 2015 as 160 million tonnes of rice. The production trend remained upward over the decades despite registering certain fluctuations. The area under rice increased from 30.81 million hectares to a high of 44.71 million hectares in 2001. The major contributor to the production has been the yield (Table 9.2).

In the entire period since independence till date (1949-50 to 2006-07), the rate of growth in production of rice is not very impressive. The area growth was as small as 0.68% per annum. The yield growth was 1.85% per annum. The production growth was an average 2.54% per annum.[6] The annual average rate of growth in rice production and productivity has not been smooth. Unlike wheat, rice production did not record a high growth. The main reason is that more than 45% of the rice area is still not irrigated. The period since independence can be divided into four sub

Table 9.2. Decadal changes in area, production and yield of rice in India[5]

Year	Area million hectares	Production million tons	Yield kg/ha	% coverage under irrigation
1950-51	30.81	20.58	668	31.7
1960-61	34.13	34.58	1013	36.8
1970-71	37.59	42.22	1123	38.4
1980-81	40.15	53.63	1336	40.7
1990-91	42.69	74.29	1740	45.6
2000-01	44.71	84.98	1901	53.6
2006-07	43.70	91.05	2084	52.6*
2007-08	N.A.	94.08	N.A.	N.A.

*Estimate.

periods; the pre-green revolution period, the green revolution period, post-green revolution period and the recent post liberalization period.

In the pre-green revolution period, especially between 1949-50 and 1964-65, rice production recorded a high growth of 3.50%. The area growth was 1.21% and the yield growth was 2.25%. Increase in irrigation, increase in yield as well as area expansion helped rice production to achieve a high growth rate even without the introduction of High Yielding Varieties (HYV) of rice.

In the green revolution period, starting in mid-sixties and ending in 1979-80, the productivity growth helped to increase the rice production. However in the seventies the rate of growth in rice production was only 1.95% per annum. This was because of the slow spread of high yielding varieties. The beginning of the green revolution in India was marked when in 1963 India requested Norman Borlaug to come to India. Two years later in 1965 the government ordered 250 tons of wheat seed from Mexico. The rice revolution began soon in 1966 when International Rice Research Institute (IRRI) released the first modern high yielding semi-dwarf variety IR8. However, India did not depend upon this introduced variety for long. Soil and location specific varieties were soon indigenously developed. Thus the HYV varieties of wheat and rice were introduced around the same time. However the spread of HYV rice lagged behind that of wheat.

The production of rice increased from about 39 million tonnes in 1964-65 to about 49 million tonnes in 1979-80. The production was fluctuating from year to year. During this period the rice production recorded a high of 52 million and 53 million tonnes only in 1977-78 and 1978-79 respectively. Thus the growth rate had decelerated in seventies compared to sixties in the green revolution period. This was because the HYV rice did not spread in the water-rich eastern regions of India.

In the post-green revolution period in the eighties, rice production recorded the highest growth in production and productivity due to the increase in the yield in the eastern zone. The eastern zone has the highest potential for rice production as the rainfall is high. However percentage of rice area irrigated is low. Special efforts were made to spread the HYV rice in the eastern states of Bihar, West Bengal, Orissa and Assam, along with measures to improve drainage and reduce water-logging. In the eighties (1980-81 to 1989-90), the rice production grew at the rate of 3.62% for the country as a whole. Eastern sector recorded a production growth of 6.51% and the northern sector recorded a growth rate of 5.03. Productivity growth was also the highest in the east and north zones at 4% (Table 9.3).[7]

Table 9.3. Zone-wise growth in production and productivity of rice during 1970-80 and 1980-90 in India[8]

| Zone | 1970-80 | | 1980-90 | |
| | Growth in | | Growth in | |
	Production	Yield	Production	Yield
East	0.10	-0.02	6.51	4.02
North	6.80	-4.10	5.03	2.88
South	2.04	-0.98	2.19	3.25
West	0.69	0.41	1.42	3.25
India	1.98	-1.10	3.63	3.25

NB: Figures are in percent.

There was a decline in the area under rice for the country as a whole in the eighties. Despite this, the production growth reached the peak. The area growth was negative at -0.41 per annum. The yield growth was high at 3.19 the highest ever for any decade. The production growth at 3.62 was slightly higher than the pre-green revolution period.

In the post-liberalization period starting with nineties, growth of rice productivity and production decelerated substantially. The production/ productivity growth trend in the 1990s was one-half of realized gains of the 1980s. The growth rate recorded for rice in the last decade of the century between 1990-91 and 1999-2000 was 1.9%. The growth in the yield of rice also decelerated to 1.27%. However, as the rice cultivation and the high yielding technology spread to rain-fed areas and semi-arid regions, there was an increase in the area under rice. The rate of growth of area for the same period was 0.62. The main reason for the deceleration in yield and productions was the stagnancy in the yield in the traditional rice growing areas and moderate gain in the yields of semi-arid areas. Yield growth was also found sensitive to weather conditions. About 47% of the rice area was rainfed in 1999-2000. Hence sensitivity to weather is not surprising.

Rice Production After 2000

The rate of growth of rice production and productivity has further decelerated since the turn of the century and area growth turned negative for rice in the liberalization era of 2000-01 to 2007-08. Fertilizer application was not as effective as before due to depletion of certain soil nutrients over time. All crops had shown deceleration in growth due to the fatigue of the green revolution. A fall occurred in the areas where soil fertility has been depleted. A negative growth rate of rice area at -0.29% per annum was recorded between 2002 and 2006. The yield growth was 2.10% per annum and the production growth was 1.75%. Production growth has been lower than that of the previous decade. The area irrigated under the crop also declined at the rate of -1.86% per annum.

It would mean that irrigated rice area is shifting to other crops even as the yields are improving in the existing rice producing areas. The increase in fallow land in the southern states also may have come from declining rice cultivation. Another important factor could be the declining profitability of rice cultivation due to escalation of labour costs and declining world and domestic prices of rice during this period.

The scenario at present is not really encouraging unless productivity gains are substantially increased to compensate for the loss in area under rice cultivation. Continuous deceleration of production has been worrying the government and therefore efforts are made to increase production and productivity. To sustain the share of rice in total food grain production, as well as to ensure sufficiency, minimum rice production and productivity required in 2006-07 are estimated at 100 million tonnes and 2450 kg/ha (based on a population growth at 1.9% and income growth at 5%).[9] As against this, the production is 94 million tonnes in 2007-08, yet to touch the 100 million mark. The average yield remained around 2040 kg/hectare in 2006-07.

Kharif vs. Rabi Rice

Many irrigated rice growing areas produced three crops of rice in the same field, in rainy season, autumn and summer. Such practices some times lead to land degradations. Similarly, rice-wheat cropping system of Punjab and Haryana damages natural soil fertility. Although introduction of pulses and leguminous crops into the cropping system could make the cropping patterns more sustainable, such measures were not adopted as required for the success.

Rabi (December-April) production of rice increased from 2.7 million tonnes in 1970-71 to 12.6 million tonnes in 2007-08. The *kharif* (July-November) production of rice increased from 39.5 million tonnes to 81.5 million tonnes over the same period. However the yields of *rabi* rice are higher than that of *kharif* rice. *Kharif* yield over this period increased from 1100 kg/hectare to 2015 kg/hectare while the *rabi* rice yield increased from 1625 kg/hectare to 3167 kg/hectare.

Wheat vs. Rice Production

The concept of self sufficiency in production and the expected demand for rice are based on the consumption requirements. Rice is the staple diet of larger population in India than wheat. Hence rice cultivation is more widespread than wheat cultivation across the country. Rice is produced in 14 major states. The production composition of rice and wheat changed over the decades as the rice production was lagging behind that of wheat and the growth slowed down decade after decade since eighties. In 1950-51 the country produced 9.75 tonnes of wheat and 20.58 tonnes of rice. Wheat production was about 32% of the superior cereals production. In

Table 9.4. Change in the relative production of rice and wheat
in India during 1950-51 and 2006-07

Year	Production	Area	Yield	% of area irrigated
Wheat				
1950-51	9.75	6.46	663	34.00
2006-07	73.70	28.17	2617	88.40
Rice				
1950-51	20.58	30.81	668	31.70
2006-07	91.05	43.70	2084	52.60

NB: Production in million tonnes, area in million hectares and yield in kg/ha.

2006-07 the country produced 73.71 million tonnes of wheat and 91.05 million tonnes of rice. The share of wheat in the superior cereal production increased to about 45%. Shift in production by about 15% has also resulted in higher consumption of wheat per capita. It is because wheat is supplied in the ration shops to low income families all over the country. Further, the price of wheat was lower than that of rice even in the open market for most part of the independent India. We can also see clearly that irrigation for rice increased only from 31.7% to 52.6% whereas that for wheat has gone up from 34% to 88% (Table 9.4).

Price Policy, Dual Pricing and Price Incentives

The aim of the price policy in India is to provide incentive prices to producers and affordable prices to consumers. Price stabilization and preventing wide fluctuations in prices are the other objectives. Thus in essence the unique price policy of India to protect both the producers and consumers is basically conflicting. The consumer's interest is to keep the prices low and the producer's interest is to keep the prices high. To achieve these objectives, India has followed the buffer stock policy of procuring rice at above the market price to give incentives and to sell the procured rice through public distribution network to the poor consumers at below market prices. To facilitate such operation, given the year to year fluctuations in production, stocks of rice are maintained by the government and increased during bumper years when prices are low. Besides, in such bumper years, government purchases more rice at a price higher than the market rate. In the lean production years or drought years, the prices are high and government has to pay higher than the market price to procure. Government procures less when it refills stocks to required levels. If sufficient rice could not be procured in a drought year, the government imports rice to augment the supplies. The entire range of costs of paying higher than market price to producers, charging a price lower than the market price to the consumers, maintaining the stocks and distributing through ration shops to the low income ration card holders are borne by the government. This outlay by the government is popularly known as food subsidy.

Despite the inefficiency of the system, the government of India wishes to continue the policy as it helps to give production incentives to producers and helps poor consumers. The popular criticism of the system is that the government is incurring huge food subsidies that could create budgetary deficits and fiscal deficits. The second problem is that the dual pricing encourages corruption through the diversion of low priced Public Distribution System (PDS) rice into the local domestic markets. The third problem is that the procurement of rice is indirect. The procurement quotas are given to the millers who are expected to give the Minimum Support Price (MSP) to the farmers. Often the collusion among the corrupt officials of the distribution network, the procurement officials and the millers results in the existing stock being shown as purchased from the millers, disposed off by the PDS and then repurchased from the millers. In such cases, both the producers and consumers lose the benefit.

The domestic price policy of procurement and minimum support price announced by the government keep the domestic market prices of rice close to the Minimum Support Price announced and provide incentives for production. However, from 2000-01 to 2004-05 the price of rice fell below that of 1999-2000 level while the cost of cultivation continuously increased incurring losses to rice growers. Lower profitability on rice cultivation caused a shift in the rice area in later years. Since 2005-06 the international as well as domestic prices of rice have increased.

However from 2001-02 to 2004-05, the country accumulated large stocks of food grain in stores of the Food Corporation of India due to high levels of procurement and low level of off-take as the open market prices fell substantially below the issue price of the PDS in some areas. At that time, government sold rice to exporters at Below Poverty Level price. This has been criticized by many. In subsequent years the stocks dwindled and the government had to import wheat at exorbitant prices.

Rice Research and Technology Development

Initiatives in the pre-Independence Period

The Imperial (now Indian) Council of Agricultural Research (ICAR) was set up on 16 July 1929 meeting one of the recommendations of the report of Royal Commission on Agriculture of 1926. With the policy guidance and financial support from the ICAR, agricultural research stations (and/or rice research stations) were established in all the provinces (and in some major princely states) in (pre-independent) India to provide better technologies to the farmers for increasing agricultural production. These research stations collected the local rice varieties of their areas, evaluated them under their local conditions and identified the best performing varieties. The then rice breeders of the country (India, Pakistan and Bangladesh) identified the cream of the rice germplasm of the country and recommended more than 445 varieties for general cultivation in preference to the local landraces. The breeding method employed was mostly pure line selection; other methods of breeding such as hybridization, mutation and introduction were rarely practiced. Use of chemical fertilizers or insecticides was negligible in pre-independence days. Hence the improved varieties of that era were bred under (and were adapted to) low input conditions only.

Research in Independent India

After the establishment of the Central Rice Research Institute (CRRI) at Cuttack in 1946 by the Government of India, rice research and training received an added impetus. The traditional rice varieties already collected and maintained by the various rice research stations in the country were screened for their yield potential. Besides, for the purpose of direct introduction of better yielding varieties into the country, many Chinese, Japanese, Taiwanese and Russian types were imported and tested. The Chinese types, first tested in Kashmir Valley prior to 1947, were found fairly successful and recommended especially for the temperate climate conditions.

With a view to improve production of cereals on an international scale after the end of World War II, the Food and Agriculture Organization of the United Nations launched a collaborative project of *japonica* × *indica* hybridization for Southeast Asian countries. The object of this project was to transfer the high yielding capacity and response to higher dose of fertilizers from *japonica* varieties into local *indica* varieties which were already well adapted to the local conditions and had tolerance to diseases and pests of the region. India participated in this project and the Central Rice Research Institute at Cuttack (CRRI) in Orissa state of India became the venue for launching this project. A parallel project of *japonica* × *indica* hybridization was also started by the ICAR with the same objectives for the Indian states. These projects, however, could achieve limited success. Only four varieties, viz. Malinja and Mashuri in Malaysia, ADT-27 in Tamil Nadu (India) and Circna in Australia were released from more than 700 hybrid combinations.

In 1960, the CRRI started another project with the same objective and continued the efforts to incorporate the high yielding fertilizer responsive characters of *japonicas* into the genetic background of Indian rice varieties at eleven state rice research stations. In this project, the objective was almost achieved when the project was abruptly ended in preference to the high yielding semi-dwarf varieties developed by the International Rice Research Institute at Manila.

282

High Yielding Varieties

The International Rice Research Institute (IRRI) was established in the Philippines in 1960. This Institute developed high yielding semi-dwarf varieties within *indica* rices. These semi-dwarf rice varieties were based on the concept of changed plant architecture and used the semi-dwarf gene available in the Taiwanese varieties Dee-Geo-Woo-Gen and Taichung (Native) 1. These high yielding varieties were highlighted during the International Rice Year in 1966 by the Indian Council of Agricultural Research (ICAR) through national demonstration trials. Meanwhile, the ICAR launched the All-India Coordinated Rice Improvement Project (AICRIP) in 1965 that helped in co-ordinating interdisciplinary and inter-institutional rice research within the country for improving production, productivity and profitability of rice in India. This was the beginning of moving towards self-sufficiency in rice production.

ICAR was re-organized in 1965 and again in 1973. ICAR spearheaded the public domain research funded by the government and was instrumental for ushering in the green revolution in India. Over the years, ICAR has developed a large research and training infrastructure to work on the production and other emerging problems confronting agriculture to meet the growing demands for food, fodder, fibre and fuel. It operates through 46 Central Research Institutes, 5 National Bureaus, 10 Project Directorates, 30 National Research Centers, 90 All India Co-ordinated Research Projects, 261 Krishi Vigyan Kendras (Agricultural Demonstration Centers) and 8 Trainers Training Centres. The scientists of these institutions have done a commendable job. Prof. M.S. Swaminathan, the architect of India's green revolution, was a scientist in the ICAR set-up since 1954. He became its Director General during 1972-79 and remained Principal Secretary to the Government of India till 1980.

A program of hybridization between semi-dwarf Taiwanese types/derivatives and the Indian rice varieties was started in 1965 with a view to incorporate the semi-dwarf plant type in the genetic background of elite varieties of India that were already bred during the pre-independence phase and had proved their potential. India operated its most intensive rice breeding program under the AICRIP and achieved remarkable success. Padma and Jaya were the first two varieties that emerged from the programme. During the span of next 35 years (1965 to 2000), 632 high yielding rice varieties were released by the Central Variety Release Committee and by the State Variety Release Committees. These varieties were bred for various ecological stress situations, or to meet the new challenges faced due to diseases and insect pests which were earlier minor but became major, or for grain quality especially the Basmati types for export.[10]

Development of Hybrid Rice

Research program was initiated in early 1970s to develop hybrid rice in the country. However, there was no breakthrough in this program for two decades. The research program was accelerated and intensified from 1989 onwards with a mission mode project. With this concerted research effort, a remarkable success was achieved within a short span of five years and half a dozen hybrid rice varieties were developed from public and private sectors. The first four hybrid rice varieties were released in the country during 1994. Subsequently, two more hybrid rice varieties were also released. By the end of 2001, a total of 17 hybrid rice varieties were released.[11]

Inputs

Fertilizer Policy and Problems

To promote the use of chemical fertilizers, the farm gate prices of fertilizers have been kept unchanged. Per hectare consumption of fertilizers has increased from 69.8 kg in 1991-92 to 113.3 kg in 2006-07 at an average rate of 3.3%. Current fertilizer policy subsidizes 15 types of fertilizers which largely provide NPK (major nutrients) by fixing maximum retail prices (MRPs).

A large proportion of fertilizer subsidy goes to the fertilizer units which is paid on a (group based) cost plus basis. Although 60% of fertilizer production is gas-based, due to non-availability of adequate natural gas, some of these units through dualistic option use naphtha, which is a costlier feed-stock. The current system has allowed the inefficient units to persist.

The current pricing mechanism of fertilizers has encouraged nutrient imbalance. There is excessive use of urea and a bias against micronutrients. As against the desirable NPK proportion of 4:2:1, the average use is 6:2.4:1. The Steering Committee of the Planning Commission has observed that "because nitrogenous fertilizers are subsidized more than potassic and phosphatic fertilizers, the subsidy tends to benefit more the crops and regions which require higher use of nitrogenous fertilizer as compared to crops and regions which require higher application of P and K". The excessive use of urea has also affected the soil health adversely.

A healthy plant growth is possible only if all the 16 nutrients are available in the soil. Besides NPK, sulfur, zinc and calcium are also required in good quantity. Other nutrients such as iron, boron, etc. are also required in small quantities but their deficiency impacts plant growth and life significantly. Micronutrients are best applied through fortification of major fertilizers. However, restrictions of MRP, as well as fixed subsidy, afford no incentive for such fortification. As a result, though NPK requirements are partly made good, micronutrient deficiency continues to affect the productivity of crops significantly.[12]

Seed Production and Distribution

Seed is the carrier of new technology for crop production and higher crop yields. It is a critical input for sustained growth of agriculture. More than 80% of the farmers rely on farm-saved seeds leading to a low seed-replacement rate. The Indian seed production has the involvement of the Central and State Governments, ICAR, state agricultural universities and the cooperative and private sector seed companies. There are 14 State Seed Corporations (SSCs) in addition to the two national–level corporations including National Seed Corporation (NSC). Though the private sector has begun to play a significant role in production and distribution of seed, particularly after the introduction of the Seed Policy of 1988, the organized seed sector, particularly for food crops and cereals including rice continues to be dominated by the public sector.

However, it is estimated that about 46% of the seed commercially sold in the country is by the private seed companies. The annual rate of growth of certified/quality seed distribution is expected to accelerate from 12.1% in 2005-06 to 18.1% in 2006-07. During 2006-07, 7383 tonne breeder seed was anticipated to be produced by the National Agricultural Research System.

International Trade in Rice in India

Exports in pre-Independent India

Indian export trade in the 19th century consisted of several items such as raw cotton, cotton yarn, cotton cloth, opium and rice. Under the impact of the global diffusion of industrialization of Europe and de-industrialization of India, the structure of Indian trade was transformed from a demand-pull type based on luxury goods trade to that of a developing country, exporting primary products, both raw materials and foodstuffs, to industrialized nations and importing their manufactured goods.[13]

Economic development and rapid industrialization of the 19th century Britain had to ensure supply of cheap food to its workers. Rice exports from India were essential to provide food grains at low prices to aid industrialization of Britain. If food prices rise, most of the income would be spent on food leaving very little to fuel the demand for industrial consumer goods. Further, India became a market for consumer goods from Britain to assist the growth in export of British industry. Hence under the then prevailing economic situation, export of rice was considered as a distress factor for India.

Exports in Independent India

Rice export was banned after independence following the policy of achieving self sufficiency. Rice was not exported from India till 1970-71. Rice production picked up in the seventies and eighties and trade was liberalized in 1990-91. Starting from 500 tonnes in 1992-93, rice exports reached a peak of about 4963 tonnes in 1999. Since then, rice exports have declined. Rice exports stood at 4088 metric tonnes in 2006. Gulf countries and Saudi Arabia are the major importers of Indian rice (Table 9.5).

At present rice exports are contributing substantially to the agricultural exports. Exports of Basmati rice are especially encouraged.[14] Basmati rice and non-Basmati rice, both are exported from India to various countries around the world. The variety Basmati is unique in its quality charac-teristics and has a good demand in international markets. But presently India is facing stiff competition with other rice exporting countries of the world.

Measures are also taken to restrict export of non-Basmati rice so that the domestic consumer is protected from shortages and high prices. Export restrictions are imposed on non-Basmati rice from time to time to maintain sufficient stocks and prevent supply drain that may lead to high prices to the domestic consumer.

"Fixing a higher minimum export price definitely helped domestic supply in January 2008. It protected lower priced rice only for government procurement. In the normal case, the same quality rice could have been exported to Africa or Asia. India raised the minimum free-on board export price thrice through October to December after relaxing a ban on non-Basmati exports to help government purchase. Base export price was raised from $425/tonne to $500 during the period. Non-Basmati rice prices range between $150 and $1200 a tonne depending on the variety and region where it is grown. India exported 3.7 million tonnes of non-Basmati rice, worth 42.43 billion rupees in 2006/07, a 39% rise in value terms over last year as government data showed."[15]

At present the international supply-demand situation is tight, despite bumper crops in India and China, due to the increased demand from Africa and Asia. The prices are likely to increase.[16]

Problems of Rice Exports

Rice exports are facing several problems in India. The major problem is the price competitiveness compared to the rice exports from Thailand, Vietnam and Pakistan. There are several export levies and taxes on rice exports in several states of India to discourage exports and to improve local availability. The cost of production is much higher in India compared to other countries. The minimum support price is high in India and the market prices often rule higher than international prices making exports uncompetitive. Many exporting countries give export subsidies to rice.

Rice mills have not been fully modernized to ensure high milling recovery and to reduce the percentage of broken rice. Lack of infrastructure facilities for export, especially port facilities and shipping facilities create bottlenecks leading to cancellation of orders by importers from other countries.

In these days Basmati rice is facing aroma problem because intensity of aroma in traditional Basmati varieties is not as high as it used to be. Basmati varieties are highly prone to lodging and lodging affects the natural grain development. In such situations, both aroma and linear kernel elongation are affected. Post-harvest handling of produce is another important aspect. Generally, farmers harvest the crop at different moisture levels and keeping the produce at higher moisture level for a long period impairs the intensity of aroma. In absence of genetically pure seed of Basmati varieties, variation in plant height, grain size and maturity of the crop is found in majority of Basmati rice fields. This is one of the major reasons for poor quality of Basmati rice.

Rice Imports

Rice imports were negligible for a long time in India. Historically, prior to liberalization in independent India, domestic prices of rice were always above the international prices giving an incentive for farmers to produce rice. Rice imports in the pre-liberalization era were controlled totally by the State Trading Corporation, a government body based on the estimated shortage in the procurement for the public distribution system. Since trade liberalization, imports are allowed by private parties. Since 1990-91 after the liberalization, the domestic prices of rice have come closer to the international prices. The imports were 102.38 thousand tonnes in 1992-93. The rice imports have come down over years and were negligible in 2006-07.

As per the World Trade Organization's guidelines, the quantitative restrictions on imports and exports have to be removed and the countries will have to move to an import and export tariff regime. India has been protecting the domestic producers by imposing high tariffs on imported rice. Thus rice imports have been prevented. Considering the high cost of production and the procurement for the public distribution system, it is not possible to protect the domestic producers without high import tariffs. The low profitability of rice has been making many farmers to shift to non-rice crops. Announcement of high minimum support price and preventing imports of cheap rice so far has been able to protect the rice producers to some extent. Import tariffs are very high for rice and wheat, making it very expensive to import for the domestic trade (Table 9.5).

When India gained her independence just after the Second World War,

Table 9.5. Export and import of rice in India during 1990-91 to 2005-06 (000 tonnes)[17]

Year	Imports	Exports
1990-91	66.04	504.99
1991-92	12.12	678.24
1992-93	102.38	580.04
1993-94	75.52	767.67
1994-95	6.99	890.57
1995-96	0.08	4914.01
1996-97	0.00	2511.98
1997-98	0.05	2389.86
1998-99	6.63	4963.59
1999-2000	34.99	1896.12
2000-01	13.20	1534.48
2001-02	0.06	2210.98
2002-03	0.87	5057.43
2003-04	0.05	3412.05
2004-05	0.00	1010.24
2005-06	0.00	4088.20

the food situation of the country was grim. In subsequent years, the country had to depend on PL 480 grants of USA for her subsistence. In 1964, the then Prime Minister of India, Mr. Lal Bahaur Shastri, gave a call to his countrymen to "miss a meal a week." Thanks to the determination of the policy makers and efforts of its rice scientists, the country is now in a comfortable situation with regard to its food production despite three-fold increase in its population and has a promising outlook as reflected in the Report of the National Commission on Farmers (2006):

> "Prime Minister has rightly emphasized the need to double annual food grain production from the present 210 million tonnes to 420 million tonnes within the next 10 years, i.e. by 2015, which is also a benchmark year for achieving the UN MDGs. This will call for producing at least 160 million tonnes of rice from 40 million ha and 100 million tonnes of wheat from 25 million ha. Pulses, oil seeds, maize and millets will have to contribute 160 million tonnes.[18] In a major new initiative, the Union Government has launched the National Food Security Mission (NFSM). The Mission aims at increasing the production of rice by 10 million tonnes during the 11th Five Year Plan ending in 2011. NFSM-Rice will be implemented in 133 districts of 12 States."

BANGLADESH
(Contributed by Mofarahus Sattar)

Rice is the principal crop of Bangladesh and has been cultivated since ancient times. The chalcolithic background of this region provides evidence of peasant groups, dating back at least 6th century BC, who were involved in rice cultivation.[19] Earliest direct evidence of rice cultivation is found from the stone inscription from Mahasthan (near Bogra) of 3rd century BC, which records instructions for the 'Mahamatra' to distribute money and 'Dhanya' (rice) from the royal treasury at the time of need.[20]

The present day Bangladesh constitutes the eastern part of Bengal province of British India. When British colonial rule ended in 1947 and India was partitioned, Bangladesh formed the eastern wing of Pakistan. In 1971, Bangladesh severed its links with Pakistan and became a sovereign independent nation.

Rice in Bangladesh Economy

Bangladesh agriculture is dominated by rice as it covers almost 75% of total cropped area. At least 65% of all employment is in rice production, post-harvest processing, storage, and handling. Contribution of agriculture to GDP is currently 23.5%,[21] while contribution of crop agriculture is 13%. Rice alone contributes 40% among crops. Over 75% of the population live in rural areas and an estimated 80% of them are directly engaged in rice production.

Geography and Rice Culture

Bangladesh is the largest delta in the world intersected by three main river systems, Ganges (Padma), Brahmaputra and Meghna, and a number of tributaries and distributaries.

Except for small hills in the north-east (Sylhet region) and south-east (Chittagong region), the country is largely flat and rises only a few meters above the sea level. The Sylhet basin with its *haor* systems and the Gopalganj-Khulna *beels* are the perennial low-lands. The largely alluvial plain receiving an average annual precipitation of 2000 mm, having an average 25° to 30°C temperature regime provide excellent growing condition for rice. It can be grown around the year as *aus, aman* and *boro. Aus* rice is generally direct seeded and grown during March-April to May-June-July. *Aman* rice is normally transplanted between July and September and harvested during October to December. *Boro* rice is usually transplanted during January-February and harvested by April-May-June. Broadcast *Aman* or *Joli Aman* is the deep water rice and usually grown between March-April and November-December and can withstand from 1.8 to 2.0 meters of flooding.

Until the end of British rule in 1947, farmers were growing traditional varieties that evolved over centuries in a biological adaptation process

Table 9.6. Local rice varieties of Bangladesh grown until advent of modern varieties

Aus	Transplant Aman		Broadcast Aman	Boro
Kataktara	Nizershail	Dudhsar	Malia bhangar	Kali boro
Hashikalmi	Latishail	Kalijira	Gabura	Tepi boro
Dular	Jhingashail	Tulsimala	Baisbish	Habiganj boro II
Shaita	Indrashail	Biroi	Pankhiraj	Habiganj boro IV
Dharial	Raghushail	Kataribhog	Bagdar	Habiganj boro VI
Marichbati	Harma	Badshabhog	Bajail	
Panbira	Tilokkachari		Laki, Goai	
Charnock	Dadkhani		Dhala aman	
Kumari	Chitraj		Lal aman	
	Patnai-23		Chaplash	

under influence of agroclimatic forces. Some of these traditional or local varieties, as mentioned in Table 9.6, were recommended by researchers and became popular.

Extent of Rice Cultivation Since 1947

The growth of rice cultivation in Bangladesh is presented below in Table 9.7, which starts from the year 1947, but proceeds with every five-year data starting 1950. Total land under rice cultivation in 1947 was 7.9 million hectares (ha) and the production was 6.7 million tons (t). The yield level was very low at 0.85 tons per hectare (t/ha) only. Although Bangladesh was reportedly self-sufficient in food until the pre-World War II period, the famine of 1943 changed the circumstances. Despite production starting to gain grounds after the war, the chaos of partition affected the growth and we see the country slowly slide towards deficiency. This trend of deficit never recovered for next four decades as growth rate of population outpaced that of production. Productivity remained low because cultivation technology was completely traditional and cultivators were engaged in the trade absolutely with a subsistence motive. Use of chemical fertilizers was practically unknown to ordinary farmers in the early 1950s.[22] The new regime of Pakistan did little to improve the situation through strengthening of either research or extension, especially when a number of non-Muslim experts left to live and work in India.[23] During the

Table 9.7. Area, production, yield of rice in Bangladesh, 1947-2005[25]

Year	Area ('000 ha)	Production ('000 t)	Yield (t/ha)
1947	7,900.00	6,720.00	0.85
1950	8,095.00	7,380.00	0.85
1955	7,800.00	6,380.00	0.80
1960	8,800.00	9,520.00	1.00
1965	9,200.00	10,371.00	1.10
1970	n. a.	11,919.00	1.11
1975	9,790.20	11,109.00	1.13
1980	10,157.40	12,539.00	1.23
1985	10,222.20	14,622.00	1.43
1990	10,411.10	17,710.00	1.70
1995	9,930.70	16,832.70	1.82
2000	10,712.96	23,067.00	2.15
2005	10,296.78	25,183.00	2.45

period between 1952 and 1962 rice production increased by only 11.72%.[24]

By late 50s and early 60s a number of organizations were set up to provide support to agriculture, especially to increase rice cultivation. Bangladesh Water Development Board (BWDB—the then WAPDA or Water and Power Development Authority) was established in 1958 to develop and operate certain irrigation projects, such as, Ganges-Kobadak Project in Jessore-Kushtia region, DND Project in Dhaka-Narayanganj, Dinajpur Thakurgaon Deep Tubewell Project, etc. Bangladesh Agricultural Development Corporation (BADC) was established in 1961 mainly to provide production inputs like fertilizer, seed, irrigation equipment, and power tillers. Rapid Soil Fertility Survey (1957-59) enabled BADC to develop balanced fertiliser application rates for rice and other crops. Over and above, Department of Agricultural Extension (DAE) started to expand its network to provide extension services. The Agricultural University (BAU) was established in Mymensingh in 1962 to have better trained agriculturists for research and extension. Despite these facts, demolition of Dhaka Agricultural Experiment Farm in 1962 without re-locating research activities to another facility or sanctioning any new farm dealt a severe blow to rice research.

After IRRI developed IR8, it was introduced by Bangladesh Academy for Rural Development (BARD) in Comilla. Slowly other HYVs like Purbachi (originally from Japan called Chen chu ai) and Pajam (originally Malinja from Malaysia) were introduced and got huge popularity. BARD trained farmers in Comilla since late 60s to apply better input standards to achieve higher yield and the result was that Comilla Kotwali Thana showed significantly higher rate of growth of rice production in Bangladesh for a long time since then.[26]

The liberation war of Bangladesh in 1971 again affected rice production that took a couple of years to normalize. Rice area expanded, according to Goletti,[27] during the 70s at 0.66% and during the 80s at -0.21%, but the HYV increment was at 5.06% and 7.28% respectively. During the same periods the rice yield growth rate was 2.16 and 2.72 respectively. The rice production growth rates, however, give a brighter view, as presented in Table 9.8. Production grew at a rate of 2.5% per annum during 1975-90, but real surge at 4.71% was registered during 1985-90.

From late seventies to the 90s and beyond, other than HYV expansion, a number of policy reforms contributed to this growth. These are described under policy reforms. But there was another dimension to it. The gradual shift of rice cultivation in favour of *boro* from *aus* is also a factor. Since rice yield is the highest during the *boro* season due to more abundant solar radiation, and as irrigation scope expanded, farmers chose to grow more

Table 9.8. Rice production and yield growth rates in Bangladesh during 1975-90[28]

Year	Annual rice production growth rate	Average rice yields (t/ha)	Annual growth rates of rice yields (%)
1975-79	1.02	1.88	1.13
1980-84	1.66	2.07	2.54
1985-90	4.71	2.37	4.63
Total 1975-90	2.50	2.12	2.30

boro rice. *Aus* crop shrunk to one-third its area from 1971-72 to 2004-05, while that of *boro* rose more than four-fold during the same period. Production of *aus* rice over the same period was reduced by half while *boro* production skyrocketed eight times, in fact also surpassing total *aman* rice production. Currently total rice area covers more than 10 million hectares producing over 25 million tons.

Demand and Supply of Rice

But population also grew during the same periods at a faster rate outpacing food production.

The population growth rates in the 50s and 60 were around 2%, in 70s at 2.32%, in the 80s at 2.03% and in the 90s below 2% per annum.[31] But rice production growth rates remained below 1% levels until about late sixties (Table 9.9). By 70s and 80s production growth rates started gaining pace as mentioned earlier, and by early 90s reached nearly 5% growth per annum. As a result, we see that Bangladesh ultimately reached rice self sufficiency by the turn of the century stepping out of all those previous years of deficit (Fig. 9.1).

Table 9.9. Population, food requirement, production and deficit/surplus during 1951-2001 in Bangladesh[29]

Year	1951	1961	1973	1981	1991	2001
Population (million)	42.45	50.90	71.48	85.69	111.50	130.00
Food requirement (million tons)	7.34	9.73	13.31	16.40	20.69	24.90
Food production* (million tons)	7.03	9.55	10.02	14.50	18.25	26.90
Deficit (million tons)	0.29	1.81	3.29	1.90	2.44	+2.00

*Includes rice and wheat.

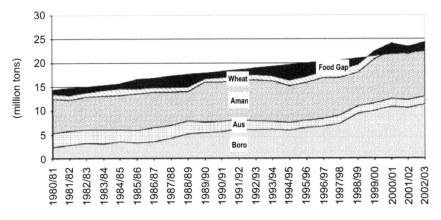

Fig. 9.1. Food grain production and food gap.[30]
Source: Dorosh, del Ninno and Shahabuddin.

Production Practices of Rice Farmers

Rice land preparation till today is largely with country plough using bullocks. The plough makes shallow furrows and hardly inverts the soil, but suffices as much of the rice is transplanted in puddled soil with very shallow submergence. Use of tractors is very rare, but power tillers are increasing. Other cultural practices of weeding, harvesting, threshing, cleaning and processing are also largely manual as labour is still abundant in rural areas. As regards fertilizer use, Samad[32] records that 'in 1951-52, the earliest year for which any data are available, about 10,000 tons of ammonium sulfate was used, almost entirely in the tea gardens. Several schemes were undertaken for popularizing fertilizer use, but progress was poor due to prejudice, apathy, and ignorance about the benefits of fertilizer application among farmers, almost total absence of a marketing infrastructure, and a lack of vigour in the government effort. Fertilizer consumption rose to only 50,000 tons by 1959-60.' However, with expansion of HYVs and reform in fertilizer distribution system in the late 70s, use of fertilizers gradually increased. However, till this day, application of recommended and balanced doses has been highly erratic. A number of researchers including the author, who conducted large agro-socio-economic surveys in 1981, 1991-1993 and 1999-2000, have reported over-use of urea and under-use of phosphate and potash. Use of herbicide is close to non-existent and use of other pesticides has increased after late 70s. Farmers usually take care of their own rice seed requirements, but the production and management of seed as a separate crop has never been in practice and they chose to keep a portion from the general produce.[33]

While by 1980 certain study areas revealed 80% adoption of modern varieties,[34] the same farmers still were transplanting rice haphazardly as late as 1982, rendering improved weeding tools unsuitable for use.[35]

Rice Research

One of the brightest aspects of rice in Bangladesh has been its historic research achievements. A chronology of important events in rice research in Bangladesh is summarized below.

1910   Rice research started in the Indian sub-continent when Dhaka farm was established.

1911   Breeding work on *aus* and *aman* started.

1922   Improvement work on deep water rice started at Dhaka.

1934   Improvement work on deep water rice accelerated with establishment of Habiganj Rice Research Station. Up to this period quite impressive progress was achieved given the limitations of facilities and expertise. Local varieties were screened to select pure lines and were recommended (a few named in Table 9.6); genetic manipulation, induction of mutation by polyploidy, etc, were taken up and some improved varieties were released.

1947   Improvement work on *boro* rice started at Dhaka.

1950   Improved methods of rice culture (e.g., line planting, increased input use) following expertise support from Japan.

1962   Demolition of Dhaka Farm to create second capital of Pakistan results in severe blow to rice research, as also no alternative land for experiment was arranged.

1970   Bangladesh Rice Research Institute (BRRI) established at Joydebpur farm, now in Ghazipur district, about 32 km north of Dhaka.

1973   BR3 (Biplab) released, the first BRRI variety to move Bangladesh rice production beyond IR8.

1975   BR4 (Brrishail) released, for many years remained very popular as transplant *aman*.

1980   BR11 (Mukta) released, became most widely grown and popular of BRRI varieties.

1994   BR29 for *boro* and BR32 for *aman* seasons released, bringing in higher productivity and shorter field duration.

2001   Brri Hybrid 1 for *boro* season released, ushering in home-grown rice hybrid seed era.

BRRI has also been continuously working on specific problem aspects of rice production, such as: flood, drought, salinity, and cold tolerance, pest resistance, integrated pest management (IPM), etc. It has identified a number of local varieties, which possess either fair or mild or substantive tolerance level to one or the other adverse condition. BRRI scientists are now trying to induce such genes into new varieties to achieve intended tolerances. In order to better suit the needs of the small and marginal farmers—who are the overwhelming majority of the farming households— BRRI is tuning transferable technologies with a whole-farm approach and therefore also collaborating its Rice Farming Systems and Adaptive Research programs with other national agricultural research organizations. It is also very important to note here that BRRI took steps in early 80s to conserve the local varieties, as they started disappearing rapidly in face of HYV expansion, and established a Germplasm Bank. It has over 7500 local varieties preserved by now in the gene bank.

Policy Reforms

BADC was established in 1961 specifically to ensure timely and adequate availability of modern inputs to farmers, which are mentioned earlier. Fertilizer was sold through excessive bureaucratic control which dampened dealer initiative. Instead of selling from over 450 Upazilas (sub-districts), BADC established 76 primary distribution points (PDPs) across the country during 1975-77, beyond when movement of fertilizers went in the hands of private wholesalers. Farm level prices were decontrolled in 1982-83. Samad[36] notes that 'During the 5-year period following the price decontrol, the marketing margin, i.e., the difference between wholesale and farm level prices of fertilizers, shrunk by approximately 29% (in deflated 1974 prices), constituting a significant gain to farmers.' Also in 1991, fertilizer import went from public-sector monopoly to private sector. By then fertilizer use increased by an average of 8.5% per annum; and reached a total of 2.3 million metric tons of fertilizer nutrients in 1994.

BADC's seed supply, comprising HYVs and local improved varieties through 22 seed multiplication farms and contract growers, accounted for near about 5% only of the total country requirement until mid-80s.[37] The remaining 95% requirement was met by farmers' own stock. Since late eighties efforts were taken to liberalize seed production and marketing to private sector and creating a level-playing field for the market actors. Seed Policy promulgated in 1993 created provisions for private sector quality seed production with Seed Certification Agency in a monitoring role for quality control. As a result BADC now supplies over 23% of all HYV rice seed (about 15-16% of total rice seed) while the private sector supplies over 95% of the hybrid seed marketed.

Pest control coverage was the responsibility of DAE and was free until 1974, when the government started taking nominal payments primarily to check abuse. Despite that farmers complained of poor and inadequate service, ordering was inappropriate, wastage was massive due to poor storage, and pest control budget was going up every year. So government transferred pest control services to the private sector, when after a year, import dropped by 50% but without decline in coverage.[38]

One of the most spectacular effects was seen in irrigation policy reforms. Between 1986 and 1988, government slowly liberalized private import of diesel engines for irrigation purposes. Finally in early 1989, restrictions on private installation of wells were removed by suspending key clauses in the Groundwater Management Ordinance. Only in the first three years between 1988 and 1990, irrigated area expanded by 3% per annum.

Famines and Disasters

The famine of 1974 in Bangladesh is one of the most tragic experiences in recent times and has been a much discussed topic among researchers examining its causes and consequences. The causes of the famine were multi-faceted, from natural to economic, social, and to political maladies.

The prelude to the famine lay in the economy still weak from the devastations of the liberation war of 1971. Budgetary deficit resulted from severe revenue shortfalls as well as from increased expenditure of establishing new government, and large deficits in bank financing hindered credit for manufacturing while private credit went increasingly to trading, construction, and commodity speculation.[39] Harvests since 1971 although started recovering from war ravages, were still lower compared to 1969-70 indexes. Post-war relief distribution had already forced the cash-short government to import on commercial terms 1 million tons of food grains during 1972-73.[40] By early 1974, Government stocks decreased drastically and public distribution through rationing and relief plummeted[41] causing serious loss of confidence in government's ability to improve the situation and rice prices to surge. Government attempts to procure foodgrains internationally met with further setback due firstly, to global food shortages pushing prices to prohibitive levels; secondly, to high demand for shipping spaces by all the other countries trying to import food—also escalating freight prices; and thirdly, to oil crisis. A heavily depleted foreign exchange reserve coupled with unavailability of short term commercial credit, seriously undermined government's capacity to procure and ship foodgrains from abroad.

The famine struck during the peak monsoon of 1974, when catastrophic flooding damaged standing jute crop and delayed *aman* rice planting. Although records available later show that in reality total rice production during 1974 was not lower than 1973, the flood critically affected the employment and income of wage labors dependent on cultivation and processing of jute—the main cash crop of the day. Delay in *aman* planting raised fears of a shortfall, fuelled further by adverse and incorrect newspaper reports, and worsened to the core by accumulation of traders' stocks and speculative hoarding by farmers, consumers, and anybody who could. Rice prices soared to record levels, increasing 250-300% compared to similar periods of previous years. By then landlessness over the past decade had swelled to a staggering 40% of the households,[42] who mostly depended on wage labor for access to food.

A very significant turn of event aggravating the crisis was postponement of committed US food-aid due in October 1973. Negotiations for the food-aid mysteriously lingered for almost nine months to fulfill this or that clause of PL-480, the US law governing such aid. Later, the aid was halted on the ground that recipient Bangladesh exported jute bags to Cuba—regarded an enemy country by the US, whose law prohibited such trading to qualify for the aid. Although Bangladesh government in desperation even scrapped the deal and stopped shipment of the jute bags, the food-aid of 250,000 tons arrived much later in 1975, when the famine had already taken its toll.[43]

It is widely reported that nearly one million people died from starvation during this famine;[44] the figure, however, is not without criticism as being highly inflated. The victims were mostly farm wage labors as the famine was a rural phenomenon, but industrial and other labors were also affected. The districts of Rangpur, Mymensingh and Sylhet were the worst hit, surprisingly the same areas as during the Great Bengal Famine of 1943 and the worst period had been July through October, largely the same as in 1770, 1866 and 1943 famines. People traversed long distances through villages, towns and to cities in search of food. Families were separated and many were totally uprooted. Distress sale of land and assets became very common. The biggest fallout of the famine culminated in the assassination of the President and his family, who founded the nation, and overthrow of his government in a military coup in August 1975. The incident however, did not bode well in the long run for the unrepentant nation.

The flood of 1987, 1988 and 1998 also caused widespread crop losses, especially *aman* rice, but government and peoples' efforts were able to avert any significant loss of lives, although livelihoods were affected. Of particular mention is the flood of 1998, when over 60% of the country remained under the deluge for three long months. Increasingly governmental capacities grew to face disasters, substantial international aid arrived, and peoples' organizations and non-governmental organizations (NGOs) worked ceaselessly to ameliorate human sufferings.

Future Potentials and Challenges

If the past is any indication of its future potentials, then rice in Bangladesh has both very bright and highly challenging prospect. The prospect is bright as yield levels have a potential to increase three times the present average at 2.5 t/ha, by expanding coverage under modern and of hybrid varieties. But the bigger challenge collaterally remains the supply and availability of adequate amounts of required inputs, particularly fertilizers and quality seed. Current HYV area covers *ca.* 65% of the total acreage producing nearly 80% of all rice. But required fertilizer quantities for a balanced application have never been available in the country. Tendency of excessive urea application, as already mentioned, creates a shortfall in demand, whereas cash-poor small and marginal farmers apply very low amounts or none. Phosphatic or potash application has always remained low, between 20 and 30% of the recommended dose,[45] and therefore, this has been one of the reasons why yield potential has never been realized. Similarly, availability of certified seed has been only 23% of total HYV requirements, as mentioned earlier. Diffusion of technology has been very slow as extension services have never been very successful.

But the aspects most difficult to improve upon are the social phenomena governing the exploitative tenurial practices, land fragmentation, scattered plots of farm holdings, and absenteeism. There is an overwhelming tendency among almost everybody to invest their savings in land. These processes also accelerated marginalization and as a result landless and marginal farmers[46] currently constitute between 52 and 58% of the rural households. Further 25 to 30% small farmers together form a colossal 77-88% of the rural households, who perpetually suffer from cash shortage that prevents them from procuring adequate fertilizers, quality seed, and for paying labors to allow timely cropping operations. Any attempt at reforming the landholding and tenurial patterns to create incentives for cultivators is bound to meet with failure as the members of the powerful bureaucracy constitute a significant section of the absentee landlords. A partial solution to this has been attempts to disburse seasonal credits for crops. But so far such credits by commercial banks and NGOs have never been disbursed beyond a very small chunk of the farm households.

Growing population and demands for housing, roads, industries, and other infrastructures are continuously encroaching upon crop lands, causing a continuous decline in cultivable land. Flood and other disasters once every 2-3 years often wash away the built-up capital necessary for furthering production growth. Increasingly a better and concerted management to tackle disasters is taking shape, but more efforts are needed.

In technological terms, achievements of rice scientists have been more than spectacular, which however remains underutilized. In order to harness full benefits of these technology a proper social, institutional and policy environment needs to be ensured. A better political understanding and knowledge-oriented leadership can effectively pave the way to overcome these challenges towards desired rice production.

PAKISTAN
(Contributed by S.D. Sharma)

When Pakistan was a part of undivided India, Punjab was considered the granary of India. In the initial years of her independence and through most of the 1950s, agricultural surpluses were taken for granted in Pakistan and the government policy demonstratively discriminated the agricultural sector in favor of the manufacturing industry. This neglect soon started to take its toll and the surplus existing at the time of inception of the country was rapidly dissipated. This made Pakistan to resort to wheat import for avoiding food shortages.

Most of the crops in Pakistan are grown in the Indus River plain in Punjab and Sindh. Rice is grown mostly in these two states. (Table 9.10) Considerable expansion and development of agriculture has taken place since the early 1960s. However, the country's potential from the well-irrigated and fertile soil has still not been fully realized. In FY 1993, agriculture, small-scale forestry and fishing contributed 25% of GDP and employed 48% of the labor force.

Wheat is the most important cereal crop of Pakistan and staple food of the majority of the people. The wheat production in Pakistan has several phases since its independence: (a) 1947-65, that is the period prior to the release of semi-dwarf wheat, (b) 1965-1972 when there was spread of semi-

Table 9.10. Percentage of rice harvested area in different
provinces of Pakistan[47]

Province	Harvested area as % of total rice area
Punjab	59.46
Sindh	32.13
NWFP	2.86
Balochistan	5.52

dwarf varieties and there was a dramatic increase in productivity, (c) 1973 to 1980s when there was less dramatic growth, and (d) declines in productivity staring in the 1980s and extending into the 1990s. Between 1961 and 1990, the area under wheat cultivation increased nearly 70% and yield increased 221% in Pakistan. In the early and mid-1980s, Pakistan was self-sufficient in wheat but by early 1990s, more than two million tons of wheat was being imported annually. With the support and incentives of the Government of Pakistan, the wheat production is picking up and for the first time in the history of Pakistan, the country exported wheat in 2000.

Rice is the second most important cereal crop grown in Pakistan. Rice yields have gone up sharply since the 1960s following the introduction of high-yielding rice varieties. Rice production increased by 40% during 1970/71 and 1992/93 period.[48] In 1992, rice was grown in about 2.1 million hectare and production was 3.2 million tons of which one million ton was exported.

Two major types of rice are grown in Pakistan: the traditional aromatic Basmati rice and the "IRRI-Pak" type, a term used for semi-dwarf high yielding rice varieties introduced initially from International Rice Research Institute (IRRI), Manila but subsequently bred in Pakistan. Basmati rice accounts for 62% of total rice area and 52% of total rice production in the country. The Basmati rice varieties are mainly cultivated in the Punjab province. There is 60% increase in Basmati area with 155% increase in its production from 1982 to 2002 in the Punjab province (Table 9.11).[50]

The Pakistan Agricultural Research Council (PARC) was formed in the mid-1970s. PARC has the authority to conduct research and has several research institutions under it including the National Agricultural Research Centre (NARC). The irrigation, land reclamation and water management research are conducted by the Water and Power Development Authority (WAPDA). There are two rice research institutes, one at

Table 9.11. Area, prodiction, import and export of rice
in Pakistan during 1985 to 2000[49]

	1985	1990	1995	2000
Harvested area (Th ha)	1863	2112	2161	2312
Yield (kg/ha)	2349	2315	2752	3027
Production (Th t)	4378	4891	5049	7000
Import (Th t)	7	25	68	NA
Export (Th t)	718	743	1852	NA

Kala Shah Kaku for the Punjab province and the other at Dokri for Sindh province. The Rice Research Institute at Kala Shah Kaku was started for the canal area of Chenab as a Rice Research Station in 1926. The Kalar tract located between the Ravi and Chenab Rivers is the home of Basmati rices in Pakistan. The Institute has been doing research and selection of Basmati types for the last eighty years. Basmati 370 was developed in 1936 and remained a popular variety of Punjab up to 1960s. From 1968 onwards, the Institute has released many strains of Basmati. In 1996, the Station released Super Basmati. The Rice Research Institute at Dokri was initially established as Agriculture Research Station in 1938. In 1950, it was converted to Rice Research Station with the opening of Sukkur Barrage. In 1973, it was upgraded to the level of Rice Research Institute.

Irrigation

During the British rule, the Punjab province had received better attention than other provinces of India. For example, the area of a district (the smallest administrative unit of the country) was much smaller than that of many other provinces (such as Bengal or Bihar) and much less in population than that of many other provinces (such as United Provinces, now Uttar Pradesh State of India). This resulted in more efficient administration of the province. Secondly, a network of irrigation canals was laid in the then western Punjab (now the Punjab province of Pakistan). In addition, the Sukkar Barrage was built on Indus River in Sindh in 1932. In total, 37,000 kilometers of canal in Punjab and 10,000 kilometers of canal in Sindh were built during the British rule.[51]

Partition placed portions of the Indus River and its tributaries under India's control leading to prolonged disputes between India and Pakistan over the use of Indus waters. After nine years of negotiations and technical studies, the issue was resolved by the Indus Waters Treaty of 1960. After a ten-year transitional period, the treaty awarded India use of the waters of the main eastern tributaries in its territory—the Ravi, Beas, and Sutlej rivers. Pakistan became entitled to use waters of the Indus River and its western tributaries, the Jhelum and Chenab rivers.[52]

After the treaty was signed, Pakistan began an extensive and rapid construction of its irrigation system, partly financed by the Indus Basin Development Fund. Several link canals were built to transfer water from western rivers to Punjab to replace flows from eastern tributaries that went in favor of India in accordance with the terms of the treaty. The Mangla Dam on the Jhelum River was completed in 1967. The dam provided the first significant water storage for the Indus irrigation system. The dam also contributes to flood control and regulates the flow of some of the link canals and to the country's energy supply.[53]

A second phase of irrigation expansion began in 1968. The key to this phase was the Tarbela Dam on the Indus River, which is the world's largest earth-filled dam. The dam, completed in the 1970s, reduced the destruction of periodic floods and in 1994 was a major hydroelectric generating source. Most important for agriculture, the dam increases water availability, particularly during low water, which usually comes at critical growing periods.[54] In addition, many barrages and canals were constructed.

In the early 1990s, irrigation from the Indus River and its tributaries constituted the world's largest contiguous irrigation system, capable of watering over 16 million hectares. The system includes three major storage reservoirs and numerous barrages, headworks, canals and distribution channels. The total length of the canal system exceeds 58,000 kilometers.

This, however, is not without any problem. The ground water table and soil salinity has increased. In late 1950s, the Government started Salinity Control and Reclamation Projects (SCRAPS) wherein large, deep tube wells were installed to control the ground water table. This has been followed by private investments in shallow tube wells. By 2004, more than 500,000 tube wells pump out 41.6 MAF of supplemental irrigation water every year.[55]

Rice as a Cash Crop

Cotton and rice are the two important cash crops of Pakistan. Cotton cultivation was developed in Punjab during the British rule for export of cotton to England and also to meet the requirement of textile mills located mostly in Bombay province (now Maharashtra and Gujarat states of India). Cotton has continued to be an export crop of Pakistan.

The oil boom in the Middle East opened greater opportunities for the export of quality rice like Basmati to that region. The production of high-grade Basmati rice and its export was promoted by the Government and its procurement price has been increased disproportionately. The government has allowed the private traders into rice export business alongside the public sector Rice Export Corporation. Rice export from Pakistan has been steadily rising since 1980s. The export of rice accounted for 6% of the total value of exports in 1991.

SRI LANKA
(Contributed by S.D. Sharma)

When Sri Lanka became independent in 1948, the country was facing immense shortage of rice. However, the British left behind a good network of roads, railways, ports and a plantation economy (The country was earning around US$ 900 million per annum from exports of tea, rubber and coconut).[56] Sri Lanka, enjoyed the advantage of offering a high price for export of rubber to the world market during the Korean War but, by the end of the war, the price of rubber dropped unexpectedly. This was mainly due to two reasons i.e. demand for rubber by western countries was meagre after the end of the Korean War and production of synthetic rubber by USA.

In 1952 Sri Lanka and China entered into a bilateral trade agreement. As per terms and conditions of this agreement, China consented to supply 80,000 tons of rice to Sri Lanka within a short period in exchange of Sri Lankan products mainly natural rubber. Primarily it was considered to be on a short-term basis but on negotiations it was later extended to a long-term trade agreement. Sri Lanka was assured of getting 270,000 metric tons of rice each year for a period of five years and China was guaranteed of receiving 50,000 tons of rubber each year for five years over the same period. This Trade Pact of Rice and Rubber agreed to between Sri Lanka and China was mostly favorable to Sri Lanka. The price of rubber offered by China was 40% higher than the price offered by the Western countries but the price of rice offered by China was less than 1/3rd of the world market price. Sri Lanka imported approximately 350,000 to 400,000 metric tons of rice to meet the requirements in early 1950s to avert a shortage of rice in the country. This agreement enlivened Sri Lanka to meet the crisis in the world market to a certain extent. This agreement was being renewed every five years till 1982.[57]

Irrigation

After 1930, the large and medium hydraulic structures were restored. The first modern river basin irrigation system, the Gal Oya scheme came up in Sri Lanka in the mid 1940s. In 1970, the Polgolla Barrage and Bowatenne reservoir and associated tunnels and trans-basin diversions were constructed and linked to cultivate 132,000 acres of existing fields in the Anuradhapura, Polonnnaruwa and Trincomalee districts. Local engineers and construction agencies were responsible for this exercise. In economic terms, it was a viable project and paid for itself in five years. In 1978 the accelerated Mahaweli project based on the Mahaweli Master Plan was launched and half the original plan saw completion in 1990. And this could provide irrigation for 450,000 acres and not 900,000 acres as envisaged originally. However, the emphasis for hydro-power generation and the neglect of irrigation for rice cultivation has led to crop failures.[58] Examining the investments made over the past half a century or so up to 1984, more than 50% of the public investment has been committed to agriculture. Of this, the lion's share has been devoted to gravity irrigation, ranging from 90% of it in 1950s to 60% in 1990. Irrigation referred to here is gravity irrigation for paddy production.

Speaking on the occasion of the 25th anniversary of the International Rice Research Institute (IRRI), the Minister of Agricultural Development and Research of Sri Lanka rightly claimed:

> Sri Lanka has a vast network of ancient large irrigation reservoirs. In addition, there are thousands of small village tanks, where rainwater is stored for rice cultivation during the dry season. Most of these small village tanks were inoperative due to breaches in bunds and silting of the tank-beds. The government's vigorous program to desilt and restore these small tanks, and the appointment of committees of local farmers who will be responsible for their maintenance and water management, have been a success, and contributed much to increased production in recent years.[59]

Research and Production

There has been a steady increase in rice production in Sri Lanka since 1940 to 1990. This rise has been attributed to increased area under cultivation, increased irrigation, improved rice varieties and availability of their seeds, increased fertilizer application, proper application of pesticides, improved milling practices and higher purchase prices for rice.[60] Over the years, rice production has a ten-fold increase over a three-fold population rise—indeed an unparalleled achievement on the part of a country over 50 years that has kept Sri Lanka's teeming millions well fed.

During the decade 1990 to 1999, rice production has further increased by about 10% and import has fluctuated between 0.5% in a good year to as high as 24% in an adverse year (Table 9.12). In 2004, about 1.8 million farmers of Sri Lanka cultivated rice; 67% of them possessed less than 0.8 ha of land. About 80% of rice land was irrigated. In that year, the country produced 2.6 million tons of rice and only 0.2 million ton was imported. Evidently, the country has the potential to be self-sufficient in its rice requirement.

Table 9.12. Production and import of rice in Sri Lanka during 1990 to 1999[61]

Year	Production (000 Mt)	Import (000 Mt)
1990	1723	116.80
1991	1625	132.90
1992	1590	237.20
1993	1747	202.80
1994	1824	58.40
1995	1910	9.40
1996	1401	338.70
1997	1522	305.60
1998	1830	167.50
1999	1921	214.20

The credit for the achievement goes to the Department of Agriculture of Sri Lanka. Since independence, the department has officially released over 50 new varieties of rice. The variety H4 bred at the Central Rice Breeding Station, Bathalagoda made a good impact on increasing rice yield. This variety revolutionized rice cultivation in the country and replaced almost 90% of traditional rice varieties grown by farmers. The key features of the variety were the high yields (4 tons per hectare compared to 2 tons per hectare with traditional varieties) and excellent palatability. The yield potential of some of the varieties released by the Department of Agriculture exceeds 10 tons per hectare. Sri Lankan farmers sow over 95% of their land with the improved rice varieties, thanks to the agricultural extension scientists and their programs enabling farmers to realize high yields. The national rice grain yield level is approximately 4 tons per hectare and the production has been around 3,000,000 metric tons annually. Sri Lanka ranks first among her neighbors in yield per hectare. In the decade of 1940s, the population of Sri Lanka was six million and rice import was 60%. In 2000, there were no imports at all despite a 19 million population.[62]

Acknowledgements

The author for Bangladesh gratefully acknowledges the help of Dr. Hamidur Rahman Molla, Senior Scientific Officer, BRRI for accessing literature from BRRI Library and Mr Nesar Uddin Ahmed, Chief Seed Technologist, Seed Wing, Ministry of Agriculture for accessing data and information from the Ministry.

References

1. As in 2002/03, Source: www.rice-trade.com/world-wide-rice-production. htm
2. Amartya Sen (1981).
3. Paul Ehrlich (1968).
4. Table 9.2 gives production in British tons whereas the production in independent India is in metric tonnes. Since 2240 pounds make a British ton and 2204.6 pounds make a metric tonne, it is reasonable to assume that the rice production doubled over a decade.
5. Source: Ministry of Agriculture and Cooperation.
6. Government of India, Ministry of Economic Survey, 2007-08.
7. Tiwari (2002).
8. Source: K.N. Tiwari (2002).
9. Tiwari (2002).
10. Prasada Rao (2004).
11. Shobha Rani *et al.* (2008).
12. Government of India, Ministry of Finance, (2008), Also see Ministry of Finance Website: http://indiabudget.nic.in
13. Kaoru Sugihara (2006).
14. Government of India (2003).
15. Sourav Mishra, Jan 2, 2008, Reuters.
16. World Grain Council Report (2007).
17. Source: Ministry of Commerce, Govt of India.
18. Government of India (2006).
19. Akmam (1991), Chakrabarti (1991).
20. Bhandarkar (1931), Barua (1934).
21. MoA (2007).
22. Sidhu and Mudahar (1999).
23. Alim (1968).
24. Alim (1968:81).

25. Sources: Alim (1968), BBS (1974, 1980, 2004), BRRI (1978), Niaz and Miller (1952).
26. Malek (1973).
27. Goletti (1994).
28. Source: Goletti (1994).
29. Source: BBS (1974, 1980, 2004), Alim (1968), BRRI (1978), MoA (2007), (Includes rice and wheat).
30. Dorosh, del Ninno and Shahabuddin (2004).
31. BBS (1974, 1980, 2004).
32. Samad (1999).
33. Sattar and Hossain (1986).
34. Hossain *et al.* (1981).
35. Sattar and Hossain (1986).
36. Samad (1999).
37. Sattar and Hossain (1986).
38. Sidhu and Mudahar (1999).
39. Hossain (1999).
40. Islam (2007).
41. Sobhan (1979).
42. Osmany (1987).
43. Islam (2007).
44. Banglapedia (2004).
45. MoA (2007).
46. Farm size in Bangladesh is defined by total land operated by a household, which differs from ownership classification and includes land cultivated by a household through ownership and/or taken through some tenurial arrangement. Accordingly, they are classified as landless (operating less than 0.02 ha), marginal (0.02-0.39 ha), small (0.40-1.01 ha), medium (1.01-3.03 ha) and large (over 3.03 ha) farmers.
47. See *website:* www.fao.org
48. Kayank (1999).
49. Source: Website: www.fao.org
50. Mahmood, Sheikh and Kashif (2007).
51. Jaffrelot and Beaumont (2004:154).
52. See website: www.photius.com/countries/pakistan
53. See website: www.photius.com/countries/pakistan
54. See website: www.photius.com/countries/pakistan
55. See website: www.unitar.org
56. See website: http://www.mediquipment.com/history.htm
57. See website: http://www.asiantribune.com/index.php?q=node/7893
58. Fernando and Denis (2002).
59. Jayasuriya (1985).
60. See website: http://oryza.com/Asia-Pacific/Sri-Lanka-Market/359.html
61. Source: Rafeek and Samaratunga (2002); For 2004 data, see website: www.irri.org
62. See website: http://oryza.com/Asia-Pacific/Sri-Lanka-Market/359.html

10 | History of Rice in Western and Central Asia

Mark Nesbitt, St John Simpson and *Ingvar Svanberg*

The Chinese envoy Zhang Qian was probably the first official who brought back reliable information about the economic and social conditions of Central Asia to the Han dynasty imperial court. Zhang visited the Ferghana valley—which he calls Dayuan—around the 2nd century BC, and gave the following description:

> "Dayuan lies southwest of the territory of the Xiongnu, some 10,000 li directly west of China. The people are settled on the land, plowing the fields and growing rice and wheat. They also make wine out of grapes. The people live in houses in fortified cities, there being some seventy or more cities of various sizes in the region."[1]

Zhang Qian is certainly one of the oldest preserved eyewitness reports we have on rice cultivation in Central Asia.

Today rice is an important food product used both as a daily dish as well as in more elaborate festival dishes in Central Asia and adjacent regions including Iran, Afghanistan, Azerbaijan, and Turkey. Various kinds of *pilovs* are an integral part of the local food culture of the Turkic- and Iranian-speaking oasis and town-dwellers of the region. In addition former nomads today eat rice as an important part of their meals. During Svanberg's rather extensive travels in Central Asia in the 1980s, he was usually served dishes based on rice and meat—i.e. various kinds of *pilovs*—when visiting Kazak, Kirghiz, Uzbek and Uighur homes. A variety of *pilovs* constituted festival dishes and treats in most Turkic homes, no matter if they were nomads, farmers or urban people. *Pilov* is actually today regarded as a kind of standard dish among these people and rice must be seen as an important contemporary staple food for many households in the region.

SCOPE

The region covered in our survey is vast, stretching from Istanbul to Chinese Turkestan (Xinjiang). Although dominated for centuries by two cultures, Turkic and Persian, countless regional culinary histories and traditions exist, in part surveyed in important papers by Zubaida and Fragner.[2] The rice cultures of western and central Asia are far less well-known than those of south and east Asia, but the literature is nonetheless immense. Our coverage is of necessity selective: detailed for the earliest history and archaeological record of rice, for medieval Arabic culinary texts (valuable evidence of Persian influence on cuisine), and for the recent history and culture of rice cultivation and consumption in central Asia. We have been able to give too little attention to the fascinating history of the irrigation technology associated with rice paddies, or to the rich literature on rice in the Ottoman Empire.

GEOGRAPHY

Rice is a highly adaptable crop. The main limiting factor for rice cultivation is its water requirement, estimated at 9000 m^3 per hectare of paddy in Iran.[3] This is all the more so in temperate areas, such as Western Asia, where rice is grown during the dry summer period in order to meet rice's temperature requirement. As abundant, standing, water is required throughout the growing season, a relatively sophisticated system of irrigation channels and bounded fields is required. The combination of abundant water, low altitude (for summer temperature) and flat land limits large-scale rice cultivation to three regions in our study area (Fig. 10.1):

- Mesopotamia—lowland Iraq and southwestern Iran.
- Southern shores of the Caspian—Iran.
- Desert oases of Turkestan—Turkmenistan, Uzbekistan.

Rice is also cultivated, on a smaller scale, in Turkey:

- Cilicia—southern Turkey. Seyhan river delta.
- Izmir—western Turkey. Büyük Menderes River (ancient Maeander).

- Marmara—northwestern Turkey.
- Black Sea—northern Turkey. Valleys to the south of the Pontic mountains, and scattered locations in eastern Turkey.

Other small areas of rice cultivation exist on the southern slopes of the Zagros mountains (accounting for about 15% of the nation's crop), in southern Iran, and in scattered small oases in Xinjiang. In all the areas listed cultivation has increased in the last 40 years, thanks to the availability of modern technology for irrigation, and increased access to farmer credit.

In Western and Central Asia rice is usually planted in the spring and then transplanted into fields in May or June, with the harvest from August to November.

Fig. 10.1. Map showing areas of rice cultivation in southwest and central Asia. Solid shading: introduction before 0 AD; cross-hatching: after 0 AD; vertical hatching: after 1500 AD. Dates are those proposed by Bertin *et al.* and are in agreement with the views expressed in this paper. Adapted from *Atlas of World Food Crops. Map 4, Rice.* By J. Bertin, J.-J. Hémardinquer, M. Keul and W.G.L. Randles, 1971, Paris: Mouton.

FOOD CULTURES OF RICE

Medieval Mesopotamia

Early medieval Islamic cookbooks give a rich insight into dishes consumed by the middle and upper classes in parts of Western Asia. These have just begun to attract serious attention from food historians but are barely known outside a handful of specialists of this period in this region. The following extracts relate to the uses of rice which are documented in these sources.

The earliest of these cookbooks was compiled by ibn Sayyār al-Warrāq in about the 940s or 950s and details 615 recipes drawn from over twenty cookbooks often written by or for caliphs, princes, physicians and leading political and literary figures. A small number of these recipes involved rice. This was typically husked white rice (*aruzz abyad maqshūr*), which is often referred to as being washed, sometimes several times, "until it is clean."[4] Rice-bread (*khubz al-aruzz*) is described by al-Warrāq as "less bloating than wheat bread." Among the recipes for alcohol-free beer (*fuqqā'*) is one where rice was substituted for bread: the type of rice is described as *ja'farī* (literally "river") which may refer to its origin in the marshes of southern Iraq.

Many of these recipes were for rice porridge. These included smooth thick rice porridges with pounded meat (*harīsat al-aruzz*), chicken breasts and optional wheat (*harīsa kānūniyya*) or shredded fatty meat and wheat, spiced with cassia and galangal (*khaytiyya*). Several varieties of coarse rice porridge are also described which were spiced with cassia and galangal, sometimes sweetened either with sugar "the way the Persians used to do" or with honey, and served in a bowl (*aruzziyyat*). Another type of coarse rice porridge involved the addition of thin slices of seasoned fried meat. A lentil, bean, chickpea and rice porridge with meat, chard roots and stalks, olive oil, and the standard early medieval spice mix of ground coriander, cumin and black pepper is described as best served in a bowl over olive oil drenched white bread and described as the *tafshīl* of Sālih bin 'Alī, a grandson of the caliph Hārūn al-Rashīd. A dish of rice porridge with shredded boiled leeks (which substituted for the meat), safflower seeds, and ground sesame and almonds is described as a Christian recipe used during Lent.

A recipe for stuffed tripe (*qibba*) is described as containing a small amount of rice. Three recipes are characterized by the combination of three types of grain and pulse, namely rice, wheat and lentils, chickpeas or beans, and either with meat or without (*muthalathāt*). Other recipes involve boiled meat and spinach (*isbanā<u>kh</u>iyyāt*) or meat and cabbage (*kurunbiyyāt*) with onion, galangal, cassia, seasoning and a little rice; boiled turnip (*<u>sh</u>aljamiyyāt*) with meatballs, onion, seasoning, spices and small quantities of chickpeas and rice; meat medleys with onions, milk, beans and rice, "the amount of which is one and a half times more than the beans [or lentils]" (*ma<u>kh</u>lūtāt*). Three thin grain stews recommended for those with upset stomachs, indigestion, gastric ailments and "good for liver, fevers, and pain in the upper gate of the stomach" relied on dry toasted rice as the key ingredient. A spicy poultry dish served with eggs and cheese, was cooked with an optional added "handful of rice and smoked strips of meat [which] would be a good thing to do, delicious and scrumptious" (*Nibātiyya* of Ishāq bin Ibrāhīm al-Mawsilī). A rack of lamb steamed in a modified chlorite cooking pot with rice and milk cooked separately in the resulting juices has the Iranian name of *dākibriyān*. Another oven-baked meat dish was characterised by the addition of beans and "a similar quantity of good quality [or Mutawakkilī] rice" (*tannūriyya*).

There were also sweet desserts. These included a recipe for moist condensed pudding (*<u>Kh</u>abīsa*) of sugar, honey, saffron, sesame oil and rice flour (*daqīq al-aruzz*) which was attributed to the Abbasid vizier Hāmid bin al'Abbas and specified as "non-Arab" (*muwallada*); a variety of oven-baked rice pudding was cooked with duck dripping (*jū<u>dh</u>āba*). A smooth rice pudding made with rice-flour, milk, sugar and fat (*bahatta*) is better known today as *muhallabiyya*, and two other versions of this are described, including one with chicken lightly seasoned with coriander, cumin, cassia and saffron, served with honey and rosewater: "the beauty of the dish is when the rice grains show through the honey". Two recipes for golden condensed pudding (*fālū<u>dh</u>aj*, from the Middle Persian *pālūdag*, literally "purified") refer to the use of thoroughly washed, dried and pounded rice which was sifted through silk or linen, mixed with honey, ground camphor and butter, cooked and served with crushed white cane sugar; one of these recipes also specifies that the cook should "choose Levantine rice (*ruzz Shāmī*) or Yemenite (*ruzz Zabīdī*). These are the best and whitest rice varieties available".

Many of these recipes are therefore porridges or desserts, only small quantities of rice are specified for most dishes and in these cases it is mainly employed as a thickening agent, and there are no references to serving meals on a thick bed of rice whereas, by contrast, many of the recipes refer to serving dishes with bread. Some of the recipes are vegetable or pulse-based and therefore probably represent common dishes elevated in status through the addition of spices: the references to adding rice may reflect courtly associations rather than common cooking traditions and it may be significant that only two of these dishes are attributed to named individuals. This is particularly clear in the case of the golden condensed puddings, where the ingredients include expensive long-distance imports (the main source of camphor was Japan and even the rice is specified as preferably coming from Syro-Palestine or Yemen, in both cases far removed from the kitchens of Baghdad), the process was labor intensive and there is gratuitous reference to straining through costly silk or linen cloth. The implication of these references is that rice was still regarded as a speciality ingredient rather than a staple. Several of these recipes are referred to in connection with Iranians or have Iranian names, and the fact that many other Abbasid court dishes also have Iranian names suggest a strong legacy from earlier Sasanian cuisine. This is made explicit in a poem quoted by al-Warrāq and recited to him by Abū al-'Abbās al-Adīb, a resident of the middle Euphrates city of al-Anbār,[5] as it associates the invention and popularity of a variety of rice porridge with members of the Sasanian royal house:

"The most delicious food one may ever eat when April the
    arrival of summer heralds,
And when kids and lamb are at their best, is *harīsa* made by
    niswān [women].
With skilled hands, tastiest *harīsa* they make, birds and lamb
    combining.
Fats and oils are added to pot, and meat and tail fat and tallow.
Then geese and quails and fair wheat and grass pea follow.
Next, milk and rice, which the miller perfectly ground,
And salt and galangal. It wearied the hands that beat and
    stirred it.
Like the shining constellations in the sky, it puts all other
    dishes to shame,
As it comes carried by the slave boys, embraced by bowl and
    tray,
Above it is a bamboo vault, which roof and walls support,
Domed and rounded. The slave boys did uncover it and offer.
Its radiance dazzles the eyes. With *murrī* [*liquid fermented sauce
    akin to soy*] brought, just what it needed.
Coveted by the hungry and the full, craved by host and guest
    alike.
Among its peers it reigns, mind and intellect clearing,
Eating it does the body good. Sāsān in his days invented it,
And Kisrā Anū Shirwān loved it.
If the famished catch sight of it, they will scramble for it."[6]

In a chapter on "Humoral properties of grains and bread made from wheat and rice", al-Warrāq states that:

> "Rice is closer to moderation with regard to heat and cold. It is not recommended for people suffering from colic because it constipates. It is very nourishing and cooking it with lots of fat will facilitate its digestion. When cooked with milk and sweetened with sugar, rice is wonderfully nourishing, healthful, and helps increase blood, so know this."[7]

Al-Warrāq's cookbook is a major mine of culinary information but it is not the only one. There are four important 13th century cookbooks. One was compiled in Baghdad in the first half of the century by Muhammad ibn al-Karīm al-Kātib al-Baghdādī.[8] The second is known as *Kitāb al-Wusla ilā l-habīb fī wasf al-tayyibāt wa-l-tīb* ("Book of the elation with the Beloved in the Description of the Best Dishes and Spices") and is usually attributed to Ibn al-'Adīm of Aleppo who emigrated to Gaza and thence to Egypt; it includes a number of North African dishes as well as recipes identified with particular regions. The third is known as *Kanz al-Fawā'id fī tanwī' al-mawā'id* ("The Treasury of Useful Advice for the Composition of a Varied Table") and was compiled in Egypt under the Mamluks; the fourth was almost certainly compiled by ibn Razīn at Murcia in Andalucia and is titled *Kitāb Fadālat al-khiwān fī tayyibāt al-ta'ām wa-l-alwān* ("Book of the Excellent Table Composed of the Best Foods and the Best Dishes").[9] Together they add a number of references to the use of rice as a culinary ingredient. These include the addition of roasted cumin and cinnamon to rice with yoghurt, and adding boiled spinach to a garlic-flavored stew of meat, rice and chickpeas. *Kitāb al-Wusla* is the most important for our purposes as it expressly refers to certain dishes which resemble what we recognize as *pilaf*: one is described as "Indian rice" which is described as being cooked in a copper pot with two and a half times the quantity of water compared to the rice; others refer to the addition of meat, fat, chickpeas or pistachios, and sweetened at the final stage with sugar and rose water.

In summary, these cookbooks describe a number of specific dishes involving rice. One has an Indian association and many have Iranian affinities or names which suggest that these may belong to a longer pre-Islamic culinary tradition centered in Iran and Mesopotamia (the latter was the seat of the political capital and at the cultural heart of Iranian empires for over seven centuries before the Islamic conquest). The care expressly taken to prepare the rice by washing and occasional references to steaming suggest that the principles of *pilaf*—namely a non-mushy dish whereby the individual grains remain separate (*mufalfal*)—were already well understood by the medieval period and possibly earlier.

Central Asia

A common dish—made in various ways among the Turkic and Iranian peoples of Central Asia—is the so-called *pilov* (Uighur *pelaw*; Uzbek *palov*; Karakalpak *palua*; Kazak *palau*; Kirghiz *paloo*, Tajik *palov*). The essence of all these dishes is a slow-cooked recipe beginning with the melting or rendering of fat from a fat-tailed sheep in a heavy cauldron, followed by the sautéing of onions and carrots, the addition of the meat and spices, and cooked on a slow simmer to blend the flavors before carefully adding the rice (or barley) on top and then carefully filled with water; the pot is then tightly lidded and allowed to slow cook until all the water has boiled away. Great care is taken to ensure that the layers of ingredients are not disturbed either through the addition of each set of ingredients, careless slopping of the water or during the cooking cycle. After cooking the rice is then ladled out and some of the spices spooned on top. An American traveler to Tashkent in the 1870s describes its preparation:

> "a quantity of mutton tallow or fat is melted in a pot, and the mutton, after being cut into pieces, is stewed in this; when the meat is cooked it is taken out, and the rice, which has been properly washed and cleaned, is put in and stewed until done; with this are mixed usually small thin slices of carrot, and the whole is turned out on a large platter, the pieces of meat and bones being placed artistically on the top."[10]

Pilov therefore has been prepared in essentially the same manner for at least 130 years in Central Asia. The name of the dish is originally borrowed from Persian (<*pīlāv*) but is probably an early borrowing into the Turkic languages. *Pilov* is considered to be derived from Sanskrit *pilaaka'* 'millet'. The dish most certainly reached Central Asia from Iran but probably originated in India. *Pilov* has been known since medieval times and is mentioned in 13th century Arabic sources from western Asia (see above), thus considerably earlier than in some previous statements that it originated in the Qajar period in Iran.[11] Today it is regarded as "the flagship of Central Asian cookery," as a couple of authors recently phrased it.[12] For the people of Central Asia and adjacent regions such as Afghanistan and Iran *pilov* has a central place in their food culture, so it can be regarded as what nutritionists sometimes call a cultural super-food.[13]

Fig. 10.2. Plov vendor at New Year festival, Russian Turkestan. Photograph taken 1865-72 by S.M. Prokudin-Gorskii. *Courtesy*, Library of Congress (LC-DIG-ppmsca-14391).

Pilov has since spread to other parts of the world. Nowadays *pilov* in various shapes and forms are found throughout many parts of Eurasia. *Pilov* thus is part of an international food culture. It entered the Russian empire at a relatively early date and is nowadays a common dish, known as *plov*, among people all over the former Soviet Union. Through the Ottoman Empire it also entered the European languages and in English is known as *pilaf*. An early Swedish encounter with the dish was through the envoy Claes Rålamb, who during his travel to the Sublime Porte in Constantinople in 1658, was served *pilou* by his Turkish hosts.[14]

Although the name indicates a long tradition in Central Asia it was until rather recently more commonly made of barley rather than rice among the oasis dwellers. It was only the rich and wealthy that could afford to use rice in their *pilov*. However, during the approximately last century or more rice has become the most common ingredient.[15]

Nowadays there are many ways of making *pilov*. There are said to be over four hundred within Uzbek cooking alone and almost endless variants reflect the availability of local seasonal ingredients, and household preference.[16] Any cookbook from the region gives a rich variety of *pilov* dishes: *pilov* made of mutton, carrots, rice and topped with cut dried apricots and apples; *pilov* with rice, mutton, dry peas, raisin and pomegranate's kernels; *pilov* with chicken (or duck, pheasant or turkey meat), rice, carrots, onions and parsley roots; etc. Every ethnic group is proud of having *pilov* of their style, especially when it comes to festival dishes, so-called *bairam pilov* ("festival pilov"). In reality there are many local and individual varieties of these recipes as well.[17]

Pilov plays an important role as ethnic food in the Central Asian diaspora communities in Turkey and Europe.[18] *Pilov* has always been an important meal for large feasts, for instance in connection with circum-cision, wedding, memorial feasts and other major events. Women now traditionally prepare the *pilov* in the home, but for public feasts—weddings, etc.—men did and do the cookery and take pride of their skill.[19] One modern writer states how,

> "In the 1960s a member of the Central Committee of the Soviet Communist Party, for whom I was interpreting, shocked his American hosts by stating that as far as he was concerned one of the most significant achievements of the 1917 Revolution was that it finally gave Uzbek women the right to make plov."[20]

Religious festivals are also celebrated with huge meals including a *bairam pilov*. Also at important religious shrines pilgrims were provided with *pilov*, prepared in giant cauldrons. The Swedish explorer Sven Hedin describes in the late 19th century how *pilov* was prepared at Ordam Pasha tomb—a famous mazar ascribed to Sayid Ali Arsland Khan—in western Taklamakan desert.[21]

Other dishes known in Central Asia are a rice soup, *shöilä*, mentioned already in the 19th century among the oasis dwellers in Turkestan and the Kazak rice bread, *muiz*. Among the Uzbeks and Tajiks there is also a kind of porridge eaten with meat, called *mashkichiri*. The Koreans, who were exiled to Uzbekistan, brought a variety of rice dishes with them as well, such as rice cakes (not sweet). Uzbek urban dwellers would eat these in lieu of *non* (bread). When they could not get *non* but rice cakes they would preface it with "now we are being like the Koreans." The Dungans, a small minority living in Kyrgyzstan, Kazakstan and Uzbekistan which are still primarily farmers growing rice, wheat and vegetables, are said to eat rice as food, but will seldom offer it to guests.[22]

Most Central Asian peoples are Muslim, and alcoholic beverages have therefore hardly any historical tradition among them, although beer, wine and hard liquor became available with the increasing influence of the dominant Russian and Chinese cultures in the 19th century. Chinese rice wine has traditionally been consumed by the Mongols—known by them as *tururgajin ariki*—and is of course readily available in contemporary Xinjiang.

Fig. 10.3. Rice field, Samarkand. Photograph taken 1905-15 by S.M. Prokudin-Gorskii. *Courtesy*, Library of Congress (LC-P87-8031A).

Iran and Afghanistan

Rice is an important crop in contemporary Iran, and it is the second main food consumed in the country. It is used in practically every meal. Most rice is produced within the country, especially in the Caspian lowlands in the north, although the urban population to a large extent nowadays consumes imported rice, especially from South East Asia. The main areas of cultivation in Iran are Mazandaran and Gilan provinces, although rice is also produced in Zanjan, Golestan, Khuzestan, Esfahan, Fars and Khorasan. In Iranian food culture, there are, in addition to various forms of *polov*, numerous other rice dishes such as rice pudding (*shir berenj*) and rice cookies (*kalucheh berenj*), and of course the rice-stuffed *dolmeh*.

As in Iran, rice is a staple eaten several times a day in Afghanistan, and the crop is widely grown in the country. It is cultivated around the central mountains where rivers and larger streams make irrigation possible. Two different kinds of rice have traditionally been grown, lok and mahin. The former is a Turkestan kind of rice, while mahin originates from India and has during the last century been grown in the provinces of

Fig. 10.4. Rice field near Resht, on the plain of the Caspian coast of Iran. Figure 242, *Persia. Geographical handbook series B.R. 525.* September 1945. London: Naval Intelligence Division.

Laghman and Qataghan. The famous *qabuli palau*, a dish made of mutton, spices and yellow rice, is nowadays a "national dish" in Afghani restaurants all over the world.[23]

Turkey

In Turkey, rice is cultivated especially in northwestern part (Marmara-Thrace) and in the northeastern part of country. Some rice is also produced in southeast Anatolia. The main rice growing provinces are Edirne, Samsun, Çorum, Balıkesir, Çanakkale, Sinop, Kastamonu and Diyarbakır. Most of the rice farms are small.

Rice is used in various dishes. In many Anatolian villages, yoghurt is regularly eaten with rice as *yoğurt çorbası* 'yoghurt soup'. *Sade pilaf* is boiled rice which can accompany many dishes (meat, fish, and vegetables). Various kinds of pilafs containing meat from chicken, lamb and other animals—such as *etli pilav* (rice with pieces of meat)—are popular in Turkish cookery. Green peppers, eggplant, tomatoes and other vegetables are stuffed with a rice mixture, and known as *dolma*. They can also be made of rice-stuffing wrapped with wine blades. Another kind of *dolma* are *midye dolması* (mussels stuffed with rice) known from the coastal areas. Various kinds of rice puddings, *sütlaç*, are popular as dessert and are also served in restaurants.[24] The consumption of rice increases as the living standard rises in contemporary Turkey. To cope with the increasing demand, rice is imported from U.S., Egypt and other countries. Rice consumption in Turkey per person is estimated around 6.5 kilograms annually (2008).[25]

DOMESTICATION AND EARLY HISTORY

It is certain that rice was domesticated somewhere in the distribution area of wild rice, which stretches from the eastern Indian subcontinent, through Indo-China to southern China. However, the number of domestication events, area(s) in which domestication took place, and the timing remain controversial.

Several types of genetic evidence point to at least two domestication events in rice. Cultivated rice can be divided into two main forms, *japonica* rices with a broad, thick grain grown in more northerly parts of China and southeast Asia, and *indica* rices with a thin, elongated grain. It has been known for many years that these are not fully interfertile, suggesting that they might arise from separate wild ancestors. Recent studies of several forms of genetic evidence support this hypothesis, although most literature still refers to a single area of origin.[26] Modern populations of *indica* rice are most closely related to wild rice populations south of the Himalayas, in India or Indo-China, and *japonica* rices are most closely related to wild rice in southern China.[27] This may be the result of independent domestication in the two areas, with a subsequent genetic history further complicated by post-domestication introgression between the crop forms of rice and the wild ancestors.[28]

Another complicating factor is the existence of less common forms of rice, which may derive from one of the two major forms, or be of independent origin. These types of rice are *aus* (from Bangladesh), *ashina* (floating rice of India), *rayada* (floating rice of Bangladesh) and *aromatic* (e.g. Basmati rice in the Indian subcontinent and *sadri* from Iran).[29] Rice landraces in Iran, the Indian subcontinent and much of Indo-China are predominantly of *indica* or *aus* races; landraces in southeast Asia are of *indica* or *japonica* races.[30] Broadly speaking, the distribution of the two main races is consistent with spread from two centers of domestication, and with rice in Iran coming from the India/Indo-China centre of origin. However, more detailed genetic characterization of Central Asian, Iranian and Turkish rice populations is necessary before genetic evidence can be used to trace their routes of spread.

As with genetics, the archaeological evidence for domestication is incomplete and controversial. Rice remains are abundant at many archaeological sites in the Yangzi and Huai rivers of southern China, at archaeological sites dating from about 10,000 BC. Wild rice was an important foodstuff for pre-agrarian foragers, hence its abundance. It has been widely suggested that the first appearance of domesticated rice is at about 6000 BC.[31] However, some criteria used to distinguish wild and domesticated grains, such as grain size and shape, are ambiguous, and the most reliable criterion, rachis scars on rachis bases, has not been widely applied. Identification of ancient grains to *indica* or *japonica* race is particularly problematic. By 4000 BC, there is unambiguous evidence for rice domestication in southern China.[32]

In the Indian subcontinent early records of rice are characterised by vague identification criteria and indirect dating.[33] Evidence from sites such as Chopani Mando (6000 BC) and Khairadih (2500 BC) must be set aside.[34] Sporadic early occurrences are likely to be of wild rice. Rice first becomes abundant at the beginning of the second millennium BC, marking its establishment as a summer crop in the subcontinent. The foothills of the Himalayas, in the eastern half of the subcontinent, and the Ganges basin to the south, lie within the distribution of wild rice, and are a potential centre for rice domestication.[35] Rice is found at numerous sites in the Ganges Basin from 2000 BC onwards. To the west of the sub-continent, rice occurs at Harappan sites on the Indus valley, such as Harappa, from 2000 BC, and soon afterwards at the site of Pirak, at the edge of the Indo-Iranian plateau (Fig. 10.6).[36]

In summary, genetic evidence no longer supports the view that rice spread to the Indian subcontinent from a single centre of origin in southern China; the earliest domesticated rice in India may have been domesticated locally. In terms of understanding the spread of rice to the Near East, the most likely starting points are in the Indus Valley, with its coastal connection to shipping routes, or on the eastern edge of the Iranian Plateau. In both areas, rice cultivation was established around 2000 BC.

In the case of Central Asia, a possible route of spread is westwards from northern China. Rice joins the millet-based agriculture of northern and northwest China about 3000 BC,[37] but then spreads slowly eastward, reaching Korea about 1000 BC and Japan about 400 BC.[38] In the light of this slow eastwards spread, any travel through the desert regions north of the Himalayas is unlikely to have been fast.

EARLY WRITTEN SOURCES AND ARCHAEOLOGY

Iron Age and Achaemenid Periods

A single rice grain is said to have been recovered from a pit at Hasanlu, in north-west Iran, and provisionally attributed to Hasanlu Period III (ca. 750-590 BC).[39] However, although referred to in a brief preliminary report, this was not mentioned in the more detailed subsequent publication of the excavated plant remains.[40] This slender piece of evidence therefore should be treated with caution until further details become available. However, Potts' suggestion that this grain could be from wild rice seems unlikely given that the nearest known area of wild rice lay some 2000 kilometers to the east of Hasanlu.

References in Iron Age texts are ambiguous. It has been suggested that rice is included in the list of products in Ezekiel (xxvii. 17), as *minnīth*.[41] The book of Ezekiel was written in Babylon in the 6th century BC. However, standard Bible encyclopaedias today consider this text as possibly corrupt, and if valid, referring to a town from which wheat was exported. Thompson translated the Neo-Assyrian term *kurangu*, used in the 7th century BC, as rice. The *Chicago Assyrian Dictionary* does not translate this term beyond cereal.[42]

One author has proposed that rice was introduced to lower Mesopotamia from India during the Achaemenid period, probably on the strength of a 4th century BC description by Diodorus (XIX.13.6) of rice among military provisions at Susa.[43]

Evidence from subsequent Parthian and Sasanian periods is more conclusive yet this has been widely overlooked within the context of the history of rice cultivation in the Near East.

Parthian Period

The earliest reliable Near Eastern archaeobotanical evidence for rice consumption derives from Susa and dates to the 1st century AD. This consists of 373 carbonised grains of short-grained rice sealed beneath roof collapse on a floor in Level 3A at Ville Royale II; associated jars suggest the means of storage.[44] Results from the 1973 Susiana Survey suggest that rice may have been cultivated during this period in the South Dez plain.

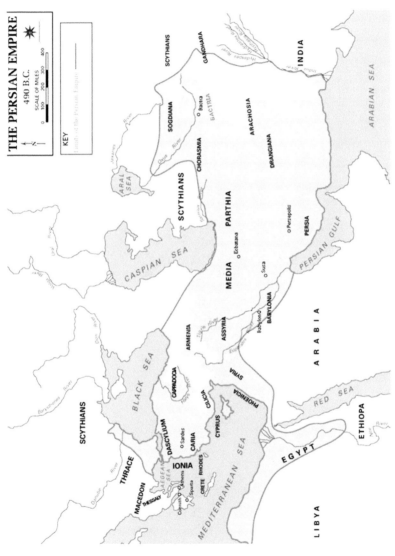

Fig. 10.5. The Persian Empire in 490 BC. With acknowledgements to the Department of History, United States Military Academy, West Point.

324

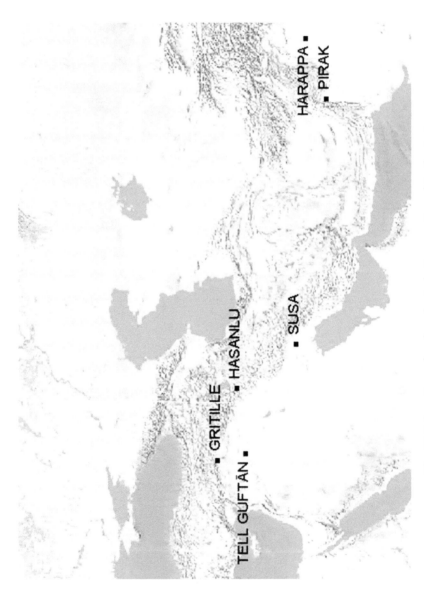

Fig. 10.6. Location of archaeological sites mentioned in the text.

This evidence is in the form of rice-hull impressions that were provisionally identified in fired bricks found on the surface of several sites here dating between *ca.* 25 BC and 250 AD.[45]

Classical and Byzantine Evidence

It is clear that rice was not grown in Mediterranean Europe in the classical period.[46] However, opinions vary as to how much of it was eaten. Dalby suggests that rice was uncommon in classical Europe, and was mainly used as a medicinal plant.[47] Rice is mentioned in medical texts by Dioscorides and Galen. Writing in about 180 AD, Galen says that "This grain is universally administered to check the stomach. It is cooked using a method similar to that employed with groats, although it is harder to digest than groats, contains less nutrition, and generally falls short of groats in culinary terms."[48] Rice was a sufficiently well-known food to be mentioned alongside barley and broad beans in the Greek manuscript *Peri Trophon Dynameos* ("On the power of food"), written in the 2nd century AD; the *Apicius* mentions the use of water in which rice has been cooked as an ingredient, suggesting that rice cannot have been too exotic in the better-off households of Rome in the 1st centuries AD.[49]

Archaeobotanical finds from the classical period are rare. Small quantities of grain are found at the Red Sea ports of Quseir al-Qadim and Berenike, in Egypt.[50] The *Periplus Maris Erythraei*, a survey of trading and sailing conditions in the Red Sea and Indian Ocean written between AD 40 and 70, refers to the export of rice from the west coast of the Indian subcontinent, by ship to Socotra and the northern coast of Socotra. From here, rice was traded to Roman Egypt. It has been suggested that the rice exported from India to the Red Sea coast was for consumption by Indian merchants resident there. Elsewhere, small quantities of rice have been found at the Roman sites of Zurzach in Switzerland, and Novaesium in Germany.[51]

Travelers with Alexander the Great, who reached the Indus in 326 BC, are quoted in Strabo's *Geography*:

> "The rice, according to Aristobolus, stands in water in an enclosure. It is sowed in beds. The plant is four cubits in height, with many ears, and yields a large produce. The harvest is about the time of the setting of the Pleiades, and the grain is beaten out like barley. It grows in Bactriana, Babylonia, Susis, and in the Lower Syria. Megillus says that it is sowed before the rains, but does not require irrigation or transplantation, being supplied with water from tanks."[52]

Aristobolus's observations appear to be of the Punjab, in the extreme northwest of the Indian subcontinent. It is unclear whether the observations regarding the other areas—Bactria (northern Afghanistan), Babylonia (southern Iraq), Susiana (southern Iran) and Lower Syria (possibly Lake Huleh, Israel)—are from Aristobolus or from Strabo himself, writing in the 1st century AD. Megasthenes is further quoted on the consumption of rice beverage and pottage, again in the Indian sub-continent.[53]

Jewish textual sources suggest that rice was a significant crop in ancient Israel by 70 AD, for example in the valley of Dan; they also refer to the cultivation of red rice in the Orontes Valley, near Antioch.[54]

Sasanian Period

Literary and historical sources suggest that rice cultivation continued in areas near the head of the Gulf during the Sasanian period. Rice-bread appears to have been consumed in Ḥozai (Khuzistan) at the head of the Persian Gulf and elsewhere in the alluvial plains of Mesopotamia, judging by sporadic references in the Babylonian Talmud, which was largely composed in central Mesopotamia between the 3rd and 5th centuries AD.[55] According to Jarir al-Tabari, rice was a local crop subject to taxation during the Late Sasanian period.[56] An Arab conquest anecdote of al-Hamadani refers to the transport of rice in Meshan during the second quarter of the 7th century.[57] In addition, rice (Middle Persian brinj) was listed as a primary ingredient of certain Sasanian desserts, including a rice-jelly served among other items as the "Dish of the King" and a type of rice-pudding known as a "Greek Dish."[58]

However, according to the Zhou shu (History of the Northern Zhou Dynasty, AD 557-581), compiled by Linghu Defen (583-666) and presented to the Chinese throne in 636, the kingdom of Bosi [Persia] lacked rice:

> "Their five cereals, fauna, and other things are about the same as those in China, except that they have no rice or millet."[59]

In the light of the sources cited above, this statement suggests that rice was considered a sweetmeat rather than a staple, except possibly among peasants in areas of southern Mesopotamia and south-west Iran where larger-scale rice cultivation was feasible.[60] The additional possibility remains that a certain proportion of rice consumed within the Sasanian Empire may have been imported by sea via the Persian Gulf from the Indian subcontinent and there is archaeological evidence from the site of Kush for the import of Indian goods—including cooking pots and glass beads—into the Gulf at this period.[61]

Archaeological (including archaeobotanical) evidence for rice during this period is surprisingly scanty. Adams and Nissen have proposed a Sasanian date for a dense network of irrigation channels to the north of Warka Survey site 265 but the date of this field-system may be considerably later than the nearby settlements and the suggested function as evidence for rice-cultivation is questionable.[62] No phytolith work has yet been done for this period and systematic flotation recovery of carbonised plant remains has only been conducted at two Sasanian sites, namely the city of Merv in Turkmenistan and the much smaller Persian Gulf site of Kush in Ras al-Khaimah. A large number of samples have been analyzed from nine seasons of excavations at Merv. These include two areas dating from the 4th to 7th centuries and consisting of Sasanian private houses, located within the citadel of Erk-Kala and the adjacent lower city of Gyaur-Kala.[63] These results clearly demonstrate that the dominant irrigated summer crop in the oasis during this period was cotton. In contrast, millet was rare and rice was completely absent.[64]

The evidence cited above therefore suggests that rice cultivation did play a role in Partho-Sasanian agriculture but that it was limited to suitable lowland areas of southern Mesopotamia and south-west Iran. The economic significance of rice therefore is likely to have been over-emphasised in earlier studies. The same may be also said for the alleged "uniformity of mode of production" and degree of centralization of Sasanian agriculture as there was undoubtedly considerable variation in the agricultural economy between different ecozones.[65]

Early Islamic and Later Medieval Evidence

There is more evidence for the range of foodstuffs produced and consumed across southwestern Asia as we enter the early Islamic period, although evidence is scanty for Central Asia. The increase in the number of lengthy written accounts, especially cookery books compiled in Baghdad, directly reflects the wider availability of paper and the creation of libraries.

The written sources of this period suggest that there was somewhat more widespread rice cultivation than previously, and that this was undertaken in humid and/or well-watered areas of north-west Afghanistan, Iran (Dailaman, Gilan, Tabaristan, Fars, Khuzistan provinces), Azerbaijan, lowland Iraq, southern Turkey (Cilicia), north-east Syria (the Nusaybin area), Palestine (the Jordan valley), Egypt (Nile valley, Faiyum), Yemen (the Tihama) and al-Andalus.[66] In addition to these regions, recent archaeobotanical evidence in the form of carbonized grains and chaff recovered from discrete samples from small rural sites along the Syrian middle Euphrates confirms rice cultivation in that region between the 9th and 12th centuries (Fig. 10.7).[67] One charred and one silicified rice glume were found in a 12th century context at Gritille, further upstream along the Euphrates in present-day Turkey.[68]

This increase in cultivation in Mesopotamia has been attributed to rising demand owing to an influx of Iranians "from Khuzistan and from the Caspian provinces [who] were accustomed to rice."[69] Rice merchants (*ar-razzaz*) are also attested from the later 9th century onwards, raising the possibility of long-distance trade to areas of consumption, beyond those of cultivation.[70] Importantly, rice appears to have been considered a luxury outside areas of production and rice-bread (*khubz al-aruzz*) was a staple among the poorer classes, notably in lower Mesopotamia, during the Abbasid period.[71] Indeed, rice-bread was only customary within Iran in rice-growing areas such as the southern shores of the Caspian, and has been more recently replaced even here owing to the use of imported wheat flour.[72]

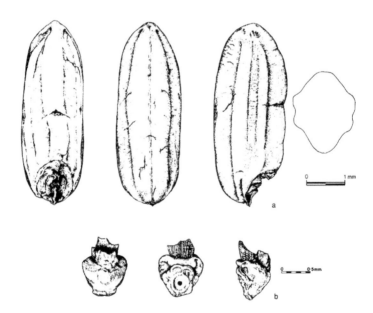

Fig. 10.7. Rice remains from 11-12th century phases at Tell Guftān, southeast Syria. a: grain; b: pedicel with lemma and palea stubs. Page 444, *Peuplement rural et aménagements hydroagricoles dans la moyenne vallée de l'Euphrate fin VIIe-XIXe siècle.* Edited by S. Berthier. 2001. Institut français d'études arabes de Damas, Damascus.

Across other parts of Western Asia, rice became increasingly available during the Ottoman period following wider cultivation, not only on state-owned lands but also in private agriculture.[73] However it is telling that detailed analysis of 18th century Ottoman court records for Damascus reveals that although rice had a high reputation comparable to wheat, it was mostly imported from Egypt and thus reserved for feasts and special occasions, whereas barley was the most commonly consumed grain (Table 10.1).[74]

Table 10.1. Amounts of grain and fodder in Damascus estates, *ca.* 1750-1767

(Grain/fodder quantity)	(in ghirara)
Barley	238.4
Wheat	180.2
Rice	56.51
Vetch	5.99
Sorghum	4.8

Source: Greehan, J. (2007:59).

Ottoman Turkey

Rice could be bought in quantity in Byzantine Trebizond in 1292, although it is unknown whether it was grown locally, or imported from the Caspian.[75] It became available as a local crop in Anatolia and Rumelia during the next centuries.[76] During the 16th century rice was grown in Anatolia, especially around Boyabad (in the sub-province of Kastamonu) and at Beypazarı. Also the area around Philippopel (contemporary Edirne) had a reputation for its rice farming. It was, according to Faroqhi, usually not practiced by ordinary peasants. Therefore rice farming was in the hand of the central administration that employed share-croppers. The sources mention a considerable division of labour, with people responsible for watering (*saka*), the cultivators (*kürekçi, rençbi*) and a foreman (*reis*). Some rice was also produced by pious foundations. The rice which was sold on the market became popular as a food item among the wealthy, while it probably was seldom consumed by peasants outside the producing areas in the 16th century Ottoman society.[77]

Still in the early 20th century, rice production was not adequate to the local demand. Marshes in which it was grown were so unhealthy, due to the malaria situation, that the government at one time forbade its cultivation.[78]

Former Soviet Central Asia and Chinese Central Asia

Although the Chinese to the east, and the Indo-Iranian peoples in the south, have a long tradition of rice cultivation, we have no information about when rice culture first reached Central Asia. We have no Neolithic record of cultivated rice from Central and Western Asia.

The recent report of rice dated to 3000 BC from the Xishangping sites in the Gansu province, northwest China, raises again the question as to whether the cultivation of rice spread westwards into Central Asia via ancient trade routes.[79] Written documents are much younger. When Zhang Qing arrived in Dayuan in the 2nd century BC—located in a region that nowadays belongs to Uzbekistan—rice cultivation was already well-developed. Zhang is probably describing agriculture with dry-land rice. The lack of data makes it difficult to speculate about an unbroken rice culture until today. During the Sasanian period, Chinese annals tell us that rice was plentiful in places like Kucha, Kashgar and Khotan.[80]

Large-scale cultivation of rice must be regarded as a relative newcomer in the Central Asian area. Travelers during medieval times have very little to say about rice cultivation although when the Franciscan friar William of Rubruck passed the Mongolian-controlled areas in 1253-1255 on his journey to the great khan Möngke he encountered a drink which was probably rice wine.[81]

Lacking archaeological evidence for Central Asian rice agriculture, linguistic evidence may provide useful insights into the background of this grain. However, the words used for rice in the various Turkic languages do not give us any proof about the origin of rice culture in western or central Asia. In the western Turkic languages, such as Turkish, Gagavuz and Azeri, the common word for rice is *pirinch*, which is of Iranian origin (< *berenj*). A variant is also found among the Turkmen in Central Asia rendered as *bürinch*. It clearly indicates that rice was introduced to the western Turkic peoples from their Iranian neighbors, i.e. from the south. Originally it is probably a loan-word from Sanskrit *vrīhi*.[82]

However, there are also Turkic words for rice that are indigenous, and they at least indicate that rice has been present in the area for a long time. The common word in Central Asian Turkic is *gürish* (Karakalpak), *kürish* (Kazak), *kürüch* (Kirghiz), *kürüch* (Tobol Tatar), *gürich* (Uzbek) and *görüch* or *gurunj* (Uighur). These words date back at least to medieval Turkic, but are probably older. Nevertheless the glosses do not give any indications as to when rice was introduced into the area. Other words for rice grain in Uzbek and Uighur are *sholi* and *shali* respectively and are also Iranian loan words (< *shālī*). Iranian influence has been very strong in Central Asia since antiquity and continued up to modern times, especially among the farmers in the oases.[83]

TOWARDS A SYNTHESIS

Rice cultivation was well established in northern China by 3000 BC and in the Indian subcontinent, including the edge of the Iranian plateau, by 2000 BC. There is no secure record of rice in the abundant archaeo-botanical record for the Iron Age Near East, or in the sparser record for Central Asia. Textual references in the Iron Age are ambiguous.

Reliable records of rice cultivation in the Near East and adjacent areas occur in classical antiquity, fairly frequently from the 1st century AD. Diodorus reports rice as a military ration at Susa in 318/317 BC. A large 1st century AD archaeobotanical find of rice at Susa, combined with evidence of rice hull impressions in contemporary Parthian bricks, points to well-established cultivation of rice in southern Mesopotamia (partly in current-day southwest Iran). Also in the 1st century AD, Strabo mentions the cultivation of rice in Susiana (around Susa), Lower Syria (perhaps the lower Orontes), Babylonia (southern Mesopotamia) and Bactria (Afghanistan). Rice was traded extensively to areas in which it was not grown: probably from the Indian subcontinent to the Red Sea in the 1st century AD, and as a medicinal plant and a luxury food in the wider Roman Empire of the 2nd century AD.

In the Sasanian Period there continue to be regular references in the Talmud, Arab and Chinese sources to cultivation in southern Mesopotamia. However—as in the preceding periods—it is possible that use was mainly in sweet desserts and elsewhere on a highly localized scale rather than being widely consumed as a staple food. Textual and archaeological data for crop cultivation in general is very scanty in the period between the fall of the Roman Empire and the time—the 10th century onwards—from which Arabic texts become abundant.

The occurrence of archaeobotanical studies and ancient texts in western Asia is highly sporadic for the many centuries between the end of the Iron Age and the coming of Islam. Nonetheless, there is sufficient negative evidence to suggest that rice cultivation was localized in the first millennium AD.

After the Islamic conquests "Rice early came to be grown in the Islamic world almost wherever there was water enough to irrigate it."[84] Rice cultivation is well documented further north in Mesopotamia, in Cilicia, the Jordan and Nile valleys, the southern shores of the Caspian, and north-central Anatolia, all before 1500.[85] The two phase model of rice establishment, with a first phase in southern Mesopotamia starting in the period 400 BC-100 AD, and a second phase linked to the expansion of Islam after 1000 AD, seems reasonable.[86]

Rice of unknown origin was available in Trebizond (Trabzon) in northern Anatolia in 1292; it might well be, as suggested by Venzke, that rice cultivation did not reach Anatolia until the Turkish conquests of the 11th century AD. By the 14th century rice cultivation was well established in the Ottoman Empire.[87]

Most surveys of rice introduction to the Near East and adjacent regions broadly agree, although this might in part be because all such studies are based on a small number of mainly classical texts.[88] There are two aspects that are still controversial; one is the reliability of Iron Age texts and the weight to be given to the possible Iron Age find of rice at Hasanlu; the other is the emphasis given to Islamic cultivation of rice.[89] To us, it appears that rice cultivation was well established in southern Mesopotamia for more than 1000 years before the coming of Islam; however, it then spread far wider in the Islamic world. We would identify a third area that is problematic: dating the first cultivation of rice in areas adjacent to Mesopotamia, for which textual and archaeological data are particularly scarce. As Bazin and his collegues point out, it would be odd if rice cultivation was established in Bactria and in southern Mesopotamia in Parthian times, but absent from the southern Caspian, a major area of rice production.[90] However, the Caspian shores were heavily forested, making the implementation of sophisticated irrigation systems complex. The beginnings of rice cultivation elsewhere in Central Asia also remain mysterious.

Ultimately the textual sources—those in which rice can be securely identified—are too laconic: they are useful for establishing presence, not absence.[91] Only archaeobotanical research at sites dating to the period between 500 BC and 1000 AD will give the desired level of detail.

ROUTES OF SPREAD

The clear relationship between Iranian *berenj*, Turkish *pirinç* and Sanskrit *vrīhi* is consistent with the westward spread of rice and rice cultivation from India to Iran and then Turkey.[92] Although sea trade was certainly responsible for distributing rice grain for consumption in the Roman Empire, to Egypt and beyond, it was not necessarily the route by which rice cultivation spread. Rice has traditionally been cultivated in irrigated areas of the Arabian Gulf, including Bahrain, but is absent from the (now relatively rich) archaeobotanical record.[93] However the absence of rice in antiquity is not necessarily significant: most of the crops transferred by sea from Africa to India are similarly absent from the ancient Persian Gulf. However, the presence of rice in Afghanistan by the 1st century AD (*fide* Strabo), in the context of early rice cultivation on the eastern edge of the Indo-Iranian plateau, suggests that rice cultivation may indeed have spread overland from India to Iran.

Why did this take some two millennia? Rice is not an unusual case;[94] after the initial dispersal of Neolithic founder crops from the Fertile Crescent, the movement of crops into the Near East was erratic. A group of woody crops—almond, date, and grape—became established in the early Bronze Age, around 3000 BC. Crops from outside the Fertile Crescent arrived later and erratically. Dates of arrival in the Near East include sesame (2500 BC), common and foxtail millet (1000-700 BC), pistachio nut (classical period?) and sorghum (Islamic period?). It could be argued that rice requires special agricultural conditions, as an irrigated crop and a summer crop. However another summer crop, millet, spread quickly in the Near East, and sophisticated irrigation techniques had been used in southern Mesopotamia since at least 3000 BC.[95] Perhaps it was simply the high, difficult relief between India and Iran that delayed the spread of rice and other Indian and Central Asian crops.

What of the thinly scattered cultivation in Central Asia? Eastern Turkestan is not an obvious route of spread from northern China, as cultivation of any kind—let alone irrigation agriculture—is so sparse. The Himalayas are a real barrier to the south. Western Turkestan, in contrast, has a long history of political and economic integration with Iran, for example during the Parthian and Sasanian Empires. Turkmenistan and Uzbekistan also contain major oases, used for large-scale irrigation agriculture since at least 3000 BC.[96] It is likely that rice spread into western Turkestan either from Iran, or at least via the Indo-Iranian plateau, at the same time that rice spread in parts of Iran and Mesopotamia.

FROM CZAR TO COMMISSAR

The Chinese pilgrim Xuanzang, who travelled across Chinese Turkestan in the 7th century, mentions rice culture.[97] Evidence from more early modern times is scant. The vast arid areas of Chinese Turkestan were unsuited to most types of agriculture but despite this the region has been able to grow an abundance of food through extensive irrigation. Thanks to the water supply and the irrigation system, local Turki (now called Uighur) farmers also sow some rice.[98] Local farmers have relied on melting snow during spring to swell streams coming down from the mountains, thus providing adequate supplies of water to feed their irrigation channels. In areas of extreme heat, where rivers run underground, the local farmers have used the *kariz* or underground wells, to tap this water supply.

Travelers in the 19th and early 20th centuries report rice cultivation from various oases in Chinese Turkestan. Although wheat, barley, maize, sorghum and millet were more important, some rice was grown in oases like Yarkand and Kashgar. Travelers reported from Yarkand, the largest oasis in southern Chinese Turkestan, that it is well watered by irrigation channels and had an extensive cultivation of rice, which was exported to other cities, like Khotan and Kashgar. The small isolated oasis of Kalpin had for instance, owing to the excellent water supply, an above average harvest of rice for the region in the early 20th century. The Manchu-speaking Xibe minority in the Ili valley has also raised wet rice with the aid of irrigation. Many rice cultivars exist in Xinjiang and at least 16 landraces are known. Despite this, rice has never been an important food crop in Xinjiang.[99]

Wet rice is incompatible with the climate and soil of northern Xinjiang. However, during the Communist take-over wet rice production was introduced with the influx of Han Chinese immigrants. They changed pastures into agricultural fields, but it was not until the 1970s that they tried to cultivate rice. The production has increased manifold since then, but still local production only reaches 25% of the total and the remainder must be imported from other provinces in the northeast of China and Jiangsu province to the east.[100] In recent years, some Han Chinese rice farmers have begun rearing fish in the rice fields, an old practice in China proper that now has been transferred into Xinjiang.[101]

Rice cultivation existed in Western Turkestan during the pre-modern time, but it was a luxury product. Bread cereals were more important. Rice continued to be a minor crop in the oasis of Central Asia. Other grains, such as wheat and barley continued to be of greater importance for centuries along with fruit crops such as grapes. With the increasing impact of the Russian imperial power in the region rice began to be more important. In the 1860s, the remaining Central Asian principalities were conquered by Russia. The final integration came in 1895. According to a traveler's report from Bokhara in the late 1890s, rice thrived only in the level areas. A Bokharan proverb said that the rice must have water three times; it grows in water, boils in water and is washed down with water. Because of the problems with malaria, rice cultivation was restricted in densely populated areas. The local Bokharan rice was small and not as white as the Persian, but it was plentiful and the country could therefore export the surplus to Persia and elsewhere.[102]

Refugees and deportees have played an important role in developing and changing the mode of agriculture in Central Asia during the history. This is particularly true for rice. After a Muslim rebellion in Kashgaria in the 1870s, a few thousand Chinese-speaking Muslims, so-called Dungans, arrived in Russian Central Asia. They settled in the Osh, Przheval'sk and Tokmak areas. Further Chinese Muslims came to Turkestan after the signing of the Sino-Russian treaty of Saint Petersburg in February 1881. They settled north of present-day Bishkek in Kyrghyzstan. The Dungan refugees settled as farmers and beside various grains such as wheat, barley, oats and sorghum, they also sowed rice, which in fact was already their main crop in the pre-Soviet period. They actually introduced the cultivation of rice to many places of Central Asia. Rice agriculture was especially a conduct for the rich Dungans. After the Dungans settled in Russian Turkestan, rice became much cheaper in many parts of the region, underlining the importance of their production.[103]

However, the Russian authorities favored the production of another plant, and they decided to use the conquered areas of Central Asia for cultivation of cotton. This had an immense impact on Central Asian agriculture and was intensified when the Soviet regime was established in the 1920s. Massive irrigation efforts were launched that diverted a considerable percentage of the annual inflow to the Aral Sea, causing it to shrink steadily. Traditional agricultural practices were destroyed by collectivization and the new crop increased in importance.[104]

While rice production was thriving in the neighboring Caspian areas of Iran and in Afghanistan, it decreased within the Soviet Union. During the 1930s the impact of Stalinist politics of deportation increased the rice production in Central Asia. Between 1937 and 1939, Stalin deported over 170,000 Koreans to Kazakstan, Kyrgyzstan, and Uzbekistan. Most of them were sent to what was then uninhabited land, land yet to be cultivated, or to former collective farms depopulated by the Kazak famine of the 1930s. Others were settled on existing Kazak kolkhozes. As a result of the relocation, seventy new Korean rice and fish collective farms were created. The deportees cooperated to build irrigation works and start rice farms.[105] Thanks to these Korean deportees, rice became an important crop in Central Asia, especially in Kazakstan. The rice grown in Central Asia belongs mainly to the sub-species *japonica* and many cultivars exist.

The presence of rice fields has, following independence and the introduction of market economy to Central Asia, increased the potential for the spread of malaria and during the last few years several cases of malaria have been reported in Kyrgyzstan. According to one report the rice-growing plots close to private dwellings, especially in the Batken, Osh, and Zhalalabad regions, lack any effective mosquito control measures and are therefore viewed as dangerous breeding grounds for mosquitoes.[106]

Rice is considered one of the most labor-intensive crop production enterprises and machinery is rarely used.

CONCLUSION

The pre-Islamic history of major Western and Central Asian summer crops—namely rice, cotton, sesame and millet—remains generally obscure. Recent research has traced the introduction and diffusion of broomcorn millet across the Near East and India during the Iron Age but the remaining crops are more heavily dependent upon irrigation.[107] The potential ecological range of these crops thus is more restricted and susceptible to changes in centralized agricultural policies. For instance, although rice has been grown earlier to this century in upland valleys of Iraq, it has been traditionally cultivated in Mesopotamia only by a small proportion of the local population in areas of the southern marshes.[108] The favored location and season was close to the lower water channels after the spring floods had receded.[109]

Although wheat continues to be the main grain crop in countries like Kazakstan, Kyrgyzstan and Uzbekistan, rice plays an important role in the economy and food culture in Central Asia. The production of rice in 2003 was estimated at 200,000 tons in Kazakstan, 18,342 tons in Kyrgyzstan, 59,415 tons in Tajikistan, 109,500 tons in Turkmenistan, and 311,200 tons in Uzbekistan. Also in neighboring Azerbaijan rice is an important crop with a production of 15,651 tons in 2003. Other countries, like Turkey, have a production of 372,000 tons and Iran produces as much as 3,300,000 tons. No reliable figures were available from Afghanistan (FAO).

Rice continues to be of great importance for the new nations in Central Asia, both for local consumption and for trade. When the UN General Assembly declared the year 2004 as the International Year of the Rice, this declaration was sponsored also by Kazakstan, Kyrgyzstan and Tajikistan, together with many of the large rice producing countries in Africa, Asia and America, which is a proof of the importance given to the rice crop in the national economies of Central Asian countries. In the light of world food shortages, particularly of rice, in the early 21st century, local cultivation of this much-appreciated grain can only increase in importance.

References

1. Ch'ien (1961:231).
2. Zubaida (2000), Fragner (2000).
3. Bazin, Bromberger, Balland and Bāzargān (1990:147-163).
4. Nasrallah (2007:98, 109, 110, n. 6, 117-118, 238, 245, 258-259, 261-268, 270-271, 296, 308, 362, 373-374, 378, 384-386, 393, 408-409, 446-447, 456).
5. This was an early Arab garrison city after the conquest. This was founded close to the pre-existing Sasanian city of Peroz-Shapur (referred to as Pirisabora by the Roman historian, Ammianus Marcellinus) and was the seat of the rabbinical academy of Pumbadita. It is referred to as Pallughtha in Syriac documents and survives today as the important Euphrates crossing point of Fallujah.
6. Nasrallah (2007:259-260).
7. Nasrallah (2007:117).
8. Perry (2007).
9. Zaouali (2007), Bolens (1990:170-171).
10. Schuyler (1876:125).
11. Räsänen (1969:385), Jarring (1964:225).
12. Mack and Surina (2005:94).
13. Fieldhouse (1986:48, 56-59).
14. Rålamb (1679:25).
15. Bacon (1980:59).
16. Visson (1999:141-159).
17. Tlemisov (1990:110-113), Makhmudov and Salikhov (1983, No. 25-28), Davidson (2006:606-607).
18. Svanberg (1989:160).
19. Bacon (1980:59), Werner (1999:47-72).
20. Visson (1999:141).
21. Hedin (1898:405), Jarring (1935:348-354).
22. Dyer (1980:42-54), Radloff (1911:1034, 2152).
23. Krochmal (1958:186-191), Ferdinand (1959:195-232).
24. Sencil (1980:12, 110-113).
25. *Turkish Daily News* 18 April 2008.
26. e.g. Glover and Higham (1996), Smith (1998).
27. Londo, Chiang, Hung, Chiang and Schaal (2006), Sweeney and McCouch (2007), Vitte, Ishii, Lamy, Brar and Panaud (2004).
28. Sang and Ge (2007).
29. Garris Tai, Coburn, Kresovich and McCouch (2005).
30. Londo, Chiang, Hung, Chiang and Schaal (2006, Supplementary Table 4).
31. Liu, Lee, Jiang and Zhang (2007).
32. Fuller, Harvey and Qin (2006).
33. Glover and Higham (1996).
34. Glover and Higham (1996).
35. Fuller (2006).
36. Glover and Higham (1996).
37. Xiao, Xin, Hong, Jie, Xue and Dodson (2007).

38. Glover and Higham (1996).
39. Rahimi-Laridjani (1988), Potts (1991).
40. Tosi (1975). The plant remains from this site were studied by Costantini but no final report was published, see Harris (1989).
41. Rabin (1966).
42. Wislon (1995), Thomson (1939).
43. Ghirshman (1978:182), cf. Potts (1991).
44. Miller (1981:137-142, cf. pp. 140-141), see also Boucharlat (1993).
45. Wenke (1975/76, figs. 3-13, cf. pp. 41, 87-88, 106-108, 120, 144-146). Note that these survey dates, as with those of Adams and others, are tentative (cf. Miroschedji, Desse-Berset and Kervran (1987:43, n. 85).
46. Sallares (1991).
47. Dalby (2003b).
48. Grant (2000): 96, Strabo also mentions rice cultivation in Babylonia and Susiana, Laufer (1919: 372), Bacon (1980:3), Täckholm and Täckholm (1941: 410-413). For a history of rice in the Near East, see Canard (1959).
49. Dalby (2003a), Dalby (2003b), Grant (2000:200).
50. Cappers (2006).
51. Reviewed by Cappers (2006).
52. Strabo 15.1.18.
53. Strabo 15.1.53.
54. Feliks (2006).
55. Newman (1932:90, 93), Oppenheimer (1983:71, 76-77), cf. also Adams (1981: 203). Despite reference in the Palestinian Talmud, rice is considered to have only played a minor role in the agricultural economy of Roman Palestine, Brosh (1986:41-56, cf. p. 44).
56. Bosworth (1999:244).
57. cf. Canard (1959:114), quoted in full by Lagardere (1996:71-72).
58. Simpson (2003), Christensen (1936:471), Morony (1984:259), Roden (1986: 31).
59. Miller (1959).
60. A different view was expressed by Laufer (1919:372-373).
61. Simpson et al. (forthcoming).
62. Adams and Nissen (1975:62), contrast Adams (1965, fig. 10). However, a considerably later date is more likely for these remains: traces of "recent" (i.e. probably 18th century or later) occupation have been observed at this site, Adams and Nissen (1975:231).
63. Nesbitt (1993), Nesbitt (1994), Boardman (1995).
64. This renders dubious an earlier report of "very large quantities" of short and long-grain rice said to have been found in Parthian mudbricks, bonding and later infilling of Room 14 of the so-called "castle" on the southern circuit of the fortifications at Erk-Kala, see Usmanova (1963:80). This construction is now regarded as being Sasanian rather than Parthian in date yet the reliability of the archaeobotanical identifications is questionable and is more likely to reflect a mis-identification of barley chaff, for which there is substantial evidence in these very bricks (Sheila Boardman pers. comm. 1998).

65. Adams (1962:116-119), Wenke (1975/76:87-88), Wenke (1987:256), cf. Neely (1974), Rahimi-Laridjani (1988:126).
66. Le Strange (1905:234), Canard (1959), Ahsan (1979:90-92, 140-142), Watson (1983:17-19), Nasrallah (2007:385), Lagardere (1996).
67. Samuel (2001).
68. Miller (1981).
69. Ashtor (1976:43).
70. Ashtor (1976:43).
71. Ahsan (1979:89-90).
72. Wulff (1966:291), Petrushevsky (1968:500-501).
73. Inalcık (1982).
74. Greehan (2007:58-59).
75. Hill and Bryer (1995).
76. Beldiceaunu and Beldiceanu-Steinherr (1978:9).
77. Faroqhi (1978).
78. Meriam (1926:88-89).
79. Xiao, Xin, Hong, Jie, Xue and Dodson (2007).
80. Laufer (1919:372), Bailey (1946).
81. Laufer (1919), Clark (1973:188), Doerfer (1963:326-327).
82. Räsänen (1969:381).
83. Räsänen (1969:311), Redhouse (1974:833), Yudakhin (1957:700), Abdurahmanov (1954:742), Baskakov (1967:886), Jarring (1964:106).
84. Watson (1983:17).
85. Canard (1959), Waines (1995), Watson (1983:16), Petrushevsky (1968:500-501).
86. Venzke (1987-92:179), also proposed long ago by de Candolle (1886:386).
87. Venzke (1987-92), Beldiceanu and Beldiceanu-Steinherr (1978).
88. de Candolle (1886), Potts (1991), Potts (1996), Rahimi-Laridjani (1988), Venzke (1987-92), Watson (1983).
89. Glover and Higham (1996).
90. Bazin, Bromberger, Balland and Bāzargān (1990).
91. Bazin, Bromberger, Balland and Bāzargān (1990).
92. Bazin, Bromberger, Balland and Bāzargān (1990).
93. Potts (1994), Tengberg (2003).
94. Zohary (1998).
95. Nesbitt and Summers (1988).
96. Nesbitt and O'Hara (2000).
97. Chang (1949:68).
98. Jarring (1998:14).
99. Golomb (1959:81), Jarring (1951), Roerich (1931:86-87), Schomburg (1928), Hua, Nagamine, Kikuchi and Fujimaki (2004).
100. Wiens (1967), Weggel (1985).
101. Ziuzhen (2003).
102. Olufsen (1911:495) .
103. Dyer, Tsiburzgin and Shmakov (1992:243-278).
104. Whitman (1956), Andersson and Svanberg (1995).

340

105. Gelb (1995).
106. Usenbayev, Yezhov, Zvantsov, Annarbayev, Zhoroyev and Almerekov (2006).
107. Nesbitt and Summers (1988), Weber (1990).
108. Johnson (1940).
109. Buxton and Dowson (1921:289, 293), Guest (1933), Bawden (1945:384), Edmonds (1958), Salim (1962:85), Thesiger (1967:64, 174-175, 193), Fernea (1970:9).

11 | History of Rice in Europe

Aldo Ferrero and *Francesco Vidotto*

Introduction in Europe

The earliest reference to the word rice in Europe can be found in the writings of the Greek tragedian Sophocles (495-406 BC) who mentioned with the name *Orinda* a cereal cultivated by a people who lived along the lower reaches of the Indus (Beluchistan, in the present days).[1] The name was the same that some ethnic groups from India used for rice. The philosopher and scientist Theophrastus, who succeeded Aristotle as head of the peripatetic school, named it *oryza*. Rice was called with different names according to the languages of the populations who used it or cultivated it. For the Illirian peoples it was *"Oriz"*, *"Tragos"*, *"Trophe"* or *"Olyra"*. *"Lyra"* was the name for the ancient Egyptians. The Arabs called it *"Eruz"* *"Arouz"* a word from which is derived *"Arroz"*, the Spanish and Portuguese name for rice.

Even if the introduction of cultivation of rice to Europe is quite uncertain there are many evidences about its knowledge as a food product. The historian Herodotus (*ca.* 480-*ca.* 484 BC) wrote of a seed, described as similar to millet that Indians used as food.[2]

Theophrastus in his *"De plantarum historia"* (the Enquiry into Plants) included the reports of Callisthenes, a relative of Aristotle, and other scholars who took part in the expedition to India undertaken from 326 BC in India by Alexander the Great, King of Macedonia, with the main task of writing the history of the undertaking. Theophrastus wrote in fact that "the plant is similar to spelt and far (both are early forms of grain) and like far its ears hang down. When boiled, it becomes visibly bigger like grain; when it stays in water, its grains separate, but not from an ear, but rather from a tuft similar to that of millet."[3] The Greek historian Diodorus Siculus (*ca.* 90 BC-*ca.* 30 BC) wrote that Eumenes, being short of wheat during the war against Seleucus in Mesopotamia, fed his troops on rice. Ancient eminent writers of agricultural subjects rarely mentioned rice in their writings. Pliny (24-79 AD) included rice in his *"Naturalis Historia"*, but giving of it a very imprecise and incorrect description. No reference to rice is given in *"De Re Rustica"* written in the 1st century AD by the Spanish agricultural writer Columellus at the time when Spain was a Roman colony.

Rice was not an ordinary food for the ancient Greeks and Romans. It was mostly known for medicinal uses. The earliest information about the use of rice for medical use are from Horace (65 BC-8 BC) the poet and satirical writer who reported that rice used for the infusion was so costly that only few Roman could afford it. In the *"De Materia Medica"* the physician and pharmacologist Dioscorides from Anazarbus (Cilicia-Turkey) (40-*ca.* 90 AD) who practised in ancient Rome at the time of Nero suggested the use of the rice infusion against intestinal pains and intestinal worms. Rice flour was also advised to prepare cosmetic potions.[4]

In the 2nd century AD, the physician Soranus of Ephesus in his *"Gynaeciorum libri"* recommended the use of rice for digestive troubles, during pregnancy.[5] From Ancient Greeks and Romans, for many centuries through the middle ages and renaissance, rice continued to be an important product in Europe for the preparation of semolina, porridges, paps, decoctions and juices to treat numerous types of health troubles and pains. There are documents stating that rice was used on 1253 AD in the S. Andrea hospital in Vercelli (Italy) for specific prescriptions of *"risum and amigdolas"* (rice and almonds). In the same period, in the book of expenses of the Savoy's Lords, it is listed, among other purchases also that of few hundred grammes of rice to prepare cookies.

The early information on the migration of rice towards West (Europe) can be found in the book *"Geographika"* by Strabo (63-24 BC), who referring to writings of Aristobolous another participant in the expedition of Alexander the Great, described with great precision the procedures of rice cultivation in the region between the estuaries of the Tigris and Euphrates, where the land for certain period of the year was waterlogged. The same Greek geographer also mentioned that the Indians produced both wine and bread from rice.[6]

In his *"Periplus Maris Rubri"* Arrian of Nicomedia (Flavius Arrianus) (95-175 AD) described the trade between Greece and Egypt, indicating that rice was at his time one of the principal products which were shipped from the Persian Gulf to the Red Sea.[7] There is a little and uncertain information on the introduction of rice cultivation in the western countries. During its diffusion from the East to the West, in an unspecified period, rice has been cultivated in Palestine at least for a certain period of time. According to the Book of Michna, rice cultivation was judged of great importance to the Jewish people.[8] What is certain is that rice cultivation began to spread particularly in Egypt most probably starting from the 1st century. Strabo during his travel in Egypt recorded to have seen rice cultivation in a desert oasis inhabited by the people named Garamanti.

It is not exactly known how and when rice started to be cultivated in Europe. It is likely that it was spread in Europe by the Arabs, who started the cultivation in 12th century in the towns of Alzira and Xàtiva in the Valencian area[9] from where rice was already exported to other countries throughout the Mediterranean basin.

Rice cultivation was limited to the Valencia region for centuries. Rice fields expanded from the late 19th century to other parts of the Iberian Peninsula, out of the Valencian area.[10] In 1880 there were already 3300 ha in the delta of the Ebro river. In Sevilla region, the most important area of rice cultivation today, rice arrived only in 1940.

In Portugal rice cultivation was probably started in the reign of D. Dinis (1279-1325), probably in swampy areas located near Montemor-o-Velhoand (along the Mondego river) and South of Tejo river estuary.[11] Several documents certify that Portuguese knew rice cultivation as early as 1446 when the Portuguese chronicler Gomes Eanes de Azurara described showing a very good knowledge on the subject, the rice growing in the Upper Guinea Coast. Starting from as early as the middle of the 16th century, Portuguese introduced in the same area and in all the West Africa the cultivation of Asian rices, that almost everywhere have replaced African *O. glaberrima*.[12]

Rice farming in France started between the 15th and 16th century in the Camargue, the delta zone of the Rhône River, the same area where the cereal is cultivated today. The spread of cultivation was favoured by the intensive works carried out to reclaim the land which was regularly subjected to floods. Most of the works were organized under the aegis of the great monasteries present in this area. In this period they dug drainage canals, cleared forests growing on the riverbanks and provided embankments to protect hollowed lands.[13]

An important document on rice cultivation in France is an edict issued in 1593 by the king Henry IV which ordered the growing of this crop in the Camargue.[14] Rice cultivation gradually spread in the following centuries in this area together with the development of the network of canals to manage water. The construction of drainage system was more and more accompanied by the creation of an irrigation network suitable to prevent soil salinization, thus making possible the cultivation of the low lying lands and providing fresh water for humans and animals. During English embargo to France, Napoleon imposed the cultivation of rice, cotton and sugar beet in the Camargue and wherever was possible. During the 19th to the mid 20th century, rice was cultivated on less than 1000 ha and it was considered a key crop for the de-salinization of the fields with the main purpose of preparing them for other crops such as grapevine or wheat.[15] However, in view of low yields, poor grain quality and high milling costs, rice cultivation was substantially discontinued by the end of World War I.[16] Rice cultivation had a substantial boost after the World War II as a consequence of the food shortages that occurred during German occupation (1940-44) and because of the progressive abandoning of winegrowing and the encouraging measures related to the Marshall Plan. In less than two decades, rice area increased from about 250 ha to 32,000 ha, meeting the entire French consumption.[17]

The period of rice introduction in Greece is not known, but the fact that some villages and towns still have names deriving from the word rice is considered a proof of its cultivation in the country since long time.[18] This crop was certainly present already in 1834, the year of the institution of the honour "For the crop of rice". The area of rice cultivation certainly broadened in 1912 after the annexation of Epirus, the region were the crop was largely grown, and its introduction towards the 1920s also in the region of Messenia. Rice has been considered a very minor crop and was cultivated only on a few thousand hectares up to the World War II. Since then its cultivation became intensive and spread in a few years on more than 20,000 ha.[19]

The introduction of rice cultivation in Italy is not much clear. After the fall of the Roman Empire, the Italian peninsula was split into many regions politically separated up to the middle of the 19th century; it was a land of invasion and conquest of Arabian and European Powers. Routes of introduction may have been different. According to some authors rice was brought to north western regions (Piemonte and Lombardia) from Spain through France by the soldiers of Charlemagne (Charles the Great) when he returned from the battles against the Moors.[20] According to others, it had been imported by Venetian merchants who had extensive trade relations with the East.[21] It might have been introduced into the southern regions by the Arabs, in Sicily during their two centuries occupation and in Naples from Spain during Aragon domination.[22] The certain proof of the cultivation in Italy witnessed by written documents dates only back to the 15th century. Notwithstanding that in Sicily rice was likely introduced during the 9th century, its initial spread and importance should be probably very limited. The first records of rice cultivation in this island date back to 1500s and describe the crop as a sporadic substitute of wheat, indicating the marginal role played by rice at that time.[23] In southeast Sicily rice cultivation was also experienced by the Jesuits. In spring 1648 abundant rains caused a great overflow in the Tellàro river, thus hampering the seeding of wheat. In that occasion, Jesuits escaped from certain famine by seeding rice. In general, rice cultivation in Sicily was very limited and reached the maximum extent (about 7700 ha) in 1855. Since then it reduced gradually and in 1912 only about 500 ha were still cultivated in the Catania and Siracusa provinces. Afterwards, cultivation declined constantly and nowadays it is practically negligible.

In a letter dated 1468, a rice grower from Pisa requested the authorization to use waters to grow rice. Three letters written in 1475 by the Duke Sforza of Milano document that twelve bags of rice seeds were exported to Ferrara with some advices to start cultivation in that area.[24] In the same century rice cultivation spread rapidly from the area around Milano to the neighbouring territories of Novara and Vigevano. The period between the end of the 15th and 16th century was characterized by an intensive investment of work and capital which led to the extension of rice cultivation in many regions such as Piemonte, Lombardia, Veneto, Toscana and Emilia.[25] In the Duky of Milano even the genial scientist and artist Leonardo da Vinci was appointed to organize and update the existing irrigation system. The success of rice was due to the high yield obtained in comparison to other cereals. Rice yields were as high as about 10-12 times the quantity of seeded grain, while that of wheat and rye was only 4-5 times.[26] An important merit to the spread of the crop has also to be attributed to the efforts of various religious orders such as the Benedictine monks who had abbeys in isolated and uncultivated places.

Rice became so popular and its cultivation so widespread and uncontrolled that restrictive actions were taken to limit rice exports to Switzerland, France, Holland and Germany.

Starting from the second half of the 16th century, the rush to cultivate rice was slowed as the crop went up against the objections of the people living near rice cultivation because of the spread of malaria. This resulted in legislation which imposed limitations both on the quantity of rice to be cultivated and security distances from urban settlements. The controversy with respect to the sanitary risks due to rice fields encouraged local authorities in several areas to prohibit rice cultivation. The history of rice cultivation in Europe has various examples of tolerance and prohibition. In the Valencia area (Spain), an edict dated 1483 that prohibited rice cultivation established that punishments for transgressors would be very severe, including the death penalty and confiscation of all belongings.[27]

In a document dated 21st September 1691, the viceroy of Sicily imposed to destroy within eight days all the rice fields included in the territory of Noto village, as they were considered to cause several infirmities.[28] Despite the low extent of rice cultivation in Sicily during 1700s and 1800s, several letters of compliant sent to local authorities prove the deep uneasiness of the inhabitants towards this crop. Similar disputes involved, in general, other crops that need the presence of stagnant water, as in the case of flax and hemp, which require water maceration to allow fiber extraction. In order to avoid the severe limitations imposed to rice cultivation, several unsuccessful attempts of cultivation without continuous flooding were made during 1827 and 1853.

Notwithstanding these issues, in the rest of Italy rice cultivation continued to spread up to the 19th century together with a contemporaneous development of networks of canals and ditches. At the turn of this century, there were over 5000 km of irrigation canals just in the area of Vercelli province (Italy). In this area it was built one of the most rational systems of the world thanks to the fact that the land was properly levelled off prior to construction.

Rice Today

Rice Area and Yield

In the European Union rice is at present cultivated on about 410,000 ha, principally located in the Mediterranean countries. It is mostly grown in concentrated areas such as the Po valley in Italy, the Rhône delta in France and the Thessaloniky area in Greece. In Spain rice cultivation is scattered in several areas such as the river delta lands of Guadalquivir river, Ebro river, Jucàr river and in other non-humid areas such as Extremadura and Aragón, as well as in Portugal, where it is cultivated in Tejo and Mondego valleys (Fig. 11.1). In some environments (estuaries, river basins, etc.), submerged rice cultivation is fundamental for a sustainable management of the wet ecosystem where the crop is grown.[29]

The two top rice producers which contribute together more than 80% of the total rice production in Europe are Italy (224,000 ha) and Spain (117,000 ha) (Table 11.1). In the period 1995-2005 some slight variations in

Table 11.1. Evolution of rice area (ha) and yield (t/ha)
in Europe in 1995-2005[30]

Country	Area (× 1000 ha)[a]		Yield (t/ha)	
	1995	2005	1995	2005
Italy	223	224	6.15	6.17
Spain	105	117	6.80	7.23
France	18	18	5.70	5.72
Greece	22	24	7.52	7.25
Portugal	24	25	6.04	4.80
Hungary		3		3.41
Romania		1		4.18
Total[a] —Weighted mean[b]	392[a]	412[a]	6.37[b]	6.43[b]
Variations 1995-2005	+ 5.1%		+ 0.9%	

the harvested area have been recorded in each country of cultivation, in relation to the market price or water availability.

In 2005 the average crop yield has been 6.43 t/ha and was quite variable among the different countries, as it ranged from 4.80 t/ha in Portugal to 7.25 t/ha in Greece. In the last decade Greece's average yield showed a slight reduction because of the unfavorable environmental conditions recorded in 2005.

Environmental Conditions

Rice fields show a wide pedological, edaphic and structural variability and a quite differentiated natural flora between the different areas, because of the old geological genesis and evolution of the soils.[31]

A frequent alternation between clayey and alluvial soils, even in short distances can also be noted. Some areas close to the river estuaries are peaty due to an ancient sedimentation of marshes. Fine textured, poorly drained with impervious hardpan or claypan soils are well suited to rice production, like those with low water permeability that are not very suitable to other crops.

In the Valencia area (Spain) rice is cultivated in marshlands of a coastal lagoon. Other lands of autochthonous origin nearest to the Alps (known in Italy as *baragge*), which were in the past woods or untilled fields, are cultivated nowadays only with rice as they are very firm and compact with little organic matter content and little fertility. The values of pH are quite variable as they can range from 4.5 to 8.5.

Organic matter content of soils from different areas of rice cultivation averages 2.0%, even if it can range from a minimum of 0.5% to a maximum of about 14% in the peaty soils.

In coastal areas (e.g. the Camargue in France, Ebro delta in Spain, Tejo estuary in Portugal, and Po Delta in Italy), soils are saline or very saline. The subsoil in the southern half of the Camargue, recently reclaimed from the sea, has a high salt content.

In spite of this large variability, all European rice is cultivated in irrigated ecosystems.[32] Most of the irrigation water comes from rivers (the Po in Italy, Ebro in Spain, Rhône in France, Tejo in Portugal, etc.) and lakes, while less than 5% of rice irrigation water is pumped from wells.

According to FAO classification, in most European areas of rice production, the primary climate is temperate-continental with a cold winter and warm summer and main rainfall occurring during the first stages of the crop growth (April-June) and the harvesting period (September-October).[33] In the Mediterranean countries, the climate is sub-tropical (Mediterranean climate) with a dry summer, warm, dry, clear days and long growing season longer than in areas with temperate-continental climate. Average temperatures range from 10 to 12°C during rice germination and from 20 to 25°C during crop flowering. This climate is favorable for high photosynthetic rates and high rice yields while its low relative humidity throughout the growing season limits the development of rice diseases. Poor crop establishment may occur following partial stand loss and seedling drift caused by cool weather and strong winds during the first stages of crop growth, especially in some areas close to the sea, such as the Valencia region.[34]

Drops in temperature can also occur during flowering (August), eventually causing serious damages from flower sterility and caryopsis abort or poor caryopsis filling.

The Camargue area is constantly swept by a strong prevailing wind ("Mistral") blowing more than 200 days a year resulting from the meeting of Mediterranean and Atlantic air masses.[35] This wind may cause rice uprooting during early crop stages and increases evaporation.

Farm Structure and Organization

Main traits of European rice farms are quite variable among countries. Nevertheless, because of the several peculiarities of the crop and in particular all those related to the cultivation under flooding conditions (such as the need of accurate soil leveling, construction and maintenance of an adequate network of canals and ditches), rice growing is usually performed in large contiguous areas by farms that tend to be highly specialized in terms of labor organization, machinery and equipments and that show the strong tendency to adopt monoculture. These conditions, that are usually the basis for the development of large holdings, have promoted in Europe a development of a wide range of average rice farm sizes. Rice farm is actually characterized by a mean area larger than that of the average farm only in Italy and France (Table 11.2), while in Greece and Portugal the rice farm size is similar to that of the average national farm. At the opposite side, it is worthy of mention that in Spain, the second European rice producing country, the rice farms are more than 50% smaller than national farm average size.

These differences among countries could be mainly related to the different geographical and historical context in which rice growing developed over the time. For instance, in Spain, the farm size is much influenced by the numerous scattered areas of cultivation and in particular by those where the agricultural activity is generally carried out in small holdings (e.g. the Valencia area). By contrast, in Italy rice cultivation was introduced and spread mainly in an area which was already characterized by the presence of large farms (e.g. the Po valley area). Differences may occur also within the same country. Even within Spain the average rice farm size spans from 4.7 ha in Valencia to more than 34 ha in Andalucía.[36]

Table 11.2. Average area of rice farms and average area of farms

Country	Average rice farm size (ha)	Average farm size (ha)
Italy[37]	48.5 (2007)	7.4 (2005)
Spain[38]	9.4 (1999)	23.6 (1999)
France[39]	221 (2000)	45.3 (2003)
Greece[40]	6.0 (2005)	4.8 (2005)
Portugal[41]	11.5 (1999)	14.75 (2005)

The mechanization of roughly all operations of rice cultivation and the spread of the monoculture led to the increase of the average farm dimensions. This trend is still continuing, driven by the reduction of the crop profitability, the increase of the working capacity per ha and the cost of the machinery.[42]

From 1980 to 2006 the mean surface per farm increased quite remarkably in all European rice producing countries (from 20 to 48.5 ha in Italy and from 1.9 to 4.7 ha in Valencia area).[43] In the same period, the number of farms showed a reduction roughly proportional to the increase of the average farm surface. For example the total number of rice farms diminished to one half in Italy and to one-fifth in Valencia area of rice cultivation. An exception is represented by France, where the number of rice farms increased steadily in the period from 1979 to 2000, growing from 89 to 193, respectivley. In this case, an increase of the total rice area had also occurred, thanks to a plan of re-launching of rice cultivation decided in 1981, with the help of the authorities.[44]

In Italy the area managed by a single worker can range from 40 to 60 ha and the farm size is then usually a multiple of that value. Many farmers, aged from 25 to 45, are graduates and there is a positive correlation between the education level and the farm size. Primary school licensed farmers manage farms with a mean surface of 49 ha, while the junior secondary school licensed farmers manage farms with a mean surface of 66 ha and the farms led by farmers with senior secondary school diploma or degree certificate have a mean surface over 100 ha.[45]

Table 11.3. Correlation between farmer's education level and farm surface in the Vercelli area, Italy[46]

Farmer educational level	Average farm surface (ha)
Primary school	49
Junior secondary shool	66
Senior secondary school	> 100
University degree	> 100

The good returns obtained from rice cultivation during the period from 1970 to 2000 encouraged the young farmers to continue their parent activity. In Spain a large proportion of rice growers are over 55 years old or retired and the oldest populations can be found in Tarragona and Valencia areas. In Tarragona several rice farmers are part-time workers, while in Valencia and Andalucía part-time farming is limited to the smallest farms.[47] According to Lima, Portugal rice growers can be grouped in specialized farmers that manage large holdings with modern means and in small producers, who often cultivate just a few or less than one hectare and are more disinclined to technology innovations.[48] This is in contrast with the Italian conditions, where the level of mechanization and cultivation technology applied in a small farm is not much different from that applied in a big farm, because of the high cost of labor and shortage of manpower.

With regard to the labor organization, it is very difficult to find reliable information sources, mainly because of the typical traits of the agricultural work: seasonal activity peaks, collaboration within the farmer family, extra time during festivities or early in the morning and late in the evening. In this regard, data from the national labor bureaus show large differences between rice growing areas in Europe. In most of the north Italian rice area, there is a prevalent presence of family members in the working population of the rice farms (about 70%; Table 11.4). This reflects a general peculiarity of the Italian agriculture: the relatively small mean farm size and the high level of mechanization limit the presence of hired regular worker, while the temporary workers, whose functions have been partially replaced by mechanization in the last decades, are employed for short periods.

By contrast, in France (the Camargue) about 73% of the working population is represented by paid workers. This proportion falls to about 35% in small and medium sized-farms, while it peaks up to 89% on 300-550 ha farms.[49]

Table 11.4. Agricultural labour employed in Vercelli area[50]

Category	Number of workers
Self employed workers	2,254
Full time workers*	370
Seasonal workers	514
Total**	3,138

*Includes the staff devoted to the irrigation network maintenance.
**One should remember that the agricultural population in Vercelli province represents about 3.5-4% of the total working population.

The rice growing farm may be directly involved in part on paddy processing. For instance, most Italian rice farms are equipped with their own drying systems. Whereas the largest proportion of rice paddy produced in Europe is processed by milling industries, a few rice-growing farms have their own milling plant for processing their production and are organized for the direct selling of the finished product. The rice-growing equipment belonging to the typical rice-farm includes also a wide set of farm implements: ploughs, harrows, furrowers, scrapers, fertilizer-spreaders, boom sprayers and many more. The adoption of external services to carry out the cultivation operations is not so frequent, but is constantly increasing.

The cost of rice production in the European rice countries is much higher than that of most other growing countries. It is about 360-420 USD/ton.[51] Main components of this cost are related to water, fertilizers, crop protection products, seed, machinery, fuel and labour.

Evolution of Mechanization

The evolution of mechanization in rice cultivation in Europe has been of particular importance.

Until the 16th century rice was primarily cultivated in marshy areas, naturally covered by stagnating water. Since that time a network of artificial canals and ditches started to be organized and rice began to be introduced in areas previously cultivated to other crops. At that time the modification of natural soil level was quite difficult because of the limited availability of motive power, which was mainly of animal origin. For this reason levee used to follow the level curves and basins had an irregular shape with an area of a few hundreds of square meters.

Mechanization of rice production started to develop from the early 1900s with the introduction of the first tractors and threshers. Mechanization led to simplification of cultural systems in rice farms and spread of monoculture in several areas of Europe. Monoculture was also favored over the time by the reduction of animal breeding, because most of the soils of rice cultivation were acidic and waterlogged and not much suitable for the production of forage and grain crops. The introduction and development of mechanization led to deep transformations both on crop management and on farm structure and facility requirements. Significant evolutions involved also cultural and social aspects related to rice economy. Since the mid-18th century until the 1960s, when herbicides started to be widely used, rice cultivation in northern Italy demanded for great employment of seasonal workers for weeding, a job that lasted for some forty days a year from spring to summer. At the very beginning, this work was not identified by the scholars as a specific operation in rice growing, also because men, women and even children took part in this work. Seasonal workers were needed also for transplanting. This practice, which was introduced in Italy from Spain in the early 1910s and lasted till the 1950s, allowed a reduction of weeding labor (from about 300-320 hours/ha to about a third).[52] A reverse side was the need of additional time for raising nurseries and transplanting and that the actual operation of placing the plants in the rice field was a much more demanding task than weeding, as it had to be performed with both hands away from the body in a bending position without the possibility to lean against one's own body. Since the 1890s, the gender ratio of seasonal workers showed a constant increase in women. It is estimated that in the 1930s in Italy, about 180,000 working women coming from many parts of the Po Valley were enrolled for transplanting and for weeding in the rice fields every spring.[53] In some areas, such as in the Valencia region (Spain), several operations such as transplanting were perfomed by seasonal working groups almost exclusively formed by men.[54] A significant peak of labour requirement was also constituted by all operations associated with harvesting, such as reaping, bundling into sheaves, transporting to farm, threshing, cleaning and drying. Thus, in addition to the recurrent presence of cattle sheds and all buildings and structures required by ordinary crops, rice farms had to be endowed with large drying floors, threshing and storage units, as well as of dormitories and other utility rooms to host the seasonal workers. Moreover, the large request of animal power for cultivation and food for workers implied spaces to raise horses, oxen, chickens and pigs. After the progressive decline of recourse to seasonal workers as consequence of mechanization and adoption of herbicides, along with the reduction of animal breeding, these buildings became unused or partially converted to shelters for machinery or to host drying plants or storehouses.[55]

Mechanization was particularly boosted starting form the early 1950s because of the progressive reduction of manpower, which was more and more attracted by better jobs offered by the industry.[56] The important changes in the social structure induced rice growers to modify rice cultivation systems by introducing new agronomical practices that mechanization and chemical industries made available in those years.

As herbicides were introduced and their use spread on rice regions, transplanting, which had been the most adopted practice for the crop establishment over about 50 years also because it favored weed identification and uprooting, was rapidly replaced by direct broadcast seeding on flooded soil. Approximately in the same period, manual harvesting was gradually replaced by mechanical harvesting. After the first experimentations of cutting and binding machines in the early 1910s, the issue of mechanizing the harvest operations remained actually unsolved until the beginning of 1950s, when combine harvesters imported from the USA and produced in Europe became available to the farmers.[57] In Spain, the diffusion of mechanical harvesting with self-propelled and auto-threshing combines started very few years later, at the beginning of 1960s.[58]

The diffusion of mechanization and use of herbicides determined a dramatic reduction of the labour requirment for main cultural operations and for transplanting, weeding and harvesting, in particular. For example, in Italy the amount of manpower required for these operations dropped from 1950 to the end of 1960s by about 95%.

For all these reasons, and also because of a favorable economical trend for rice marketing, the machinery fleet of the rice farms changed abruptly starting from the 1950s. The complete mechanization of rice production in Europe eventually turned in a remarkable way towards the super mechanization and the consequent under-utilization of machinery and facilities. This trend, which has been recorded during the last decades in the farm machinery fleets of all European rice farms, is probably related to the relatively limited farm size and to the scarce adoption of external services for the cultivation operations. The first factor mostly depends on social and historical backgrounds, while the small amount of cultivation operations carried out by external services can be explained mainly by the necessity of performing the operations timely, sometimes in difficult field conditions, without being subjected to delays due to the service provider time-table. Starting from the beginning of this decade, the need to rationalise the rice production costs led to a decline of the investments in the agricultural machinery. At present, the main demand of new equipments is for the replacement of the old and outdated ones. This inclination involves also the new combines: frequently in the farms over 150 ha a single new combine replaces two older and obsolete machines. In this farm type, the mean use of tractors is less than 350 hours/year, while the use of combines is about 150 hours/year.

The current situation of the rice farm mechanization in Europe is quite difficult to depict as detailed information are available only for some areas.

According to a recent survey carried out in the Vercelli area, one of the most representative for rice cultivation in Italy with a total of about 85,000 ha, during the 1950s and 1960s, about 50% of the majority of new tractors had a power ranging from 60 to 100 kW, while starting from the 1970s most of this machinery was over 100 kW (Table 11.5). The number of less powerful tractors decreased gradually becoming meaningless in the last years.[59] The number of new tractors registered every year showed the strongest increase during the 1980s and it is now quite similar to that observed in this period. The total number of tractors still operating in the Vercelli area is around 5,300, giving an area per tractor rate of about 16 ha/tractor. In Valencia area (Spain) the number and the power of tractors in use increased quite remarkably in the last years. In particular, the percentage of tractors with more than 75 kW increased from 9.4% in 1985 to 23.8% in 2005 (Table 11.6). As a consequence, in this region the area per tractor rate decreased from 36 ha/tractor in 1985 to about 14 ha/tractor at the present.

The portioning out of the tractor fleet in age classes is quite similar in Vercelli and Valencia areas (Figs. 11.3 and 11.4). A certain amount of

Table 11.5. Number of new tractors registered the Vercelli area
(north-west Itlay) in the last 20 years[60]

Power class (kw)	1960s	1970s	1980s	1990s	2000-2003
0-30	16	16	5	6	3
31-45	40	25	3	12	3
46-60	64	36	9	10	7
61-75	80	96	34	20	12
76-100	96	271	227	57	9
> 100	57	922	1416	1213	549
Total	353	278	1694	1318	583
Average new registrations per year	35	28	169	132	146

Table 11.6. Number of tractor in use in Valencia in the last 20 years[61]

Power class (kw)	1985	1990	1995	2000	2005
0-30	30	32	33	60	61
31-45	66	84	105	116	125
46-60	115	147	174	207	254
61-75	77	98	112	152	186
76-100	28	65	87	127	165
> 100	2	9	8	11	31
Total	318	435	519	673	822

tractors is outdated, as from 30 to 33% of the tractors are more than 25 years old. Old machinery is commonly maintained to alleviate activity peaks. Farmers frequently consider it more convenient to maintain two extra–tractors than employing one more laborer.

On the other hand the presence of a significant number of recent and powerful tractors (over 100 kW), is related to the current evolution of soil cultivation, which tends to replace the traditional plowing with minimum tillage techniques, employing combined cultivators (e.g. disk + rigid tines) that must operate at high speed (2.3-2.5 m/s). Such a change demands a further increase in the use of very powerful farm implements.

Presently, the number of combines in the Vercelli (Italy) and Valencia (Spain) areas is of about 1000 and 230, respectively, giving an area per combine ratio ranging from 59.5 to 67.4 ha/combine. The distribution of the combines in age classes is similar between the two areas, with a higher presence of old machines (more than 35 years old) in Valencia (Figs. 11.3 and 11.4).

The adoption of big combines, with a great working capacity, allowed to reduce the harvest period, but imposed to adapt the drying capacity in order to complete, within few hours, the drying of the rice harvested daily. At the beginning of 1950s, the first available combines allowed to harvest about 2.8 tons of paddy/h, while the machines that are presently used have a harvesting capacity of about 11 tons/h, and can reach up to 20 tons/h if they are fitted with stripper headers.[62] The harvesting time was considerably reduced from 35-40 days in the 1950s to the current 20 days. Since the 1970s, the drying plants which demanded a continuous monitoring (dynamic—open type) have been gradually replaced with facilities which demand control and monitoring in the final stage only (batch drier or dynamic drier with recycle type); frequently, advanced monitoring systems were adopted with totally automatic processes. Dryers operate with ambient or hot air obtained from oil-fired or methane-fired burners. Drying plants are commonly associated with storehouses, sometimes equipped with refrigerating devices to preserve rice from fermentations and from insect attacks, to keep rice until the market is favorable for selling. In Valencia, almost all of the grain harvested is artificially dried through large plants available at cooperatives and rice grower associations, and through the facilities of about 20 individual farmers.

Consumption and Market Policies

Rice is not the staple food for most of the European population; nevertheless, rice consumption in the continent has increased in the last years due to immigration and diversification of the diet of the Europeans. In the last decade, rice consumption has significantly increased in all European countries, both rice producers (Southern Europe) and non-rice producers (Northern Europe). It is expected that this trend will continue in the coming years, particularly in northern European countries.[63]

In 2005, milled rice consumption in rice-growing countries ranged from 4.9 to 7.1 kg/capita/year, except in Portugal, where rice is considered as a staple food for some population groups,[64] accounting for an average individual consumption of 17.1 kg/capita/year (Table 11.7). In Southern European countries, about 80% of the consumed rice belongs to japonica-type varieties (mainly medium and long A type) and 20% to indica-type varieties.[65]

Rice produced in southern Europe is processed by local milling plants. Very often, local and "specialty" varieties are gaining significant importance for small and medium farms in local markets. For example, varieties like "Carnaroli" and "Vialone nano" in Italy or "Senia", "Bomba", "Bahia", in Spain have a good appreciation in local markets, also thanks to the promotion of their quality through the attribution of APO (Appellation of Protected Origin) and the direct selling of rice by rice growers, who directly process their own rice production on the farm with small rice milling plants.

Table 11.7. Consumption of milled rice in European countries (kg/capita/year)[67]

Country	1995 (kg/capita/year)	2005 (kg/capita/year)
Denmark	2.3	4.9
France	3.4	4.9
Germany	2.7	3.4
Greece	4.9	6.1
Hungary	4.1	4.5
Italy	4.9	5.4
Norway	2.5	4.3
Poland	1.8	2.3
Portugal	15.1	17.1
Romania	2.5	5.6
Spain	6.1	7.1
Sweden	4.5	7.2

Quite different is the situation in North-European countries (UK, Denmark, Germany, Sweden). About 85% of rice consumption is related to indica-type varieties that are mainly imported from the USA, Thailand, India and Pakistan and processed by local milling plants. To meet the demand of indica-type varieties, about two decades ago, European rice growers were encouraged to increase the cultivation of long grain rices suited to temperate climatic conditions. Some of these varieties are characterized by a long cycle and exposed to the damages caused by low temperatures which occur particularly at night during the flowering period.[66]

The growing presence of Asiatic communities, primarily in western European countries, has remarkably favored the increase of the demand for scented rice varieties (Basmati type).[68] It is expected that this trend will be maintained in the years to come because of the expanding migration from Asian and North African countries, the growing interest in ethnic cuisine and the significant reduction of the import duties.

Following the international trade agreements, the prices paid to European rice growers showed a sharp reduction over the last 10 years (Table 11.8), and gradually approached the international prices. The return from rice production remained substantially unchanged thanks to the subsidies to compensate price reductions. After the introduction in 2003 of the Common Agricultural Policy, subsidies were not any more linked to the production and were in part devoted to environmental protection.

As a consequence of the Uruguay Round agreement, rice imports from third countries increased remarkably. Tariff reduction policy started to be applied with a cut of 20% in 2006, 59% in 2007, 80% in 2008 and total market liberalization from 2009. In the meantime a duty-free quota based on previous exports to the EU has been established with an increase by 15% each year until 2009 when all tariffs and quota will be removed.

Table 11.8. Price to rice grower of paddy rice in 2005 (USD/t) in European countries and other outstanding rice producing countries[69]

Country	1995 (USD/t)	2005 (USD/t)
France	564	255
Greece	741	191
Italy	477	270
Portugal	571	242
Spain	479	238
Average Europe	566	239
Egypt	207	221
USA	202	155
Brazil	193	128
Thailand	165	164
India	150	150

The enlargement of the EU resulted in a remarkable market increase of the rice grown in the area.[70] In 2006 the additional consumption due to the 10 new countries which entered the EU was about 250,000 tons, mostly from indica-type varieties, equivalent to about 10% of the total European production.

Management Decisions

Cultural Practices

In Europe rice is principally grown as mono-crop. It is directly seeded from mid-April to end-May and harvested from mid-September to end-October. Main agricultural practices carried out in rice cultivation in Europe are included in Fig. 11.5.

During all or almost all cycles of cultivation, rice fields (basins) are maintained flooded mainly to protect the plant from the low temperatures, avoid fast temperature variations and limit weed growth.

Rice fields are with 2 to 10 ha in size and are usually laser leveled in order to maintain a uniform and shallow layer of water with a maximum level variability of ± 2 cm in each cultivation basin. This results in better rice germination after flooding and, at the same time, in limited weed growth.

Plowing depth does not exceed 20 cm, in order not to limit efficiency of tractors in the submerged rice-field during the operation of fertilizer application and herbicide spraying.

To avoid water losses soil is often compacted through several passes with tractors equipped with special rollers and cage-shaped wheels, during tillage operations (puddling).

More and more frequently plowing operations are replaced by minimum tillage normally carried out using chisels suitable to operate at 20 cm. This trend in soil tillage is mainly due to (1) labor cost reduction, (2) improvement of efficiency of dry seeding technique on non-submerged soils, and (3) application of the false seeding technique (stale seed bed) through stimulation, before rice-seeding, of the germination of weed seeds, especially those of weedy rice.

Seeding is generally done by broadcasting seed (often pre-soaked) using centrifugal spreading devices on fields submerged with a 5-10 cm water layer.

About 70% of the European rice area is cultivated with long and medium-grain japonica-type varieties.[71] The remainder is cultivated with long-grain indica-type varieties. The cultivation of indica-type and early-maturing varieties has significantly increased over the last years in all European rice-growing countries. Most indica-type varieties are short with a low growth and moderate competitive ability towards weeds, especially weedy rice plants. There is a growing interest towards early high yielding varieties with a cycle of 130-145 days as they allow to seed from mid-May and then escape negative effects of the low temperatures of April and August months, during which the delicate phases of emergence and flowering, respectively, occur. These varieties also allow the application of the weedy rice control to be carried out before rice planting.

In about 40,000 ha, mostly in Italy, in less compact and more permeable soils, seeds are drilled to dry soil in rows. The rice is generally managed as a dry crop until it reaches the 3-4 leaf stage. After this period, the rice is flooded continually, as in the conventional system. In these conditions, rice has no competitive growth advantage over weeds, which can compete with the crop from the beginning of stand establishment.

The main benefits of drill seeding are water saving, prevention of alga propagation, seeding before water availability (even in early April), prevention of seed drift due to wind during seeding, better rice rooting, a higher resistance to lodging as well as to pests. The most important disadvantages are related to a lower tillering rate, a greater competition by the weeds and a risk of nitrogen loss due to the important nitrification process that can occur during the pre-flooding period.

Through the centuries a dense network of canals has been built originating from rivers. These canals deliver water to the rice fields while at the same time allowing for correct draining so as to prevent soils from becoming marshy.

Once it has been dammed by weirs, water flows by gravity into the canals and then into the fields, commonly applying a flow-trough systems where water flows from uphill to downhill basins. By this system it is estimated that irrigation water is used at least twice. In some regions of Spain, Italy and Portugal water may be supplied to each basin through a head ditch.

Water requirements for rice cultivation can be very different, depending on the soil conditions. Total seasonal volumes of water supplied can vary from about 17,000 m^3 per ha in the heavy soils up to 42,000 m^3 per ha in the sandy soils. To reduce water requirements soils are compacted through several passes with tractors equipped with special rollers and cage-shaped wheels during tillage operations. On average water losses in most rice fields due to percolation evapo-transpiration (calculated on a period of 150 days of flooding) are about 3000 and 4500 m^3/ha respectively.[72]

In each cultivation area, irrigation associations take care of the measurement, delivery and management of water flow, in order to support the needs of the rice farms.

Water is maintained at depth ranging from 5 cm during the early stages to 12-15 cm from tillering to early ripening and drained 2 or 3 times over the growing season. The higher water level during flowering prevents rice from temperature variations and from incurring pollen sterility.

Rice fields are drained several times during growing season to improve seedling rooting, prevent or control the formation of algal scum, spray foliar-absorption herbicides and apply fertilizers.

One of the most discussed aspects concerning rice management is that of the fate of the straw after crop harvesting. Some rice farms burn crop residues in the autumn or the following spring, before starting a new rice cultivation. Since the 1980s more and more farms bury straws turning them into the soil during plowing operations, usually in the autumn, in compact clayey soils or in spring on sandy or half-mixed soils.

As a general rule, nitrogen is supplied at 80-120 kg/ha, 50% in pre-planting and 50% in post-planting, using urea or other ammonium-based fertilizers.[73] Phosphorous and potassium are supplied in the pre-planting stage at 50-70 and 100-150 kg/ha, respectively.

Rice is affected by the attacks of weeds, diseases and insects. In European rice fields weeds are considered the worst noxious organisms affecting rice production. It has been estimated that without weed control, at a yield level of 7 to 8 t/ha, yield loss can be as high as about 90%.[74]

In Europe, herbicides account for more than 80% of the total consumption of pesticides utilized for crop protection with a total spending of about € 110 million/year.

The major weed problems in the European rice fields are aquatic species or species that tolerate flooding. In relation to their relevance in terms of competitivity against the crop and to the practices which are adopted for their control, main weeds can be grouped as follows:[75] (1) *Echinochloa* species (*E. crus-galli*, *E. crus-pavonis*, *E. oryzoides*, *E. erecta*, and *E. phyllo-pogon*); (2) *Heteranthera* species (*H. reniformis*, *H. rotundifolia* and *H. limosa*); (3) *Alisma* species (*Alisma plantago-aquatica* and *A. lanceolatum*) and sedges (*Cyperus difformis*, *Bolboschoenus maritimus* and *Schoenoplectus mucronatus*); (4) various weedy rice biotypes; (5) weeds of the drill seeded fields; and (6) other weeds of secondary importance, occasionally representing an issue in restricted areas (e.g. *Leptochloa* species).

Weed management currently rely on strategies based on a combination of herbicide application with appropriate agronomic practices such as soil leveling, accurate water management and variety choice.

In case of infestations characterized by the presence of weedy rice together with other species, weed control programs are principally aimed at controlling weedy rice. These practices are mostly performed in rice pre-planting either with an antigerminative herbicide such as flufenacet or pretilachlor applied about one month before rice planting, or with mechanical means (harrows or cage wheels) or systemic graminicides (dalapon, cycloxydim, glyfosate) to destroy weedy rice seedlings grown after stale seedbed application. In both cases, other treatments are usually required to control the other species.

When the pre-planting treatments against weedy rice are not sufficiently satisfactory, rice growers frequently carry out an additional intervention against this weed at rice flowering time. With a few weedy rice plants per ha, the control of the weed is usually done manually. In the case of high infestations, the weed is devitalized by applying systemic herbicides (glyphosate) with wiping bars, provided that the weed plants are taller than those of the crop.

Particularly promising for weedy rice control appears to be the Clearfield technology based on the planting of a rice variety tolerant to imazamox, an imidazolinone herbicide with a wide spectrum of activity that also includes weedy rice plants.

In case of absence of weedy rice, in general, one treatment is done in pre-emergence mainly against *Heteranthera* spp. (with herbicides containing oxadiazon) and other one or two from 10 to 40 days after crop emergence against *Echinochloa* spp., sedges, alismataceae and other species (with herbicides containing molinate, thiocarbazil, propanil, chyalophop-buthyl, profoxidim, azimsulfuron, bispyribac-sodium).

When rice is directly seeded in dry soil, in general two or three herbicide interventions are required. The first treatment is commonly carried out in rice pre-emergence with pendimethalin or clomazone to control *Echinochloa* spp., *P. dichotomiflorum* while others, done in post-emergence, are targeted to control sedges, alismataceae and other species.

Major diseases in the European rice fields are caused by parasitic fungi such as: *Magnaporthe grisea* (anamorph *Pyricularia grisea* [= *P. oryzae*]), the agent of rice blast disease, *Cochliobolus miyabeanus* (anamorph *Helminthosporium oryzae*), the agent of the brown spot disease, *Gibberella fujikuroi* (anamorph *Fusarium verticillioides* [= *F. moniliforme*]), the agent of the bakanae disease.[76] The rice blast disease, in particuar, is considered the most serious disease of cultivated rice worldwide.[77] As occured in other rice rowing areas, historically the disease had driven the abandoning of most of the varieties cultivated until the beginning of 1900. Most applied fungicides applied against blast disease in Europe are tricyclazole, azoxystrobin, pyroquilon, isprothiolane, prochloraz, tebuconazole. Especially against the blast disease, two treatments can usually be carried out: one when the first leaf spots appear at the panicle formation (or a little later); the second one at the booting stage (neck blast), a little before panicle exsertion (heading). Sometimes it is suggested to do the second treatment only.

Invertebrate pests are not a major problem to rice production in Europe. Some, such as crustaceans, insects and worms can affect the establishment phase of the crop. The most dangerous among crustaceans is *Triops cancriformis* (tadpole shrimp) which feed on young rice seedlings, uprooting them and making water muddy.[78] Water turbidity results in a lower photosyntehic activity and growth reduction of the young plant. Immediate seeding after flooding is the best agronomic system to prevent tadpole shrimps damages. The Louisiana red swamp crayfish, *Procambarus clarkii*, native to the south-central United States and north-eastern Mexico is in progressive spread in European ponds and streams since its introduction in Spain in 1972.[79] This cambarid, which represents an important issue in Portugal, may interfere with rice cultivation both because of its intense feeding activity (mainly on recently germinated plants) and as a consequence of burrowing, which can cause severe water leakage.[80] Main insect dangerous in the early crop stages is *Hydrellia griseola* (rice leafminer). This insect can be controlled by lowering the water level to favor plant growth. Among the insects which can particularly affect rice plants during late vegetative stages, *Chilo suppressalis* (stem borer) and *Eysarcoris inconspicuus* (called "pudenta" in south of Spain) can have a major impact. The most visible symptom of the stem borer is that of the "white head". In the area of higher risks of attack (Estremadura, Spain) this pest is mainly controlled by using lures in traps or mating disruption products. Recently, *Lissorhoptrus oryzophilus* (rice water weevil) has been reported in Italy and its diffusion is expanding.[81]

Between mid- and late-August, rice fields are drained for harvesting; in this period, rice plants have passed the caryopsis formation stage and entered the dough grain stage.

Harvest time is determined by the average grain water content that has to reach an optimum average of 21-22% in the case of the indica-type varieties and an average of 24-25%, in the case of japonica-type varieties.[82]

Cultivation Issues

Main constraints that affect rice productivity are related to poor crop establishment, water availability, weed resistance, rice grain quality, environmental and public opinion issues.

Very frequently rice emergence rate does not exceed 30-40% of the planted seeds. This is primarily due to a combination of low temperatures at planting period with the anaerobic conditions of the submerged soil. This in turn can result in a delay of the heading stage with a high risk to incur in temperature below 14°C at flowering stage and in a risk of devitalization of the pollen. To overcome poor crop establishment rice can be planted, whenever possible, in dry soil. Great emphasis is also given to the development of new varieties characterized by early vigor and good tolerance to anaerobic conditions[83] and low temperature during germination.

Main water problems include uneven distribution, nitrate and pesticide pollution, water logging in heavy soils and the increasing costs of irrigation systems.[84] It has been calculated that about 20% of the losses due to evaporation, leaching and other inefficiencies can be saved through a more careful use of the water. The planting of rice in dry soil followed by the flooding only from the stage of 3-4 leaves of the plant as well as the planting of short-cycle and high yielding varieties could successfully lower the amount of irrigation water used. Referring to these issues, agronomists are paying particular attention to developing management strategies which can increase the efficiency of irrigation systems and limit water pollution. A consistent reduction of water can be obtained by introducing profitable varieties suitable to discontinuous irrigation in the European climatic conditions.

The repeated use of herbicides with the same mechanism of activity, in particular the inhibitor of the enzyme ALS, has led to the spread of weeds resistant to these products. Since 1994, lack of control has been reported for *Alisma plantago-aquatica* L. and *Schoenoplectus mucronatus* (L.) Palla.[85] It is estimated that weed resistance to herbicides is, at present, spread over about 30,000 ha out of the total European rice area. Main techniques currently adopted by rice farmers to tackle the problem of herbicide resistance are the rotation of herbicides and the application of mixtures of herbicides with different modes of action. In order to give successful control of ALS-resistant *Alisma* and sedges, the farmers are increasingly using hormonic herbicides such as MCPA, a widely used herbicide before the introduction of the ALS inhibitors. Crop rotation, which is the best preventive and curative method to face herbicide resistance, has no good chance to be adopted by farmers, both for technical (soil suitability for other crops), economic, and organizational reasons.

According to the European regulation, paddy rice of standard quality have to be sound and fair marketable, free of odor, with a wholly milled rice of 63% in whole grains (with a tolerance of 3% of clipped grains) and a maximum moisture content of 13% (Table 11.9). Paddy should not show abnormal color (green, chalky, striated, spotted, stained, yellow, amber).

Grain fissuring may result in important negative effects during techno-logical processing and it is often due to overexposure of mature paddy to fluctuating temperature and moisture conditions. Cracks in the kernel are the most common cause of rice breakage during milling. Milling degree is influenced by grain hardness, size and shape, depth of surface ridges, bran thickness and mill efficiency. Producers are paid less for broken kernels than for whole.

Table 11.9. Grain type categories and other quality components of paddy rice[86]

Grain shape			Other quality components of paddy rice	
Type	Length (mm)	Length/Width ratio	Chalky grains*	1.5%
			Grains striated with red*	1.0%
Long	Long A > 6.0	> 2.0 < 3.0	Spotted grains*	0.5%
	Long B > 6.0	≥ 3.0	Stained grains*	0.25%
Medium	> 5.2	< 3.0	Amber grains*	0.05%
Short	< 5.2	< 2.0	Yellow grains*	0.02%

*Percentage of wholly milled rice.

The growth of mosquitoes and concern for the spread of malaria was a major reason causing the restriction in rice production in the past. Recently, there is a significant increase of concerns related to negative effect of rice production on the environment especially the emission of methane gases which cause global warming and the harmful effect of pesticide application on the agricultural bio-diversity in rice-based production systems, as well as on quality of air, surface water and groundwater. These new concerns may lead to further restrictions in rice production in the continent. Integrated management systems for efficiency in input utilization including the use of water, need to be promoted in rice production in the continent. Also, the promotion of agricultural bio-diversity in rice-based production systems such as rice-livestock, rice-other crops is desirable.

Prospects for the Sustainability of Rice Production

Rice is not a major food crop in Europe but it plays an important role in the agricultural economy and conservation of the environment. Rice cultivation is fundamental for a sustainable management of the wet eco-systems[87] (estuaries, river basins, albuferas, etc.). Many European rice fields are located in natural parks or environmentally protected areas. In these areas, rice allows a diversification of agricultural production and landscape maintenance.

Rice production in the continent has been present since its introduction in spite of the low rate of rice consumption and a number of agronomic, economic and social constraints.

Rice cultivation in Mediterranean climate areas also has to face strong competition with a worldwide supply and a local market more and more characterized by a demand of speciality and quality rices.[88]

The sustainable development of rice in the European countries will most likely be related to the implementation at farm level of methods of production which allow to increase rice yield and enhance quality by fostering water and fertilizer efficiency, minimizing the use of crop protection products and introducing new varieties.

A better agronomical management of the crop could allow to bridge the gap between the yield obtained by the currently planted varieties in the experimental and in the farm fields.[89] In other areas of rice cultivation (e.g. Australia and Egypt), it has been demonstrated that much of this gap could be reduced by applying Rice Integrated Crop Management (RICM) systems.[90] The development and dissemination of RICM systems in Europe could also help lower production costs and minimize environmental impact of agricultural practices.

Important advances are expected in development of new varieties with better traits such as, for instance, dwarf height, improved nutrients response, shorter growing cycle, tolerance to pests, diseases, drought and salinity. Most of the achievements will likely be boosted by the recent success in rice genome mapping.

The adoption of hybrid rice technology would be another major step in raising the yield potential. Hybrid rice varieties, which are cultivated on a large scale since many years in China and are being developed in Egypt, have demonstrated to provide an yield advantage of 15-20% over the existing high-yielding varieties. The adoption of these varieties in the European countries, however, still needs technologies to increase the F_1 yield and lower, consequently, the cost of the hybrid seed. Promising results with these varieties have already been obtained in Spain and Italy.

A crucial aspect of the success in rice research is collaboration among the few scientists who are working in the European rice Centres. Sharing the research programmes is a valuable way to address the issue of the

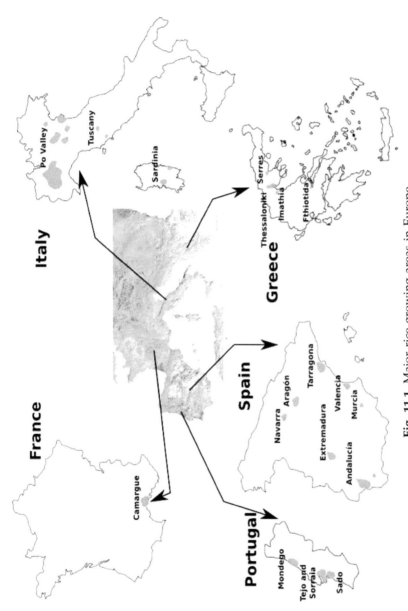

Fig. 11.1. Major rice growing areas in Europe.
(Color image of this figure appears in the color plate section at the end of the book).

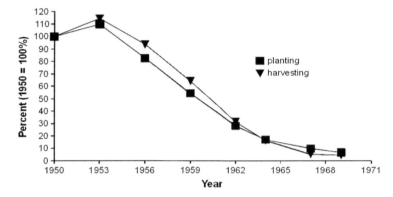

Fig. 11.2. Manpower requirements for planting (transplanting or seeding) and weeding and for harvesting in Italian rice fields in the period 1950-1969 (as percentage of average values recorded in 1950).[91]

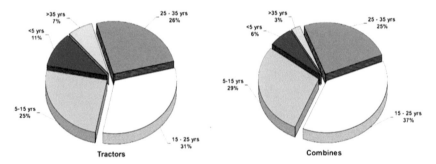

Fig. 11.3. Age class distribution in the tractor and combine fleets in the Vercelli area, Italy.[92]

(Color image of this figure appears in the color plate section at the end of the book).

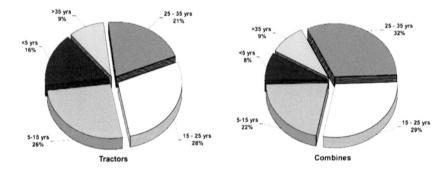

Fig. 11.4. Age class distribution in the tractor and combine fleets in the Valencia area Spain.[93]

(Color image of this figure appears in the color plate section at the end of the book).

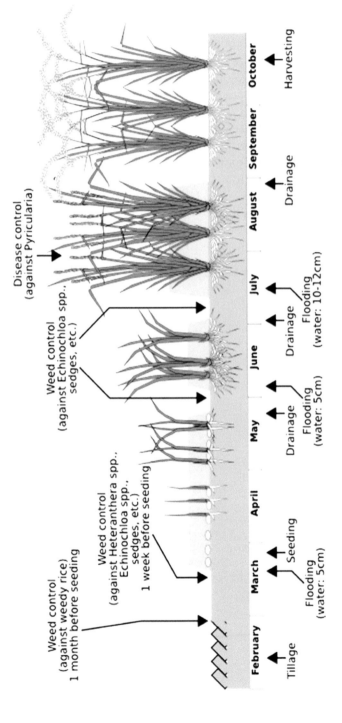

Fig. 11.5. Scheme of main agronomical practices carried out during rice cultivation. (Color image of this figure appears in the color plate section at the end of the book).

scarce financial resources and take advantage from the transfer of information and advanced research methodologies which result in a shortening the time needed to solve problems.

Several research projects carried out in Europe and the Mediterranean regions have been fostered by Medrice, the FAO Inter-Regional Cooperative Research Network on rice in the Mediterranean Climate areas. This organization was created in 1990 to promote scientific exchanges among rice scientists working in the Mediterranean areas and in other regions with a Mediterranean climate. Research Institutions from sixteen countries participate at present in Medrice: Bulgaria, Egypt, France, Greece, Hungary, Iran, Italy, Morocco, Portugal, Romania, Russian Federation, Spain, Turkey, UK, Ukraine and Uzbekistan. Some of the important subjects considered in the collaborative research include resistance to blast, stemborers and diseases, quality and competitiveness of European rice, control of red rice and cataloguing of rice genetic resources in the region.

References

1. Marinone (1992).
2. Marinone (1992).
3. Marinone (1992).
4. Tinarelli (2006).
5. Tinarelli (2006).
6. Mastrangelo (1988).
7. Blandi (1827).
8. Mastrangelo (1988).
9. Piqueras and Boira (2005).
10. Piqueras and Boira (2005).
11. Vasconcellos (1953).
12. Linares (2002).
13. Barbier and Mouret (1991).
14. Audebert and Mendez del Vilar (2007).
15. Barbier and Mouret (1993).
16. Crist (1960).
17. Audebert and Mendez del Vilar (2007).
18. Anagnostopoulos (1954).
19. Ntanos (2001).
20. Morbello (1952).
21. Mastrangelo (1988).
22. Angelini (1936).
23. Braudel (1982).
24. Motta (1913).

25. Morbello (1952), Motta (1913), Della Chiesa (1777), Betti (1783).
26. Actis Caporale (2005).
27. Clavera (2004).
28. Polto (2004).
29. Ferrero and Nguyen (2004).
30. FAOSTAT (2007).
31. Grant (1964).
32. Garrity et al. (1986), IRRI (2007).
33. FAO (1996).
34. Carreres (2007).
35. Barbier and Mouret (1992).
36. Navarro et al. (1997).
37. ENR (2007), ISTAT (2007).
38. MAPA (2006), INE (2007a).
39. Audebert and Mendez del Vilar (2007).
40. Ntanos (2001), GSNSSG (2005).
41. INE (2007b), GPP (2007).
42. Finassi and Ferrero (2004).
43. Finassi and Ferrero (2004).
44. Audebert and Mendez del Vilar (2007).
45. Finassi and Ferrero (2007).
46. From Finassi and Ferrero (2007).
47. Carreres (2007).
48. Lima (1997).
49. Audebert and Mendez del Vilar (2007).
50. Data 2003; National Bureau for Social Security; from Finassi and Ferrero (2007).
51. AIDAF VC/BI (2003).
52. Crocioni (1973).
53. Imbergamo (2007).
54. Cinotto (2007).
55. Crocioni (1973).
56. Herruzo et al. (1992), Finassi and Ferrero (2007).
57. Finassi (2001).
58. Carreres (2007).
59. Finassi and Ferrero (2007).
60. From Finassi and Ferrero (2007).
61. From Carreres (2007).
62. Finassi (2001).
63. CEC (2002).
64. Lima (1997).
65. Ferrero (2007).
66. Ferrero and Tabacchi (2002).
67. FAOSTAT (2007).
68. Faure and Mazaud (1996).
69. FAOSTAT (2007).
70. Chataigner and Salmon (1996).

71. Ferrero and Nguyen (2004).
72. Ferrero *et al.* (2001).
73. Bocchi (1996), Audebert and Mendez del Vilar (2007).
74. Oerke *et al.* (1994).
75. Pirola (1968), Batalla (1989), Tabacchi and Ferrero (2006), Ferrero and Vidotto (2007).
76. Luppi *et al.* (2000), Aguilar Portero (2001).
77. Lupotto (2004).
78. Aguilar Portero (2001).
79. Barbaresi and Gherardi (2000).
80. D'Agaro and Sparacino (2007), Gherardi *et al.* (2007).
81. Lupi *et al.* (2007).
82. Luppi *et al.* (2000).
83. Ferrero *et al.* (2001).
84. Ferrero *et al.* (2001).
85. Busi *et al.* (2004).
86. Ferrero (2007).
87. Greenland (1997a, 1997b), Nguyen and Ferrero (2006).
88. Van Tran (2001).
89. Van Tran (2001).
90. Clampett (2001).
91. Adapted from Crocioni (1973).
92. From Finassi and Ferrero (2007).
93. From Carreres (2007).

12 | History of Rice in Africa

Abd El-Azeem T. Badawi, Milad A. Maximos and
Ibrahim R. Aidy, R.A. Olaoye and *S.D. Sharma*

This chapter deals with the history of rice in Africa. The great desert of Sahara divides the continent into two broad regions. In the north of Sahara desert, rice has been grown mainly in the valley of River Nile in Egypt. In sub-Saharan Africa, rice was traditionally cultivated in West Africa and in the Madagascar Island in East Africa. In West Africa, rice is grown mainly in the valleys of river Senegal and river Gambia, in the inner delta area of river Niger and around Lake Chad. In the east Africa, the only country with a long history of rice cultivation is the Madagascar Island. The history of rice cultivation in Tanzania is very recent.

In Egypt, rice cultivation was introduced by the Arabs. In West Africa, two species of rice are cultivated: an indigenous African species (*Oryza glaberrima*) that has been cultivated there since three millennia and the Asian species (*O. sativa*) that has been introduced there about 500 years back. In East Africa, rice was brought to Madagascar by the Polynesian settlers more than a thousand years ago. In the rest of the sub-Saharan countries, rice has been a recent introduction and hence will be out of our purview.

NORTH AFRICA
(Contributed by Abd El-Azeem T. Badawi, Milad A. Maximos and Ibrahim R. Aidy)

Early History of Rice in Egypt

Rice was unknown to Egyptians of ancient times as there is no archaeological evidence of its existence during the rule of Pharaohs and it does not find any mention in the Bible. After the rule of Pharaohs, Egypt was ruled by the Greeks and then by the Romans until the Arabs reached there with their new religion Islam. It was only after the conquest of India by Alexander who brought rice to the knowledge of Egypt and Europe.

Alexander the Great arrived in Egypt in 332 BC. He founded Alexandria in 331 BC as a gateway to his motherland, Greece. Alexandria became a major center for trading. The Ptolemies ruled Egypt up to 30 BC when they were succeeded by the Romans.

Diodorus Siculus described rice as a crop of India. Theophrastus, Aristotle's successor, mentioned that rice was cultivated in Egypt; probably there was an attempt for acclimatization of rice as a crop in Egypt by Ptolemy Soter who was keen to promote agriculture in Egypt.[1]

Egypt was under Roman occupation from 30 BC for the next six hundred years. Egypt was considered the granary of the Roman Empire, sending food to Constantinople, Rumelia, Anatolia and Syria. Considerable amount of rice was grown in the Roman Syria, Palestine and also in western Asia Minor. By contrast only four Egyptian papyri mention rice. However, in view of the lack of papyri from Lower Egypt, one cannot rule out the possibility of rice being grown in Lower Egypt.[2] The Romans took the fullest advantage of the fact that Egypt was directly accessible to Rome and also to South Asia by sea route. Though spices were the main items that Romans imported from India via Egypt, rice was also imported as one of the items.

In 642 AD, Egypt witnessed the beginning of its Arab and Islamic epoch. Successive Arab rulers governed Egypt as part of their Islamic empire and at various junctures established it as the center of power of their empire. The Arabs introduced rice to Egypt around 600 AD probably from Mesopotamia. Since then rice has played a role of paramount importance to the Egypt coming second after wheat in terms of contribution to foreign exchange earnings.

By the time of the reign of Abbasid Caliph, Al-Mansur (754-775), rice cultivation had already spread to the province of al-Fayyum where enough water was available in seasons other than the flood of the River Nile.[3] This province maintained its lead in rice cultivation for many centuries. For example, in the 10th century, Muqaddasi[4] writes, "Al-Fayyum is a district for fine rice and excellent linen production." In the 12th century, al-Idrissi mentions that the fields of al-Fayyum are fertile and rice cultivation is popular among its farmers.

The Arabs were followed by the Mamluks who ruled Egypt from 1250 to 1517 AD. During the Mamluk period rice cultivation in al-Fayyum declined and the lead was taken over by the districts of Lower Egypt. In Mamluk days, rice was grown in the region of Damietta, Rashid and Manzala i.e. the region north of Damanhur and Mansura and manifested an early form of capitalist activity since the merchants exporting rice financed its production in the region. They invested in maintenance of canals and employed labourers seasonally.[5] According to al-Nabulusi,[6] in the later half of the Ayyubid period, rice cultivation in al-Fayyum was abandoned in favour of the more profitable sugar cane cultivation. He also informs that, in 1296 AD, the summer crops like rice and sugar cane were damaged due to hot winds.[7] The Mamluk reign ended with the Ottoman conquest of Egypt by Sultan Selim.

Rice fitted very well in the cropping scheme of Egypt. Whereas wheat, bean, berseem (Egyptian clover, used as fodder for animals) were grown as winter crops; rice, cotton and sorghum were cultivated during summer.[8] Rice and cotton have been important crops for Egypt and have been foreign exchange earners for centuries.[9]

Egypt was an Ottoman province from 1517 to 1914 though it gradually slipped out of the Ottoman control and had virtually followed an independent policy since 1805 when Muhammad Ali was appointed as its governor. In the 16th and 17th century, rice was mainly exported to Ottoman Empire where it was considered a luxury item. In fact, there was an embargo on export of rice outside the empire. Evliya Celesi who was very close to the Sultan Murad (r. 1623-1640) has recorded an interesting incidence that shows how important rice was for the Ottoman Empire. During a ceremonial public procession, first the royal guards and then the police were to march; then the butchers and then only the Egyptian merchants. This was not acceptable to the Egyptian merchants, so they appealed to the Sultan and said,

> Our rice is fine and white and cooks nicely especially that is grown in the regions of Maricala, Damietta, Faraskur and Birimbal which has a wonderfully fragrant scent when cooked with butter. Before the Prophet there was no rose-water, rice, bananas or *abdallawi* [Hamar: jujube], so the creation of rice is one of the miracles of the Prophet . . . . My padishah, the goods arriving in galleons provide the public treasury an annual revenue 11,000 purses from customs duties. As a matter of justice, we ought to have precedence in the Mediterranean procession and the butchers ought to come after us.[10]

Then, addressing the butchers, they said, "But tell us this, O band of butchers, what benefit and return do you offer to the public treasury?" The

Sultan felt that the appeal by the Egyptian merchants was justified and decided in favor of the Egyptian merchants.

Towards the end of the 17ᵗʰ century, the Ottoman central government was losing its authority over Egypt. The governors appointed by the Ottoman Emperors had little control over the Mamluk households and locally based amirs. The Bedouins in the north and the Hawwara shaykhs in the south were equally disobedient. Plagues and pestilence were also frequent. But despite all these odds, the Egyptian economy was strong.

During the 18ᵗʰ century, Egypt was dominated by Mamluk-merchant alliance and its trade was mainly confined to the Ottoman territories only. But trade with Europe was vital especially for acquiring modern weaponry and other finished products. The elite invested in export crops like rice and sugarcane. Sheikhs and rich Mamluks planted these cash crops in the Delta and the Said. The Syrians who had become money lenders tried to divert rice export to Europe for better profits though Egyptians rioted and continued to riot periodically.[11]

> Throughout the 18ᵗʰ century, Egypt remained the richest province of the Ottoman Empire. Its cornucopia of agricultural surpluses sustained not only its own population but the inhabitants of Mecca and Medina; it sent quantities of rice, wheat and sugar to Istanbul and rice, fruits and wheat to ports of eastern Mediterranean . . . . By the second half of the century, many products particularly rice were diverted from Ottoman ports to Europe.[12]

On the eve of the French invasion in 1798, Egypt annually sent 40,000 to 50,000 ardebs (7.4 to 9.2 million litres, @ 184 litres per ardeb) to the holy cities in Arabian Peninsula and 30,000 ardebs of rice to Syria.[13]

Rice in Egypt in the 19ᵗʰ Century

Muhammad Ali (1805-1848) was appointed as governor of Egypt by the Ottoman Emperor in 1805. He, however, followed an independent policy and modernized Egyptian agriculture. He introduced various measures with regard to land ownership, land taxation, banking, irrigation and land reclamation. He expanded the area under cultivation especially the area under export crops such as cotton, rice, sugarcane and indigo.[14] It was during his reign that long staple cotton was grown for the first time in 1821. He built dams, barrages, canals and promoted industrial development of Egypt. Foundation of the delta barrage in 1847 was a landmark in the field of irrigation in Egypt. It assured supply of irrigation water to farmers at any time of the year. Rice cultivation was expanded from Damietta to its neighboring provinces of Gharbiyya and Daqahliyya which became major rice growing areas and benefited from the Duwaida and Buhiyya canals (each 500 km long), the Mansuriyya (350 km) and Sharqawiyya (400 km), all in the province of Mansura. Gharbiyya province acquired three new canals with a total of 1790 km. The rest of the Delta which had acquired *saifi* canals turned to rice and cotton cultivation. A total of nearly 98,000 feddans were now allocated to rice cultivation. Rice production was so important that to encourage its production, the rice-growing areas were exempted from conscription; they were subject to conscription only later when the need of military men outweighed the demands of rice production. He attempted to increase agricultural exports and to use the profits to create industries. By the 1840s, Egypt's major trading partner had become Europe and Egypt had become a part of the world trading system.[15]

Khedive Ismail who came to power in 1863, carried the developments further. A school of agriculture was started in 1867. Suez Canal was inaugurated in 1869. The Ministry of Agriculture and Department of Agriculture were created in 1875. A law was passed in 1890 determining the legal situation of irrigation, dams, barrages and canals and the responsibilities of the government, irrigation inspectors and farmers were fixed. After the construction of the Aswan dam in 1898 and a number of barrages in the delta region, the water table in the delta area increased by three to four meters; hence drainage scheme was taken up in 1911. Rice cultivation area increased from 12,091 thousand feddans in 1899 to 242,367 thousand feddans in 1913.

It would be wrong to assume that everything was fine with the Egyptian farmers in the 19th century. The farmer had usufructory right only on a piece of land; he had to work on the land of the *multazim* (tax farms) as a leaseholder or sharecropper or córvee or wage laborer for his subsistence. He had to grow the crop as determined by the state, sell to the state at a pre-fixed price and pay tax in the form of a fixed produce (usually wheat). In the Lower Egypt, where rice was a monopolized crop, the farmer had to sell all his produce at a price fixed by the state and buy from the market at the market rate for his own requirement. Women and children were not spared from the coerced labor. In the late 1830s, the state did cede the monopoly but continued to control the production and price indirectly.[16]

History of Rice in Egypt in the 20ᵗʰ Century

The rice belt has been restricted to the northern half of the Delta where the Ministry of Irrigation guarantees a special irrigation regime. On the other hand, soils of the Northern Delta include large areas with various levels of salinity in which rice is grown as a reclamation crop to help leach and lower their salt content.[17] Outside the rice belt, rice is grown on a limited scale in the southern Delta and middle Egypt. An estimated one million feddans (0.40 million hectares) is planted annually with rice. This area has further increased during the last five years. New rice varieties bred in Egypt have played a significant role in boosting the yield of rice in Egypt.

Rice, which is the preferred food for most Egyptians, contributes about 20% to the per capita cereal consumption. Its impact on the economy lies in the fact that rice occupies about 22% of the cultivated area in Egypt during the summer season, thus consuming about 18% of the total water resource. Rice farming also engages one million families, which corresponds to about 10% of the Egyptian population. Rice is also one of the most effective and profitable means of reclaiming hundred of thousands of feddans of saline lands.[18] Moreover, rice is an important export crop (Table 12.1).

Table 12.1. Egyptian rice export during the last three decades

Year	1970	80	78	91	92	93	94	95	96	97	98	99	2000	2001
Export (1000 t)	400	25	70	200	177	133	300	350	253	360	280	320	332	775

In 1933, the elevation of the Aswan Dam was increased by eight and a half meters so that the highest level of water in front of the dam reached 122 m and irrigation of four million feddans could now be possible. The construction of Aswan High dam was a landmark in the history of Egypt. It was started in 1960 and finished in 1970. The basic purpose was to make round the year cultivation in the Lower Egypt possible and to reclaim more area for cultivation. It also generated electricity. By 1974, i.e. just four years after the dam's completion, the cost of the construction was realized.

In 1955, President Nasser used rice and cotton export to Czechoslovakia to acquire arms for the Egyptian army. Rice production in 1988 was 4.5 m metric tons; it increased to 5.8 m metric tons in 1999 and to 6.0 m in 2000.[19]

Role of Rice Varieties Bred in the 20th Century

Before 1917, rice was grown mainly as a reclamation crop, so no attempt was made to improve rice varieties. Rice farmers cultivated and collected scattered non-pure rice varieties with unknown botanical description. Some of these varieties had long growth duration (210 days) and were planted in April as a summer crop while other varieties with short growth duration (75-90 days) were planted in August as a Nili crop during the flooding of River Nile.

The importance of rice as a major economic crop, nevertheless, encouraged the Ministry of Agriculture and Land Reclamation (MOALR) to initiate a research program in 1917 for improving rice yield and quality.

Varietal Improvement and Yield Progress
during the Past 85 Years (1917-2001)

Rice varietal improvement efforts started in 1917 after the establishment of the Plant Breeding Department under Ministry of Agriculture. The successive progress made in the varietal improvement efforts during the past 80 years was the result of the initiation of systematic breeding research (1917-34), the release of the improved varieties (1935-54), the release of high-yielding Nahda variety which brought about major progress in rice production (1955-74), the period of stable rice yield and production (1975-86) and the period of a great strides in rice production in Egypt (1987-2001).

Phase I (1917-1934)

Yields of non-improved varieties during the initial years of this period (1920-1924) were low (Fig. 1). At that time, dominant non-improved varieties included Ain-Elbent, El-Phino, El-Agami, El-Nobari, El-Sabeeni and El-Fayoumi.[20] In 1917 the Ministry of Agriculture imported samples of 250 rice varieties from different parts of the world, such as Spain, Italy, USA, Japan, China and India. Yield evaluation showed that *japonica* varieties were more productive in Egypt than *indica* varieties.[21] During this period, the Ministry released several new varieties selected from the Yabani and Agami strains using individual plant selection. Yabani 2 and New Yabani (selected from Yabani strains) and Nabatat 1 and Nabatat 2 (Nabatat Asmar) (selected from Agami strains) were the most popular varieties. As a result, average yield increased to 3.2 tons/hectare (t/ha) (1.3 t/feddan) at the end of this period.

Phase II (1935-1954)

During this phase, rice yield and production increased by about 25% (3.1-3.8 t/ha) largely due to the release of another group of new improved varieties and the adoption of transplanting method for rice in Egypt. During the period 1935 to 1940, hybridization as well as pure line selection within the relatively narrow base of imported *japonica* resulted in the release of Yabani Pearl, Yabani Montakhab 5 and Yabani Montakhab 7. Yabani Montakhab 7 was released in 1952 to replace Yabani 15 and Yabani Pearl which were highly susceptible to blast disease. In 1953, Giza 14 (derived from the crosses between Yabani Pearl and Iraki 16) was released to Egyptian farmers as a high-yielding variety. Incidentally, Giza 14 was the first released variety[22] which was developed through hybridization in Egypt.

In addition, some other rice varieties combining special characteristics were also developed for specific use. Among them was Agami M1 (a variety with high salinity tolerance) which was released in 1952 for cultivation in the saline soils of Egypt. Yabani Momtaz and Sabeiny Abiad were released as early varieties for cultivation in the Nili season at Fayoum governorate.

Phase III (1955-1974)

Rice research in Egypt gained momentum during this phase as rice yield increased by about 40% (from 3.7 to 5.3 t/ha). High production characterized this phase as a result of two major factors: the doubling of area planted to rice following the completion of the Aswan High Dam and the release in 1955 of high-yielding variety Nahda derived from a pureline selection.[23]

In 1958, high yielding Nahda occupied about 95% of the rice area in Egypt. Nahda maintained its popularity and dominance for nearly 20 years.[24] Besides its high yield potential, Nahda was also characterized by high milling percentage, less breakage and good cooking quality. Although Nahda was highly resistant to blast when it was first released, it became susceptible to the disease later.[25]

Three other varieties viz Arabi, Giza 159 and Giza 170 were also developed and released during this phase.[26] Arabi, a long-grained variety developed from a cross between Java 3 and Yabani Montakhab 3 was released in 1958 mainly for export purpose. Arabi, however, had limited success because of its low yield and poor milling out turn.

Giza 159 was released in 1964 with particular advantage of being more salt-tolerant than Nahda. It was derived from a cross between Giza 14 and

Agami M1. It had better yield, milling out-turn and grain appearance than the old salinity tolerant variety Agami M1. Giza 159 was grown in about 5% of rice area in Egypt during this period. Giza 170 was developed from a cross between Nahda and Giza 14 and was released in 1970. Although slightly superior to Nahda in terms of yield and blast resistance, Giza 170 was not grown on a large scale basis.

Phase IV (1975-1986)

In 1975, highly resistant and equally productive varieties, Giza 171 and Giza 172, were developed after Nahda variety became increasingly susceptible to blast. These two varieties occupied more than 95% of the rice area in Egypt by 1980. Stable production and yield characterized this period probably because Giza 171 and Giza 172 offered no marked genetic yield advantage over Nahda as evidenced by the national average yield which scarcely changed from 5.3 to 5.6 t/ha.

Giza 171 (a cross between Nahda and Calasy 40) and Giza 172 (a cross between Nahda and Kinmaze) were released for cultivation in 1977. They out-yielded Nahda by about 10% and were resistant to blast races prevalent in Egypt at that time. They were also distinguished by short grain *japonica* type like Nahda.

Giza 180: Collaboration with the International Rice Research Institute (IRRI), which started in the late 1960s, led to the introduction of hundreds of new short statured lines. Of these, IR 1561-228 proved superior to Nahda because of its high yield, early maturity and blast resistance. These lines were developed in 1974 and were named as Sakha 1 and Sakha 2 respectively. Sakha 1, later renamed Giza 180, was released in 1977 but because the cooking quality of Giza 180 was not acceptable to Egyptian consumers and also because the rice mills in Egypt were not equipped for long grain milling, this variety was grown on a limited basis.

Reiho: This Japanese variety introduced in 1972, was found to meet the intensification requirements. In 1978, rice breeders in Egypt focused on the development of varieties combining non-lodging and dwarf habit to make them responsive to heavy doses of fertilizer application, mechanical harvest and early maturity and thus match crop intensification patterns. Reiho was released in the year 1983. Highly accepted by rice growers and consumers, Reiho by 1984 occupied about 30% of the total rice area of Egypt. Unfortunately, Reiho was severely affected in the same year by new races of blast disease, and hence was prohibited ever since.

IR 28: To slow down the spread of new blast races whose out-break started in 1984, a new variety called IR 28 was released in 1985. IR 28 was an IRRI long-grain short-statured variety highly resistant to blast. Its high yield potential and earl maturity enabled IR 28 to occupy about 10% of rice area in Egypt within two years after its release. However, the area planted with IR 28 gradually declined in the following years as its cooking qualities was not accepted by the Egyptian consumers.

Phase V (1987-2001)

There has been significant increase in the production of rice in Egypt during the last 15 years (1987-2001). This has been due to the increase in yield per unit area as well as increase of area under rice cultivation.

In terms of yield per unit area, the national average recorded a 30% increase in rice productivity compared with that of 1975-86 period. Average yield amounted to 5.6 t/ha during 1975-86, 7.3 t/ha during 1987-97 and reached 9.3 t/ha in 2001.

Total production during this phase increased from 2.3 million tons (1975-86 average) to 3.9 million tons, an increase of 1.6 million tons representing a 70% increase in rice production. The outstanding increase in rice production was attributed largely to improved rice productivity per unit area and expanding rice cultivated area. Although rice area was restricted by the Ministry of Irrigation to 1 million feddans annually (0.4 million hectares), it exceeded this limit during the past 10 years. Major reason for the increase was preference of farmers to grow rice rather than cotton or maize in the summer seasons. Farmers considered rice as more profitable because of its high productivity and reasonable price.

Progress achieved during the fifth phase was mostly brought about by the development and release of promising varieties with high yield potential, early maturity and high blast resistance. Currently the following varieties are widely grown in Egypt.

Giza 181: Released in 1988, this *indica* type was introduced from IRRI as IR 1626-203 and was derived from a cross between IR 28 and IR 22. Characteristics include high yield potential belonging to the highest group (> 4.5 t/feddan or > 11 t/ha), 145-day growth duration, 95-cm plant height with erect flag leaves, blast resistance and response to high level of nitrogen fertilizer (60 kg/feddan or 150 kg/ha). Its long grains have white bellies and good appearance. Amylose content is intermediate (20%) compared to *japonica* rices. Giza 181 has excellent cooking quality. It was grown on a limited scale in Egypt.

Giza 175: The origin was a breeding line (1394-10-1) selected from the local top cross between IRRI varieties and the local variety Giza 14. This variety combined features of both *indica* and *japonica* such as short-grain *japonica* type and short stature (95), short growth duration (135 days) and high blast resistance transferred from *indica* type. Milling out-turn is 69% with high amylose content (28%). It was recommended for general cultivation in 1989 but was registered as a new variety Giza 175 in 1991. Although Giza 175 was acceptable to rice farmers for its high yield potential and other agronomic characteristics, it was less acceptable to Egyptian consumers for its unsatisfactory cooking and eating qualities. Giza 175 had now been replaced by Giza 178.

Giza 176: Released in 1991, *japonica* type, Giza 176 was developed from the local cross Calrose 76/Giza 172//GZ 242. It has high yield potential (3.5-4.5 t/feddan or 8-10 t/ha), 145-day total growth duration and 100-cm plant height. It became susceptible to blast disease after 1993 when its growing area increased. Optimum level of fertilizer input is 40kg N/feddan (100 kg N/ha); increasing the N level enhances blast infection. It has short grains, 70% milling out-turn, 19% amylose content and has excellent cooking quality. It currently occupies about 15% of the rice-growing area in Egypt.

During the period (1996-2001), early maturity was given top priority in breeding programs in addition to other characteristics such as blast resistance, salinity tolerance and grain quality. Early maturity is important for stabilizing yield, economizing water use and increasing farmer's income through cropping intensity to get at least one cut more of berseem clover which corresponds to 1 ton of rough rice. Breeding efforts had led to the development and release of early-maturing varieties widely adopted by farmer.

Breeding for Short Growth Duration

Besides growing long duration high yielding varieties, increased crop intensification approach was practiced worldwide to increase food production. Major emphasis was, therefore, placed on developing improved varieties with short growth duration. Use of early-maturing varieties helps to save 20-25% of irrigation water by shortening growth duration by 25-30 days. It allows the planting of the following winter crop, particularly clover (berseem), one month earlier than the traditional period. This helps increase the number of clover cuts and helps reduce exposure of the crop to biotic and a biotic stresses. Finally, using early-maturing varieties will help bring about major changes in the Egyptian cropping pattern in the future.

Because yield is the primary consideration in varietal development programs, during the selection process, only short-duration entries that matched the yield potential of medium-growth duration varieties were included. Success of this program lies in the selection of genotypes with rapid vegetative vigor at earlier growth stages and high harvest index.[27]

Growth duration of Egyptian traditional rice varieties ranges from 160 to 180 days. Breeding progress made so far, indicates the possibility of developing rice varieties with short growth duration (125-135 days) while retaining high yield potential.

The major constraint found in early-maturing varieties is their low leaf area index due to their low-tillering ability. Two solutions are possible: one is to change the cultivation pattern through close spacing or by introducing direct seeding methods. The second is the transfer of high tillering ability genes through breeding.

Early-maturing varieties Giza 177 and Sakha 102 were developed in Egypt and were released in 1994. Both have growth duration of 125 days.

Recent Developments in Rice Production

Further increase in rice production per unit area was achieved through the rice crop management program including varietal improvement, optimization of cultural practices and integrated pest management. Rice production policy aimed to maintain self-sufficiency with a modest surplus for regional export through sustaining production growth and shortening the rice growth duration to minimize water requirements and increase crop intensification.

After the implementation of the rice crop management program, the average annual value of rice from 1994 to 1997 amounted to about one billion Egyptian pounds representing about 12% of the average annual value of plant production and about 8% of the total value of agricultural production.

Because of the fertile soils of the Nile Delta, high intensity of sunlight, few diseases and insect pests, warm weather and good irrigation system, rice yields in Egypt (9.3 t/ha in 2001) had been among the world's highest.

The national average yield of rice increased from 5.71 t/ha during the base period 1984-1986 to 7.28 t/ha in 1990. It reached 8.20 t/ha in 1995 and 9.28 t/ha in 2001 (Table 12. 2).

Studies on yield potential (yield gap) showed that the yield of demonstration fields using the package of recommended technology and improved high-yielding varieties averaged 10.56 t/ha during 1988-2001 (Table 12.3). Yields of the demonstration fields exceeded national average yield (8.03 t/ha) by about 32%. For 2001 season potential yield ranged from 7.92 to 13.49 t/ha (Table 12.4).

Table 12.2. Annual averages of rice area, production and yield
with responsive indices during 1984-2001

Year	Area		Production		Yield (national average)	
	Hectares (1000)	Index	Tons (1000)	Index	Tons/ha	Index
1984-86*	420	100	2.400	100	5.71	100
1987	414	99	2.413	100	5.83	102
1988	360	83	2.182	91	6.06	106
1989	412	98	2.668	111	6.47	113
1990	435	104	3.167	132	7.28	127
1991	454	108	4.411	142	7.51	132
1992	411	122	3.914	163	7.66	134
1993	538	128	4.154	173	7.71	135
1994	558	133	4.424	184	7.93	139
1995	587	140	4.821	200	8.20	143
1996	589	140	4.927	205	8.35	146
1997	672	160	5.752	240	8.56	150
1998	515	123	4.450	185	8.64	151
1999	655	156	5.757	240	8.79	154
2000	659	157	6.003	250	9.11	160
2001	563	134	5.225	228	9.28	163

*Base period.

Table 12.3. Progress in rice yield from 1988 to 2001

Year	Demonstration fields	National average
1988	9.9	6.1
1989	10.5	6.5
1990	10.4	7.8
1991	10.6	7.6
1992	10.6	7.7
1993	10.3	7.7
1994	10.5	7.9
1995	10.5	8.2
1996	10.3	8.4
1997	10.3	8.6
1998	10.3	8.6
1999	10.8	8.9
2000	11.5	9.1
2001	11.4	9.3
Mean	10.56	8.03

The outstanding yield increase was brought about by the release and spread of new short-duration, high-yielding varieties, the transfer appropriate technology to the farmer, monitoring of production constraints during the season, coordination among various agencies involved in rice production, as well as solving the problems of the farmers through the National Rice Campaign.

Table 12.4. Yield of demonstration fields, 2001

Variety	Yield (t/ha)		Remarks
	Average*	Range	
Giza 177	10.26	7.92-12.65	Japonica
Giza 178	11.28	8.52-12.94	Indica/Japonica
Sakha 101	11.45	9.52-13.49	Japonica
Sakha 102	9.97	8.10-11.19	Japonica
E. yasmin	9.88	7.92-10.00	Indica

*National average: 9.3 t/ha.

Water Saving

Rice is adapted to grow in flooded soil. Though it also grows well in non-flooded soils under upland conditions, rice is a semi-aquatic plant that requires more water than other crops.

Under Egyptian conditions, rice is one of the major water-consuming crops and continuous flooding is the only method used for irrigation. Our share in River Nile water is insufficient for both reclaiming and irrigation purposes. Limited water resources and considerable increase in population has forced research workers to find ways to economize the water use without any loss in grain yield.

Research has shown that rice can grow under shallow water far better than under deep submergence. Shallow water increases water temperature during the day and lowers it during the night, thus allowing more tillering and better growth for rice. Most Egyptian rice varieties produce higher grain yield when soil water content is kept near saturation throughout the season to simulate continuous flooding.[28] This indicates that better yield does not necessarily require standing water on the soil surface.

Because the water resources in Egypt are limited to 55.5×109 m^3 a year (in addition to the tremendous population increase), production has to be increased and irrigation water has to be well-managed. Finding ways to save irrigation water is, therefore, of utmost importance.

Prolonging irrigation intervals for six days at any of the growth stages is another means to save irrigation water under rice conditions. This has demonstrated non-significant reduction in grain yield.[29] Giza 178 can tolerate withholding of irrigation water at least twice: the first period during the vegetative growth (4 weeks after transplanting) and the other during ripening stage. This variety is suitable for areas suffering from irregular irrigation at the terminals. Ibrahim and his associates noted that drought-tolerant variety showed significant difference in yield due to irrigation intervals ranging from 6 to 10 days while Giza 172 showed no significant reduction in yield when irrigated at 6 days intervals during the first month after transplanting.[30] Irrigation intervals can be extended to ten days until the end of the growing season.

More efforts have to be devoted to educate the rice farmers about the importance of saving irrigation water through improved irrigation methods suitable for newly developed rice varieties. Intensive research to developed more short duration varieties with high yield potential and resistance to insects and diseases is also needed.

National Strategy for Rice Production Improvement by 2007

Rice production in Egypt during the Second Five Year Development Plan (1987-1992) increased by about 42% from 2.4 million tons during the base period (1984-1986) to 3.4 million tons in 1991. Rice production further increased by about 49% because of the higher productivity (8.5 t/ha) achieved in 1997.

To keep pace with the 2.7% annual population growth and the yearly exports of about 500,000 tons, rice production has to increase from 3.1

Table 12.5. Rice production strategy (2002-2007)

Period	Area		Yield		Production	
	1000 (feddan*)	%	*Ton/ feddan*	%	*Million tons*	%
1984-1986 (base period)	1000	100	2.4	100	2.4	100
1989-1991	1033	103	3.0	125	3.1	130
1992-1997	1200	120	3.5	146	4.2	175
1998-2001	1423	142	3.8	158	5.4	225
2002-2007	1100	110	4.2	175	4.6	192

*1 feddan = 0.42 ha.

million tons (average of 1989-1991 period) to 4 million tons by the year 2000 (about 30% increase) (Table 12.5). To achieve this goal, a multi-pronged strategy has been developed as a part of the Third Five Year Development Plan (1992-1997). The strategy basically aims at increasing productivity per unit of land, water, and labor as well as increasing farmers' income from rice based cropping systems.

WEST AFRICA
(Contributed by R.A. Olaoye)

The region generally identified as West Africa stretches from the Atlantic Ocean in the west to the Lake Chad in the East and spans for about 3000 km from west to east. In the north, the region is bound by the Sahara desert which is not very habitable. The south of West Africa is limited by the Bight of Benin of Atlantic Ocean. The southern coastal stretch is covered by the tropical rain forest which was not very habitable in the past because of tse-tse fly and also because it was difficult to clear the dense vegetation during those days. The Savana belt limited by the desert in the north and the tropical rain forest in the south was comparatively hospitable and hence has been an arena of much human activity in the past as well present.

This broad savanna belt is covered with fertile soil but crops must be grown during the short rainy season during the summer. The desert edge of the savanna is known as the *Sahel* (an Arabic term for "coast" since the Sahara is viewed as a "sea" of sand), has fluctuating rains and is semi-desert. Some years it could be farmed; some other years it could be fit for grazing only and some years it had to be abandoned.

An important and positive feature of the region is the presence of three rivers, namely, Niger, Senegal and Gambia along with their tributaries and the sweet-water lake, Lake Chad. The basins of these water bodies provide vegetation for the wild life, grasses for domestic animals and water for the cultivation of crops as well as for human life. These rivers and the lake also provide fish and easy transport of men and material by boats. The River Niger is the longest river in the region. It originates in the highlands of Fouta Jallon of Guinea and flows eastward. As it starts from the region of tropical rain forest, it carries enough water and carries silt along with it for onward journey. In the middle of its course, it forms many islands which are called inland Niger delta. It then takes a southern turn and ultimately empties its water in the Bight of Benin after running a course of 4000 kilometers. Near the inland delta, it meets its tributary Bani River. Beyond the inner delta it meets River Sokoto and River Benue in

Nigeria. The River Senegal rises in the same highland of Fouta Jallon but takes a westward course, traverses a distance of 1600 kilometers and drains in the Atlantic Ocean. The River Gambia runs parallel to River Senegal and is better navigable. The Lake Chad is the eastern limit of West Africa. The lake was much larger in pre-historic times and was a major centre of human activity but has now shrunk to mere 2500 square kilometers.

The people of West Africa neither developed a system of writing nor borrowed one from their (distant) neighbors until the Arabs reached this area from the north by trans-Saharan land routes in the 9th century and Europeans from the west and south by sea route in the 15th century and lent their writing systems. Written records are, therefore, available only after the Arabs started visiting these lands. The historians have tried to reconstruct the history of the past based on archaeological findings (which are not many), local oral traditions and linguistic, anthropological and sociological studies.

River Niger may be called the lifeline of the people of Savanna region. Every year in the months of August and September, about 50,000 square kilometer of alluvial plain of inland Niger delta is inundated with floodwaters. As the water recedes from the floodplains, marshes, swamps and grassland are formed in succession. They create a very favorable environment for the fishers to catch fish, herders to graze their animals and farmers to grow many crops including rice.

According to anthropologists, when desertification of Sahara started around 3000 BC, the people of Sahara started moving southward. Their main destinations were Senegambia, inland Niger delta and Lake Chad. They were initially pastoralists; agriculture developed later.

Prof Roland Porteres of Paris has studied the biological diversity of rice and other crop plants of West Africa and has provided a comprehensive picture of the domestication and spread of rice in this region.[31] According to him, the African rice was first domesticated in inner delta region of River Niger around 1500 BC. In course of time, it spread upstream and downstream of River Niger and also along the tributaries of this river. Then the crop spread along the River Senegal and River Gambia. It also spread eastward to Lake Chad basin. The type of rice that the people of Inland Delta developed was "floating" type that could grow in deep water. This type of rice spread to riparian situations of Gambia and Senegal and also to lacustrine situations of Lake Chad. The upland types suitable for swidden cultivation were developed in the highlands of Guinea. These upland types spread eastward to similar ecosystems in Guinea, Sierra

Leone, Liberia and Ivory Coast.[32] According to Harlans, the African rice could have been domesticated in more than one location.

Archaeological Evidences

Archaeological evidences suggest that people of Savanna region were herding cattle, sheep and goats in the second millennium BC. They started cultivating crops towards the end of the second millennium BC or in the beginning of the first millennium BC. The earliest evidence of crop cultivation is available in present-day Ghana, Burkina Fasso and Nigeria. The earliest crop seems to be pearl millet followed by sorghum and rice as evidenced from the excavations at Jenne Jeno.[33] The tools that these early farmers used were made of stone only. There is no evidence of the use of copper or bronze in subsequent periods. Use of iron appears for the first time in the middle of the first millennium BC.[34]

Archaeological excavations at Jenné-jeno suggest that, during the period 250 BC and 400 AD, the people of this area followed a mixed economy of hunting wild antelopes and water fowl, fishing in the river Niger, herding domestic cattle, sheep and goats and cultivating domestic rice, some domestic millet, sorghum and wild *Brachiaria*. According to oral tradition, these people had migrated from the plateau of the present-day Mauritania and settled between the River Senegal and inner Niger delta during the end of second millennium BC and the beginning of the first millennium BC.[35]

Advent of Arabs

Before the Europeans reached West Africa by sea route, the only contact of the people of West Africa with outside world was across Sahara desert. It is likely that trade across the Sahara desert existed long before the camels were brought to North Africa in the 3rd and 4th century AD and certainly before the arrival of Arabs and Islam in the 7th century. Arabs and Berbers intensified and expanded the trading networks across Sahara and along with them came the preachers of Islam. Arabs were attracted due to the gold, ivory and slaves of West Africa and the people of Sahel region needed salt (from Saharan salt mines), clothes, metals and arms from the north and the east. The rulers of West Africa also needed the horses from the Arabs to assert their authority.

The Arabo-Berber traders and Islam reached this region around 900 AD. According to Arab geographer, Abu Ubayd al-Bakri, there was a powerful kingdom named Ghana between 900 and 1100 AD and the area was habited by Soninke people. The king had a large army and his wealth came from taxes levied on salt, copper and gold passing through his territory. Since the climatic conditions of the this region has been fairly constant during the last three thousand years and the region experienced a comparatively wetter period during 700 AD to 1100 AD,[36] it can be assumed that production of rice and other food grains must have been sufficient in this period to sustain the population of this kingdom.[37]

The next great empire of the region was the Mali Empire that flourished from the early 13th century to 17th century. This empire was established by Sundiata Keita. Sundiata's Empire stretched from the Atlantic coast south of River Senegal in the west to Gao in the east of the middle Niger bend. His kingdom included cities like Jene, Timbuktu and Gao on the River Niger and the salt mines of Tagaza in the north (in Sahara desert). Slaves were used to clear farmlands and grow crops such as millet, sorghum, rice, beans, cotton, etc. Cattle, sheep, goats and chicken were bred. Many small kingdoms owed allegiance to Sundiata and paid tributes in the form of rice, millet, lances, arrows, etc. The next great king of this Empire was Mansa Musa. His pilgrimage to Mecca in 1324-25 became famous for his display of wealth during his stop-over in Egypt.

When the Mali Empire became weak in the middle of the 15th century, the Songhai people who had settled along the middle region of River Niger asserted themselves and in course of time conquered all the territory that was under Mali Empire. Their military superiority was due to their horsemen and a fleet of war canoes by which they could move very swiftly along the River Niger. Askia Mohammed Toure, who was a Soninke and a devout Moslem, extended his territory through *jehad* farther east into the Hausa land near Lake Chad. He revived Timbuktu as a centre of Moslem scholarship and trans-Saharan trade. The king had his own royal farms in the Niger flood plains and also in the Songhai heartland. Cloth was woven from the local cotton. Gold, kola nuts, and slaves were traded for salt, cloth, cowries, and horses.[38]

Many Arab travelers such as Az-Zuhri (12th century), Al Damishqi, Ibn Battuta (14th century), Marmol (16th century) and Leo Africanus (early 16th century) visited and described the life at Timbuktu and Gao. From their accounts, it appears that wheat, barley, sorghum and rice were grown and eaten in this region. Among vegetables, legumes (possibly kidney beans), pumpkins, cucumbers and melons were grown. Fish was eaten daily. Cattle, sheep and chicken were domesticated. Hippopotamus was hunted in the River Niger for its meat. Salt was, however, dear and came from Taghaza mine in Sahara.[39]

The *Kano Chronicle* informs us that during 1500-1800, in the north of

Nigeria, the Hausa farmers were cultivating not only millet, sorghum, beans, rice, cotton, henna but also maize, peanut and tobacco i.e. the crops that must have been newly introduced from the Americas by the Europeans.[40] The overall picture that emerges for the Sahelian Africa during the historic period until the advent of Europeans was a period of agricultural self-sufficiency and prosperity.

Rice (*O. glaberrima*) that was cultivated in the West Africa before the arrival of Arabs was of African origin. It was domesticated in West Africa from its wild progenitor, *O. barthii* that grows wild in the savannah. That some of the tribes such as Jola [Diola] had gained expertise in rice cultivation has been attested by the European travelers who visited them later.

> As Robert Baum points out, "the Diola are considered the best wet rice cultivators in West Africa" and the evidence for their long-standing cultivation of rice in the well-watered coastal plain bisected by the Casamance River stretches over two millennia. The Diola even sold rice paddies to ransom relatives who had been captured by slave raiders before they were sold into slavery. The Baga-Sitem in the Rio Nunez region, a subgroup of the Baga peoples who generally inhabited mangrove islands located between Guinea-Bissau and Iles de Los and who were said by one traveler in 1793 to be "very expert in cultivating rice and in quite a different manner to any of the nations on the Windward Coast," were unusual for neither holding slaves nor selling them. [p. 18][41]

The Arabs introduced rice cultivation that was already being grown in Mesopotamia and Egypt in Sicily and Spain. It has still not been ascertained if they introduced the Asian cultivated rice in West Africa and if they did not, why not.

The Colonial Period

The Portuguese were the first maritime Europeans to navigate along the West African seacoast, explore its hinterland and trade with the people. By the middle of the 15th century, they had established trading posts along the sea coast from Senegal to present-day Ghana and their cargo included rice, copal, civet, palm mats, sacks and carved ivory objects. According to their records, the people residing in the coastline between Casamance and Sierra Leone practiced wetland rice cultivation and used waterways for transport and communication. The main crop along the Guinea coast was plantain, banana, yam and rice. The Portuguese were quick to introduce new crops such as maize, cassava, peanut and tobacco from the New

World. The Portuguese maintained their maritime supremacy for the next one hundred and fifty years until the Dutch, British and French joined the race.[42] The military prowess of these maritime powers could easily subdue the local rulers and occupy the territories. As a result, the West Africa became ultimately a mosaic of French and British possessions cutting across the ethnic or cultural identities of the people.

These European rulers exploited their possessions for their own commercial benefits. As the farmers of West Africa practised subsistence agriculture, there was not much to be gained from it. There was not much of gold and ivory either left for exploitation. The next cheaply available commodity was the people who could be enslaved and used for plantations in their newly discovered continent America. Rice plantations in South Carolina and Georgia of United States and in Maranhao in the northeast of Brazil were based on cheap African slave labor and Bance Island of Sierra Leone and Isles de Loss were the ports where slaves were assembled and were exported to the New World. According to Judith Carney, rice cultivation in North Carolina and Georgia (USA) was based on the technology of rice cultivation carried by these African slaves from the west coast of Africa[43] but "a close look at the slave trade from an Atlantic perspective suggests not a shred of hard evidence that the rice culture of South Carolina, Georgia and Amazonia was any more dependent on skills imported from Africa than were its tobacco and sugar counterparts in the Chesapeake and the Caribbean respectively."[44]

The European powers also introduced many new crops into West Africa such as maize, peanut, tobacco and cassava from Americas and cotton and sugarcane from Asia. They introduced Asian rice (*O. sativa*) from their Asian colonies for its cultivation in their possessions in Africa. However, they were more interested in the commercial and plantation crops than in any of the cereals except in so far as these (cereal) crops were necessary to meet the local requirements. For example, the British introduced peanut cultivation in Gambia and coffee plantation in Tanzania and the French introduced cotton cultivation in Mali and coffee plantation in Madagascar.

After the World War II, the African colonies became a liability and the British and the French felt that it would be more sagacious to grant them independence. As a result, most of the British and French colonies attained their independence during 1960s. But this did not result in their political, social or economic emancipation. Military dictatorships, ethnic conflicts and economic mismanagement hindered the progress of many of these countries. Besides, natural calamities have taken their own toll. For example, beginning in the late 1960s, the Sahel region suffered from a prolonged drought, famine and loss of life. Drought and famine affected the Sahel region again in the mid-1980s and early 1990s. This resulted in meager water supply for irrigation and shattered agricultural economy of the region. This also caused deforestation, desertification, increase in population and migration of the rural population to towns and cities. The African states, however, are gaining political maturity and people social awareness. The international organizations and developed countries are realizing that African nations, if helped, will prove a grand asset for the world community and, if not, a great liability.

Let us examine the recent history of rice production in Mali and Senegal which had long history of rice cultivation and in Nigeria which has the highest rice production (Table 12.6) in West Africa. It is encouraging that

Table 12.6. Average milled rice production (thousand tons) in major rice producing countries of West Africa for selected periods

Country	1970	1980	1990	2001-2005
Cote d'Ivoire	290.58	355.06	453.54	444.60
Ghana	52.87	48.10	125.20	177.18
Guinea	266.70	411.24	633.95	770.41
Mali	122.39	147.15	377.07	569.13
Niger	20.47	38.09	42.35	49.23
Nigeria	357.24	1177.95	2086.25	2103.41
Senegal	62.20	94.56	116.98	140.33
Sierra Leone	356.73	329.84	255.06	336.06

Source: Africa Rice Center (WARDA) (2007).

Table 12.7. Self sufficiency ratio of major rice producing countries of West Africa by decades

Country	1980	1990	2001-2005
Cote d'Ivoire	0.54	0.74	0.42
Ghana	0.47	0.61	0.33
Guinea	1.39	1.23	1.28
Mali	0.82	1.13	0.80
Niger	0.55	0.65	1.10
Nigeria	0.87	0.93	0.64
Senegal	0.22	0.20	0.18
Sierra Leone	0.96	0.65	0.79

Source: Africa Rice Center (WARDA) (2007).

there has been definite increase in rice production in all the countries except in Cote d'Ivoire and Sierra Leone which had been afflicted with civil strife. But mere increase in rice production would not indicate the position of a country unless the increase in population of the country and increase in rice consumption is also taken into account. The self-sufficiency ratio is a better indicator of rice production with regard to its consumption in the country (Table 12.7).

Mali

The northern 60% of Mali is desert or semi-desert and hence not suitable for agriculture. After independence, the country was under the dictatorship for 23 years; it returned to democratic government in 1991.

In 1932, the French authorities created Office du Niger (ON) as a public enterprise in Mali to focus on rice and cotton production. After independence, it became a Malian public enterprise that controlled all aspects of land and irrigation, management of input and output, marketing, extension, etc. By 1970, ON gave up cotton and concentrated on rice only as rice production had stagnated and area under this crop could not expand as per objective. In 1980, the process of liberalization was started and the role of ON was gradually restricted. By 1995, the role of ON was limited to land management and extension, input supply and credit to private sector only. The adoption of the market economy paid dividends: the use of fertilizer increased, area under transplanted rice increased, more area was planted under new and better-yielding rice varieties. In 1992-93, the area under transplanted rice was only 50%; by 1998-99, it was almost 100%. Between 1992-93 and 1998-99, the yield increased by 27%.[45]

Senegal

Certain types of rice were available in Senegal even in pre-colonial times, but rice was not a staple food. Rice has become an integral element of the Senegalese diet since the colonial era only.

Rice is grown in the flood plains of the Senegal, Saloum, and Casamance rivers but is mainly produced along the Senegal River Valley where irrigation water is readily available. It is estimated that approximately 45% of the harvested rice area in Senegal is irrigated from recessional floods along the river banks, and another 45% is irrigated by controlled flooding. In the southern Casamance region, rice is produced mainly by recessional irrigation along the ephemeral flood plains, and this rice is used mostly for home consumption. Rice is cultivated on small landholdings and large-scale agriculture is very limited, with the exception of some large-scale rice and sugar production at the Richard Toll irrigation scheme located near the Senegal River delta. The Senegalese government is attempting to reduce its dependency on rice imports by sponsoring dam construction and irrigation projects, especially along the Senegal River.

During the colonial period, rice was imported in large quantities in order to keep food prices under control and at the same time promote groundnut production as an important cash crop. Senegal was already exporting groundnut oil to Europe in pre-colonial times; after colonization, groundnuts became Senegal's main export product. A further factor encouraging rice imports was that in the French colonies of Indo-China, rice production and export were under French control. That region exported the bulk of the rice coming to the West African colonies.

During the first two decades after independence, this import policy remained unchanged, and there was a continued interest in exporting groundnuts and importing rice. However, rapid industrialization appeared to offer better prospects for economic development. Policies which aimed to promote local food production through price increase of imported food encountered stiff opposition. Besides, another major factor was that the preparation of rice dishes takes far less time, fuel wood and labour than preparing Senegal's traditional millet dishes. In urban households in particular, these inputs—time, fuel and labor—are scarce. This goes a long way towards explaining why high rates of urbanization in the West African states are generally accompanied by increased rice consumption.

The strategy for industrialization, Senegal's high level of urbanization (around 50%) and the promotion of imports combined with the positive image of rice as a foodstuff created a new pattern of consumer behavior based around rice as the staple food. Most of the demand for rice was—and still is—met by imports. Today, rice imports amount to 7-8% of total imports and is posing a major burden on the country's trade and foreign exchange balance. When the old model of industrialization at the expense of agricultural development proved to be non-viable, intensive efforts to encourage local food production began in the early 1980s.

A series of droughts occurred in the 1970s, leading to a permanent drop in rainfall and harvest yields. The situation was exacerbated by changes in the global vegetable oil markets which greatly reduced the profit margins for exported groundnut oil.

The progressive shift towards local food production from the early

1980s therefore took place. The development strategy focused primarily on rice cultivation based on large-scale irrigation. Over recent decades, more than 60% of investment in the agricultural sector has been channeled into irrigated agriculture, especially in the Senegal River Valley.

Irrigation

The Senegal River is 1800-km long and begins in Mali at the confluence of two rivers, the Bafing and the Bakoye rivers, some 1000-km inland from the Atlantic Ocean. It flows across the western part of Mali and then defines the border between Senegal and Mauritania before discharging into the ocean. Downstream from Manantali Dam, its main tributaries are the Falémé, Kolimbiné, Karakoro and Gorgol rivers. The Organization for the Development of the River Senegal (OMVS), formed by Mali, Mauritania and Senegal, developed a plan to construct two dams, the Diama and Manantali dams, along the Senegal River to increase irrigated area by 340,000 hectares, supply fresh drinking water and industrial water, produce hydroelectric power and improve seasonal river navigation. The Diama Dam, located 23-km upstream from the river's mouth, was completed in 1986 and its primary function is to prevent salt-water intrusion from the Atlantic Ocean. The second multi-purpose dam, Manantali Dam, is located approximately 1200 km. upstream, and was built by the OMVS on the Bafing tributary of the Senegal River in Mali. The Manantali Dam is a 66-meter high dam and was completed in 1988, although no turbines were installed. The Manantali Dam evens the flow of the Senegal River and the dam has a maximum capacity of 11 billion cubic meters, much more than the Diama Dam. The Manantali Dam stabilizes the water level along the stretch forming the border between Mauritania and Senegal, thus facilitating navigation and improving irrigation potential.

Construction of these dams created an additional 240,000 hectares of land which can be irrigated on the Senegalese side of the Senegal River. This gave the country the potential to diversify its crop base and increase food production. Operation of the upstream dam has also reduced annual floods along the floodplain, where an ancient and productive form of recessional irrigation has been practiced for hundreds of years. Recessional irrigation is still practiced along these flood plains for an estimated 50,000 hectares, but simulated flood waters from the Manantali Dam do not carry the same load of nutrient-rich silts which are deposited on the fields after the flood waters retreat.

Improvement in Rice Production

Compared with the downturn that had occurred during the previous decades, a slight increase in output was achieved, but the overall growth trend during the period 1980-1996 stood at just 3.1% per year, barely exceeding the rate of population growth. The development of key production factors—harvest yields, number of harvests per year, water consumption and the expansion of irrigated areas—lagged far behind Senegal's potential. An important factor was the ineffective role of the state in all areas of production development, starting with research into irrigation management, agricultural extension, the granting of loans and delivery of inputs, extending to processing and marketing. The other factors were disputes with the neighboring states and conflict of interest among the stake holders (such as farmers, fishermen and animal herders).

Between early 1994 and mid-1996, the economic parameters for food cultivation changed radically. The currency was devalued by 100%, imports were liberalized and privatized and the role of the rural development associations was reduced to a handful of core tasks such as the granting of loans, agricultural extension and providing support to private producers and processing companies. The transfer of responsibility for rice imports from the state to the private sector proceeded smoothly despite fears to the contrary, although over time a significant concentration process has occurred. With the liberalization of imports, the state's levies and quota system were abolished and the basic tariff now stood at around 15%. Rice imports increased dramatically as a result. In all, imports of broken rice have increased from around 400,000 to almost 900,000 tonnes since 1994. The high levels of food imports, especially broken rice, into Senegal undoubtedly pose a major problem for local food producers. They have created consumption patterns which local agriculture now finds almost impossible to satisfy while foreign producers can meet this demand at very low cost with agricultural by-products.

Notable increases in production were not achieved in the rice sector until 1999. In cereal production as a whole this took until 2002. Due to the major fluctuations in production, it is still impossible to say whether this marks the start of a long-term trend.

Rice Imports

Since independence in 1960, rice consumption in Senegal has increased by almost 1000% in just four decades, currently standing at around one million tonnes. Rice consumption now exceeds 70 kg per capita per year and, since the 1970s, has replaced millet as the most important staple food. In urban households, rice accounts for 54% of cereal consumption and 18% of total household spending. In rural regions, the figures are 24% of cereal consumption and as much as 25% of total household spending.

Senegal currently imports more than three-quarters of its rice for domestic consumption. Rice imports into Senegal consist almost entirely of broken rice. Senegalese consumers have developed a marked preference for broken rice, with the result that in 2003, broken rice costs 10-20% more than locally produced whole rice. In addition, many rice farmers have difficulty selling their crop, because the markets are inundated with cheaper imported rice from Asia.

Nigeria

There were Islamic movements in the early 16th and 17th centuries in the regions of Futa Toro leading to dispersal of people to different parts of West Africa, particularly Nigeria. Uthman dan Fodio and most of his followers, who were Fulani splinter groups from the earlier Islamic movements, launched *jihad* in northern Nigeria. It is plausible that the West African indigenous rice got into Nigeria through the migration of these people.

The northern communities of Birnin Kebbi, Sokoto and Nupeland pioneered the cultivation of the crop. Probably, the cultivation of the crop started in these regions in the 16th century.[46] In Kebbi and Sokoto regions, cultivation was based on cultural practice of *fadama* whereby the crop thrived not on rainfall but on wet and marshy lands. In the Middle Belt, around the rivers Niger and Benue basins, the Nupes were, and still are, prominent producers of rice, especially the native rice, *Oryza glaberrima*. The prominence of rice production in this region is due to the presence of large bodies of water from rivers Niger, Benue, Kaduna and small rivers and streams such as the Gbako, Yenmi, Gazum, Eche, Duku, Kampe, Nalli, Egwa, etc.[47] The availability of water in this region ensures that large areas of land remain marshy throughout the year. In effect, the states of Niger, Kwara and Kogi covered by this region are among the largest rice production areas in Nigeria. The prominence of rice production here is manifest in the local words: *raisi tapa*, which means the rice of Tapa (Nupe)—a term which has generally been adopted by some communities in Nigeria.[48]

The Asian rice (*O. sativa*) did not get into Nigeria until 1920 when it was introduced by the British colonial Department of Agriculture.[49] Unlike the indigenous species which has red grains, *Oryza sativa* has white grains. In addition, the Asian species has many varieties that are suitable for different ecosystems like shallow or deep swamp and upland cultivation.

The most widely-cultivated shallow swamp varieties of *O. sativa* are BG79 (introduced from British Guiana in 1921), MAS 2401 (which originate in Indonesia and is used in irrigation schemes by the Nupes and Yoruba both in Niger and Kwara States) and D114 (grown in Southern Zaria and the Jerre Bowl in Borno State).[50]

The deep swamp variety, Maliong, was introduced from Thailand in 1957 and it is now mostly grown in deep flooded areas in the Rima valley in Sokoto State, the Illushi area of Edo-Delta State and among the communities in Ebonyi State.[51]

The best-known upland variety of *O. sativa* is Agbede 16/56. This is noted to have been selected from a complex variety of upland types at the Federal Rice Research Station in 1956 and widely grown throughout Nigeria, including by the communities in and around Ibadan and Egba/Ijebu in the southwestern part of Nigeria.[52]

Rice production in Nigeria has been substantially boosted by the government agencies and international organizations. In Kwara State, for instance, the Shonga Agricultural Project of the State Government is a bold policy for rice production which has become a model for other States in Nigeria.[53] Likewise, government establishments such as the Upper and Lower Niger Basin Authorities have jointly and severally encouraged rice production in the country. The Upper Niger Basin Authority in conjunction with other agricultural agencies takes charge of rice production in Northern Nigeria. The Lower Niger Basin Authority, on the other hand, superintends over similar functions in southern Nigeria.[54] Thus the blending of the indigenous cultural practices with other more recent techniques ensures that a large number of communities are today engaged in rice production in Nigeria. In this regard there are production areas such as Sokoto, Kebi, Zaria, Kano, Maiduguri, Kaduna and Lake Chad Basin in the North; Bida, Minna, Lafiagi, Patigi, Ilorin, Makurdi, Otukpo and Lokoja in the Middle Belt; Ibadan, Ijebu-Ode, Sagamu and Abeokuta in the South West; and Abakaliki, Ikwo, Ohaozara and Ivo in the South East of Nigeria.

Production Techniques

The cultivation of rice in Nigeria is done in four main types of environments, namely rain-fed uplands especially in the southern part; tidal fresh-water mangrove swamps in the southeast; naturally flooded areas such as the *fadamas* of northern region and irrigated lands. As these environments are found in many parts of the country, rice is cultivated in virtually all ecological zones. This is because the conditions of high temperature ranging from 20°C to 38°C during the growth period of the crop and long period of sunshine are available in most parts of the country.[55]

Unlike more established cereals such as maize, guinea corn and millet, rice is traditionally produced in small scattered areas which are isolated from each other. This is due to the fact that the traditional cultural practices of cultivation, land management, cropping patterns and storage have significant influences on the decisions of farmers concerning different aspects of agricultural production. The most common traditional modes of production are the swamp and *fadama* rice cultivation. The swamp rice is of shallow, medium, deep and floating varieties. It has to be flooded for a period ranging from 60 to 90 days during growth.[56] This is because rice is unique among other cereals which can germinate and thrive on flooded terrain. Thus in Nigeria, the areas along rivers Rima, Kaduna and Siroro and around the Chad Basin in the North offer favourable ecosystems for swamp rice production. Similarly, in the Middle Belt, Niger, Kwara, Kogi and Benue States utilize the flood and water from the rivers Niger and Benue and their tributaries to produce a large quantity of rice all the year round.

In the same vein, the fresh water areas of the southeast in Ebonyi, Imo, Bayelsa and Akwa Ibom States are known for the swamp rice production. This is possible because of a combination of factors. First, the region has a bimodal rainfall and experiences long rainy seasons with more than 1800 mm of annual rainfall starting as early as the in the month of March each year.[57] In addition to the well-spread over annual rainfall, most communities in the south-south and southeast have mangroves and creeks where water is available for rice production.

Among the Hausa/Fulani in the North, *fadama* rice production is a common agricultural practice. The practice is encouraged by the presence of lowlands which are naturally flooded. When the rain-induced flood is at low ebb, there is always the advantage of rivers, streams, lakes, etc from which water is drawn to wet the rice fields. In places like Sokoto, Kebbi, Bornu and Nupe land, *fadama* rice cultivation is very popular. Studies have shown that there are areas in the south where *fadama* rice production is done. For instance, in spite of the fact that upland rice is the most popular practice in the region, it is not unusual to find *fadama* rice cultivation around Ibadan, Iwo, Ikire, Gbongan, Owode, Ifo and Odeda to name but a few.[58]

Upland rice is grown in the areas that are far away from river basins, streams, and tributaries. This category does not depend on flooding or irrigation. Instead, it thrives on good soil and adequate rainfall. Although good soil type exists especially in the South of Nigeria, the less-than-required and non-consistent rainfall makes upland rice less practicable. Instead, the rainfed lowland rice assumes prominence since most areas have one type of alluvial plain or the other.

Today, technology-driven production system has, however, considerably raised the level of commercial rice production in Nigeria. Apart from what the government is doing, there is also the contribution of international institutions to rice production in the country. Institutions such as International Institute of Tropical Agriculture (IITA), in Ibadan, Nigeria, and African Rice Center (WARDA) in Cotonou, Republic of Benin, have generally raised the level of agricultural production and particularly that of rice crop. Through its policy of New Rice for Africa (NERICA), for instance, WARDA has introduced new techniques and varieties of rice in many African countries including Nigeria.[59]

The modern production techniques have certain features. These include the introduction of new varieties, the use of tractors in cultivation of land, system of irrigation, the use of pesticides, insecticides, fertilizers, weed control, chemicals, harvester and milling machines to name but a few.[60]

The effect of this is that both the government and people are sensitized on the need to utilize the new techniques in agricultural production, as manifest in the efforts of several State Governments towards establishment of commercial agriculture with a focus on rice production. Kwara State is today noted to be one of the leading producers of rice in Nigeria, if not in West Africa.

Importance of Rice in Nigeria

In the Hausa community of Northern Nigeria, for instance, rice from the early period had been both for local consumption as well as trade and became known as *shinkafa* or *tuwo shinkafa* while among the Nupes, for example, rice on the menu is called *jankafa buboci*, and in Yoruba land, it is called *iresi*.[61]

It should be noted that before 1960 when Nigeria regained its independence, rice was not particularly popular in southern parts of the region. At that time, the notion was that rice was a food for lazy people. Farmers in those regions preferred the cultivation and consumption of root crops like yam, cassava, coco-yam, etc that provide heavy carbohydrate and hence energy required for long and tedious farming activities.

The unpopularity of rice then earned its derisive description as *onje tanpepe* i.e. food for the ants and birds.[62]

All these however, began to change in the post-independence period. On the 1st of October 1960—the day Nigeria secured independence from the British colonial rule—the food for the celebration of the occasion was rice. The government provided free rice meal for schools, public establishments and corporate organizations throughout the country. The event marked a high rise in the taste for rice meal among Nigerians in general. Moreover, rice was discovered to be a food that could be easily prepared and consumed by family members, individuals and at social functions. Apart from easy preparation, rice can be readily packed as take-away and fast food. In this regard, it is a choice food at occasions like those of marriage, naming, house-warming, traditional and religious festivals and other public and private functions. Indeed, for the children, rice is a must-eat food. It is common in homes for children to pester their parents when they yearn for rice. Thus rice, which was initially rejected in some regions and served only as an occasional food, is today one of the most sought after food in those areas.

Besides being a staple food, rice is also a commercial commodity. In Nigeria, the traditional production goes hand in hand with modern production methods. Rice is now available in large commercial quantities for buying and selling activities. Again, the traditional and modern varieties complement one another in making rice available and affordable for both the poor and the rich. More importantly, the government involvement in commercial farming has provided the required synergy for rice production on large commercial basis. Such involvement has, in fact, put states like Sokoto, Kebbi, Niger, Ebonyin and Kwara as exporters of rice to other African countries.

Nigeria, like most others in the sub-region depends on Asian countries (like India, Thailand, Taiwan and Korea) for the supply of rice. Currently, Nigeria is packaging an agreement with South Korea on rice, especially in the areas of production technology, varieties resistant to biotic and non-biotic diseases and storage techniques.

International Efforts for Rice Production in West Africa

Efforts have been (and are being) made to improve agriculture in general and rice cultivation in particular by the African governments and international organizations. West African Rice Development Association (WARDA) was created in 1970 by eleven West African countries with the assistance of the United Nations Development Programme (UNDP), the

Food and Agriculture Organization of the United Nations (FAO) and the Economic Commission for Africa (ECA). Now known as Africa Rice Center, it is one of the 15 international agricultural research centers supported by the Consultative Group on International Agricultural Research (CGIAR) and plays a much wider role throughout sub-Saharan Africa. Its research and development activities are conducted in collaboration with various stakeholders, primarily the National Agricultural Research Systems (NARS), academic institutions, advanced research institutions, farmers' organizations, non-governmental organizations and donors for the benefit of African farmers. Earlier, this Centre was working from Bouaké in Côte d'Ivoire but because of civil strife there, it is now working from Cotonou (Benin) since 2005.

In 2006, Sub-Saharan Africa has been importing 50% of rice it needs.[63] The leading African rice-importing countries are Nigeria (16%), South Africa (11%), Senegal (9%), Côte d'Ivoire (8%), Sierra Leone (4%), Ghana (4%) and Burkina Faso (3%). It is projected that these countries will continue to import rice in the near future. These countries have suitable ecologies and water bodies to support rice production.[64]

In recent years, the African Rice Center has developed NERICA rice varieties which are the derivatives of the hybrids between the Asian (*O. sativa*) and the African (*O. glaberrima*) cultivated rice. These varieties combine the desirable traits of both the species and are better adapted to different ecosystems of Africa. The NERICA rice varieties offer great hope to the next generation in Africa.

In West Africa, except Sierra-Leone which can boast of some level of rice production and reserve, other West African countries are experiencing sharp shortfall which makes them resort to importation. The main reason for this situation is that the area under rice production is generally far below the area that should be available for its cultivation. To achieve self-sufficiency, the area under rice cultivation has to be increased. Through assistance from WARDA, most countries of the region are taking advantage of NERICA rice varieties and, as a long term measure, are going for large processing plants, use of solar energy (due to poor electricity) to power machines and for irrigation purposes and, of course, knowledge acquisition in storage and preservation/general capacity building.

EAST AFRICA
(Contributed by S.D. Sharma)

By about 900 AD, Arab, Persian and Indian traders had settled on the east African coasts and were exchanging cloth, beads and metal goods for ivory, gold and African slaves. By 1506, Portuguese controlled most of the east coast of Africa. In 1698, the Portuguese were expelled by the Arabs. Rice was introduced in the eastern coast of Africa by the Arabs, Persians and Indians in or after the 10th century. The main country on the east coast of Africa where rice good get a foothold was Tanzania only.

Tanzania

Before 19th century, millet and sorghum were the staple food in Tanganyika (i.e. the mainland of Tanzania). Since the 19th century, maize slowly replaced the millets and the sorghum. At present, maize is the staple food of the majority of the people. Rice and wheat are consumed by the middle and the upper class that form a small percent of the population. Tanzania was exporting maize in 1950s and 1960s. Tanzania imported large quantities of food for the first time in 1972 and continued until 1985. These imports consisted mostly of maize. Now rice is grown in many areas of Tanzania; in 95% area it is grown as a rainfed crop.

In the 1870s, Joseph Thompson (the young explorer of 19th century who discovered Lake Victoria) was extremely impressed by agricultural practices in the Rufiji river delta and flood plains. Much food—rice, maize, beans, and sesame seed—was exported from the villages of the Rufiji Delta to Zanzibar; this area was referred to as "Little Calcutta" because of its large rice export industry. Before Tanzania became a German colony in 1884, the Rufiji river flood plains and delta were important areas for rice production.[65]

After the First World War, Tanganyika came under British control until it became independent in 1961. In 1964, Tanganyika and Zanzibar united to become the present-day Tanzania. Among the countries of East Africa, the rice production of Tanzania is next only to that of Madagascar (Table 12.8) and is nearly self-sufficient with regard to its rice requirement (Table 12.9). The only country which had a long history of rice cultivation in east Africa is Madagascar.

Table 12.8. Average milled rice production in major rice producing countries of East Africa for selected periods

Country	1970	1980	1990	2001-2005
Madagascar	1333.37	1460.22	1655.21	1942.52
Tanzania	184.05	329.54	446.29	456.97

Source: Africa Rice Center (WARDA) (2007).

Table 12.9. Self sufficiency ratio of major rice producing
countries of East Africa by decades

Country	1980	1990	2001-05
Madagascar	1.16	1.25	1.13
Tanzania	0.94	1.03	0.86

Source: Africa Rice Center (WARDA) (2007).

Madagascar

Madagascar is located in the Indian Ocean 350 miles off the eastern
coast of Africa just south of the equator. It is 1680 miles (2735 km) long
and 350 miles (570 km) wide. Steep mountain ranges run parallel to the
east coast of the island. Its climate is tropical. The island receives 30-35
cm rainfall mostly during December to April.

According to 1984-85 census, only 5.2% of Madagascar's total land
area was under cultivation, the average farm size in the country was only
of 1.2 hectares and less than 2 million hectares were permanently
cultivated. However, agriculture employed almost 80% of the labor force
in 1992 and constituted 33% of the GDP in 1993 and provided 80% of the
country's export. In 1992, rice production occupied about two-thirds of
the cultivated area and produced 40% of total agricultural income.
Cassava, maize and sweet potato were the other crops in that order of
importance. Coffee, cotton, cloves and vanilla were the commercial crops
cultivated in Madagascar.[66]

There are four major rice growing zones in Madagascar:[67]

Hauts-Plateaux	300,000 ha of rice land
Lac Alaotra	90,000 ha of rice land
Marovoay	30,000 ha of rice land
Somangoky	10,000 ha of rice land

In Madagascar, the central highlands are the densely populated region
of Madagascar. It is inhabited by the Merina group in the north and
Betsileo group in the south. The Betsileo people are the most efficient rice
farmers in the country. They practise irrigated paddy cultivation in their
terraced lands in the valleys in the southern portion of the central
highlands. The dryland crops are grown only on lands which cannot be
converted to wetland paddy cultivation. The people of Merina group who
reside in the north of the central highlands also practice paddy cultivation
in irrigated fields but their technique is less advanced and intensive. Rice

is the dominant crop here also but more dryland crops are grown in this region.

In the eastern coast, the Betsimisaraka and Tanala peoples also practise irrigated rice culture wherever possible but slash-and-burn method is still used in this region even though it is now declared illegal. Only short duration varieties are grown in such drylands.

It is likely that Polynesians from southern Borneo migrated to Madagascar and settled there around 200 to 400 AD along with their blow pipes, rice and rice culture.[68]

The people of Madagascar remained undisturbed until the arrival of Arab traders in the 9th century. Swahili speaking traders settled in the northwest coast and established trading posts in northwest of Madagascar and traded for rice, ivory and slaves. The central highlands were settled by the 1200 AD. By 15th century the population of Madagascar was still mobile and there were empty lands when the Portuguese, the first European traders, reached the island in 1506. By the middle of the 18th century, especially after 1724, Madagascar was one of the main exporters of rice and slaves. The powerful king of Madagascar was receiving his tribute as slaves, cattle and rice. In early 19th century, Arab dhows used to make regular trips to Madagascar to buy rice, millet and young men and women.[69]

The 19th century saw the rise and collapse of a centralizing Merina economy. By the 1820s, the Merina had conquered the eastern parts of the island and gained access to the coastal trading system. But by the middle of the 19th century this was weakening. In this century, there was a demand for slave labor in the neighboring plantation economies of Reunion and Mauritius and beyond (such as in the Cape and the Swahili coast). An indigenous plantation economy and the spectacular growth of rice cultivation emerged.

> Before the mass arrival of Europeans on the island in 1860, the Malagasy had a virtually self-sufficient economy. The bulk of the population was engaged in agricultural production. Rice, the staple food of the island, was produced in large quantities to the extent that the Malagasy became specialists in its production—Rice and beef became important items of export.[70]

By 1890, the Merina economy collapsed and the French occupied the island. In the first decade of the 20th century there was "gold rush" and there was influx of settlers for gold and for coffee and vanilla plantations. During the second and third decades, the economy depended on export of rice. There was migration from the east and southeast to western region of the island to introduce irrigated rice cultivation in this sparsely populated region. There was a severe drought during 1917-18. In 1930s, coffee and clove became the main export items.[71]

During the Second World War, "the island was valued by France as a source of cheap rice, manioc, butter beans, skins and beef extract."[72] After the war, the country gained independence in 1960.

In the initial years after its independence, coffee, vanilla and sugar were exported but the economy based on the export of these cash crops could not be sustained because of the international competition and falling prices of these commodities.

Table 12.10. Area and production of rice in Madagascar[73]

Year	Area (ha)	Production (t)
1905	300,000	400,000
1936	500,000	680,000
1944	540,000	640,000
1951	700,000	1,025,000

During 1970s, the area under rice expanded at the rate of 3% per year but the production increased by less than 1% per year. As a result, Madagascar became a net importer of rice beginning in 1972. By 1982, it was importing nearly 200,000 tons of rice per year i.e. about 10% of the total domestic rice crop.

Until 1982, rice research in Madagascar was being looked after by the Institut de Recherche Agronomiques des Tropicales et de Cultures Vivrieres (IRAT) based at Paris. In 1982, with the cooperation and participation of International Rice Research Institute (IRRI), Food and Agricultural Organization (FAO), World Bank and some other national and international organizations it was reorganized and revitalized with a determination to achieve self sufficiency by 1990.[74] In recent years, the country has achieved near self-sufficiency though there was a severe cyclone in 1994 just when the rice crop was ready for harvest. Similarly, in December 2004, the Indian Ocean *tsunami* caused heavy damage to the island and, in 2006-07, there was a series of cyclones that damaged to the rice crop.

References

1. Hunt (1842).
2. Scheidel (2001:81).
3. Sato (2005).
4. Muhammad ibn Ahmad Shams al-Din Al-Muqaddasi (also known as Al-Maqdisi) (945-988) was a notable medieval Arab geographer, author of *Ahsan at-Taqasim fi Ma'rifat il-Aqalim* (The Best Divisions for Knowledge of the Regions). He published his masterwork forty years later after extensive travels.
5. Sayyid-Marsot (1984:154).
6. Abd al-Ghani al-Nabulusi (1641-1731) was a sufi visionary of Ottoman Syria.
7. Sato (2005).
8. Taylor and Francis Group (2002).
9. Taylor and Francis Group (2002).
10. Dankoff (2004).
11. Sayyid-Marsot (1984:14).
12. Crecelius (1998:67-68).
13. Scheidel (2001).
14. Sayyid-Marsot (1984).
15. Sayyid-Marsot (1984).
16. Tucker (1985).
17. El-Tobgy (1976).
18. Wally (1989).
19. Taylor and Francis Group (2002).
20. Refaie (1965).
21. Sidky (1988).
22. Balal (1989).
23. Momtaz (1989).
24. Balal (1989).
25. Balal (1989).
26. Balal (1981).
27. Khush (1990, 1993).
28. RRTC (1996).
29. Nour *et al.* (1994).
30. Ibrahim *et al.* (1995).
31. Porte'res (1970).
32. Porte'res (1970).
33. Connah (2004).
34. Connah (2004).
35. McIntosh (n.d.).
36. McIntosh (n.d.).
37. McIntosh (n.d.).
38. Eltis, Morgan and Richardson (n.d.) see website: www.uga.edu/colonialseminar/EltisEssay.pdf
39. Maclean and Insoll (1991).
40. Ogot (1999).

41. Eltis, Morgan and Richardson (n.d., p. 18), see website: www.uga.edu/colonialseminar/EltisEssay.pdf
42. Ogot (1999).
43. Carney (2004).
44. Eltis, Morgan and Richardson (n.d., p. 43), see website: www.uga.edu/colonialseminar/EltisEssay.pdf
45. Chohin-Kuper, Kelly and Mariko (2001).
46. Agboola (1979).
47. Jiddah (2002).
48. Oral Interview, Madam Alimat Ajagbe, C. 67, rice trader, Ibadan, 14th June, 2007.
49. Agboola (1979).
50. Agboola (1979).
51. Agboola (1979).
52. Agboola (1979).
53. Information Bulletin, Kwara State Ministry of Agriculture, No. 147, Passim.
54. Olumese (1988).
55. Agboola (1979).
56. Agboola (1979).
57. De Datta (1981).
58. Oral Interview, Ramonu Jimoh, C. 65, rice farmer, Ikire, 3rd July, 2007.
59. Africa Rice Center (WARDA), Annual Report 2005-2006, pp. 8-9.
60. Agboola (1979).
61. Oral Interview, Nuhu Abubakar, C. 72. rice farmer, Lokoja, 9th July, 2007.
62. Oral Interview, Nuhu Abubakar, C. 72. rice farmer, Lokoja, 9th July, 2007.
63. Africa Rice Center (WARDA) (2007).
64. See website: www.warda.org/publications.
65. Maghimbi (2007).
66. See website <www.wildmadagascar.org/overview/loc/>
67. IRRI (1985).
68. Bellwood (1970).
69. Ogot (1999).
70. Ade Ajayi (1998).
71. Fremigacci (1985).
72. Fremigacci (1985).
73. IRRI (1985).
74. IRRI (1985).

13 | The Rice Industry of the United States

Peter A. Coclanis

To paraphrase Churchill's famous quote regarding Russia, the U.S. rice industry is a riddle wrapped in a mystery inside an enigma. The industry *qua* industry is over three hundred years old, but few Americans know much about it. The U.S. is a major rice producer and one of the leading exporters of rice, but per capita consumption is very low, and Americans get only about 3% of their calories from rice (as opposed to two-thirds or even three-fourths in parts of Asia).[1] Although rice is generally associated with small-scale production and heavy labor requirements, the U.S. rice industry can be characterized as agribusiness *par excellence*: There are relatively few producers, most of whom are very large, and rice is the most-capital-intensive crop grown in the United States. These large-scale producers, not surprisingly, have long had a good bit of political influence in the U.S., and the industry, which is concentrated in a handful of states, has long been heavily protected and/or subsidized. Riddles, mysteries, and enigmas abound, then, but how did the U.S. rice industry develop in this way.

Unlike its venerable history in many parts of Afro-Eurasia, rice cultivation in the Western Hemisphere, including the area that later became the United States, dates back only five hundred years or so. The cereal arrived in the Western Hemisphere along with Europeans and Africans as part of the so-called Columbian Exchange of plants, animals, germs, and peoples that began in the late 15th century CE. Two species of domesticated rice—*Oryza sativa* and *Oryza glaberrima*—made it across the Atlantic. *O. sativa* first originated in Asia, of course, while *O. glaberrima* originated in West

Africa, where it was fairly widely grown by the time the Columbian exchange began. Of the two species, *O. sativa* became the more important by far in the Western Hemisphere—it alone was commercialized—although it should be noted for the record that small amounts of *O. glaberrima* were grown for home consumption by African populations in various parts of the Americas over the centuries.[2]

With adequate water and sun, rice can be grown virtually anywhere, and in the Western Hemisphere it was in fact tried in seemingly unlikely places such as the Central Andes. In North America specifically there were early experiments in Spanish Florida in the 16th century and in the English colony of Virginia in the 17th century, but the first important rice-producing complex in the Americas did not develop until the early 18th century with the rise of the industry in the English colony of South Carolina. To be sure, well before that time small quantities of rice were consistently being grown in Brazil, and in the circum-Caribbean area, but in such places the cereal was produced primarily for home consumption rather than as a market crop. In South Carolina, too, both Europeans and Africans likely grew small quantities of rice for home consumption from the first decade of settlement in the 1670s, but it was not until the early 18th century that sufficient stocks of labor, capital, entrepreneurship, and "local knowledge" were in place, and infrastructure and marketing networks sufficiently developed to support and sustain significant levels of production for commercial purposes. Not until then, in other words, can we speak of a rice industry *qua* industry in the Western Hemisphere.[3]

The precise origins of rice cultivation in South Carolina are mired in controversy. Until the 1970s Europeans—and Europeans alone—were viewed as the progenitors of rice production in South Carolina, and various "foundation myths" were often invoked to explain whence rice cultivation began. One such story involved a Captain Thurber, who in the mid-1680s supposedly brought rice (the legendary "seed from Madagascar") to the colony after a voyage into the Indian Ocean, and Europeans and European-Americans in South Carolina ostensibly took it from there. Over the past few decades, however, another "origins" story has gained great currency, this one according primary responsibility for the introduction of rice into South Carolina to Africans rather than to Europeans. A number of scholars now claim that enslaved Africans, who in earlier accounts had provided little more than the brute labor necessary for rice cultivation, transferred both knowledge about risiculture and production technology from West Africa to South Carolina where in time it was appropriated/expropriated by Europeans and European Americans. According to proponents of the Africa-centered "Black rice" view, various

groups in West Africa, particularly in the Senegambia, had been cultivating rice for centuries, and many of these peoples ended up as slaves in
South Carolina. Moreover, early cultivation technologies and labor practices in South Carolina seem to have shared many characteristics with
technologies and practices in West Africa. Certainly, the evidence adduced
by proponents of the "Black rice" thesis has now convinced almost everyone that Africans contributed more than brute labor to the beginning of
rice cultivation in South Carolina. Still, a number of questions remain.[4]

Many Europeans were familiar with both rice and rice production by
the late 17th century, for example, and a number of scholars have recently
pointed out that the timing of the "Black rice" story does not always hold
up well. In particular, they point to the fact that only a small percentage of
the Africans who ended up in South Carolina were from rice-producing
areas in West Africa and most of those entered South Carolina after rice
cultivation was well established in the colony. Moreover, there is a big
difference in any case between small-scale production of rice by Africans
(or anyone else) for home use and the creation of a rice industry. Sorghum
is another cereal brought from Africa to the Americas, but it never developed into an "industry" of note in the early modern period. Whichever
group in fact has the best claim for initiating cultivation in South Carolina,
clearly, there were several possible routes of transmission and, just as
clearly, there is a big difference between rice production and a rice
industry. Regardless of the exact origins of rice cultivation in South
Carolina, that is to say, the industry that developed was informed by
European and European American aspirations, capital, and entrepreneurship, along with African labor and technology.[5]

Given how heated the debate over the origins of American rice cultivation has become, it is interesting and not a little ironic to note that we
know very little about early production techniques or even sites. Some
"wet" rice may have been grown from the start, but most scholars believe
that for a time at least most of the rice grown in South Carolina was grown
"dry"—without any irrigation, that is to say—on relatively "high" (non-
swamp) ground in the area known as the low country, the easternmost
third of the modern state of South Carolina. Throughout its history in
South Carolina, most (and until the Civil War, almost *all*) rice cultivation
took place in the low country. This said, production sites were distributed
quite unevenly within this large area of roughly 13,400 square miles or
about 8.6 million acres. Never in South Carolina's history—not even at
the peak of production in 1859, when South Carolina produced about 119
million pounds of clean rice (almost 64% of total U.S. production) were

much more than 100,000 acres planted in rice, and it is likely that this estimate is high.[6]

However murky the early history of rice in South Carolina, it is clear that a rice industry *qua* industry had begun to develop in the low country by the 1720s. Production became more routinized and systematic, and we can document the fact that production in the area centered on various and sundry inland, freshwater swamps. After such swamps were drained, rice was planted on these sites, and rudimentary water-control mechanisms—impoundments and the like—were constructed nearby to ensure adequate supplies of water at appropriate times in the growing season. Such sites remained central to the emerging industry until the second half of the 18th century, when production shifted again, this time from inland, freshwater swamps to drained swamps located on, or adjacent to particular stretches of South Carolina's principal tidal rivers.

The so-called tidal rice zone, which was to constitute the heart of the low country rice industry until its demise in the early 20th century, developed within severe geographical and hydrological constraints. The idea behind tidal cultivation was to harness river tides to draw water onto and off of diked rice fields, which allowed for more effective irrigation and reduced the amount of labor necessary during the growing season. The trick was to find sites where tidal action was strong enough as to raise and lower water sufficiently to allow for commercially viable rice crops without getting too close to estuaries or the coast where the water was brackish and salty and would injure or kill the plants. Over time, such sites were in fact found, and tidal cultivation became centered along narrow bands stretching from about ten miles inland to twenty miles inland on the five major tidal rivers in the low country: the Santee, the Cooper, the Ashley, the Combahee, and the Savannah. As rice cultivation spread to other parts of the South Atlantic coast—beginning with Georgia and North Carolina in the 1750s—similar concerns held sway, and production became concentrated along similar bands of tidal rivers, stretching north to south from the Cape Fear River in southeastern North Carolina, along the major tidal rivers of Georgia (the Savannah, the Ogeechee, the Altamaha, the Satilla, and the St. Marys) down to the St. Johns River in northeastern Florida.[7]

The heart of the United States rice industry lay in the South Atlantic region from the early 18th century until the late 19th century. As suggested above, beginning in the second half of the 18th century, the great majority of the rice produced in this region was produced using tidal-irrigation technology in narrow bands of land in tidal zones. For a complex set of reasons to be discussed below, beginning in the 1880s the center of the

U.S. rice industry shifted to the "Old Southwest"—Louisiana, Texas, and Arkansas—and a bit later to California as well. Commercial rice production in the South Atlantic region collapsed completely in the first decade of the 20th century, but for well over a century, the South Atlantic region—South Carolina and Georgia in particular—for all intents and purposes constituted the American rice industry. For this reason, let us look at cultivation in this area more closely.

To point out that rice production imposes heavy labor requirements under most cultivation regimes is to state the obvious. Such requirements are particularly arduous, however, when the cereal is cultivated under so-called wet conditions as was the norm first in South Carolina and, later, in other parts of the South Atlantic rice region. Moreover, because populations living and working in the swamps of this region were also subject to a variety of serious mosquito-borne diseases—malaria, most notably—morbidity and mortality levels in rice areas were extremely high. In light of the rigorous labor demands and unhealthy conditions associated with rice cultivation, it is not surprising that it proved difficult to attract sufficient numbers of free laborers to meet the needs of the industry as it developed. The African slave trade to North American venues was well established by the time the rice industry emerged in South Carolina, and from the start African and African American slaves provided the vast majority of the labor required in the industry. As we have seen, some slaves from West Africa were already familiar with risiculture, and Africans and African Americans had higher degrees of inherited and acquired immunities to certain mosquito-borne diseases than did populations from other available labor pools (European, European American, and Native American). These considerations, along with the general acceptance of slavery amongst the European and European American populations in the rice region and the relative costs of recruiting/retaining slaves versus free laborers, help to explain why Africans and African Americans, working as slaves until 1865 and as free or quasi-free laborers thereafter—constituted most of the labor force for the South Atlantic rice industry from the time of its establishment until the time of its death.[8]

To say that rice cultivation in the South Atlantic region had heavy labor requirements is to understate things grossly. Indeed, from today's perspective it is difficult to appreciate just how much labor was involved in establishing and maintaining the "wet rice" regime in this region. Although the amount of land devoted to rice in this region never comprised more than about 160,000-170,000 acres, the amount of forest clearage, water engineering, bund/embankment construction and maintenance, and field preparation required to render and keep this land suitable for commercial rice production was staggering. Before any type of agricultural activity could be pursued, swamp lands had to be drained, cleared, and leveled, of course, but that was just the start. Once these preliminary tasks were completed, wet-rice cultivation, particularly tidal cultivation, entailed the construction (and subsequent maintenance) of the elaborate system of irrigation works needed to regulate the diurnal flow of tidal waters onto and off of the fields. For many larger units of production—unlike rice production in many parts of the world, most of the rice grown for commercial purposes in the South Atlantic region was produced on large plantations—this meant miles of bunds and embankments, ditches and culverts, with assorted floodgates, ditches, and drains. Clearly, thousands upon thousands of miles of "riceworks" were constructed over time in the region, the vast majority of which construction was done by African and African American laborers.

Furthermore, unlike cereals such as wheat, rice was generally cultivated in the South Atlantic region as a row crop, needing a considerable amount of attention during the growing season. Thus, rice workers in the South Atlantic region had to endure a rigorous year-long labor regimen as well: Preparing or repairing irrigation facilities and other "riceworks", readying the fields for cultivation, planting, periodic flooding/draining, hoeing, harvesting the crop, processing it, preparing it for market, and transporting it for further milling or for sale. Although the use of nurseries and the practice of transplanting young rice were not part of the South Atlantic cultivation regime, very heavy inputs of labor and units of energy were nonetheless expended in rice production in the region. The vast majority of the human labor inputs was provided by Africans and African Americans. To be sure, draft animals supplemented human labor in some stages of the production sequence, and inanimate power sources—wind, and, later, steam, in addition to water—were employed as well, particularly in milling. But, clearly, human labor was always paramount.[9]

One other feature of the labor system in the South Atlantic rice region merits our attention, to wit: Its manner of organization. Unlike the case in almost all other parts of the American South, under slavery, the labor system in the South Atlantic rice region—both in rice and in other activities—was dominated by the so-called task system rather than the so-called gang system. According to the latter system, slaves, working in groups, were subject to close monitoring, and worked over set periods of time (typically, or at least proverbially, from "sunup to sundown"). The task system, on the other hand, was quite different. In this scheme, a slave was responsible for a set amount of work—a daily "task" or set or tasks, as it were. Once complete, that slave was relieved of any further labor obligations to the master, and for the rest of the day he or she could do as he or she so chose, which often meant pursuing economic activities on his/her own account. Not only did this system facilitate the accumulation of small amounts of property by slaves working on their own time, but, because the close minute-by-minute supervision associated with the gang system was absent in the task system, slaves working under tasks generally achieved greater independence, autonomy, and "space" within the context of slavery. Indeed, the hold of the task system in the South Carolina low country and other parts of the South Atlantic rice region was so strong that even after emancipation, black workers, now free, often contracted with planters and landowners not for a calendar year or even for a growing season, but for specific, mutually-agreed-upon tasks such as field preparation, sowing, hoeing, harvesting, and the like.[10]

What about the efficiency of rice production in the South Atlantic region? Did slave labor and the task system provide the basis for a high-productivity agricultural regime? Such questions, particularly the former, have led to lively (and sometimes acrimonious) debates among economic historians of the United States. Although most scholars working on the efficiency of slave labor in the United States have focused on slaves involved in cotton production rather than rice production, some work has been done on rice. This work is often quite technical in nature, but it is reasonably clear from such work that rice production on large slave plantations in the rice region of South Carolina and Georgia was relatively efficient in comparison to other producing areas in the world in the 18th century and in the 1800-1860 period. We know, for example, that daily work requirements, annual yields per acre, and annual yields per laborer increased for much of this period, at least until the 1840s or 1850s, and total factor productivity (TFP)—measured in standard Cobb-Douglass form—in South Carolina and Georgia in 1859 was greater than TFP in Lower Burma in the 1880s. Because many other considerations, often difficult or impossible to measure—soil fertility/exhaustion, variation in microclimates, weather shocks, the quality of the capital stock, biological innovation, etc.—likely impacted productivity as well, we must interpret the above findings cautiously. Nonetheless, it seems fairly safe to conclude that the slave-labor regime in the rice region was hardly retrograde.[11]

In American history, the South Atlantic rice region—the South Carolina low country and coastal portions of Georgia in particular—is considered the quintessential plantation zone in the southern part of the United States. That is to say, the area is often viewed as having been completely dominated during slavery times by large units of production, worked by sizable numbers of slaves, performing highly specialized agricultural tasks related primarily to the production of either rice or another secondary staple. Such units or production ("plantations") are said to have been highly capitalized, marked by economies of scale, and owned, if not directly operated by free white men and women of extraordinary wealth. This conceptualization is not wrong, but it is incomplete. To be sure, as a group the rice planters of the region were often quite wealthy, which is not surprising given the fact that the minimum number of slaves needed to be considered a "planter" was typically put at twenty, and slaves were expensive. Clearly, a sizable number of planters (and planter families) in the rice region lived very well in the style associated in the American popular imagination with rural aristocrats or grandees. This said, it is important to note both that rice plantations, even large ones, were not always profitable (particularly in the decades prior to the Civil War) and that many small farmers in the South Atlantic rice region also grew some rice. As suggested earlier in this essay, rice production was concentrated in certain parts of the region, and most agriculturalists in the region did *not* grow rice. For example, in the South Carolina low country and coastal Georgia, the heart of the South Atlantic rice region, in the year 1859 there were 4126 agricultural units of at least three acres in size, the minimum size threshold for farm status in the U.S. at the time. Of these units, only 1608—less than 40%—grew any rice at all. Rice, then, may have been the quintessential plantation crop in the South Atlantic region, indeed, in the American South, but the region cannot be reduced entirely to great rice planters, huge agricultural estates, worked by dozens, if not hundreds of African and African American slaves.[12]

Within the South Atlantic rice region, South Carolina was always the leading producing area. Indeed, South Carolina was the greatest producer in the Western Hemisphere until the 1880s, when the state was surpassed in production by Louisiana. Systematic data on total rice output in the U.S. are lacking until 1839, when the federal census began to compile reasonably complete crop output series. From the late 17th century on, however, we do have good data on rice exports, which, when employed carefully, allow us to speak with some confidence about the development of the industry over time. Thus, we find that rice exports from "Carolina"—the colonies of South Carolina and North Carolina combined—averaged 268,602 pounds between 1698-1702, growing to over 30 million pounds annually between 1738-1742 and to over 66 million pounds annually between 1768-1772 on the eve of the American Revolution. The vast majority of rice exports from "Carolina"—roughly 90%—originated in South Carolina during the late colonial period, but it should be noted that a small proportion originated from Georgia and the Cape Fear region of North Carolina.[13]

But for small quantities of rice grown in Louisiana, almost all of the rice grown in North America during the 18th century originated in the South Atlantic rice region, particularly in South Carolina and Georgia. The rice-producing areas in these two states remained the largest producing—and exporting—areas by far until well after the American Civil War. Export series and scattered data on output suggest strongly that U.S. rice production grew steadily over the course of the of the eighty-five year period between the American Revolution (1776-1783) and the American Civil War (1861-1865), but that rice exports from the U.S. stagnated after about 1800 or so. Throughout this period, South Carolina dominated both production and exports, constituting roughly three-quarters of total U.S. rice production in both 1839 and 1849, and about 64% of a much higher total in 1859. Georgia's share of total U.S. production grew from a little over 15% in 1839 to 18% in 1849 and to fully 30% in 1859. Total U.S. production in these three census years amounted to about 81 million pounds of clean rice in 1839, 144 million pounds in 1849 and about 187 million pounds in 1859. Over the course of this eighty-five year period, the share of rice output that was exported shrank significantly, from about 90% at the time of the Revolution to between 30% and 40% in 1859.[14]

The rice export and output figures above at once capture and symbolize a fundamental fact about the U.S. rice industry in the period from the American Revolution to the advent of the Civil War: Over the course of this period the U.S. position in international rice markets became increasingly threatened and, in many of the most important venues, untenable. In the 18th century, when a very high proportion of the rice produced in North America was exported, the principal markets were in Europe, particularly northern Europe, where rice was used as a cheap and versatile source of bulk calories, complementing, supplementing, or, as relative price considerations dictated, substituting for more desirable foodstuffs, most notably, small grains. Much of the American rice imported into Europe was consumed by *lumpen* groups of one sort or another— soldiers, sailors, orphans, the poor—while some found industrial applications or was used as animal feed.

Prior to the entry of rice from North America into European markets, the northern Italian states of Lombardy and Piedmont supplied most of the Continent's demand for rice. During the second half of the 18th century, however, South Carolina and Georgia had supplanted Italian rice in a number of the key northern European markets, and rice from the South Atlantic rice region dominated these markets for the rest of the century.

The outward expansion of European capitalism changed things in the 19th century, however. By the last decade of the 18th century, in fact, European merchant capital had begun to redirect and reroute sufficiently large quantities of much cheaper rice from South Asia—from Bengal primarily— as to make inroads into the major U.S. markets in Britain and on the Continent.[15] Indeed, it is interesting and important to note—as political economist David MacPherson did in 1795—that rice was the first "neces- sary" sent to Europe from Asia, all previous trade consisting of articles and products "rather of ornament and of luxury than of use."[16]

The penetration of Asian rice into European markets continued and intensified in the 19th century. During the 1800-1860 period—a period in which European commercial penetration into South and Southeast Asia deepened considerably—Asian rice-exporting areas (India, Java, and Burma in particular) came to dominate the major European markets. As a result, U.S. rice was increasingly forced out of these markets and into either lesser markets on the Continent, markets in the Caribbean (especially Cuba), or into the domestic market, which grew slowly. Thus, we find that even though European demand for rice grew dramatically in the 1800-1860 period—the result in large part of population growth, rising per capita income, and urbanization—U.S. rice exports to Europe were far lower in the 1850s than they had been in the 1790s. Indeed, *total* U.S. rice exports were lower in the 1850s than they had been in the last decade of the 18th century, further proof that North American rice was becoming uncompetitive as global market integration proceeded, and as Asian exporters of cheap rice were rendered into viable suppliers to the West.

The basic scenario described above intensified considerably with the coming of the Civil War, provoking a true crisis in the American rice industry. The gradual drop-off in the South Atlantic rice region's ability to export to major markets—and by the late 1850s even to produce rice profitably in many cases—accelerated sharply after the war. Indeed, not only did most markets for U.S. rice collapse after the war, but the U.S. actually became a large net importer of rice until well into the 20th century. Until recently, scholars have attributed the declining fortunes of the South Atlantic rice region almost entirely to domestic causes, particularly those affecting rice supply. Thus, they have focused on factors such as the emancipation of slaves and the postbellum transformation of agriculture in the region, on wartime destruction to rice infrastructure, to the dearth of capital for rebuilding such infrastructure in the aftermath of war, and on destructive late 19th-century weather events (particularly a number of hurricanes) that wreaked havoc on the rice region.

If such scholars have correctly identified *some* of the factors that affected the South Atlantic rice region in negative ways, thereby impeding its ability to adjust, they have nonetheless framed their analysis too narrowly in both spatial and temporal terms. In so doing they have missed the forest for the trees, neglecting crucial considerations such as the long-term decline in the U.S. rice industry's competitive position *vis-à-vis* Asian suppliers, the hugely important mid-19th-century improvements in transportation and communications—steam shipping, the Suez Canal, and transmarine undersea cable come immediately to mind in this regard— that served to link Europe and Asia more closely together, and the intensification of the European imperial project in South and Southeast Asia after 1840 or so, which worked greatly to increase the flow of foodstuffs and raw materials from these parts of Asia to Great Britain, the Continent, indeed, to the Western Hemisphere as well.[17]

To put things bluntly, the South Atlantic rice industry was already deeply troubled by the 1850s, and the region would have collapsed—at least as a significant exporter—with or without the Civil War. In international context, the region was a relatively high-cost producer of a basic commodity well before the war, and the postbellum changes mentioned above exacerbated its disadvantageous cost structure and competitive position in rice. International competition, moreover, did not merely cost the region its export markets, but gradually its domestic markets as well. As we shall see below, beginning in the mid-1880s the locus of domestic production in the United States shifted decisively to southwestern Louisiana and soon after to other parts of the "Old Southwest"—to southeastern Texas and east central Arkansas specifically—where a new industry, based on entirely different production technologies, was born. The new rice industry was much more efficient than was that of the South Atlantic, and the South Atlantic rice industry shrank in the 1880s and 1890s before collapsing completely in the first decade of the 20th century. In 1909 the region accounted for only about 3% of the 658.4 million pounds of clean rice produced in the United States. A decade later, it accounted for only 0.6% of the 1065.2 million pounds produced. By the end of the 1920s, the industry in the South Atlantic region was completely gone, leaving a decidedly mixed legacy behind. The region, built for and by rice, was once one of the wealthiest parts of America—the free population in one part of the region (the South Carolina low country) was by far the wealthiest in British North America on the eve of the American Revolution—but once the rice industry collapsed, the region lost its economic *raison d'être*. Other than rice, there were few viable economic alternatives in the swampy region, and the area, burdened by history—or, more to the point, by slavery and war—lacked the human capital and financial resources to adjust to its new circumstances. As a result, after lapsing into decline, it remained an economic backwater, mired in poverty and social pathologies of one type or another, until the second half of the 20th century, when new possibilities began to emerge and things at long last began to change. By that time, rice producers in the Old Southwest, along with producers from California, were once again major players on the international stage.[18]

Small amounts of rice had been grown in Louisiana since the early 18th century, shortly after the French arrived in the area. Most of this rice was produced in small plots of wet ground near the Mississippi River in the southeast part of colony/state. Production increased slowly after the U.S. gained control of the area with the Louisiana Purchase (1803), and by 1859 the area accounted for about 3.4% of total U.S. rice production.

The disruptions and economic dislocations associated with the American Civil War affected different parts of the South in different ways. The South Atlantic rice region was hurt badly, as we have seen, but the modest rice industry in southeastern Louisiana fared better. Louisiana fell to Union forces early in the war, which minimized war-related destruction in the state, and, after the war, many agriculturalists in southeast Louisiana shifted production from their traditional staple, sugar, to rice, which was less labor-intensive, thereby decreasing both labor needs and the degree to which white planters and farmers would have to engage with, much less depend upon newly-freed African American field hands, who were perceived as being overly "independent", often unruly, and relatively costly.

Taking advantage of the window of opportunity opened by the disruptions in the South Atlantic rice region, rice producers in southeast Louisiana expanded production significantly in the 1860s and 1870s. In 1869 rice production in Louisiana amounted to about 15.8 million pounds (up from about 6.4 million pounds in 1859) and, a decade later in 1879 production reached about 23.2 million pounds. In 1869 Louisiana accounted for 21.5% of total U.S. rice production, and in 1879 about 21.1%.[19]

By the early 1880s, then, rice producers in southeastern Louisiana had developed a significant presence in the U.S. rice industry. Cultivating rice in irrigated swamps near the Mississippi, Louisiana producers grew the cereal in much the same way as did their competitors in the South Atlantic region, who by the 1880s were getting back on their feet. By the middle of that decade it seemed as though the U.S. rice industry had two bases, then, both of which produced wet rice in irrigated, embanked paddies, and both of which relied predominantly upon sizable numbers of African American laborers. At about that time, however, a new rice industry began to develop in another part of Louisiana, which industry was based on very different cultivation practices and technology. In time the new industry, sited initially on the prairies of southwestern Louisiana, would dramatically transform rice production in the United States. Indeed, the assumptions and guiding principles behind the so-called prairie rice revolution of the 1880s created the intellectual and operational contexts for the American rice industry as we know it today.

Stepping back and abstracting a bit, what those responsible for the "prairie-rice revolution" did in essence was to attempt to meet low-cost Asian competition by increasing productivity through the substitution of capital (particularly mechanized equipment and sophisticated water-pumping schemes) for relatively expensive, often scarce, and at times unreliable labor. Like most successful revolutions, this one owed its success in part to antecedent developments and institutions—most notably, the existence of a relatively sophisticated network of rice-marketing channels and rice-milling facilities in New Orleans—which those that made the revolution put to good use. Over time, the "revolu-tionaries"—aided by stepped-up biological innovation, the development of additional marketing institutions and channels, and robust govern-mental support—rebuilt the American rice industry, and, in so doing, rendering it competitive with competitors from Asia among other places. And everything started because of the ready availability of large expanses of cheap, fertile prairie lands in southwest Louisiana in the 1880s, because of the promotional efforts of railroads, speculators, and outside entre-preneurs, and because of immigration into the area, mainly from the Midwest and Great Plains.

As in other parts of the United States at various points in time, land settlement, promotion, and speculation were hopelessly intertwined in the southwestern part of Louisiana in the 1880s. This part of the state had traditionally been under populated and underdeveloped, and in the decades after the Civil War both corporations (particularly railroads) and enterprising individuals stepped up efforts to encourage and promote development of the area. Their promotional efforts were at once aggressive and broad based, with much of their attention focused not on relatively poor southerners, but on wealthier populations—successful farmers and developers—from other parts of the United States.

To cut to the chase, the convenient, if not harmonic convergence of interests in the mid-to-late 1880s and early 1890s amongst promoters of the Southern Pacific Railroad (which had the right of way through south-western Louisiana and southeastern Texas), a phalanx of aggressive land agents, brokers, and speculators, and a sizable cadre of relatively affluent, entrepreneurially-inclined small-grain and corn/hog farmers from the Great Plains and Midwest led to rapid immigration into, settlement of, and farm building on the prairies of southwestern Louisiana.

When these immigrant farmers got to southwestern Louisiana, they found prairies, as they had expected, but prairies unlike those in Illinois, Michigan, Iowa, Kansas, South Dakota, and other parts of the Midwest and Great Plains. In southwest Louisiana, the subsoil clay was almost impervious to water, holding water atop it until and unless it was removed. Once drained, though, the prairie soil was quickly found to be rich and cultivable. Moreover, unlike the soft, mucky soils in the South Atlantic rice region, the prairie soil was sufficiently dense and firm to hold the heavy capital equipment employed in Midwestern agriculture—steam plows, cultivators, and the like. All well and good, then, but the erstwhile Midwesterners were quick to learn something else as well about their new home: the climate and soil were not well-suited for the production of the crops they expected to cultivate, which is to say, wheat and corn. What it was well suited for was rice, although it took a bit of time, a lot of aggressive promotion, and some investment in substantive agricultural research and demonstration work to make the case.

Much of the credit for both the prairie-rice revolution and the early growth of the industry in the "Old Southwest" has traditionally gone to a handful of energetic promoters, developers, and agricultural scientists— S.L. Cary, Jabez B. Watkins, and especially Seaman A. Knapp—who, considered together, helped to get farmers into the area, got them onto the land, and, most important for our purposes, got them into rice. Knapp's role was particularly instrumental in the revolution *per se*, for he not only did the key scientific/demonstration/promotion work necessary to prove that the commercial production of rice made sense on the prairies of southwest Louisiana, but, along with a few others, he was one of the first to see (and thence demonstrate) that the capital-intensive, labor-saving cultivation and harvesting technologies employed in wheat could be transferred to rice.[20]

The above spadework, as it were, was done mainly between the mid-1880s and the early 1890s. Within a few years private irrigation and canal companies, employing relatively sophisticated pumping equipment, were also established in the area, which further differentiated the emerging agricultural regime in southwestern Louisiana not only from that of the South Atlantic rice region but also from every other cultivation system in the American South. With its immigrant agricultural entrepreneurs, its relatively large, capital-intensive, heavily mechanized production units, and its private irrigation companies, southwestern Louisiana by the mid-

1890s was well on its way to becoming the first thoroughly modern, proto-agribusiness regime in the South, which regime was based on rice. This process accelerated further—gathering steam, so to speak—by the turn of the century, by which time the region's transportation infrastructure, particularly rail connections, had improved markedly, its irrigation and pumping schemes had become more elaborate, its farmers more interested in, and influenced by science, and its millers, merchants, and bankers had developed greater milling capacity and financing capability.

All of the above developments and processes were made manifest in the increase in rice production in Louisiana and the growing relative importance of Louisiana in the U.S. rice industry. Already by 1889, for example—shortly after the prairie-rice revolution had started—Louisiana's share of the U.S. rice crop of 128.6 million pounds was 58.8%, a huge jump in relative terms since 1879. Such trends continued and accelerated in the 1890s. U.S. rice output nearly doubled between 1889 and 1899—from 128.6 to 250.3 million pounds—and in the latter year Louisiana accounted for fully 69% of the total. One should note here as well that during the 1890s rice production in southeastern Louisiana, along the Mississippi, faded out, meaning that by the end of the decade nearly all of Louisiana's rice was being produced on the prairies in the Southwest.[21]

Like many revolutions, that on the prairies in southwest Louisiana in the 1880s and 1890s, could not be contained, spreading in the first decade of the 20th century first to neighboring southeastern Texas, before jumping to east central Arkansas slightly later. The revolution jumped further still in the second decade of the 20th century, all the way to California, in which state the Sacramento Valley in particular began to develop as a center of production. In all of these areas, the cultivation systems that emerged were based on the model and logic of that pioneered in southwest Louisiana, to wit: capital intensity, mechanization, sophisticated irrigation/pumping schemes (including electric), agricultural science (with greater and greater emphasis over time on biological innovation).

The spread of the revolution led to the rejuvenation of the U.S. rice industry. By 1909, total U.S. rice production reached 658.4 million pounds, over 90% of which came from Louisiana (49.6%) and Texas (41.2%), with fledgling producer Arkansas already accounting for another 5.9%. Expansion and further diffusion were characteristic of the second decade of the 20th century as well. With demand for foodstuffs of all kinds growing as a result of the First World War (1914-1918), rice production in the U.S. increased substantially during that decade, rising to 1065.2 million pounds by 1919. In that year Louisiana was still the largest rice-producing state by far, accounting for 45.3% of the U.S. total. California, whose first commercial crop was grown in 1912, had already become the second-largest producer by 1919, accounting for 19.6% of the U.S. rice crop. Arkansas wasn't far behind at 19.2%, and Texas was fourth with 15% of the total.[22]

The "revolutionary" developments that occurred in southwestern Louisiana, southeastern Texas, east central Arkansas, and in the Sacramento Valley of California between the mid-1880s and 1920 created the template for the modern rice industry in the U.S. Moreover, the template created in that period in many ways still informs the American rice industry today, which industry, by the way, is still centered largely in these areas. Although the rice "story" differs in details in each of these areas—the California "story" in particular—the basic strategy followed and the results that ensued are pretty much the same everywhere. In each of these places, commercial growers, whether owner-operators or tenants, pursued and promoted self-consciously modern, scientific approaches to rice cultivation, approaches predicated on size, scale, capital intensity, mechanization, and sophisticated irrigation technologies. In so doing they grew rice not in small paddies or even quarter-acre rice "squares," as had been done in the South Atlantic rice region, but in large, flat, well-drained fields similar in many ways to those found in the wheat and corn zones of the Great Plains and Midwest. Unlike the South Atlantic rice region, labor was minimized in the "Old Southwest" and in California, and African Americans, by and large, were marginal to the industry, but for some seasonal harvest labor. The new industry took shape—and gained momentum—because it offered a way, perhaps the only way, for U.S. producers to compete with low-cost rice produced in Asia, where labor costs were much less than in the United States. The strategy pursued began to succeed at least to some degree by the second decade of the 20th century, when the U.S. produced its biggest rice crops ever, when growers in the Old Southwest and California made high profits, and when the U.S. became a net rice exporter for the first time in over a half century.[23]

During the interwar years the "new" U.S. industry continued to develop along the lines outlined above. New types of machinery and more elabo-rate irrigation and pumping systems were constantly being employed by growers, and greater and greater attention was placed on scientific breeding and the development of new seed varieties. Whereas the South Atlantic rice industry made do for the most part with two long-grain varieties—"Carolina White", supplemented later by "Carolina Gold"—growers in the first decades of the 20th century, working in the wake of the rediscovery of Mendel by soil scientists in the U.S., experimented widely with different varieties, and grew a range of long, medium, and short-grain rices for different markets. Over time, a rough geographic division of labor came about, as it were, with growers in Louisiana, Texas, and Arkansas mainly cultivating long and medium grains, and California growers specializing on short-grain varieties, developed originally from Japanese lines.[24]

At the same time principals engaged in the industry—growers, millers, brokers, agents, exporters, etc.—sought to render themselves more effective by establishing a variety of professional trade and marketing associations which, in aggregating individuals, would, it was hoped, allow for better market coordination and greater political influence. The latter—political influence—was becoming increasingly important in American agriculture, as the role of the U.S. government in the economy became greater and greater. To be sure, rice had long been a "political" crop in the U.S., enjoying tariff protection against imported rice even before the American Civil War. But in the 1920s and 1930s—in large part because of the hard times facing U.S. farmers producing other crops—principals involved in the rice industry used their professional organizations to ensure that the rice industry got its share, indeed, *more* than its share of governmental assistance and aid. Thus, the origins, particularly in the 1930s, of the panoply of subsidies, tariffs and non-tariff barriers, marketing initiatives, and export-promotion schemes have been more or less in place, for better or worse, from the time of the New Deal down to the present day.[25]

Both the increased role of government and the growing political influence of the rice industry are evident in American rice-export patterns. Once the U.S. rice industry was restructured and relocated, thereby becoming more efficient and more competitive at least in some international markets, efforts to promote exports were stepped up considerably. The home market for rice, while growing, was still relatively small, and various governmental agencies and sundry industry groups pushed exports as a way of reducing domestic supply and sustaining domestic price levels. The problem was to find buyers for U.S. rice in a world awash with cheap rice from Southeast Asia. In the 1920s most exported rice was shipped to Europe, but some was also shipped to Latin America, and (occasionally) to Japan. One strategy pursued by exporters early on was to focus on "political" markets—U.S. dependencies such as Hawaii and Puerto Rico, for example—and, increasingly from the late 1930s on, on quasi-dependencies such as Cuba, where deals were made to allow more Cuban sugar into the U.S. so long as American rice received preferential tariff treatment in Cuba.[26]

During World War II and in its immediate aftermath, surplus U.S. rice found extensive use in food-deficit areas as a form of American foreign aid, which policy became formalized—and well-nigh institutionalized—in 1954 with the passage of U.S. Public Law 480 (P.L. 480). According to this famous piece of legislation, surplus crops (including rice) being stored by the government could be released for use both to feed needy populations in the U.S. and to sell in less-developed countries. Because the law provided that international sales would be made in the currency of the country doing the buying rather than in U.S. dollars as was normally the case, such sales were designed both to offer somewhat of a subsidy to the LDCs doing the buying and to get rice of surplus crops without affecting domestic price levels adversely. A 1959 amendment to P.L. 480, which created the so-called Food for Peace program, allowed the U.S. to offer LDCs long-term, low-interest loans to purchase surplus U.S. foodstuffs. Rice figured prominently in this program from the start, a testament at least in part to the political clout of the increasingly powerful "rice interests" in the United States. And, politics being politics, such programs have only grown in number over time.[27]

Over the past fifty years or so, the U.S. rice industry—and market—by and large have continued down the path described above, albeit with a few interesting twists and turns. Production remains highly concentrated in a few areas: East central Arkansas, the Sacramento Valley of California, and southwest Louisiana/southeast Texas, of course, but now also in areas bordering the Mississippi River in southeast Missouri, Arkansas, Mississippi, and Louisiana. Arkansas remains the leading producing state, followed by California, then Texas. Scale remains characteristic of the U.S. rice industry. There are not many rice farms in the U.S.—just 8046, according to the most recent Census of Agriculture (2002)—but the average size of U.S. rice farms is much greater than for most other field crops. According to the same agricultural census: the average for rice farms in 2002 was 397 acres compared to 269 acres for wheat farms, 228 acres for soybean farms, and 196 acres for corn farms. Moreover, the vast majority of the acres in rice in the U.S. is planted on the largest farms, and this situation is not new. Indeed, as early as 1964, almost 90% of rice acreage in the U.S. was planted on farms with more than 100 acres in rice.[28]

U.S. rice output remains at very high levels: the 2007-2008 crop is estimated at about 198 million hundredweights of rough rice or—at a conversion ratio of 1 hundredweight of rough rice equal to .032 metric tons of milled rice—about 6.336 million metric tons of milled rice. This total is up a bit from 2006-2007, but still about 15% lower than the record U.S. rice crop of 2004-2005. If output is slightly below our all-time production peak, yields per acre in 2007-2008 are at a record high of 7247 pounds of rough rice (about 4471 pounds of milled rice per acre). As usual, yields were highest in California at 8350 pounds of rough rice (about 5152 pounds of milled rice) per acre.[29]

There have been some interesting developments in the market for U.S. rice as well. The domestic market for U.S. rice has doubled in size over the past twenty-five years, and today more than 50% of the U.S. crop is marketed domestically. Rice imports have grown significantly in recent years as well, particularly of high-quality long-grain varieties—Thai jasmine rice and Indian/Pakistani basmati rice—as well as medium/short-grain varieties from China for the Puerto Rican market (which is counted as part of the U.S. market). One other change in the U.S. market should be noted as well: In a reversal of historical trends in the West, the income-elasticity of rice in the U.S. has risen significantly over the past few decades: Wealthier groups in America now consume more rice, *ceteris paribus*, than do groups in more modest circumstances. Herein lies a good part of the explanation (along with increased Asian immigration) for the surge of imports of jasmine and basmati rice in the U.S market.[30]

The U.S. remains a prominent exporter of rice today, generally accounting for about 12-13% of the total world trade in recent years. In 2007 the U.S., coming in at about 3.2 million metric tons of milled rice, ranked as the fourth-largest exporting nation in the world, after Thailand, Vietnam, and India, and the U.S. is projected to place third or fourth in 2008. As in the past, the majority of U.S. exports go to countries in the Western Hemisphere—Mexico has been the biggest importer of U.S. rice for some time now—with considerable quantities also being shipped to Japan, to various parts of the Middle East, and, via numerous food-aid programs, to Africa.

Although the U.S. is a large exporter, it does not have the ability to shape the world rice market, which is thin—only about 5-7% of total world production is traded internationally, less than half the percentage in other small grains such as wheat—shadowy, highly protected, and characterized by poor and/or asymmetrical information. The recent run-up in world prices, for example, was due as much to such market imperfections, as to changes in supply and demand.[31]

It is, of course, always difficult for historians to make predictions, especially about the future! This is particularly so in the case of the U.S. rice industry, an industry which we said at the outset of this essay is a riddle wrapped in a mystery inside an enigma. Although rice is by value the eighth-largest crop grown in the U.S., the U.S. accounts for only about 1.5% of the rice produced in the world. Rice-cultivation systems in places such as the Grand Prairie of Arkansas and the Sacramento Valley of

California are arguably the most advanced, capital-intensive sites of cereal production in the world, replete with sophisticated irrigation works, laser-guided field prep, aerial seeding and spraying, self-propelled tractor/cultivator/combines employing GPS technology, computers, and delivery systems developed by the U.S. National Soil Tilth Research Laboratory in Ames, Iowa to regulate and vary the amount and composition of fertilizer dropped every few inches based on soil needs, past yields, and future potential. In such systems, yields of 6000 pounds of milled rice per acre are common, and two workers can easily cultivate 1000 acres or more. It takes this kind of "high modernist" system for U.S. producers to compete in the world rice market, and, even with this system, rice remains a highly subsidized/protected crop at home. Rice is quite profitable, generally speaking, to the small number of growers in the U.S., but the industry uses so much water that some prominent critics, particularly in California, question not merely its viability but the advisability of continued rice production in that state. Environmental issues are not the only ones clouding the future for the U.S. rice industry: The manner in which agricultural protection/subsidy questions are resolved in ongoing WTO negotiations will surely weigh heavily in this regard as well. Ironically, although rice has been grown in this country for well over three hundred years, and although it has long been commercially important, its profile is so low that relatively few Americans know about the issues the industry currently faces, much less care about its fate.[32]

References

1. Maclean, Dawe, Hardy and Hettel (2002:7).
2. Coclanis (1993a), Coclanis (2006a).
3. Coclanis (1993a), Coclanis (2006a).
4. Coclanis (2006b), Wood (1974), Littlefield (1981), Carney (2001), Coclanis (2002), Eltis, Morgan and Richardson (2007).
5. Coclanis (2006b), Wood (1974), Littlefield (1981), Carney (2001), Coclanis (2002), Eltis, Morgan and Richardson (2007).
6. Coclanis (2006b), Coclanis (1989:142, 255-256), Chaplin (1993).
7. Gray (1958:1:279-280) Coclanis (2006b), Coclanis and Marlow (1998), Coclanis (1993a), Stewart (1996), Coclanis (2004), Edelson (2006).
8. Wood (1974), Coclanis (1989), Morgan (1998).
9. Wood (1974), Coclanis (1989), Morgan (1998), Coclanis (2006).
10. Wood (1974), Coclanis (1989), Morgan (1998), Coclanis (2000), Edelson (2006).
11. Coclanis (1989), Coclanis and Komlos (1987).
12. Coclanis (2006b), Coclanis (2009).

13. Coclanis (1989:82-83), Coclanis (2006b).
14. Coclanis (1989), Coclanis (1993a), Coclanis (2006a), Coclanis (2006b).
15. Coclanis (1989), Coclanis (1993a), Coclanis (2006a), Coclanis (2006b).
16. MacPherson (1805:4:362). Note that MacPherson made the observation cited in the text in 1795, though his study was not published until 1805.
17. Coclanis (1989), Coclanis (1993a), Coclanis (1993b), Coclanis (2006a).
18. Coclanis (1989:142), Coclanis (1993a).
19. Coclanis (1989:142), Coclanis (1993a), Daniel (1985), Dethloff (1988), Coclanis (2006b).
20. Coclanis (1989:142), Coclanis (1993a), Daniel (1985), Dethloff (1988), Coclanis (2006b).
21. Coclanis (1989:142), Coclanis (1993a), Daniel (1985), Dethloff (1988), Coclanis (2006b), Post (1940), Phillips (1954).
22. See the works cited in Note 19. Also see Spicer (1964), Pudup and Watts (1987).
23. Coclanis (1989), Coclanis (1993a), Daniel (1985), Dethloff (1988), Spicer (1964), Bleyhl (1955), Pudup and Watts (1987), Hill *et al.* (1992).
24. Daniel (1985), Dethloff (1988), Bleyhl (1955), Pudup and Watts (1987), Hill *et al.* (1992).
25. Dethloff (1988), Moore (2000).
26. Dethloff (1988), Bleyl (1955:70-75), Moore (2000).
27. Dethloff (1988).
28. U.S. Department of Commerce (1975:2:515), U.S. Department of Agriculture (2006:3-6), U.S. Department of Agriculture (2007), U.S. Department of Commerce (1964:378).
29. U.S. Department of Agriculture (2007:2-9).
30. U.S. Department of Agriculture (2006), U.S. Department of Agriculture (2007), Shortridge and Shortridge (1983).
31. U.S. Department of Agriculture (2007:88-92), U.S. Department of Commerce (2007:539), Coclanis (2008).
32. U.S. Department of Agriculture (2007), U.S. Department of Commerce (2007:539), Hart (1991:301-314), Griswold (2006), Scott (1998).

14 | History of Rice in Latin America

J.A. Pereira and *E.P. Guimarães*

To best understand the historical process that culminated with the transference and adaptation of plants and animals from the oriental to the occidental hemisphere, an important chronological mark was the discovery by the Portuguese navigator Vasco da Gama of a new route connecting directly Europe to India. It was the first large stride toward globalization with repercussion to the New World that surpasses the first human trip to the moon.

Regarding the introduction of the first rice seeds in the American continent, the event occurred just before the great Portuguese conquest when during his second trip in 1493 the Genovese Christopher Columbus brought a few seeds despite the fact that these did not germinate.[1]

History of Rice in Hispanic Countries other than Brazil

The history of introduction of rice in Latin America starts in Southern Europe, more precisely in Spain, Portugal and Italy. These countries were important points of departure in the routes of all seeds that, despite their remote origins in the far orient, ended up in the American continent mainly during the 16th century but continued up to the end of the 19th century.

The first contact of the occidental man with the cultivated species of rice (*Oryza sativa* L.) was probably in Persia (present day Iran) as a consequence of Alexander the Great's invasion of India. Current consensus is that the seeds of this cereal were taken to Alexandria (Egypt) and to Damascus (Syria) from where they were later transported by the Arabs to Spain around the year 1150 and to Portugal in the next century. In this latter country, rice was cultivated during the reign of D. Diniz (1279-1325) only.

Approximately three centuries after its introduction to the Iberian Peninsula, in the 15th century, rice also began to be cultivated in Italy. Initially it was commercialized as a special product: sometimes as a medicinal food, sometimes as a simple condiment.[2]

It is desirable to mention that, prior to the 18th century, types of rice that were cultivated in Europe were quite popular and received their names after the regions where they were cultivated. In that era, the most celebrated rice types were the "Valencia rice" (from Spain), "Venice rice", "Nostrano-de-Novara rice" also known as "Nostrano" "Novara" or "Novarese" and "Ostiglia rice" (which had excellent acceptance in the Amsterdam marketplace).

The "Novara rice" was taken to South Carolina in the United States where its cultivation started in 1694[3] and was later known as "Carolina rice". In fact, it was the "Novara" variety[4] that was known as "Carolina rice" until the 19th century in southern Europe. This information helps confirm the great influence of Italy on the American rice industry since its beginning especially in the United States, Brazil and Argentina.

In the case of Hispanic America, despite the fact that the first seeds were brought by Christopher Columbus to the Hispaniola Island (present day Dominican Republic), history accounts for the cultivation of rice in Colombia, more precisely in the Madalena River valley around 1580. In Mariquita, for example, rice was already produced in abundance in those days despite employment of the same methods that were used in the cultivation of wheat suggesting that it was grown as an upland crop. In Antioquia, rice was introduced by the Jesuits in the middle of the 18th century and its production was concentrated in the region of Saint Jerome.[5]

According to that author,[6] the first incidence of growing irrigated rice in Colombia was in the Cauca valley in the region of Guacari where until 1930 the varieties of choice were "Japanese" and "Bomba". However the most widely cultivated variety was known as "Guacari". From that point on the "Guacari" variety was substituted by the "Guayaquil", a variety of rice that was brought to Ecuador by Rafael Madriñan. About the "Guacari" variety there is information that it could have been the same variety "Honduras" cultivated many years earlier in practically all countries of Central and South America.

According to Jennings,[7] commercial-scale production of rice started around 1908 and a curious fact became permanently associated with the history of rice in Colombia especially in the region of Llanos Orientales. During the presidency of Rafael Reyes, a decision was taken to liberate Bogota from a certain number of criminals and to transfer them to the agricultural colonies established in that region. In the proximity of one of those colonies, the soil was tilled and, as soon as the rains started, the rice seeds were sown. The harvest was very successful and it also contributed to combat the traditional melancholy of the inmates.

In that country, the cultivation of rice became so important that while a national caucus of rice producers was being held in 1952 in Cali, it was agreed that May 28 should become the Rice Day in honor of the day on which five years earlier was founded the National Federation of Rice Farmers—FEDEARROZ.[8] At present the main regions that produce rice in Colombia are Alto Madalena, Llanos Orientales, Caribe Humido and Caribe Seco.

In more recent times, Colombia has had a great importance for all Latin America because of the contributions of its National Center for Agricultural Research as the pioneering official organization for the genetic improvement of rice in the region. Its current headquarters are located in the city of Palmira. This organization would become the embryo for the current International Center for Tropical Agriculture (CIAT).

The research activities started there in 1930 with the goal of improving the existing rice varieties. In its initial years, the efforts were focused on the multiplication of the variety "Fortune" which was developed in 1911 by selection from the variety "Pa Chiam" at the Louisiana Agricultural Experiment Station at Crowley (USA). The original genetic material had been brought from Formosa (Taiwan) by the United States Department of Agriculture in 1905. At a later date, the variety "Fortune" was replace by "Bluebonnet 50" and "Zenith" both also originated from the United States.[9]

The importance of CIAT for the Latin-American rice industry has in no way been less than that of the International Rice Research Institute (IRRI) during these years. Its participation was vindicated after the development of the variety "Cica 8" (the most well-known variety of rice in the continent), "Metica 1" and many other varieties that are still grown by farmers or have been used as parental stock in rice breeding programs.

In Peru the cultivation of rice appears to be as old as in Ecuador, Colombia and Brazil. There is precise record that the first seeds of white rice came to northern Brazil[10] from the Vila de Santiago de Jaen de

Bracamoros in the province of Mainas in the year of 1761. This white rice too had its origin in Spain just as all the other varieties cultivated in the country had until 1927. It was then that the Northern Agriculture Experiment Station was created and direct importation of varieties from Asia and the United States started.[11]

According to Bruzzone,[12] the year of 1945 was marked as the beginning of artificial crossing of rice varieties in Peru. In 1968, the National Rice Program was created and six years later the first national variety, a semi-dwarf type ("Inti"), was released. Currently the main varieties used by Peruvian rice farmers are "Viflor", "Amazon", "Pitipo", "Alto Maio", "Capirona", "Huallaga", "IR 43", "Taymi" and "Amor".

In Hispanic America, however, singular importance must be attributed to Ecuador. During colonial times, it was a center of dispersion of rice seeds coming from rice fields in Murcia (Spain) to South America.[13]

The history of rice in the south of the American continent, especially in Uruguay and Argentina, is relatively recent if compared with the rest of the Latin-American countries. Its introduction dates back to the middle of the 18th century. Speaking of Uruguay, however, the first reference to rice cultivation is from 1869 while the first commercial operation was established in 1919 in Santa Rosa do Cuareim (present day Bela União). In 1936, with a cultivated area of only 4735 ha, the country obtained its first surplus which met the country's small internal demand and helped the country to embark on rice production for export.[14]

As rice has become an important crop in recent years, two events became mile-stones in the history of rice in Uruguay: first was the organization of rice farmers to form the Rice Growers Association which became effective from February 8, 1947, and the second was the creation of the Eastern Experiment Station at Treinta y Tres. The first major impact of this second initiative was realized when the yield of rice·increased considerably due to introduction and large scale adoption of the North American variety "Bluebelle".

In Argentina rice cultivation was introduced by the Jesuits in the province of Missiones from where it spread to the provinces of Tucuman, Jujuy and Salta reaching later Entre Rios which is currently the largest rice producing province. In Latin America as a whole, the decade of 1930s represented a period of large advances in agriculture since it marked the onset of systematic research in crops including rice. The credit for early efforts for controlled hybridization in rice also goes to Argentina as the agronomist Julio Hirchhorn (1895-1974) used this technique for the first time in the province of Entre Rios in 1934.[15]

The need to employ this technique arose when Julio Hirchhorn, after studying the characteristics of more than 1000 native and exotic varieties, concluded that there was no genetic material capable of exhibiting the desired agronomic characteristics for the conditions of Entre Rios. The varieties "Victoria", "Chacarero" and "Cumeman" were the first varieties developed through hybridization using Italian germplasm.[16] These became very popular in the entire Argentina.

According to the current information, the rice was first cultivated in Cuba around 1750 and up to 1967, all the rice varieties of that country originated in the United States. In that period, the main varieties in use were "Honduras" ("Zayas Bazan"), "Fortune", "Rexoro" ("Rexora") and "Bluebonnet 50".[17]

Following the example of what happened with all the Latin American countries from the second half of the decade of 1960, the Institute of Rice Research was established and an official program of genetic improvement of rice was initiated by introducing varieties and segregating lines from the breeding programs of IRRI in the Philippines and CIAT in Colombia. As a result of this work, the Cuban rice producers gained access to the varieties "IR 8" and "Cica 4" and, as a result of the first artificial crosses in rice, developed the variety "Amistad 82" in that country.

In Venezuela, where rice constitutes one of the most important cereals for the diet of its people, breeding programs started in 1943 only with the introduction of the variety "Zenith" from the United States. Most of the rice germplasm used up to the end of the next decade came from the United States only. In addition to "Zenith", the varieties "Bluebelle", "Saturn" and "Dawn" were also introduced among others. In the decade of 1960, with the introduction of improved varieties from IRRI and CIAT, the varieties "IR8" (called then "the Philippino miracle"), "Cica 4" and "IR22" were released. From that point, varietal improvement of rice was taken up in the states of Guárico and Portuguesa where 85% of rice production is now concentrated and new rice varieties like "Ciarllacen", "Araure 1", "Araure 2", "Araure 3", "Araure 4", "Cimarrón" and "Palmar" were regularly released.[18]

In the Dominican Republic also, until the decade of 1960, North American rice varieties dominated. The Experiment Station at Juma was established in 1963 and ten years later varieties like "Juma 51", "Juma 57", "Juma 58", "Toño Brea", "Mingolo", "Ingles Largo", "Juma 66", "Juma 67" and "IDIAF 1" were released by artificial hybridization.[19]

At present, Brazil is responsible for more than half the annual rice production in Latin America followed by Colombia, Peru, Ecuador,

Uruguay and Argentina in decreasing order of importance of this cereal. No country other than Brazil has an annual production of more than one million ton (Table 14.1). Therefore the rest of this chapter will be devoted to the history of rice in Brazil only.

Table 14.1. Cultivated area (1000 ha), production (1000 ton) and productivity (kg/ha) of rice in Latin America in 2004

Country	Area	Production	Productivity
Brazil	3732	13356	3580
Colombia	517	2663	5150
Peru	318	1473	4630
Ecuador	350	1300	3710
Uruguay	190	1250	6580
Argentina	172	1060	6160
Cuba	205	716	3490
Venezuela	135	700	5190
Dominican Republic	107	640	5980
Bolivia	142	424	2980
Panama	86	255	2970
Nicaragua	94	242	2570
Costa Rica	53	222	4180
Mexico	51	192	3790
Chile	25	115	4620
Paraguay	28	105	3820
Haiti	53	105	2000
Latin America	6258	24818	3965

Source: IRRI (2006).

History of Rice in Brazil

Presently, there is no doubt that the cultivated rice was first introduced in Brazil in the first half of the 16th century though this topic was very polemic among Brazilian historians for many years. There are records from at least that time that rice was cultivated in the Brazilian seaboard more precisely at the Captainship of Ilhéus in the present day state of Bahia and in Iguape, the ancient capital of the Captainship of São Vicente (modern day state of São Paulo).[20]

The most important original document known in Brazil is the book "*A Descriptive Treaty of Brazil*" written by Gabriel Soares de Souza, a

Portuguese plantation master, who established himself at Recôncavo da Bahia between 1570 and 1587. A verbatim transcription of the extremely important report attributed to Gabriel Soares de Sousa about the introduction of rice into Brazil:

> ". . . . Rice thrives in Bahia better than in any other known place because it is seeded in swamps and in dry land, there is no doubt that even when the soil is poor we are almost certain to be surprised from each cultivated *alqueire*[21] forty to sixty *alqueires* are harvested of rice of quality and beauty comparable to that from Valencia. After cleaning the field where rice was produced a fresh crop emerges even without reseeding other than the kernels naturally fallen during harvest. The seeds of rice from Cape Verde were taken to Brazil, its litter if fed to horses makes them fall to glanders and if they eat in abundance they die from it."[22]

At the Captainship of Ilhéus, rice was cultivated first in Valença in a forest area that had been cleared of wood and harvested with the intention of building ships for the Royal Navy.[23] In Iguape, rice became an agricultural produce of great importance in the early years of its colonization. In this regard, Friar Gaspar da Madre de Deus in his book *"Memoirs for the History of the Captainship of São Vicente"* asserts that in 1557 paddy rice was commercialized in that location for Rs. 50 (fifty réis) per *alqueire* suggesting that rice was being cultivated in the coastal area of that Captainship in the 1550s.[24]

The earliest available record testifies that in 1530 a ship carrying rice seeds departed from the Island of Santiago in the archipelago of Cape Verde with Brazil as its destination. Other similar trips occurred in the subsequent decades and, in many of these cases, seeds were delivered in the present day state of Bahia.[25]

It is opportune to stress that due to the unique climate of Cape Verde, other commonly cultivated cereals like wheat and barley that were introduced by the first colonizers into the archipelago, adapted poorly to the local conditions. Santiago, the first island of Cape Verde to be populated (along with Fogo Island), served for many years as an initial crop experiment station for several species that ended up being introduced into Brazil. Despite most of the plants cultivated in Cape Verde originated from Portugal, the origin of the most important of those crops can be traced back to tropical America and Asia.[26]

In the context of the available documents on the history of Brazil, it is difficult to prove that rice was cultivated in the country before 1530. It was during that decade that the formal process of territorial division of Brazil into hereditary captainships began: a colonization system that was successfully used by the Portuguese in the archipelagoes of Açores and Madeira but with less success in Cape Verde and Angola.[27]

By this time, Martim Afonso de Souza, the first who received a captainship, had already received the hereditary captainship in Brazil (São Vicente). However, no agricultural tools could have arrived from Portugal in the first half of the 16th century since, until that time, there was no plausible reason for a harvest either for internal consumption or for exporting in a time when communication between Brazil and Portugal was rare and difficult.[28]

This situation lasted at least until 1549 when the king of Portugal D. João III, after realizing the failure of his colonization plan, decided to change the system that had been adopted and named Tomé de Souza as the first Governor-General of Brazil effectively starting the Portuguese colonization of Brazil. Brazilian historians consider that the decades of the first half of the 16th century were lost, thus it would be advisable to consider that the cultivation of rice in Brazil had started after this period.

It should be added that, at the onset of the 17th century, rice was already acknowledged "as the provision that occupied second place" among the foodstuff consumed by the inhabitants of Brazil.[29] Also, in the northern region where the colonization process started only after a century after the colonization of the south, rice was first introduced around 1650.

It is important to stress that for more than two centuries, the rice cultivated in Brazil was not white rice but red rice known by the Portuguese and by the Spanish by the names of *red rice, land rice* or *Venice rice.*[30]

Red Rice, Land Rice or Venice Rice

The name "red rice" comes from the fact that the grains of these varieties present a red-colored pericarp. It can be easily explained as a genetic trait that is controlled by the gene *Rd* in chromosome 1 and gene *Rc* in chromosome 7.[31]

The name "land rice" probably refers to the system used for cultivating this rice i.e. upland. This name was probably used to differentiate it from the red wild rice frequently found in those days in swamps and flooded areas. In Portugal, the Portuguese agronomists[32] referred to a rice variety as "land rice" as late as in the 1940s. In Brazil, the rice variety "Mineiro"[33] is known as "land rice" in the state of Minas Gerais.[34]

The "Venice rice" was one of the first awnless rice cultivated in Italy.

The old Italian varieties "Novara rice" and "Ostiglia rice" had long awns like all the primitive varieties of rice.[35] The "Venice rice" received its name from the city of Venice which was a world clearinghouse for spices for many centuries and whose port had great influence on the commercial routes between the Orient (from where rice was brought) and the Occident. It should be noted that Venice is located at a short distance from the triangle Novara-Vercelli-Pavia in the Po River Valley which is the territory where nearly most of the Italian rice is produced even in modern times.[36] It is believed that "red rice" was cultivated in this area for many years before being taken to Portugal and from there to Brazil.

Another strong reason to believe this fact is that in the 18th century Portugal continued to be a regular importer of rice from Venice to fulfill its internal demand.[37] It was also at this time that Thomas Jefferson a plantation owner and later U.S. Ambassador to France and then President of the United States introduced the first seeds of upland rice from Italy to his country.[38]

In this regard, an interesting article was published in Portugal dealing with the political, economic and maritime ties between Portugal and the Most Serene Republic of Venice in the 16th century[39] which provide some evidences that reinforce our conviction that the so-called "Venice rice" was introduced into Portugal and thereafter in the 16th century to Brazil, more precisely in the Captainship of Ilhéus and later in the Captainship of Maranhão in the 17th century.

The year in which it was introduced to Brazil is not known with absolute certainty but the oral tradition informs us that the "Venice rice" was the same rice that was cultivated in Valença which was at that time a part of the Captainship of Ilhéus in present day state of Bahia.[40] From this information, it is reasonable to infer that it could have been brought around 1550. In the previous year, the Portuguese Crown had terminated the hereditary captainship system. Thus, at least officially, the Captainship of Ilhéus no longer existed. It is interesting that, even after being the cradle of rice cultivation in Brazil, the Captainship and later the state of Bahia were never prominent as a great rice producer in the country.

However, it can be affirmed that the major importance of the "Venice rice" in the context of the Brazilian rice industry was recognized in the state of Maranhão where it was the only cultivated rice until 1765. Some Brazilian authors[41] maintain that the entry of this rice in that state happened by intervention of the Azores colonizers. Their arrival in northern Brazil thus would have occurred after the introduction made from the archipelago of Cape Verde to Bahia. In the main historical works about the history of Maranhão, the Azorian colonization in the 17th, 18th and 19th centuries[42] is well documented but there is no record about the introduction of rice seeds by those peasants who originated from the middle of the Atlantic Ocean. Unfortunately, the primary archival sources of the above-mentioned authors are still not known.

Based on this information and assuming that the "Venice rice" was actually introduced in Maranhão (or in the state of Pará) by Azorians, the first seeds would have arrived between 1619 when the first peasants arrived from the Azores in 1676 in which case, after landing in Pará, one of the first activities of the Azorians was to cultivate rice, tobacco, cocoa and sugar cane.

There are other relevant pieces of information from those years that are worthy of being mentioned. One of them refers to the production of rice in the town of Alcântara which was at that time an important settlement established by the Portuguese in Maranhão. It was found in a will dated 1712 that, among property items left by a lady called Isabel Mendes, there were nothing less than 60 *alqueires* of rice priced at Rs. 24,000 (twenty-four thousand réis).[43] Another historian of Maranhão mentions that in 1738 a gentleman named Antônio de Oliveira Pantoja obtained authorization to install two mills in Maranhão, one of those for rice.[44] Also in 1757, rice was listed as one of the food products consumed in a Jesuit school and it was produced in the large farms that were owned by the religious congregation in northern Brazil.[45] This information makes it reasonable to assume that this rice was "red rice", "land rice" or "Venice rice".

Generally speaking, some Brazilian historians have provided confusing reports about "red rice", "land rice" and "Venice rice" calling this ancient rice a *native* rice perhaps because this rice, like the *Brazilian wild rice* (also called "red rice"), presents a red colored pericarp. While the "Venice rice" belongs to the species *Oryza sativa* L., a species cultivated in Europe during Brazil's colonial years, there are several species of red wild rice (*Oryza glumaepatula, Oryza grandiglumis, Oryza latifolia* and *Oryza alta*) that are genuinely Brazilian.[46] These native species were used in the diets of natives and of the poorest members of the colony.

Nevertheless, one must admit that there is no definite proof that allows establishing the precise year in which "red rice", "land rice" or "Venice rice" arrived in northern Brazil but it is perfectly acceptable that this introduction would have happened in the 17th century between the year 1619 and 1649. This conclusion is partly corroborated by Jennings[47] who believes that large-scale introduction of rice in the tropical Americas had started in the 17th century.

White Rice or "Carolina Rice"

White rice also known as "Carolina rice" was introduced into Brazil at the end of the colonial era as a product that would be exported to Portugal since red rice was not widely accepted as a food by the Portuguese cuisine. At that time, Portugal was undergoing a chronic deficit in availability of grains especially wheat—a staple item of the Lusitans—and consumption of rice was gaining importance. This food item had been traditionally imported from Venice and Genoa (Italy) and later brought from South Carolina of United States.[48]

Thus the name "Carolina rice" referred to the state of South Carolina in the United States where it had been cultivated since 1685[49] and had flourished among the prosperous society between the 18th and 19th centuries.[50] The rice plantations drew on labor from western African slaves.[51]

The earliest known reference to white rice in Brazil was discovered by the naturalist Alexandre Rodrigues Ferreira (1756-1815) who found a collection of orders and resolutions written by the then captain-general of the state of Maranhão and Grão-Pará João Pereira Caldas who was also the governor of Maranhão state between 1772 and 1780. It was noted by João Pereira Caldas that, in the year of 1761, a João Batista da Costa sent to the governor of the state of Maranhão and Grão-Pará Manuel Bernardo de Mello e Castro (1759-1763) in Belém a small sample of white rice seeds. This rice would have been obtained in the town of Santiago de Jaen de Bracamoros which belonged to the province of Mainas in Peru.[52] It was evidently the rice brought by Spanish settlers and it was believed that it was the famous "Valencia rice" named after a region of southern Spain.

It is believed that Governor Manuel Bernardo de Mello e Castro with the help of some peasants personally distributed these first seeds of white rice in northern Brazil with the express recommendation that they be immediately seeded. The harvest of this rice reached the table of that authority in the same year of 1761. This white rice was so widely accepted that after only two years it had replaced a portion of the area that was previously cultivated with red rice in that region.

These first references have an inestimable historical value since they confirm that one of the first events of introduction of white rice in Brazil is due to a Spanish colony. They also explain the great importance at that time of the Amazon River as a natural communication channel between Brazil and the Spanish colonies in South America especially Quito (in present day Ecuador) and Peru.

Although it went down in Brazilian history that the introduction of white rice occurred through Peru, the second cycle of the rice culture in

Brazil started only after the introduction of the other white rice that is the "Carolina rice". This is because some information is available about rice production in Rio de Janeiro in the year 1761. In that year, rice was exported for the first time from Brazil to Portugal in bags, barrels and leather pouches but the entire lot was pilfered before it reached its destination. In 1763, when the Brazilian capital was shifted from Salvador to Rio de Janeiro, more than 1200 barrels of milled rice were shipped, most of that amount was taken to Bahia to supply the ships of the Portuguese king.[53]

In the world scenario, there was a crisis in the supply of cotton and rice as a consequence of the war for the independence of North American colonies and of the Napoleon wars in Europe. This enabled the creation of some wealthy centers, the most important one being in Maranhão where the most dynamic commodity of its economy was rice followed by cotton. The main buyers were the English who were in conflict with the North Americans.[54]

Actually rice and cotton played such a towering role in the economy of Maranhão that the wealth created on the basis of these two agricultural commodities led to the construction of the building complex in the historical center of São Luís do Maranhão (currently considered by UNESCO a World Heritage Site).

It can, therefore, be stated that the introduction of the "Carolina rice" ushered the second economic cycle of the rice crop in Brazil. It is particularly true for the state of Maranhão. Before 1765, there were attempts to start cultivation of rice but it was only from that year when the captain José Vieira da Silva, one of the administrators of the Commerce Company of Grão-Pará and Maranhão, brought seeds of the rice from Portugal and started expansion of the cultivated area of white rice.

Shortly after, it was the same José Vieira da Silva who installed the first rice mill and also brought from Portugal lieutenant-colonel José de Carvalho who was considered one of the top specialists on the subject. José de Carvalho arrived in 1766 bringing with him the necessary equipment and installing the first "processing mill" which consisted of a water mill at the margin of the Anil River. Thus in the year of 1767, Maranhão was able to export its second harvest (225 *arrobas*[55]) to Portugal.

Stimulated by the growing acceptance of white rice in Portugal, the Commerce Company of Grão-Pará and Maranhão created three more mills in the year of 1771 and worked toward the installation of ten more mills.[56] At that time, the demand was so high that the Company signed a contract with Lourenço de Castro Belfort (1708-1777), an Irishman settled in

Maranhão, to supply 1000 *quintais*[57] (58.744 tons) of rice every year.

It is worth mentioning the content of the letter dated December 24, 1770 sent from Portugal by the minister Martinho de Melo e Castro to the governor of Maranhão, Joaquim de Melo e Póvoas especially the fragment that reads: "Regarding rice, it has caused me great admiration of this year's great harvest that Lourenço Belfort and his farmers harvested 10,500 *alqueires* (142,443 liters) and that not only this large amount but also the harvest of other farmers remained unpurchased by the Company neither in its entirety as they should nor in part."[58]

To have a better idea of the importance that the introduction of white rice had for Brazil and in particular to Maranhão, it is enough to mention that this event was awarded the "standard of glory" an honor of those times.[59] The sons of Lourenço de Castro Belfort sued for the right to have their father acknowledged as the introducer of the "Carolina rice" in Brazil giving rise to a lengthy dispute between them and the son of José Vieira da Silva. The final decree favored the capitão José Vieira da Silva since he had been the first to successfully introduce white rice to Maranhão.

There is also a record of another verdict in which the Prince Regent of Brazil, D. João (the future D. João VI) in January 28, 1800 appointed Manoel Corrêa de Faria as Scribe of the Legislative in acknowledgement of the fact that his father had initiated the cultivation of Carolina rice in the town of Alcântara.

Carolina rice had thus excellent adaptation to the climate and soil conditions of northern Brazil. Despite the creation of great wealth that followed its introduction, there was resistance from some farmers who did not want to adopt white rice in lieu of Venice rice. To get rid of the white rice, they consumed as much of it as they could because, if they save it for seed and grow the next crop, there would be more of white rice.[60]

Several historians[61] affirm that as a consequence of the attitude of several farmers of Maranhão not to replace Venice rice by Carolina rice, the governor issued an edict on November 29, 1772 that the cultivation of any rice other than Carolina rice was forbidden using the following language:[62] ". . . . This rice culture is of much benefit to His Majesty [D. José I] and thus I expect that all those who wish to show to be good vassals that they apply themselves with abandon to the cultivation of this rice with the certainty that all that is be produced will be placed because even if much is produced much is needed. And let it be known to all how prejudicial to this new enterprise and to its commerce are the land rice or rices other than Carolina rice [the farmers are forbidden to cultivate those under the following penalties]: (1) *Free men*—One year of reclusion and payment of a fine of Rs. 100,000 (one hundred thousand réis) half of the amount being destined for public services and the other half to the denouncing party, (2) *Slaves*—Two years wearing a *calceta*[63] with spankings interpolated during this time, (3) *Indians*—Just two years wearing a *calceta*."

According to Viveiros,[64] the prohibition issued by the governor Joaquim de Melo e Póvoas in 1772 remained effective for more than 120 years but it is certain that his edict left a lasting legacy up to the present day because Venice rice disappeared from Maranhão for ever. However, it spread to other regions of Brazil where there was no such restriction to its cultivation. It is still grown in other parts of Brazil and is preferred by farmers of semi-arid regions.

It is remarkable that rice has become an important element in Brazilian daily life especially in the north where it is part of the regional culture. For instance, in the states of Piauí and Maranhão, there is an old tradition that the best date to sow the seeds of this cereal is December 13, the day of Saint Lucy's Feast (according to the Roman Catholic Church). Farmers' lore says that "rice sown on Saint Lucy's day has high yield, eschews plagues and even the birds respect it."[65]

Between 1768 and 1777, rice was the only cereal exported by Brazil. The export volume from Maranhão (the largest producer of the colony) alone was of 428,310 *arrobas* (6287 ton); it was superior even to the cocoa exports from Pará (419,689 *arrobas* or 6161 ton), cotton exports from Maranhão itself (238,519 *arrobas* or 3501 ton) and coffee exports from Pará (41,961 *arrobas* or 616 ton) according to data summarized by Dias.[66]

Simultaneous with its introduction in northern Brazil, white rice also started being cultivated in Rio de Janeiro. In 1781, the yield of the rice crop in Brazil was already enough to meet all its internal demand and export the surplus to Lisbon from where it was re-negotiated to Hamburg, Rotterdam, Genoa, Marseille and other large European consumer markets.[67] At that time, rice gathered so much importance that it figured second only to sugar on the list of exports from the entire Colony.

After these golden years, starting from the second decade of the 19th century, production of the Brazilian rice started to decline. Production was reduced to such an extent that it became necessary to import great volumes of milled rice from India for internal consumption. According to Marques,[68] the Indian rice was better than the rice from Maranhão in appearance and grain size but the Brazilian rice was superior in terms of its cooked volume and for the fact that the grains were "gummier".

Because of the shorter distance between the production center and

Portugal when compared to the remainder of the country, the rice produced in northern Brazil in the 18th century particularly in Maranhão and in Pará received higher historical relevance than the rice produced in the south of the colony. The fact is that the rice crop had already gained some importance in Rio de Janeiro in this same period. Holanda[69] mentions a document from 1767 that deals with the difficulties found in loading the ships destined to Portugal with products originating in the Captainship of São Paulo. It reads that "if there were rice mills here as they are in Rio de Janeiro (. . . .) then it would be easier to have something to be loaded . . . ." The same author, however, estimated that the installation of the first mills in Rio de Janeiro were quite recent since in 1753 a gentleman named Antônio Francisco Marques Guimarães who lived there had applied for a license to build a rice milling factory and in the next year requested for the "penalties be established for those who transgress the privilege that he had received."

About rice cultivation in Rio de Janeiro, there are records of a certain number of producers by the year 1759. For example, a gentleman named Antônio Lopes da Costa had built a mill in Sítio do Andaraí Pequeno with 10-year exemption from tax payment from the Portuguese government with the pledge that the production would be exported to Lisbon. By the way, his mill owned 24 slaves and 16 mules with a weekly production capacity of 500 *arrobas* (7343 kg) of milled rice.[70]

All the rice produced in Latin America up to the end of the 19th century was obtained from an ecosystem known as humid swamp that did not require irrigation. Irrigation of rice crop is a recent history in Brazil; it started in 1898. This technology was first used in the southern region of the country, more precisely in the state of Rio Grande do Sul. After this, the Brazilian geography of rice production has undergone remarkable transformation: the northern region especially the state of Maranhão lost its leadership that it had held for two centuries to the southern region represented by Rio Grande do Sul.

In this connection, there is a record that Lúcio Cincinato Soveral, the postmaster of the city of Pelotas, had sown a plot of irrigated rice in that city when the cultivation of the crop was starting at the margins of several creeks in that state. However, according to important local sources,[71] the first large irrigated rice field in Rio Grande do Sul was only sown in 1904 and in the next year there were 100 hectares of this crop more precisely in the Progresso Farm in Gravataí. By the way, this is the same area where later the Rice Experiment Station of the Rice Institute of Rio Grande do Sul (IRGA) was built.

In the early years of the 20th century, the city of Pelotas gained recognition as the great producer of rice in Brazil. A local producer Pedro Luiz da Rocha Osório was able to expand his production to such an extent that he became known the "King of Rice". This businessman exported with great success to neighboring countries his rice brand POB (Pedro Osório Brilhado[72]). In 1918, he personally introduced some Japanese rice varieties like "Japanese Pragana" a variety that became one of the most cultivated in the state of Rio Grande do Sul.

In an attempt to organize the rice industry in Rio Grande do Sul, a group of businessmen founded the Syndicate of Rice Producers of Sul on June 12, 1926. Within two years of its inception, the Syndicate introduced the varieties "Blue Rose" (considered at the time to be the best rice in the world) and "Long Grain Edith" from the United States. After this major achievement, the Syndicate was renamed on May 31, 1938 and finally received its current name Rice Institute of Rio Grande do Sul (IRGA) on June 20, 1940.[73]

The first efforts at genetic improvement of rice in Brazil started only in 1937 at the Agronomic Institute of Campinas (IAC) with its main objective to develop varieties that could be cultivated without irrigation.[74] It is interesting to note that the second half of the 1930s marked the permanent entry of the country in the era of genetic improvement of rice through state sponsored initiatives in São Paulo (IAC), Rio Grande do Sul (IRGA) and Minas Gerais (College of Agriculture of Viçosa) as well as through federal initiative (Ministry of Agriculture). The first steps had thus been taken to create and equip research institutes in all the major regions of Brazil.

The first phase of rice breeding in Brazil corresponds with the years preceding 1938. It was characterized by the utilization of the existing genetic variability found in the local varieties as well as some introductions made from other countries. It was in 1937 that the researchers Hilário da Silva Miranda and Emílio Bruno Germek (both from IAC) initiated the selection of varieties in the state of São Paulo. They selected the variety "Iguape" from which later "Iguape Cateto" and "Iguape Agulha" were isolated by secondary selection. These were the first varieties obtained by a rice breeding program in Brazil.

In February 1938, at IAC, the first hybrids were produced by crossing local and exotic varieties like "Nira", "Iola", "Matão", "Fortune" and "Golden Precocious".[75] It was the same IAC that made a great contribution to Brazilian agriculture with the release of the variety 'IAC 1246' in 1958. This variety covered roughly 60% of the country's upland rice area between 1964 and 1973.[76] Besides, for many decades, IAC became the only institute for genetic improvement of rice varieties for upland condition. By breeding and releasing many "IAC" varieties, it became the major player in Brazil's rice production at least until the middle 1980s.

Genetic improvement of irrigated rice started almost simultaneously at São Paulo and Rio Grande do Sul. In the latter state, the Rice Experiment Station (EEA) was created in 1939. Then the Agronomic Institute of the South (IAS) and later the Research and Agricultural Experimentation Institute (IPEAS) that is a subsidiary of the Ministry of Agriculture participated in the research efforts. This last institute located in the city of Pelotas became responsible for executing research in other southern states under its administration. The research stations that function under its administration are the Experiment Stations of Pelotas Bagé and Passo Fundo in Rio Grande do Sul, Rio Caçador in Santa Catarina and Ponta Grossa and Curitiba in Paraná.[77]

Until the 19th century, Italy was the major contributor to the rice germplasm in Brazil. In the first half of the 20th century, the leading role was played by the United States, notably the Louisiana Agricultural Experiment Station at Crowley. The contribution of this Experiment Station started with the visit of the IRGA agronomist, Bonifácio Carvalho Bernardes (1900-1969) who was at Crowley for two years before returning to Brazil in 1939. As a result of his trip, IRGA released the varieties "Caloro", "Colusa", "Blue Rose", "Blue Rose 155", "Early Prolific", "Zenith", "Nira", "Fortune" and "Rexoro"—all of them of North American origin[78]—for the benefit of the gaucho[79] producers.

Following the example set by Bonifácio Carvalho Bernardes in the decade of 1930 that offered clear benefit of rice production in the Brazilian south, especially in the state of Rio Grande do Sul, another agronomist (Demóstenes Silvestre Fernandes) who was employed by the Ministry of Agriculture in the state of Maranhão was also sent to the same Crowley facility and to farms in Texas to study the irrigated rice culture during 1944 to 1945. As a result of his study trip, he brought back 25 varieties for his state. Two of these varieties were known in the United States as "table rice": "Rex" and "Fortune". His collection was sown in the experimental field of Barreiro on the left bank of the Mearim River in the city of Arari Maranhão. It was the first attempt for cultivating irrigated rice in northern Brazil. It is noteworthy that the variety most cultivated in Maranhão nowadays is still the "Lajeado" variety derived from one of the original table rice varieties brought by Demóstenes Silvestre Fernandes, probably the "Rex" variety.[80]

To date, the major difference between rice production in the north and the south of Brazil is that the north still produces upland rice as a subsistence crop with minimal use of technology while the south cultivates irrigated rice in commercial scale with high technological level. In the second half of the 20th century, Rio Grande do Sul has assumed the position of the leading rice producer in the country and currently contributes half of the rice volume produced in the country.

Concluding Remarks

The first seeds of rice in Latin America were introduced right after the onset of colonization. Evidences point that 500 years ago, rice was introduced in Dominican Republic, Brazil, Ecuador and Colombia just within a gap of few years only.

There is almost a consensus among historians that the Latin American rice industry became economically organized from the middle of the 18th century only after the introduction of the "white Carolina rice". Up to that point, rice cultivation was exclusively a subsistence activity and the rice that was grown, at least in the Portuguese America, was the red rice (also called "Venice rice" and "land rice").

The modernization of Latin American rice industry is very recent. This phase started only in the early years of the 20th century with the simultaneous emergence of irrigation, chemical fertilizers and introduction of better yielding varieties especially from the United States and Japan. Prior to this era, the entire rice germplasm available came from Spain and chiefly from Italy. Even 100 years ago, rice productivity in several Latin American countries was very low barely touching 1500 kg/ha.

Starting in the decade of 1960, the Latin American rice industry enjoyed a spectacular evolution in terms of productivity and grain quality. Although numerous institutions have been involved in this process, one cannot deny that the development of semi-dwarf varieties by IRRI and CIAT and the replacement of tall rice varieties with these high yielding semi-dwarf types in a short span of time have played a major role in this transformation. This fact alone has, in some countries, accounted for five-fold increase in productivity levels that was attained a century earlier. Considering that in the same time span human population has grown almost in the same proportion and that human migration has occurred from rural to urban areas and thus has become more demanding, the increased productivity and production of rice has probably contributed a most relevant chapter in the history of rice not only in Latin America but in the entire world.

References

1. González (1985).
2. Ciferri (1960).
3. Silva (1950a).
4. Brasil (1910).
5. Jennings (1961).
6. Jennings (1961).
7. Jennings (1961).
8. Anonymous (2006).
9. Melhoramentos da rizicultura no Rio Grande do Sul (1946), Jennings (1961).
10. Ferreira (1885a), Ferreira (1885b).
11. Bruzzone (2004).
12. Bruzzone (2004).
13. Jennings (1961).
14. Asociación Cultivadores de Arroz (2006).
15. Nívia (1991).
16. Marassi (2004).
17. Suárez et al. (2004).
18. Ramon (2004).
19. Moquete (2004).
20. Holanda (1947, 1957), Sousa (1974), Gandavo (2001).
21. Alqueire is unit of dry volume that is approximately equivalent to 13.566 liters or 0.38497 U.S. bushels.
22. Sousa (1974:86). Glanders is a contagious and destructive disease especially of horses caused by a bacterium (Pseudomonas mallei syn. Actinobacillus mallei) and characterized by caseating nodular lesions especially on the respiratory mucosae and lungs that tend to break down and form ulcers.
23. Faculdade Zacarias de Goes (2005).
24. Madre de Deus (1975).
25. Duncan (1972), Carney and Marin (1999), Carney (2001).
26. Brito (1966).
27. Boxer (1981).
28. Luís (1956).
29. Brandão (1997).
30. Amaral (1940), Silva (1950a, 1950b, 1955), Pereira (2002a, 2004).
31. Sweeney Thomson Pfeil and McCouch (2006).
32. Vasconcellos (1949), Silva (1955a), Silva (1956).
33. The word "mineiro" is the gentile form used to designate the state of Minas Gerais.
34. Silva (1950a).
35. Ciferri (1960).
36. Novelli (1918), Morais (1960), Gonçalves (1964), Milano (2003).
37. Dias (1970), Carreira (1988).
38. Rasmussen (1975).
39. Oliveira (2000).
40. Faculdade Zacarias de Goes. (2005).

41. Freitas (1919), Viveiros (1928).
42. Pereira (2002b).
43. Viveiros (1999).
44. Coutinho (2005).
45. D'Azevedo (1901).
46. Black (1950), Rangel (1998), Pott and Pott (2000).
47. Jennings (1961).
48. Dias (1970), Carreira (1988), Gonçalves Souza and Resende (1989).
49. González (1985).
50. González (1985).
51. National Geographic (1994), Carney and Marin (1999), Carney (2001).
52. Ferreira (1885a), Ferreira (1885b).
53. Amaral (1940).
54. Ribeiro (1995).
55. Approximately 3304 kg. The *arroba* is an ancient unit of weight used in Portugal and in Brazil.
56. Viveiros (1928).
57. The *quintal* (plural *quintais*) is an old Imperial weight measure, also known as *hundredweight*.
58. Marques (1970:91).
59. Viveiros (1928), Gaioso (1970), Viveiros (1992), Viveiros (1928).
60. Viveiros (1928).
61. Amaral (1923), Viveiros (1928), Dias (1970), Marques (1970).
62. Dias (1970:435).
63. An iron ring worn on one ankle that was usually attached to an iron ball.
64. Viveiros (1992a).
65. Iglésias (1958).
66. Dias (1970).
67. Dias (1970), Boxer (1981), Carreira (1988).
68. Marques (1970).
69. Holanda (1947).
70. Varnhagen (1975), Santos (1979).
71. Sindicato Arrozeiro do Rio grande do Sul (1935), Melhoramentos da rizicultura no Rio Grande do Sul (1946).
72. The verb "brilhar" is derived from "brilhado" and means "to shine".
73. Sindicato Arrozeiro do Rio grande do Sul (1935), Bastos (1964, 1966), Massera (1983).
74. Germek and Banzatto (1972).
75. Germek and Banzatto (1972).
76. Germek and Banzatto (1972).
77. Instituto de Pesquisas e Experimentação Agropecuárias do Sul (1967).
78. Bernardes (1947), Bernardes (1960), Pedroso (1989).
79. The word "gaúcho" spelled in Portuguese with an accent in the letter "u" is the gentle form used to refer to the state of Rio Grande do Sul.
80. Personal communication (Manlio Silvestre Fernandes, Professor at Federal Rural University of Rio de Janeiro and son of the agronomist, Demóstenes Silvestre Fernandes).

15 | History of Rice Marketing

Sushil Kumar and *Jabir Ali*

'What do we have for the dessert?' asked a Swedish friend of mine[1] Liane Ruus, whom we had invited over dinner at our place in Toronto. My wife served *kheer* (the traditional Indian rice pudding) which she had cooked for around 3-4 hours on low heat. As a natural consequence, the discussion now tilted towards history of rice and various recipes for cooking it. We were pleasantly surprised to learn that Swedish prepare rice pudding exactly the way Indians do. Liane told that not very far in history, Swedish housewives used to cook rice in milk on very low heat for hours together. 'But these days due to paucity of time and fast paced life style, no one in Sweden has the time and patience to simmer rice pudding; and availability of concentrated milk preparations in the market has replaced the traditional cooking method of rice,' lamented Liane. This is exactly what has happened to *kheer* in India. We tried to understand the underlying basis of such strong commonalities in rice cooking across two diverse and distant cultures. No doubt, history of rice has many unexplained stories attached with it.

INTRODUCTION

Throughout history of mankind, rice has been one of man's most important staple foods. Archeological evidence suggests that rice has been the staple food for more than 5000 years.[2] Today, rice is eaten by more than two third of the world population, especially in Asia where 70% of the world's poor live.[3] This unique grain is grown in more than 100 countries on every continent of the World. Historically, across cultures, rice has been documented as a source of food and for tradition and ritual as well.

It is indeed a daunting task to chart precisely the labyrinthine course of historical trade that flows within world rice market. Historians, however, with the help of extant data, scattered widely in customs records, toll registers, and commercial manuals and journals, tried to track commodity flows within this market.[4] Driven by natural factors, technological innovations, national policies, and various exogenous market forces, rice trade pattern underwent radical transformations since it first started to be traded in ancient times.

In this chapter, an effort has been made to explore world rice trade and track trade patterns right from ancient period to 2005. After brief description of world rice market characteristics, the chapter goes on to discuss world rice market in Ancient World, Medieval World, Colonial Era, 20th century and finishes with first decade of 21st century.

World Rice Market: Characteristic Features

Though rice forms the staple food of over two third of world's population, as an item international trade and commerce, it is of secondary importance.[5] However, domestic policies, related to rice production and trade, have always been of great strategic importance across the world from a broader political perspective. In most countries, the governments exert rigid control on the rice trade,[6] which has a strong impact on the structure of the market. As a commodity market, the world rice market has historically been described as an 'unstable' and 'thin' market.[7] As compared to other grain markets such as wheat and maize, the rice market shows geographic concentration in production, low domestic price elasticity of demand, uncertain supplies in the world market, low world stockholdings[8] and volatility in world prices.

World Rice Market: Structure

Structurally, the market is essentially a thin[9] residual market.[10] Thinness of the markets has resulted in unstable trading patterns as the participants vary the volume of trade considerably from year to year. The rice market is 'residual market' as the ratio of trade to production is very low. Bulk of rice production is concentrated in Asia and coincidently, most countries in this continent have very high consumption of rice owing to large population size and rice being the predominant staple food for large section of the population. Low trade to production ratio implies that fluctuations in production have large scale impacts on the volumes traded in world market. Throughout the history of rice trade, the market has witnessed and continues to witness extreme volatility in the volume traded by each individual trading country.[11] This leads to high transaction/search costs[12] which in turn makes the market further thinner. In world grain market, unlike that for wheat and maize,[13] there is no 'central' market price for rice. And, this results in higher search costs and dissuasion of countries with temporary surpluses from uploading their surpluses onto the world market.

Relative high search costs and volatility in traded volumes forced importing countries to invest in increasing productivity of rice through introduction of new technologies and increase in rice acreage so as to minimize their dependence on rice imports. Consequently, it is the importing countries which have been very aggressive in adoption of High Yielding Varieties (HYVs) or improve their irrigation facilities and supplies of other inputs for rice production. Siamwalla and Haykin[14] point out that 'importing countries, over time, have been gaining in comparative advantage relative to exporting countries, so that the growth in the volume of trade tends to lag behind the growth in production.'

Market Segmentation

Depending on the length of grain, rice is classified in three categories in the international market: short, medium or long grained. It is the long- and medium-grained rice which dominates the production, consumption and trade of rice.[15] Similarly, the market also distinguishes between *japonica* and *indica* rice. The share of *japonica* rice in the world trade is very low.[16] Further, the market is also segmented on the basis of grade i.e., percentage of the grain that is broken. As compared to other grains, wide variations in prices are reported across various qualities and grades of rice. One of the major reasons for significant price differentials is the end use to which the grain is put to. Unlike other grains which are used both as human as well as animal feed, rice is almost entirely used for human consumption, and hence aesthetic elements play very important role in determination of price.

Conduct of World Rice Market Players

The atypical structure of the world rice market influences the conduct of participants in this market. The weak price discovery mechanism makes an individual participating country design self-sufficiency policies that can help curtain off the domestic rice market from the world rice market. Governments' policies focus on expansion of irrigation and fertilizer and credit subsidies in order to increase domestic rice production and attain self sufficiency and reduce reliance on world rice market.[17] Secondly, most countries, especially developing Asian countries, try to pursue policies which ensure stability of domestic rice prices even when the world has moved rapidly in the direction of free trade in the post WTO era.[18] Low trade to production ratio in major rice producing countries implies balance between domestic demand and supply of rice which indicates non-integration of domestic prices to the international prices. Through trade controls and restrictions, most countries have been more or less successful in maintaining stable rice price regimes in their domestic markets even in the wake of high degree of price instability in the world rice market. A study by Timmer and Falcon indicated that domestic real prices vary considerably across different countries.[19] Similarly, in a study of 13 Asian countries, when domestic real rice prices were correlated with international real rice prices, the median value of the correlation coefficients was found to be only 0.28.[20]

Market segmentation based on different qualities and grades further fragment an already thin market. In such a segmented and thin market, temporary or new exporters face stringent entry barriers as many of them do not possess the facilities to ensure desired standardization. Siamwalla and Haykin reported that in the years of excess production, countries, irrespective of their size, preferred to stock the surplus rather than export it.[21] In a perfectly competitive market, such storage behavior would be treated as irrational.[22] However, throughout the history of world rice market, there have been evidences of few imperfections because of which the countries resorted and continue to resort to such irrational behavior.

WOLRD RICE MARKET IN DIFFERENT PERIODS

Ancient World

Historians believe that rice was first domesticated in the area covering the foothills of Eastern Himalayas (North-Eastern India) and stretching through Burma, Thailand, Laos, Vietnam and Southern China.[23] In the East, where the rice plant originated, things are far different today.

In the ancient world, rice trade was negligible as this was an era of peasant economy. Much of the rice produced was used for local consumption and the remaining was exchanged for other food and non-food items in a 'barter system'. There are stray references to trade in rice. In this period, rice along with other commodities like wheat, clarified butter, sesame oil, cotton cloth, girdles, and honey was imported to Egypt. Some documents[24] mention voyages made by merchants carrying these commodities in ships to various market-towns in Egypt. And, in exchange, the merchants used to take a great quantity of tortoise-shell.

The period mainly saw the spread of rice cultivation to different regions and different ecosystems. Beginning in China and the surrounding areas, its cultivation spread throughout Southeast Asia, India and Sri Lanka. It was then passed onto Iran, Mesopotamia, Egypt, Spain and Italy.

Medieval World

During medieval period, major rice trade was among Asian countries. For this period, very scant and fragmented information is available and that too is derived from accounts of foreign travelers.[25] In many cases their accounts were subjective, one-sided, and aimed at the readership in their own country of domicile.

Though the 17th century was a period of growth in most commodities, there was trade in rice in eastern India from areas of surplus production to deficit areas independent of the European presence and this dates from well before their presence.[26] Areas of surplus production existed all along eastern Indian coastline (e.g., Bengal lowlands, Orissa, Godavari and Krishna delta, Thanjavur delta). There are indications that surplus rice from these areas was exported short distances to neighboring deficit areas in India and long distances to remote markets in Central, Western and Southeast Asia. The longer distance trade embraced other wide variety of places also. There was a regular and consistent export of rice to the island of Ceylon (Sri Lanka) which, from an exporter of rice, became a massive importer since the 13th century due to important changes in settlement patterns and demographic movements.[27] When price of rice in Ceylon was not very attractive, some Bengal vessels continued their voyage to Maldives in expectation of higher prices.[28]

Some historical accounts indicate that the diffusion of Asian *sativa* rice to east Africa occurred between 8th and 12th century (in the pre-Columbian exchange). This fact gets confirmed as Egyptian hieroglyphics do not mention rice among the cereals planted along the Nile floodplain.[29] Historians relate onset of rice cultivation in eastern Africa with population movements between the 8th and 12th centuries. These movements brought Islam to coastal Kenya as well as the migration of peoples from Malaysia and Indonesia to the unpopulated island of Madagascar.[30] While there are sufficient indications that Muslim traders and the Malagasy people introduced Asian *sativa* rice to eastern Africa, no evidence indicates the dispersal of the cereal overland to West Africa prior to the arrival of European mariners.[31] In the inland delta of the Niger River, there was already a well developed rice cultivation system and robust regional trade, as reported by Muslim scholars reaching the western Sudan from North Africa in the 11th century.[32] This indicates that domestication of *glaberrima* rice in West Africa was established centuries before Asian *sativa* arrived in East Africa.[33]

There are accounts of rice trade in the West well before it was produced in that part of the world.[34] Rice to the West was imported from Syria, Persia, and India by both the Greeks and the Romans. Most scholarships are of the view that cultivation of rice in Europe began with the Moors' incursion into Spain in the 8th century and the concomitant Turkish thrust into southeastern Europe and the Balkans.[35] Rice cultivation spread to other parts of Mediterranean Europe from these points of entry, and Europeans continued, intermittently, to import rice from the East throughout the medieval period.

Colonial Era

In 15th century, with onset of rice cultivation in northern Italy, European rice industry, which had been importing rice from the east throughout medieval period, attained significance in economic terms.[36] Surplus production from northern Italy was exported both to parts of the peninsula and to other parts of Europe during 16th and 17th centuries.[37] However, rice trade remained of limited importance till modern period when Europe's outward thrust altered economic geography of rice supply which eventually helped transform demand for rice in the West.[38]

In the 17th century, rice cultivation began in Carolina when a vessel from Madagascar visited Charlestown and delivered a small package of rice seed.[39] Some historians attribute onset of rice cultivation in Carolinas to arrival of rice delivery during the 1690s when a *Portuguese* vessel arrived with slaves from the east, with a considerable quantity of rice, being the ship's provision.[40] South Carolina, with Charlestown as a trading port, very soon developed as a major centre of rice export in the West. In fact, almost revolutionary innovations occurred in the cultivation, milling and marketing of rice over the next few decades following its introduction into South Carolina. Realizing that successful rice production and trade requires not only land and labour but also sophisticated milling technologies and facilities, the General Assembly in 1691 granted a patent to Peter Jacob for a pendulum engine to "huske rice."[41] By 1774, South

Carolina had the highest per capita wealth among the white population of the mainland colonies, and much of it came from rice.[42] Eighteenth century's increased trade in rice linked these states with markets in the northern continental colonies. Peter A. Coclanis has argued that sales to these markets integrated South Carolina and Georgia into the Atlantic economy through the pull of increased European purchasing power. Diversity in rice markets across regions and legal complexities led to development of international cooperation between merchants and correspondents in various ports.

Two major impediments to growth of rice trade in the early decades of 18th century—both in West as well as in East—were large scale problem of pirates and political interferences.[43] Authorities took drastic measures against pirates. Drayton reports that 'forty pirates were hanged on one occasion in Charlestown, and buried on White Point, below the high water mark, thus easing at least one of the restraints on developing foreign commerce.'[44]

The other constraint on trade—the British Navigation Acts of 1660 along with amendments of 1661 and 1663—was rather difficult to confront. The 1661 amendment required that certain goods and commodities, including sugar, tobacco, indigo, and cotton must be exported from the colonies only to England. Rice and molasses were added to the lists in 1704.[45] This provision protected the British and American traders from competition from the Dutch merchants and guaranteed the colonists a market for the products, sometimes at lower than world prices.[46] The amendment of 1663 further strengthened control of British over colonial import business by requiring that all European goods destined for the colonies, even on English-built ships, must transship through England. In order to avoid illegal trade and ensure collection of taxes, Act of 1673 required that all duties or taxes be levied at the port of clearance, rather than the port of arrival. Though laws were intended to facilitate trade, these did create serious obstacles to expanding rice trade. Rice shipments were first brought to England which consumed almost no rice but re-exported to Europe and Mediterranean countries. The transshipment and layover in English ports added considerably to the cost of shipment and losses incurred from spoilage.[47]

Commercialization of rice cultivation had significant influence on socio-cultural practices, including slavery, in South Carolina. With conversion of agriculture from subsistence to commercial system, slavery also began to expand. As rice export from South Carolina increased from 15,550 cwts in 1710 to 65,603 cwts in 1770, so did the number of slaves from 5,700 to 75,200 in the same period (Table 15.1).

Table 15.1. Slave population and rice export from South Carolina[48]

Year	Slave population ('000)	Rice exports ('000 cwts)	Export per slave (cwts)
1710	5.7	15.55	3.92
1720	11.8	62.28	8.48
1730	22.0	161.09	8.88
1740	39.2	305.90	8.23
1750	39.0	304.08	9.02
1760	57.3	399.02	11.50
1770	75.2	656.03	12.77

Source: Nash, 1992.
Notes: Figures are five year averages, of which the year quoted is the middle year.

The consistent increase in demand for rice, more technological innovations and growing importation of slave into America led to spread of rice cultivation along the coast of South Carolina into Georgia and along Carolina rivers, during 18th century. There was considerable demand for rice in northern Europe in 18th and 19th centuries which proved central to the development and working of Western rice trade. In addition, substantial demands for rice existed in other areas such as North America, southern Europe, the West Indies, and Brazil.[49] In northern Europe, Germany was the major centre of rice consumption where rice shipments entered from such ports as Amsterdam, Rotterdam, Hamburg, Bremen, and Danzig. Some other countries in north Europe which consumed significant volumes of rice included France, Belgium, and Netherlands. There was intermittent demand for rice in Scandinavia as well as Baltic region during this period. On the other hand, Britain had limited internal consumption of rice during 18th century but it increased in the 19th century.

Economic dynamism in Europe and consequent consistent increase in demand of rice resulted in British colony of South Carolina as the West's leading rice exporter. This colony, along with neighboring Georgia, which soon became a major exporter as well, possessed conducive environment for rice cultivation and requisite human and non-human resources which help explain why rice proved successful there at that time.[50] Sufficient African and Creole labour force which was cost efficient, disciplined, and had prior experience of cultivating rice (either *O. sativa* or the related African species, *O. glaberrima*) was available to undertake production of rice in these areas. Other advantages enjoyed by these areas over rice producing areas in the East included low transport costs, availability of

460

Table 15.2. Rice exports from South Carolina[51]

Year	'ooo pounds
1698	10
1699	131
1700-09	630
1710-19	2,952
1720-29	9,659
1730-39	20,939
1740-49	31,519
1750-59	37,867
1760-69	58,705
1770-74	77,179

Source: Adpated from Dethloff (1982).
Notes: Except for 1698 and 1699, all other figures are average for the period mentioned.

market information, and high degree of integration with the world economy. The enormous world demand and attractive prices led to continuing increase in production and growth in exports of rice from South Carolina. The region exported 10,000 pounds of rice in 1698 which rose to 394,000 pounds in 1700 and 76,265,700 in 1774 (Table 15.2). Almost all rice imported into England, Wales and Great Britain in 18th century was sourced from South Atlantic areas, while imports from other rice growing regions saw decline in this century (Tables 15.3 and 15.4)

However, this trade pattern started showing signs of change with the onset of 19th century. During first half of 19th century, South Carolina-Georgia rice region, despite producing more rice than ever before, saw decline as a supply source for the European and other markets. Tables 15.4 and 15.5 illustrate this pattern very clearly. The share of North America in volume imported in Great Britain fell from 90% at the end of 18th century to 20% during 1831-40 (Table 15.5). On the contrary, share of Asia in total rice imported in Great Britain which was meager 8% during 1791-1800 saw a steep rise to 78% during 1831-40. The stagnation and ultimate demise of South Atlantic rice industry was chiefly the result of competition from Bengal, Java, Lower Burma, Siam, and Indonesia. Some segments of the European market were also captured by Italian rice industry.

The island of Java witnessed rapid increase in rice exports to Netherlands, other European countries and America in first half of 19th century due to promotion of market integration by British. Total annual export

Table 15.3. Rice imports into England, Wales and Great Britain, 1696-1808[52]

	1696-00	1701-10	1711-20	1721-30	1731-40	1741-50	1751-60	1761-70	1771-80
Total import ('ooo cwts)	30.65	60.17	259.02	850.89	1617.93	1495.09	1758.95	3037.04	2434.40
Imports by Region (% of total)									
The "Thirteen Colonies"	13.95	89.19	94.79	99.06	99.79	99.35	99.3	99.13	98.57
Other North America	0.01	0	0.01	0	0	0	0	0.01	0.47
West Indies and Central America	0.16	1.14	4.14	0.75	0.19	0.2	0.66	0.57	0.64
South America	0	0	0	0	0	0	0	0	0
Southern Europe and Atlantic Islands	83.23	8.88	0.72	0.12	0	0.37	0	0.17	0.1
Northern Europe	1.32	0.08	0.11	0	0	0.02	0.03	0.01	0.18
Africa	0	0	0.06	0.01	0	0	0	0.1	0.02
Asia	1.32	0.71	0.17	0.05	0.01	0.07	0	0	0.02

Source: Coclanis (1993).

Table 15.4. Rice imports into Great Britain, 1772-1808[53]

	1772-80	1781-90	1791-00	1801-08
Total import ('ooo cwts)	2004.77	1344.61	1983.63	1281.53
Imports by Region (% of total)				
The "Thirteen Colonies"	98.31	98.4	89.31	67.02
Other North America	0.51	0.01	0.32	0.07
West Indies and Central America	0.8	0.72	0.05	0.12
South America	0	0	0	1.47
Southern Europe and Atlantic Islands	0.12	0.4	0.26	0.12
Northern Europe	0.22	0.04	1.04	0.24
Africa	0.02	0.04	0.26	0.1
Asia	0.02	0.38	8.76	30.85

Source: Coclanis (1993).

Table 15.5. British rice imports by area, 1831-1850 (clean rice equivalents)[54]

	1831-40	1841-50
Total import ('ooo cwts)	3723.03	8312.83
Imports by Region (% of total)		
Northern Europe	0.12	0.18
Southern Europe	0.12	0.62
North America	20.52	14.26
West Indies and Central America	0.00	0.09
South America	0.19	0.85
Africa	0.69	0.23
Asia	78.36	83.77

Source: Coclanis (1993).

from Java and Madura which was 189,312 pounds (clean rice equivalent) in 1825 increased to 87,728,992 by the year 1856.[55]

In order to ensure consistent supply of rice to the Western markets in the wake of inconsistent role Java was playing as rice exporter, British started transforming lower Burma from a 'closed, under-populated "natural" economy into what would soon become the greatest rice exporter in the world.'[56] During 1860s and 1870s, very extensive flood control works were carried out in great Irrawaddy Delta of lower Burma which brought million acres of land lying waste under rice cultivation. Unprecedented increase in rice production during this period put Burma ahead of her three rivals, Siam, Indo-China and Korea, in terms of rice export.[57] Mean yearly rice export from Burma to Asia and other countries was 907,000 metric tons during the period 1872-1881, and 2,411,000 metric tons during 1902-1911. Thus, Burmese agriculture was put on the path of commercialization which has been hitherto unknown in the country.

Other area which gained importance during 19th century as source of rice supplies to Western market was Bengal region, which again was the result of strong market interventions by the British. Aggressive entrepreneurship of British merchant capital coupled with complementary action on the part of Indian merchant capital succeeded in redirecting a portion of Bengal's agricultural production, including rice, to Europe.[58] In the year 1861-62, Bengal was exporting 341 thousand metric tons of rice to the foreign market of which 38% was to the Western markets comprising United Kingdom, France, Germany, North and South America, and West Indies. In the next decade or so, though the total export remained the same, the share of Western markets in this export fell to 25%.

Between 1861 and World War I, Italian rice industry not only successfully survived the challenge from the East, but showed significant growth. However, volumes of rice exports from Italy and other Western producing areas such as Spain were insignificant in comparison to those from the East.[59]

The major factor behind this radical shift in source of rice supply to Europe was extensive British rule in India, Singapore, and Hong Kong in the 19th century which led to development of a vast free-trade area in the East and global market integration. Asian rice was being imported in Europe earlier also, but the British economic and political control of Bengal intensified the shipments of rice in the late 18th and early 19th centuries.

What were the reasons for emergence of Southeast Asia as a dominant supplier of rice to the Western markets and demise of U.S. rice exports in the 19th century? Though this question is impossible to answer categorically, but Peter A. Coclanis suggests some possibilities. One reason was easy and sufficient availability of relatively cheap land and labour in Bengal, Java and Lower Burma. Family labour in rice producing households in these regions was often completely unremunerated. Agro-climatically also, Southeast Asian region was almost perfectly suited for rice cultivation, which led to higher productivity. In addition, the indigenous population of Southeast Asia was sufficiently experienced and relatively efficient in rice cultivation. Coclanis and Komlos[60] in their study on comparison of rice cultivation between Southeastern United States and lower Burma in 19th century suggest that the 'total factor productivity of Burmese peasant producers, measured in standard Cobb-Douglas form, was roughly at par with that of slave labourers on South Carolina and Georgia rice plantation. Even under stringent neoclassical assumptions, then, Burmese rice producers were competitive with U.S. producers in the 19th century.'[61]

European political and military powers in the 19th century also played their role in enhancing competitiveness and attractiveness of Southeast Asia as rice supplier vis-à-vis other potential supply sources. In 19th century, European political and military power was in an expansionary mode in Southeast Asia. Accompanied with this power came better administrative, fiscal and tariff innovations which supplemented and complemented the natural competitiveness of Southeast Asian regions in rice supply vis-à-vis their counterparts in the developed parts of the world.[62] During this period many other transportation and communication improvements took place which further facilitated global integration of Southeast Asian markets in general. On the logistic front, technological and organizational improvement in transoceanic shipping resulted in lower cost of carriage and insurance. Steel-hull shipping, compound and triple-expansion steam engines, and laying of submarine telegraph cable between Europe and Southeast Asia eased the rice trade between rice growing countries of the East and the European and other Western markets. Establishment of rice futures markets in Rangoon contributed further to the growth in trade as the 19th century drew to close.

Thus, in the 19th century the United States, once a great rice exporter to Europe and other importing countries, turned a major importer and remained so until well into the 20th century, as Table 15.6 demonstrates. According to Coclanis, this occurred despite increased duties imposed on imported rice.[64] The traditional rice growing states namely South Carolina and Georgia, which at one point were extremely wealthy due to booming rice industry, resulted in becoming one of the poorest parts of the country.

During the late half of 19th century, the rice cultivation centered on monsoon Asia—India, Burma, the Malayan peninsula, Java, Siam, Indo-China, China and Japan. In the early 1860s, Bengal became the chief rice exporting region and retained this position till 1867-68 when Burma took over its position as chief exporter of rice.[65] Ceylon was the major importer of Bengal's rice through the port of Calcutta. Burma rice was mainly exported to Britain from where it was re-exported to Europe mainly for non-food usage such as making of starch and spirit. On the other hand,

Table 15.6. Rice trade in the United States, 1875 to 1910[63]

Year	Imports (million pounds)	Exports (million pounds)
1875	59.4	0.3
1880	57.0	0.2
1885	81.1	0.2
1890	68.4	0.4
1895	141.3	0.1
1900	93.6	2.9
1905	43.4	4.9
1910	82.7	7.0

Source: Coclanis (1993).

rice from Siam and Cochin China was not able to move further to the West, except Japan, China, Java, and the Straits Settlements, due to quality deterioration in the long voyage. Though Burma's rice was usually used for non-food purposes in Europe, during the natural calamities and at the time of scarcity, it formed a natural reservoir for the entire rice-consuming regions of the world. For example, in the time of great scarcity of rice in China in 1863 to 1865, huge quantities of rice from Bengal and Burma were exported to China for human consumption.[66]

The region of Indo-China and Malay, thanks mainly to about 300 years of European political and mercantile penetration, came on the map of world rice trade during 19th century. The rice production no longer merely supplied the staple food for the local population, but became a major source of revenue through export duties and tariffs. Singapore emerged as the key distribution centre in the world rice trade. Rice from Siam, Burma and French Indo-China was first brought to Singapore for further distribution to the Dutch East Indies and the Malay Peninsula.[67] For the Far East, Hong Kong acted as a major redistribution centre; however, there is lack of statistics for this centre.[68]

The above description presents the dominant rice trade pattern prevailing in the second half of the 19th century and to some extent for early part of 20th century. Siam and Indo-China were exporting their rice to China and the Far East. On the other hand, Bengal was a major supplier for the regions to the west of the Malayan Peninsula under normal conditions. Burmese rice continued to meet the demand of European markets for non-food purposes, except during famines and other natural calamities when it was supplied to Asian countries.

Twentieth Century (First Half)

From the beginning of the 20th century, the Asian markets, dominated by India and Ceylon, started to takeover as the more important market destination for Burma's rice, which till the end of 19th century was mainly serving non-food demands of Europe. So much so that by 1910, supply of rice from Burma to the Asian markets exceeded the supply from Burma to the whole of Europe. One factor which might have led to shift in this pattern was that Burmese rice, instead of being put to non-food purposes such as source of starch and cheap alcohol, started becoming a staple food.

In the period of de-globalization between 1914 and 1950, the export of rice witnessed many upheavals. The damage and disintegration caused by two World Wars, the economic depression in the 1930s and volatility of international markets throughout the 1920s, clearly influenced the domestic rice policies resulting in inconsistent patterns in the global rice trade.

Before the World War I, rice exports from Burma, Siam and French Cochin-china accounted for over 90% of the total world rice export. Table 15.7 illustrates that Burma with 2.3 million tons of rice export during 1911-1914 contributed 62% to the total export from Southeast Asia followed by 22% by French Cochin-china and 16% by Siam.[69] Of the total exports, around 60% was imported in East Asia while the remaining in India, Europe, and some other destinations. Burma, in comparison to Siam and Cochin-china, was exporting much greater proportion of its rice to Germany, Austria-Hungary, Turkey, and Holland. All these countries were closed to British trade during the war years and hence, when government controls and trade embargos were imposed in the post war period, Burma was significantly affected.[70]

Table 15.7. Distribution of Southeast Asian rice exports 1911-1914[71]

	East Asia	Other	Total
Burma	1188 (49.7)	1201 (50.3)	2389
Siam	564 (80.5)	137 (19.5)	701
French Cochin-china	639 (73.7)	228 (26.3)	867
Total	2391 (60.4)	1566 (39.6)	3957

Source: Kratoska (1990).

Southeast Asian Rice Crisis of 1919-1921

Kratoska (1990) gives detailed description of the rice crisis faced in Southeast Asia during 1919 to 1921. Southeast Asia faced unprecedented rice shortages from 1919 through 1921 due to poor rice harvests and speculative buying. Governments in Burma, Malaya, East Coast Residency of Sumatra, and India were forced to impose embargo on rice exports and other controls so as to keep the domestic rice prices within reach of local consumers. Malaya, East Coast Residency of Sumatra and Ceylon, where large scale labour force was involved in production and export of non-edible and edible primary products,[72] were major importers of rice from Burma, Siam and Indo-China for meeting food demands of the labour force. Until the First World War, rice in these exporting countries was readily available at relatively inexpensive prices; therefore, the importing Southeast Asian countries did not face much problem in meeting their rice demands as Britain's naval dominance in the Indian Ocean and South China Sea ensured free flow of trade.[73]

In 1919, realizing that production of rice would not match the demand, controls were imposed on the rice industry in Burma; and in 1920 rice trade throughout British Empire in Asia was put under controls. Through these controls the governments aimed at maintaining domestic prices and increasing output. However, due to peculiar features of the rice industry, administrators faced many problems in enforcing comprehensive systems of controls. Large number of small rice cultivators, dominance of Chinese rice traders through networks, and conflicting goals of controls system made imposition of controls cumbersome and the costs of enforcement very high.[74] During the war years, Chinese merchants, by offering advances on more advantageous terms and by allowing cultivators to sell at the time of their own choosing, increased their share of the rice trade in the Burmese rice industry. The main source of profit for advancers and speculators lay in seasonal variations of rice prices because cultivators lacked storage space or financial resources to hold their crops off the market during the harvest season leading to glut in the market.

The 1919 embargo on export of Burmese rice by the Government of India, which had administrative control over Burma, prohibited direct shipment of rice from Burma to Sumatra. However, traders were free to send rice to Delhi through Penang. The major importing countries which suffered extensively because of trade embargo in Burma were Straits Settlement, Malaya Peninsula, and Ceylon. Owing to heavy demand in India, exports to Strait Settlements and Malaya were curtailed by about 50% in 1919.

With the control on Burmese rice, demand for Siamese rice went up leading to tripling of its price[75] while Burmese rice, sold at controlled prices, was often unavailable. In fact, in June 1919, riots broke out in Penang and Province Wellesley over high cost of rice. In response to this all importation and internal distribution of rice in Malaya was brought under government control using war power legislation passed in 1917. Similarly, in Siam also, heavy international demand of rice had driven prices up to unprecedented levels—more than six times the normal prices —forcing the government there to increase the money supply by more than 90% within one year. In July 1919, Siamese government prohibited the export of rice except under license by issuing a Royal Decree.

Rice production in India during this period was not sufficient to meet its domestic demand, but in normal years India exported several hundred tons of rice to European markets, and in turn imported cheaper, lower quality rice from Burma. In 1911 and 1912, India exported net 340 thousand tons of rice every year, but 1913 onwards, the country became net importer of rice. In 1919, India also imposed an embargo on the export of rice from the country with the logic that this would make India largely independent of supplies from Burma.

Under a new system of control put in place in Burma in 1920 to deal with the importing countries on a government-to-government basis, the importing countries were granted allotments and then submitted tenders subject to a minimum price.[76] As per this new scheme, Burmese government was to supply rice to India at cost of purchase from Burmese millers and at concessionary rates to territories elsewhere within British Empire. Ceylon objected to this arrangement as it was to pay a price which was 75% higher than what India was paying. The Viceroy argued that Burmese farmers should not be expected to subsidize other colonies. However, with interventions from Colonial Office with the argument that imported rice in Ceylon met food needs of large Indian population there, the price was brought down by 20%. Request from Malaya, that it also should be supplied rice at the same concessionary rate on the ground that the country was facing depressed prices for its principal exports, tin and rubber, was refused. There was no control on the domestic prices assuming that maximum price for export sales would hold down prices within the country. However, market prices of rice in European markets at that time were over three times the controlled prices within Burma, making the scheme financially very attractive. Some forecasts indicated that Government of India stood to realize profits in the range of 25-30 million pounds by exporting surplus rice to the European markets. This profit, after long negotiations, was to be given to the Burmese government for investment in agricultural projects.

In 1921, Burmese rice production went up by 20% as compared to that in 1920. Realizing that trade embargo and government controls were working against the interests of Burmese consumers as well as the farmers, the Government of India, in December 1921, lifted all restrictions on the export of rice from Burma. By 1922, availability of rice became abundant and the prices went down. Learning from the rice shortage during this period, Japan started depending on rice imports from within the Japanese empire.[77] After 1920, Japan's export of rice from Korea and Taiwan went up substantially and from 1928 onwards, very little supplies in Japan came from sources outside the empire.

World Rice Markets during and after Great Depression of 1930s

The Great Depression of 1930s, which lasted for almost three years starting in 1929, caused heavy fall in prices of primary products especially staples including rice. Starting in 1929, prices of most staples started showing decline. Between 1929 and 1934, rice price in India declined by 52%[78] compared to 55% for oil seeds and 51% for raw cotton. Since 1931, South India had been importing vast quantities of cheap rice from Siam and French Indo-China. Influx of this rice further depressed the already declining prices of rice in the domestic markets. Madras Presidency made representation which led to imposition of duty on rice imports.

Wide disparity between prices of primary products and finished goods affected the barter terms of trade. This resulted in increased numbers of rent disputes and high debt burden for the farmers. In order to avoid worsening of situation, government made partial remissions of land revenue.

There was phenomenal increase in area under rice throughout South-east Asian countries since the end of 19th century. However, not much effort was made to improve marketing of rice or to introduce new techniques and technology for rice cultivation in these rice producing countries. Though the rice growing peasants were hard-working, they could not reap adequate benefits of their hard work because of backwardness in their techniques of rice cultivation. Consequent to market depression and government protectionism in the deficit countries, the rice exporting countries realized the importance of technical progress in improving the quantity and quality of rice.[79] Many countries put in place institutional mechanism or improved the existing systems to focus on research and development in rice cultivation and marketing. In India, the Imperial

Council of Agricultural Research was asked to focus their research efforts on improvement of rice production techniques and marketing methods in the country.[80]

Southeast Asian rice producing countries dominated the world rice market up to World War II with 80-90% world rice exports with 23% being intraregional trade during interwar years.[81] Until World War II Thailand, Indo-China and Burma dominated as exporting countries and most exports within Asia were to India, China, Hong Kong and Japan. In transition from subsistence agriculture to commercial agriculture in Southeast Asia, cash loans to the farmers were extended by Indian Chettyars and Chinese moneylenders.[82] Chinese merchants dominated the rice trade as middlemen. However, due to adverse impacts of large scale borrowings on local rice peasants, the Thai government in late 1930s started imposing restrictions, including export licenses, in order to reduce dependence of rice cultivators on Chinese middlemen. So much so that a domestic company, the Thai Rice Company which was established in 1938, was given preferential rates on government owned rail-roads and subsidies in marketing and storage facilities. In Indo-China, the Chinese merchant did face breakdown during the depression period, however, they quickly regained their positions after 1934. In contrast to Thailand, the governments in Indo-China and Burma did not introduce market reforms to reduce dependence of their rice cultivators on Chinese borrowings.

In 1939, Japan and Korea faced drought, which coupled with other economic factors, forced Japan to look beyond its empire for supply of rice. With the control of Indo-Chinese region[83] coming under Japan from France, Japan got control over about one fourth of the annual rice surplus produced by countries of the Indo-Chinese and Malay Peninisula (known as Further East to European traders). In addition, Japan got a priority claim on rice in Thailand because of its loan agreement with Thailand.

World Rice Markets during and after World War II

During World War II, apprehending the rise in prices and supply shortages of rice, trade restrictions were imposed by most of the Southeast Asian rice producing countries. The first country to impose restrictions was British Malaya, the single rice importing country of the region. Burma and Thailand followed suit. However, Indo-China could manage its trade in rice during war years without any restrictions because of its special relations with Japan and China.[84] The Thailand's rice market was fully monopolized by Japan.

During war years, production of rice throughout 'Further Asia' remained more or less constant. Loss of European markets and shortage of international shipping during this period did not adversely affect the rice markets in Southeast Asia due largely to increasing demands from Japan and China.[85] However, after the war, exports from Indo-China and Burma declined while Thailand succeeded in maintaining the volume of its exported supplies.

Despite the postwar expansion of the world market, Southeast Asia's share in world trade kept declining after the World War II up to the late 1970s, when Thailand started expansion of its rice trade.[86] There are many complex reasons for this decline. Cultivation of rice no more retained its high profitability due to heavy taxations on rice exports, especially in Burma and Thailand, the two major Southeast Asian rice exporting countries. The change in food preferences from rice-based food items to wheat-based food items at the global levels also contributed to decline in rice export from the southeast region. In addition, technological innovations led to higher productivity and profitability of wheat cultivation over rice. Increased production of wheat followed by lower prices led to replacement of rice on cereal markets outside Asia.[87]

Hence, the competitive advantage the Southeast Asian rice producing countries enjoyed up till the end of the war period started getting eroded in the post-war period. Developments in mechanization, new irrigation projects, intensification of rice cultivation and improved methods of plant breeding in the western world started to affect the comparative advantage in new and profound ways.[88]

Twentieth Century (Second Half)

Though rice markets have been in existence for centuries, rice accounted for a small fraction of the total trade in cereals (under 10%) even in the second half of 20th century. Most rice producing countries continued to domestically consume major portions of their rice production due to increase in population coupled with improved purchasing power, leaving very small portion for export to deficit countries.

In the second half of the 20th century, the world rice market continued to be characterized by thinness, volatility, segmentation, geographic concentration, and above all, market distortions. The market remained unstable for much of this period with prices volatile[89] and supplies uncertain. However, the world rice market underwent continuous evolution and change on many accounts during this period.[90] These changes have affected both the level and the variability of world rice prices. Major area

of concern for most of the rice producing countries had been price stabilization in their domestic markets. The countries adopted different policies
to curtain off their domestic markets from the impact of changing world
prices.

Rice Production during Second Half of Twentieth Century

Table 15.8 shows top ten rice producing countries in the world from
1961 to 2006. Total production of these ten top rice producing countries
accounted for around 87% of the world rice production. Such high share
of the top ten countries indicates that rice production continued to be
geographically concentrated, especially in South Asian countries. With
over 50% combined share in world rice production, China and India were
consistently the top rice producers in this period. On regional level, it was
the Asian region which dominated the world rice production accounting
for over 90% of the world production throughout this period (Table 15.9).
America, with around 5%, and Europe, with less than 1% shares in the
total world rice production, lagged far behind the South Asian rice
producing countries in terms of total production. Though total yearly
production of rice in the world increased from 264.5 million tons during
1961-70 to 604.08 million tons during 2001-06, the Decadal Annual
Compound Growth Rate (ACGR) of world rice production continued to
consistently decline between 1961 and 2006. ACGR, which was 3.80 for
the decade ending 1970, fell to 2.66 in the next decade and during 2001-
06 it further declined to 1.86.

Share of export in total production in each of these ten countries was
still meager, except in Thailand which exported 37.26% of total yearly rice
production in the country during 2001 to 2005. For remaining countries,
this figure remained below 5%, with Bangladesh consuming almost all its
rice production domestically. Total yearly export of rice from these
countries during 2001 to 2006 was 13.5 million tons of milled rice, 3.8% of
their total yearly production in milled rice equivalent.

Rice Trade during Second Half of Twentieth Century

Ten leading rice exporting countries with their average annual export
volumes and respective share in world rice export are given in Table 15.10.
The export of rice continued to be highly concentrated, most of it from
South Asian countries. Total yearly export of milled rice from these ten
countries was 6.29 million tons during 1961-1970 which tripled to
18.85million tons during 2001 to 2005. Throughout the second half of 20th

Table 15.8. Top ten leading producers of rice in the world, 1961 to 2006 (million tons)[91]

1961-70		1971-80		1981-90		1991-2000		2001-06	
Country	Production	Country	Production	Country	Production	Country	Production	Country	Production
China	87.88 (33.2)	China	130.58 (37.1)	China	173.30 (37.2)	China	190.96 (34.1)	China	177.43 (29.4)
India	54.94 (20.8)	India	68.90 (19.6)	India	92.84 (19.9)	India	121.67 (21.7)	India	129.87 (21.5)
Japan	17.10 (6.5)	Indonesia	23.42 (6.6)	Indonesia	39.02 (8.4)	Indonesia	49.00 (8.8)	Indonesia	52.76 (8.7)
Bangladesh	15.80 (6.0)	Bangladesh	18.06 (5.1)	Bangladesh	23.11 (5.0)	Bangladesh	29.12 (5.2)	Bangladesh	38.66 (6.4)
Indonesia	14.34 (5.4)	Japan	15.36 (4.4)	Thailand	19.07 (4.1)	Viet Nam	25.95 (4.6)	Viet Nam	34.82 (5.8)
Thailand	12.07 (4.6)	Thailand	14.93 (4.2)	Viet Nam	15.93 (3.4)	Thailand	22.13 (4.0)	Thailand	27.95 (4.6)
Viet Nam	9.24 (3.5)	Viet Nam	10.89 (3.1)	Myanmar	14.01 (3.0)	Myanmar	17.38 (3.1)	Myanmar	23.69 (3.9)
Myanmar	7.74 (2.9)	Myanmar	9.50 (2.7)	Japan	13.44 (2.9)	Japan	12.34 (2.2)	Philippines	14.03 (2.3)
Brazil	6.38 (2.4)	Brazil	7.85 (2.2)	Brazil	9.48 (2.0)	Philippines	10.50 (1.9)	Brazil	11.49 (1.9)
Republic of Korea	5.01 (1.9)	Republic of Korea	6.69 (1.9)	Philippines	8.65 (1.9)	Brazil	9.89 (1.8)	Japan	10.85 (1.8)
World Total	264.53	World Total	352.34	World Total	466.40	World Total	559.88	World Total	604.08

Note: Figures in parentheses are percent share of world rice production.
Source: FAOSTAT (http://faostat.fao.org/default.aspx)

Table 15.9. World rice production by regions, 1961 to 2006 (million tons)[92]

	1961-70	1971-80	1981-90	1991-2000	2001-06
Africa	6.06	7.89	10.24	15.48	18.65
America	13.35	18.62	23.91	28.28	33.76
Asia	243.33	323.58	429.41	512.00	547.62
Central Asia	0.00	0.00	0.00	0.70	0.62
Eastern Asia	112.05	155.37	196.70	213.08	197.20
Southern Asia	77.71	96.75	128.13	165.73	186.27
South-Eastern Asia	52.48	69.12	101.75	131.79	162.83
Western Asia	0.44	0.45	0.46	0.50	0.70
Europe	1.59	1.79	2.07	2.97	3.31
Oceania	0.21	0.46	0.77	1.14	0.74

Source: FAOSTAT (http://faostat.fao.org/default.aspx)

century, almost 90% of world rice export was from these ten leading exporting countries, which illustrates the kind of concentration the rice market had been during this period.

For most of the second half of 20th century, the major rice exporters in the world market have been the nations of South Asia: Thailand, China, India, Burma, Pakistan, and Vietnam. During the 1950s, Burma and Thailand dominated world rice exports, Cambodia also being an important player.[93] From 1960 onward, Pakistan emerged as another dominant rice exporting country in South Asia. In Europe, Italy and Egypt also became major players in the world rice market, Uruguay joined in 1990s. The US retained its position as a major rice exporting country. More important, for leading export countries, overall share of exports as a percent of domestic production was not very high throughout 1960 to 2005, indicating thinness of rice market. The thinness and concentration of world rice markets implied that changes in production or consumption in major rice-trading countries would have an amplified effect on world prices. In general, share of export in total domestic production was relatively high in countries of the West: USA, Italy, Egypt, and Australia. In contrast, though maximum concentration of rice production was in South Asian region, share of export in domestic rice production continued to be low due to high population and rice being the major staple in the food basket for majority of the population. For example, China was the leading producer of rice throughout the second half of 20th century, but its export did not constitute more than 2% of its domestic production. However, Thailand had been able to export fairly large chunk of its domestic production—17% during 1961 to 1970 and 36% during 2001 to 2005. Pakistan

Table 15.10. Top ten leading rice exporting countries, 1960-2005 (000' tons)[94]

1961-70			1971-80			1981-90		
	Export	Share of export in total production (%)		Export	Share of export in total production (%)		Export	Share of export in total production (%)
Thailand	1391.95 (20.11)	16.96	Thailand	1824.90 (22.25)	17.98	Thailand	3999.38 (36.77)	30.84
China	1316.41 (19.02)	2.20	China	1736.30 (21.17)	1.96	USA	1907.77 (17.54)	42.85
USA	1220.60 (17.64)	49.86	USA	1540.98 (18.79)	44.17	Pakistan	1066.65 (9.81)	31.83
Myanmar	1097.86 (15.86)	20.86	Pakistan	687.69 (8.39)	24.72	China	895.32 (8.23)	0.76
Egypt	441.32 (6.38)	31.05	Myanmar	483.77 (5.90)	7.49	Italy	473.71 (4.36)	64.03
Pakistan	270.51 (3.91)	17.00	Italy	327.82 (4.00)	51.95	Myanmar	456.47 (4.20)	4.79
Cambodia	254.07 (3.67)	13.99	Egypt	223.66 (2.73)	13.83	India	393.09 (3.61)	0.62
Italy	114.90 (1.66)	24.85	Korea, DPR	215.42 (2.63)	4.74	Viet Nam	264.43 (2.43)	2.44
Singapore	98.73 (1.43)	na	Australia	179.98 (2.19)	60.29	Australia	233.18 (2.14)	46.59
Guyana	87.08 (1.26)	57.69	Japan	129.75 (1.58)	1.24	Korea, DPR	178.00 (1.64)	11.84
Total	6293.43 (90.94)	7.76	Total	7350.26 (89.63)	5.69	Total	9868.00 (90.72)	4.39
World Total	6920.43		World Total	8200.57		World Total	10876.86	

Contd.

Table 15.10 continued

	1991-2000		2001-05	
	Export	Share of export in total production (%)	Export	Share of export in total production (%)
Thailand	4863.14 (30.88)	32.32	Thailand	6809.17 (33.89)	35.83
India	2090.01 (13.27)	2.53	India	3794.49 (18.89)	4.30
USA	1696.83 (10.77)	30.52	Pakistan	2114.99 (10.53)	42.60
Pakistan	1539.55 (9.78)	36.78	USA	1794.08 (8.93)	27.36
China	1516.99 (9.63)	1.17	China	1540.35 (7.67)	1.28
Viet Nam	1084.35 (6.89)	6.15	Viet Nam	1139.79 (5.67)	4.81
Italy	505.22 (3.21)	55.40	Egypt	559.44 (2.78)	13.53
Australia	459.71 (2.92)	60.15	Italy	537.97 (2.68)	56.31
Uruguay	338.82 (2.15)	57.54	Uruguay	386.96 (1.93)	51.31
Argentina	250.19 (1.59)	40.93	UAE	177.33 (0.88)	na
Total	14344.80 (91.08)	5.56	Total	18854.57 (93.85)	7.01
World Total	15749.09		World Total	20089.43	

Note: Figures in parenthesis are percent share of world rice export.
Source: FAOSTAT (http://faostat.fao.org/default.aspx)

performed exceptionally well in the rice export market since 1960 with its share of export as percent of domestic production rising consistently from 17% during 1961 to 1970, to as high as 42% during 2001 to 2005 (Table 15.10).

During this period, agriculture dominated the economy of leading rice producing countries in South Asia, and rice dominated the agricultural sector. Furthermore, since some of these countries exported large volumes of rice to world markets, these countries needed to be commercially oriented and participate actively in the world rice market as reliable suppliers. Not only was rice an important share of the economy, but it was also a key source of foreign exchange earnings and government revenue.

In contrast to rice export market, the import side had comparatively more players. In the export market, top ten exporting countries accounted for over 90% of world export, while the ten leading importing countries imported less than 50% of total rice imports in the world (Table 15.11). Import patterns indicate prevalence of wide geographical spread as well as inconsistent behavior of the importing countries. Indonesia was the major importer of rice during most of the period of second half of 20th century. During 1971 to 1980, its mean yearly import was 1.4 million tons which fell to 0.7 million ton during 1981 to 1990, rising to 1 million ton during 1991 to 2000. Vietnam was another Southeast Asian country which had very high imports till late 1970s. Bangladesh was consistently importing rice, mainly from India, to meet its domestic food demands. In 1990s and 2000, African countries—Nigeria, South Africa and Cote d'Ivoire—emerged as major importers of rice from the world rice market with rising rice consumption, primarily in urban areas. The case Cote d'Ivoire is illustrative of this point. Between 1984 and 1989, the country spent an average of U.S. $93.7 million in rice imports which increased to U.S. $107 million in 1997, making Cote d'Ivoire the largest rice importer in West Africa.[95] In the same period, Middle East also became major destination for world rice supplies.

Rice Market Distortions and Trade Negotiations

In the post-World War II, many countries, particularly the war-torn economies of Western Europe, were primarily concerned with the issues of economic recovery. An urgent need was being felt to liberalize international trade and to correct the large overhang of protectionist measures which were introduced by different nations in 1930s and remained in place since then. In January 1948, the General Agreement on Trade and Tariff (GATT) came in force to ensure free trade across the world as a

Table 15.11. Top ten leading rice importing countries, 1960-2005 (000' tons)[96]

1961-70	Import	1971-80	Import	1981-90	Import	1991-2000	Import	2001-05	Import
Indonesia	715.18 (10.72)	Indonesia	1397.51 (17.50)	China	688.24 (7.00)	Indonesia	1004.47 (7.17)	Nigeria	1226.76 (6.88)
Viet Nam	709.38 (10.63)	Viet Nam	454.61 (5.69)	USSR	503.29 (5.12)	China	778.95 (5.56)	Iran	941.77 (5.28)
India	506.35 (7.59)	China	405.55 (5.08)	Iraq	472.93 (4.81)	Saudi Arabia	768.71 (5.49)	Philippines	837.15 (4.69)
Sri Lanka	459.28 (6.88)	USSR	376.10 (4.71)	Saudi Arabia	419.46 (4.27)	Unspecified	747.35 (5.34)	Saudi Arabia	821.51 (4.61)
China	412.21 (6.18)	Bangladesh	321.04 (4.02)	Nigeria	368.09 (3.75)	Iran	589.17 (4.21)	China	782.46 (4.39)
Bangladesh	339.17 (5.08)	Sri Lanka	319.53 (4.00)	Senegal	347.83 (3.54)	Brazil	527.36 (3.77)	Bangladesh	775.04 (4.34)
Japan	331.75 (4.97)	Malaysia	225.47 (2.82)	Côte d'Ivoire	329.09 (3.35)	Malaysia	484.39 (3.46)	South Africa	702.68 (3.94)
Malaysia	294.39 (4.41)	Senegal	215.67 (2.70)	Malaysia	313.43 (3.19)	Philippines	471.11 (3.36)	UAE	597.31 (3.35)
USSR	273.41 (4.10)	Cuba	209.78 (2.63)	Indonesia	388.43 (2.94)	Bangladesh	453.72 (3.24)	Unspecified	576.76 (3.23)
Singapore	256.36 (3.84)	Nigeria	205.91 (2.58)	Bangladesh	255.45 (2.59)	UAE	442.63 (3.16)	Côte d'Ivoire	566.58 (3.18)
Total	4297.48 (64.41)	Total	4131.16 (51.74)	Total	3985.25 (40.55)	Total	6267.85 (44.77)	Total	7828.02 (43.88)
World	6671.77	World	7985.13	World	9827.00	World	14001.47	World	17838.56

Source: FAOSTAT (http://faostat.fao.org/default.aspx)

solution to the problems of postwar unemployment and economic instability. The scope of this trade treaty was significantly less than the aborted International Trade Organization (ITO), originally conceived by the US. Issues like employment, development, restrictive business practices, commodity agreements or labor standards were not covered under GATT. The treaty limited itself to broad measures among governments and did not concern itself with domestic practices.

Under the auspices of GATT, multilateral trade negotiations, or 'trade rounds', started to be organized on various issues related to international trade liberalization. GATT did not mention agriculture specifically, except at few places, as a sector requiring special treatment due to its unique features. This meant that agriculture trade was to be treated like trade in other goods. However, discussions on agriculture trade were brought in all successive 'trade rounds'. The European Community, in the early sixties, put together its Common Agriculture Policy (CAP) aimed at self-sufficiency in food production. The Community successfully achieved the goal of self sufficiency through adoption of very generous support mechanisms, first at the level of production and then by subsidizing sale on the world market. The market forces did not allow withdrawal of supports even after self-sufficiency was achieved. The United States, Canada and Japan also started following the similar path of supporting their farming communities. In order to counter European Community for their trade practices, the United States instituted the Export Enhancement Program (EEP) which subsidized U.S. exports of several commodities.[97]

In the series of 'trade rounds' held under the auspices of GATT, the Uruguay Rounds (UA), initiated in 1986, were the most extensive. It took almost eight years before these rounds concluded in 1994 with establishment of World Trade Organizations (WTO). One major reason that it took so long for the countries to conclude negotiations was the difficulties in reaching an agreement on agriculture. It was the round in 1994 which finally brought agriculture in forefront of GATT by means of Agreement on Agriculture (URAA). The basic provisions of URAA, applicable to agricultural commodities including rice, are: reductions in domestic support, improvements in market access and reductions in export subsidies.[98]

Despite various provisions under URAA for creating level playing field for the rice exporting and importing countries, the world rice market continued to be highly distorted throughout the second half of 20th century, partly because of the high degree of domestic support, government market interventions and export subsidies.

Comparison of taxation or protection policies in rice producing developing countries with those in the developed nations indicates very high divergences (Table 15.12). Poor countries of South Asia such as Thailand, Vietnam, and India were following the policies of dis-protection of their rice sectors, the rich countries of East Asia (Japan and Korea), Europe, and the United States had been heavily supporting their rice producers.[99] Japan provided support to the extent of 80% of the total value of rice produced in the country. European Union and the USA reduced their levels of support to rice between 1986 and 2002, but these were still very high. Developing countries, on the other hand, were imposing taxations on the rice industry. Data on Product Support Estimates (PSE)[100] for rice in Table 15.13 shows that up to 1997, India, for most of the period, was following the policy of disprotection of its rice producers. However, the situation changed after 1997, when India adopted protectionist interventions in its domestic rice market. Vietnam transitioned from communist disprotection of rice producers to providing net support to the sector. Indonesia, on theother hand, provided consistent support to rice in the period 1985-2003 with few exceptional years in which it disprotected the rice producers.

Table 15.12. Producer support estimates (%) for rice in OECD[101]

Country	Value (million US $)			% value of output		
	1986-88	1995-97	2002-04	1986-88	1995-97	2002-04
Rice						
Australia	11	11	5	17	5	6
European Union	471	368	389	60	34	35
Japan	199,68	23,987	16,398	84	81	83
United States	868	312	677	52	15	33
Total Agriculture						
Australia	1,321	1,306	1,068	8	6	4
EU	101,672	117,615	114,274	41	34	34
Japan	48,976	59,269	46,924	61	57	58
United States	36,390	26,304	40,409	22	12	17
Support to Rice as % of Total Support to Agriculture						
Australia	0.8	0.8	0.5			
EU	0.5	0.3	0.3			
Japan	40.8	40.5	34.9			
United States	2.4	1.2	1.7			

Source: OECD, PSE/CSE database 2004.

Table 15.13. Producer support estimates (%) for rice
in South Asia, 1985 to 2003[102]

Year	China	India	Indonesia	Vietnam
1985		-4.5	35.5	
1986		17.9	34.6	-16.6
1987		12.9	6.2	20.3
1988		-14.4	-3.4	67.9
1989		-15.0	-9.5	9.0
1990		-11.5	8.1	-60.0
1991		-2.8	7.1	-34.9
1992		-33.6	15.1	-27.3
1993		-4.3	20.7	-27.2
1994		-35.0	23.0	-26.7
1995	-14.9	-5.8	18.1	-23.7
1996	-33.6	-52.5	13.4	4.8
1997	-5.1	-22.7	14.2	17.1
1998	3.3	6.8	-72.1	18.6
1999	-10.8	-3.9	20.0	29.0
2000	4.4	22.1	22.6	27.6
2001	11.3	10.0	21.4	31.0
2002		28.6	35.8	15.1
2003			37.2	

Source: Orden *et al.* (2007).

These various types of supports provided at different stages of production to trade have led to trade distortions under WTO regime.

In 1990s, intellectual property rights became a key issue in almost all economic activities including agriculture. However, despite a number of multilateral treaties on Intellectual Property Rights finally culminating in TRIP, there were number of cases of violations of IPR not only by developing nations but by some developed ones too. In case of rice, there were many controversies over ownership of certain varieties. Basmati rice in particular, has been a controversial issue of intellectual property. This variety is an aromatic variation of rice, grown mostly in India and Pakistan.[103] It has been under cultivation in foothills of Himalayas for centuries. Today, Basmati rice forms major share of the rice exported from India and Pakistan throughout the world where countless number of people enjoy its characteristic flavor and excellent cooking quality. This is also a source of export revenue for India.

482

No one imagined that a grain which formed the staple food for more than two thirds of the world population, would be an object of "intellectual property" within the framework of the Agreement on Trade-Related Aspects of Intellectual Property Rights (TRIPS) in 1990s.

References

1. First Author.
2. Chandola (2006).
3. UNDP (1997).
4. Coclanis (1993).
5. Commodity Trade and Price Trends (1981).
6. Falcon and Monke (1979/80).
7. Falcon and Monke (1979-1980), Monke and Pearson (1991), Siamwalla and Haykin (1983).
8. Jayne (1993).
9. Siamwalla and Haykin (1983) characterize a thin market as one with relatively high transaction costs, more specifically, with high search costs.
10. Introduction of High Yielding Varieties (HYVs) in the major rice growing countries in 70s and 80s has further contributed to making it more of a residual market.
11. Siamwalla and Haykin (1983), Dawe (2002).
12. Siamwalla and Haykin (1983) report brokerage fee in rice as high as 5 or 10% which is not the case in wheat world trade.
13. For sugar, there are London daily prices or New York world market price. Similarly, for wheat, maize and soybean, we have Chicago Board or Trade futures prices.
14. Siamwalla and Haykin (1983).
15. Siamwalla and Haykin (1983).
16. Dawe (2002).
17. Dawe (2002).
18. Dawe (2001).
19. Timmer and Falcon (1975).
20. Commodity Trade and Price Trends (1981), Washington DC: IBRD (1981).
21. Siamwalla and Haykin (1983).
22. Johnson (1982).
23. Chang (1985).
24. Ancient History Sourcebook: The Periplus of the Erythraean Sea: Travel and Trade in the Indian Ocean by a Merchant of the 1st Century (http://www.fordham.edu/halsall/ancient/periplus.html)
25. Dasgupta (2000).
26. Arasaratnam (1988).
27. Arasaratnam (1988).
28. Governor and Council of Ceylon to Governor General and Council, 24 November 1683, K.A. 1272 f121.

29. Chevalier (1932).
30. Carpenter (1978).
31. Carney (2001).
32. Tymowski (1971), Lewicki (1974).
33. McIntosh and McIntosh (1981).
34. Coclanis (1993).
35. Warmington (1974), Pounds (1973).
36. Bullio (1969).
37. Dollinger (1970).
38. Coclanis (1993).
39. Coclanis (1993).
40. Collinson (1766).
41. Dethloff (1982).
42. Morgan (1995).
43. Dethloff (1982).
44. Drayton (1808).
45. Dethloff (1982).
46. Harper (1935).
47. Dethloff (1982).
48. Source: Nash (1992).
49. Coclanis (1993).
50. Coclanis (1993).
51. Source: Adpated from Dethloff (1982).
52. Source: Coclanis (1993).
53. Source: Coclanis (1993).
54. Source: Coclanis (1993).
55. Coclanis (1993).
56. Coclanis (1993).
57. Andrus (1936).
58. Coclanis (1993).
59. Barbiero, "Reassessment of Agricultural Production in Italy." For statistics on Italian rice production and exports between 1861 and 1920, see Istituto Centrale di Statistica [ISTAT], Sommario di Statistiche Storiche dell'Italia 1861-1975 (Rome, 1976:76, 118).
60. Coclanis and Komlos (1987).
61. Coclanis and Komlos (1987).
62. Coclanis (1993).
63. Coclanis (1993).
64. Coclanis (1993).
65. Cotton (1874).
66. Around 369 thousand tons of rice from Bengal and 99 thousand tons from Burma were taken to China in these three years to meet the shortages (Cotton, 1874).
67. On the supply side, Siam was the predominant supplier followed by Burma and French Indo-China. While on the demand side, Dutch East Indies was the predominant consumer followed by Malay Peninsula (Latham and Neal, 1983).

68. Latham and Neal (1983).
69. Urickizer (1941).
70. Rangoon Gazette, 16 Feb. 1917, in NARS RG 166. Entry 5 Box 326 file Burma 1917-1933: Rice.
71. Kratoska (1990).
72. Non-edible primary products produced in these countries were tobacco, rubber and tin while edible products mainly included tea and coffee.
73. Kratoska (1990).
74. Kratoska (1990).
75. Young to S of S for Colonies, 28 Dec. 1918, in C0273/470/162384 (Gov.).
76. Tel. Rev. Secy, Burma, to SG, India, R&A 30 Dec. 1919.
77. Kratoska (1990).
78. Thomas (1935).
79. Thompson (1941).
80. Thomas (1935).
81. Van der Eng (2004).
82. Thompson (1941).
83. Indo-Chinese and Malay Peninsula region comprised three greatest rice exporting countries in the world—Burma, Indo-China and Thailand—and one importing country—Malaya.
84. Thompson (1941).
85. Thompson (1941).
86. Van der Eng (2004).
87. Van der Eng (2004).
88. Coclanis (1993).
89. Dawe (2002) identifies three distinct phases of price changes in world rice market in second half of the 20th century. From 1950 to 1964, prices were relatively stable; from 1965 to 1981, prices were substantially more variable; and finally, from 1985 to 1998, world prices were relatively stable once again. From 1999 to 2001, prices of rice plunged once again.
90. Dawe (2002).
91. Source: FAOSTAT (http://faostat.fao.org/default.aspx)
92. Source: FAOSTAT (http://faostat.fao.org/default.aspx)
93. Dawe (2002).
94. Source: FAOSTAT (http://faostat.fao.org/default.aspx)
95. Becker and Diallo (1996).
96. Source: FAOSTAT (http://faostat.fao.org/default.aspx)
97. Wells (1994)—*InterParks Digest*.
98. Yap (1996).
99. Gulati and Narayanan (2002).
100. According to Orden *et al.* (2007) aggregate measure of support (AMS) primarily included production-related (amber box) domestic support commitments under the WTO. PSE, on the other hand, is a broader measure of the transfers to farmers from border protection and domestic policy interventions. It is defined by the OECD as "an indicator of the annual monetary value of gross transfers from consumers and taxpayers

to agricultural producers, measured at the farm gate level, arising from policy measures that support agriculture, regardless of their nature, objectives or impacts on farm production or income" (OECD 2002:59). Thus the PSE spans all of the categories of support policies (amber, blue, and green boxes) reported to the WTO.

101. Source: OECD, PSE/CSE database 2004.
102. Source: Orden *et al.* (2007).
103. Chandola (2006).

16 | A Century of Rice Breeding, Its Impact and Challenges Ahead

Gurdev S. Khush and *P.S. Virk*

Rice, mankind's most important food crop has been improved since its domestication about 8000 years ago. Constant human selection for improved traits has modified domesticated rice varieties from their wild progenitors so much so that domesticated rices can no longer survive in the wild state. The simple acts of reaping and sowing, for example, are selective. Primitive humans may not have known it, but they started the first rice breeding programs when they began to grow rice plants for their own use. Most farmers have a keen eye and sensitive feeling for plants. Millions of rice farmers have applied this keen insight and sensitivity for thousands of years to select better varieties.

Selection was first practiced on the variable and heterogeneous wild and semi-wild populations, which must have narrowed genetic variability. However, several mechanisms in primitive agriculture, such as the introduction of varieties from one region to another and occasional natural crosses between local and introduced varieties enhanced variability for further selection. Natural crosses between the domesticated crop and the weed complexes were another source of variability. The third source of variability was the varietal mixtures that primitive agriculturists grew as a protection against disease epidemics. Occasional intercrosses between component varieties gave still more variability. This conscious and unconscious selection by humans led to the development of over 150,000 land race varieties which were grown around the world till the dawn of scientific plant breeding after the rediscovery of Mendel's laws of inheritance.[1]

Scientific rice breeding can be divided into three phases: (1) selection phase, (2) hybridization phase, and (3) green revolution phase.

SELECTION PHASE (1901-1949)

The selection phase of rice breeding started in early part of last century when rice breeding stations were established in most of the rice growing countries in Asia (Table 16.1). Subsequently the number of rice breeding stations increased greatly in order to develop rice varieties for ecoregional adaptation. Early selection work was limited to purification of land races by removal of off-types in the varieties popular with farmers. The next step was purification of land races through single plant or mass selection to develop pure line varieties. Thus a large number of pure line varieties such as Ptb varieties at Rice Research Station Pattambi (now in Kerala), Co varieties at Paddy Breeding Station Coimbatore and Adt varieties at Paddy Breeding Station at Adutturai, both in Tamil Nadu, were developed.

Table 16.1. First rice breeding stations established in different Asian countries

Country	Place	Year of establishment
India*	Dacca	1911
Pakistan	Kalashahkaku	1926
Burma	Mandalay	1907
Sri Lanka	Batalagoda	?
Malaysia	Krian	1915
Indonesia	Bogor	1905
Philippines	Maligaya	1902
Thailand	Rangsit	1916
Cambodia	Battambang	1928
Vietnam	Phu My	1920
Laos	Salakom	1956

*United India.

A limited amount of hybridization was carried out during this period. As an example variety Co 25 was developed during late 1920s from a cross between blast resistant variety Co 4 and a popular variety Korangu Samba.[2] Similarly, several varieties with regional adaptation in Indonesia were selected from a cross of Tjina and Latisail made in 1929 at Bogor.[3] However, most of the varieties developed before the Second World War were pure line selections. Breeders concentrated on selecting long duration varieties for the monsoon season which occupied the largest land area. Their perception was that longer duration varieties were better yielders.

Since no fertilizer was used, varieties with nitrogen responsiveness could not be developed. Breeder's effectiveness was limited by the variable conditions under which rice is grown, limited germplasm variability, inadequate research facilities, lack of trained personnel and failure to recognize the importance of interdisciplinary approach to crop improvement. A review of trends in the area, production and yield of rice from 1934 to 1960 shows that moderate increase in production in major rice producing countries of Asia was primarily due to increase in area planted to rice but the change in yields was negligible (Table 16.2).

Table 16.2. Annual growth in population, rice production, area and yield in different countries of Asia, 1934-38 to 1956-60

Country	Annual growth rate			
	Population	Rice production	Rice area	Rice yield
Japan	1.3	1.1	0.1	1.0
South Korea	1.8	0.5	-0.5	1.0
Taiwan	2.8	1.6	0.8	0.8
Burma	1.1	-0.6	-0.9	0.3
Cambodia	2.1	2.7	2.3	0.4
Sri Lanka	2.4	3.2	1.1	2.1
India	1.7	1.0	1.2	-0.2
Laos	2.5	2.5	1.8	0.7
Malaysia	2.3	2.0	0.9	1.1
Pakistan	1.2	0.8	1.0	-0.2
Philippines	2.1	2.1	2.0	0.1
Thailand	1.9	2.0	1.8	0.2

From Parthasarathy (1972).

HYBRIDIZATION PHASE (1950-1960)

Immediately after World War II, the shortage of food supplies and the eminent threat of population increase directed world attention towards finding ways to increase the production of most important staple food of Asia. The establishment of the International Rice Commission (IRC) in 1949 within the framework of FAO was the first effort in international cooperation in rice breeding.[4] The first meeting of the Working Party of IRC held at Rangoon in 1950 emphasized that the primary aim of rice improvement was increased yield. It was recognized that low yield was due to limitation of the varieties under cultivation, susceptibility to diseases and insects, long growth duration and lodging and narrow adaptation. These observations pointed out the need for breeding for early maturity and lodging resistance.

The nucleus of international cooperation started with the cataloging of major rice varieties of the world, the establishment of centers for maintaining their seed stocks and for exchange of seed. Two training courses on rice breeding were held at the Central Rice Research Institute (CRRI) at Cuttack, India. One or two trainees from most of the countries of tropical Asia participated in the courses held in 1952 and 1955. These courses focused on selection procedures, field plot techniques and principles of genetics and breeding. The IRC Working Party on rice breeding held eight meetings from 1950-1959. There was consensus amongst participants that *indica* rices were not responsive to fertilizer. Thus *indica-japonica* hybridization program was initiated to transfer nitrogen responsiveness from *japonica* rices to *indica* rices. All the countries of tropical Asia participated in this program by sending seeds of their best varieties for crossing with *japonica* rices at CRRI. *Japonica* parents were early and flowered in 58-70 days at Cuttack, while *indica* varieties took 95-100 days.

F_1 hybrids were grown at CRRI and F_2 seeds were sent to collaborating countries for generation advancement and selection. This project had a very limited success because; (1) breeders knew little about the type of plants to select, (2) no country except India conducted experiments to determine response of final selections to different fertility levels as compared to *indica* parents, and (3) early yield evaluation was done on single plant basis where tall *indica* types crowded and outyielded shorter segregants. Only in India and Malaysia early maturing nonseasonal commercial varieties derived from this project were released for cultivation. In India ADT27 was recommended for early monsoon, short growing season (Kuruvai) in Tanjore Delta of Tamil Nadu. In Malaysia, Malinja and Mashuri were released for irrigated second crop season. Later on Mashuri spread widely in rainfed areas of Andhra Pradesh and Bihar states of India, Terai region of Nepal and became popular in Bangladesh and Myanmar during 1970s and 1980s.

GREEN REVOLUTION PHASE (1960-2000)

As mentioned in the previous sections little progress was made in increasing the yield potential of rice. 1960s was a decade of despair with regard to the world's ability to cope with the food-population balance, particularly in the tropics. The cultivated-land frontier was closing in most Asian countries, while population growth rates were accelerating owing to rapidly declining mortality rates resulting from advancements in modern medicine and health care. International organizations and concerned professionals were busy organizing seminars and conferences to raise awareness regarding ensuing food crisis and to mobilize global resources to tackle the problem on emergency basis. In a famous book entitled *Times of Famine* published in 1967, Paddock brothers[5] predicted, "Ten years from now, parts of the underdeveloped world will be suffering from famine. In 15 years, the famine will be catastrophic and revolution and social turmoil and economic upheavals will sweep areas of Asia, Africa and Latin America."

Thanks to the widespread adoption of "green revolution" technology, large scale famines and social and economic upheavals were averted. Between 1966 and 2000, the population of densely populated low income countries grew by 95% but rice production increased 130% from 257 million tons in 1966 to 600 million tons in 2000. In 2000 the average per capita food-grain availability was 20% higher than in 1966. The technological advance that led to the dramatic achievements in rice production over the 40 years was the development of high yielding varieties of rice with following traits.[6]

Yield Potential

Increase in yield potential resulted from a reduction in plant height through incorporation of *sd1* gene for short stature. This led to improvements in harvest index (grain-straw ratio) as well as to increase in biomass production. Conventional varieties of rice are tall and leafy with weak stems and have a harvest index of 0.3 that is 30% grain and 70% straw. They can produce a total biomass of 10-12 tons. Thus their maximum yield potential is about 4 tons. When nitrogenous fertilizer is applied at rates exceeding 40 kg/ha, these varieties tiller profusely, grow excessively tall, lodge early and yield less than under lower fertilizer inputs. The improved varieties on the other hand have a harvest index of 0.5. Because of short stature, they are lodging resistant and responsive to fertilizer inputs. Their biomass can be increased to 18-20 tons/ha. Thus their maximum yield potential is 9-10 tons/ha. This improvement in the harvest index was the single most important architectural change in the rice varieties (as well as wheat) that led to doubling of their yield potential.[7]

Shorter Growth Duration

Most of the traditional varieties of rice grown in the tropics were photo-period sensitive and took 150-200 days to mature. They were suitable for growing a single crop of rice during monsoon season in Asia. New varieties on the other hand, are photoperiod insensitive and can be planted at any time of the year. Moreover, their growth duration has been reduced to 110-115 days. The availability of short duration varieties has led to increased cropping intensity. Many farmers now grow two crops of rice where only one was grown before or even two crops of rice and another upland crop in between. IR36, the first high yielding short duration variety was accepted on a very wide scale and was grown to about 11 million hectares during 1980s. Most of the improved varieties now grown in Asia and elsewhere mature in 110-120 days.

Multiple Disease and Insect Resistance

The varietal composition and cultural practices for rice have changed significantly during the green revolution era. Farmers have adopted improved cultural practices, such as application of more fertilizers and establishment of higher plant populations per unit area. Availability of short duration, photoperiod-insensitive varieties, coupled with the development of irrigation facilities, has enabled farmers in tropical Asia to grow successive crops of rice throughout the year. These conditions are conducive for the multiplication of disease and insect organisms. Several outbreaks of brown planthopper and viruses occurred during 1970s. Therefore major emphasis was put on incorporation of genes for disease and insect resistance into improved germplasm. In most of tropical and subtropical Asia, five diseases (blast, bacterial blight, sheath blight, tungro and grassy stunt) and four insects (brown plant hopper, green leafhopper, stemborers and gall midge) are of major importance. At IRRI major efforts were directed to develop germplasm with multiple resistances to diseases and insects. A large number of germplasm collections were screened and donors for resistance were identified.[8] Using these donors improved germplasm with resistance to four diseases and four insects was developed. First variety with multiple resistance, IR26 was released in 1973. Earlier IR varieties such as IR5, IR8, IR20, IR22 and IR24 were susceptible to most of the diseases and insects. All the IR varieties released subsequent to IR26 have multiple resistance (Table 16.3). These varieties have as many as 20 parents in their ancestry which has helped restore genetic diversity on farmer's fields. Large scale adoption of varieties with multiple resistance has prevented the occurrence of epidemics of diseases and insects.

Table 16.3. Disease and insect reaction of IR rice varieties

| IR variety | Reaction[a] | | | | | | | | | |
| | Blast | Bacterial blight | Grassy stunt | Tungro | GLH[b] | BPH[c] biotype | | | Stem borer | Gall midge |
						1	2	3		
IR5	MR	S	S	S	R	S	S	S	MS	S
IR8	MR	S	S	S	R	S	S	S	S	S
IR20	MR	R	S	MR	R	S	S	S	MR	S
IR22	S	R	S	S	S	S	S	S	S	S
IR24	S	S	S	S	R	S	S	S	S	S
IR26	MR	R	MR	MR	R	R	S	R	MR	S
IR28	R	R	R	R	R	R	S	R	MR	S
IR29	R	R	R	R	R	R	S	R	MR	S
IR30	MS	R	R	MR	R	R	S	R	MR	S
IR32	MR	R	R	MR	R	R	R	S	MR	R
IR34	R	R	R	R	R	R	S	R	MR	S
IR36	R	R	R	R	R	R	S	R	MR	R
IR38	R	R	R	R	R	R	R	S	MR	R
IR40	R	R	R	R	R	R	R	S	MR	R
IR42	R	R	R	R	R	R	R	S	MR	R
IR43	R	R	S	S	R	S	S	S	MR	R
IR44	R	R	S	R	R	R	S	S	MR	S
IR45	R	R	S	S	R	S	S	S	MR	S

Contd.

Table 16.3 continued

IR variety	Blast	Bacterial blight	Grassy stunt	Tungro	GLH[b]	BPH[c] biotype			Stem borer	Gall midge
						1	2	3		
IR46	R	R	S	MR	MR	R	S	R	MR	S
IR48	R	R	R	R	R	R	R	S	MR	-
IR50	MS	R	R	R	R	R	R	S	MR	-
IR52	MR	R	R	R	R	R	R	S	MR	-
IR54	MR	R	R	R	R	R	R	S	MR	-
IR56	R	S	R	R	R	R	R	R	MR	-
IR58	R	R	R	R	R	R	R	S	MR	-
IR60	R	R	R	R	R	R	R	R	MR	-
IR62	MR	R	R	R	R	R	R	R	MS	-
IR64	MR	R	R	R	R	R	MR	R	MR	-
IR65	R	R	R	R	R	R	R	S	MS	-
IR66	MR	R	R	R	R	R	R	R	MR	-
IR68	MR	R	R	R	R	R	R	R	MR	-
IR70	R	S	R	R	R	R	R	R	MS	-
IR72	MR	R	R	R	R	R	R	R	MR	-
IR74	R	S	R	R	R	R	R	R	MR	-

[a]S = susceptible, MS = moderately susceptible, MR = moderately resistant, Reactions were based on tests conducted in the Philippines for all disease and insects except gall midge conducted in India.
[b]GLH = green leaf hopper.
[c]BPH = brown plant hopper.

Grain Quality

Grain quality of rice is evaluated relative to several consumer-oriented criteria.[9] Most consumers in tropics and subtropics prefer long or medium long, slender and translucent grains. Higher milling recovery is a universal requirement and is, to some extent, dependent on size, shape and amount of chalkiness in grains. Cooking quality and palatability is another factor that is very important to consumers and is determined to some extent by the amylose content and gelatinization temperature of starch. In tropics and subtropics varieties with intermediate amylose and intermediate gelatinization temperature are preferred. Improvement of milling recovery and grain appearance received immediate attention. The early varieties, such as IR5 and IR8 have poor grain quality. They have bold and chalky grains of poor appearance that frequently break during milling. In addition they cook dry because of high amylose content and thus have poor consumer acceptance.

All the IR varieties released after IR5 and IR8 have slender and translucent grains and have good milling recovery. However, improvements in cooking quality were only achieved slowly primarily due to the fact that all the donors for disease and insect resistance used in the hybridization program had high amylose content and low gelatinization temperature. IR64 is the first IR variety, released in 1985, that has a desirable combination of intermediate amylose content and intermediate gelatinization temperature. It also has long slender and translucent grains with high milling recovery. In addition, cooked rice of IR64 is highly palatable. Not surprisingly, therefore, IR64 has been widely accepted as a high quality rice in tropical and subtropical Asia. It replaced IR36 during 1990s and has been planted to about 10 million hectares annually. Twenty-three years after its release it is still planted to large areas in India, Indonesia, Philippines and Vietnam.

Tolerance to Abiotic Stresses

Large areas of land otherwise suitable for growing rice remain un-planted because of severe nutritional deficiencies and toxicities of their soil. A vast majority of rice soils have varying levels of alkalinity or salinity. Even well-managed rice lands suffer from mild nutritional deficiencies or toxicities. For example, zinc deficiency in rice soils is a common problem in many countries. Several improved varieties have moderate to high level of tolerance to several nutritional deficiencies and toxicities. IR36, for example, has a tolerance to salinity, alkalinity, peatiness and iron and boron toxicities. It also tolerates zinc deficiency. Similarly IR42 has a broad spectrum of tolerance to many soil problems.

Use of IRRI's Germplasm Internationally

From the inception of IRRI's rice improvement program, the germplasm was shared with national rice improvement programs. Seeds of donor varieties, early generation breeding materials, fixed elite lines and named varieties were sent to national program scientists at their request and through the International Network for Genetic Evaluation of Rice (INGER) nurseries. The seeds of breeding materials were sent to 87 countries irrespective of geographic location and ideology. IRRI even shouldered the cost of shipment. The materials were evaluated by local breeders for adaptation to local conditions. Some were released as varieties and others were used as parents in breeding program. Thus 328 IR breeding lines have been released as 643 varieties in 75 countries. Numerous IR varieties and breeding lines have been used as parents in breeding programs all over the world. During 1970s and up to 1980s many IR varieties and breeding lines were released directly by national breeding programs. However, as the national breeding programs became stronger, IR lines were used primarily as parents in local breeding programs. It is estimated that 60% of the world rice area is now planted to IRRI-bred varieties or their progenies.

IMPACT OF THE GERMPLASM IMPROVEMENT PROGRAM

The impact of germplasm improvement spearheaded by IRRI popularly described as green revolution has led not only to major increases in food production but also to improved socio-economic conditions and environmental sustainability.

Impact on Food Grain Production

The gradual replacement of traditional varieties of rice by improved ones, together with associated improvement in farm management practices, has had a dramatic effect on the growth of rice production, particularly in Asia. Farmers harvest 5-7 tons of paddy rice per hectare from high yielding varieties as compared to 1-3 tons with traditional varieties. Since 1966, when the first high yielding variety of rice was released, the rice harvested area has increased only marginally, from 126 to 152 million hectares (18%), whereas the average yield has increased

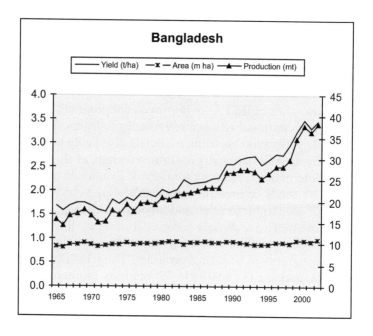

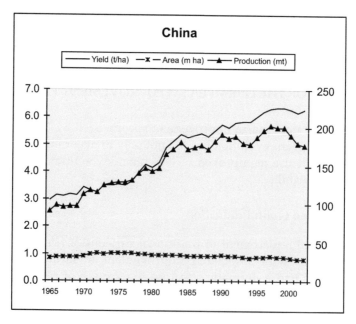

Fig. 16.1. Rice production (mt), area (m ha) and yield (t/ha) from 1965 to 2002 in major rice growing countries (Bangladesh, China, India, Indonesia, Malaysia, Myanmar, Nepal, Pakistan, Philippines, Sri Lanka, Thailand, Vietnam).

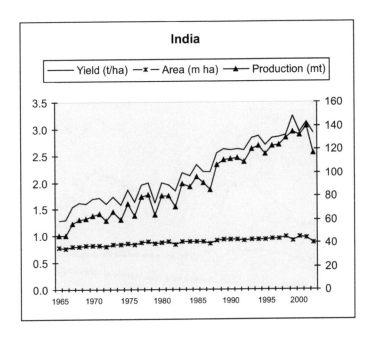

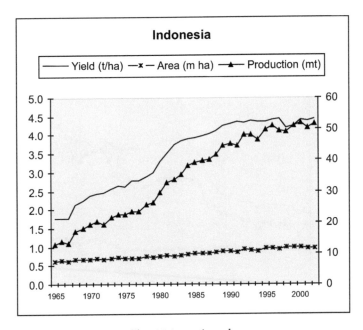

Fig. 16.1 continued.

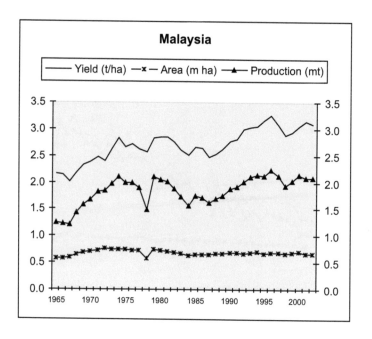

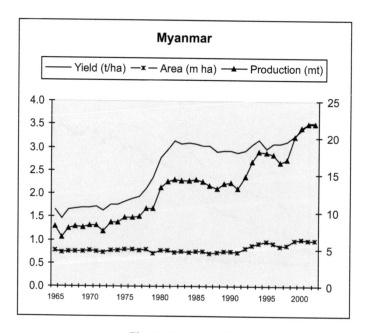

Fig. 16.1 continued.

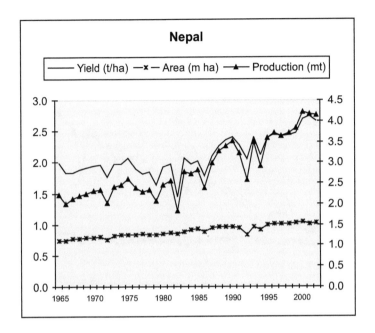

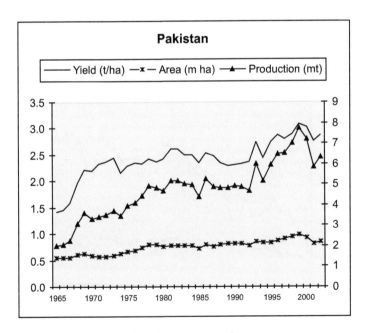

Fig. 16.1 continued.

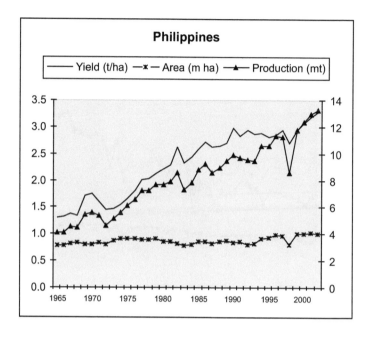

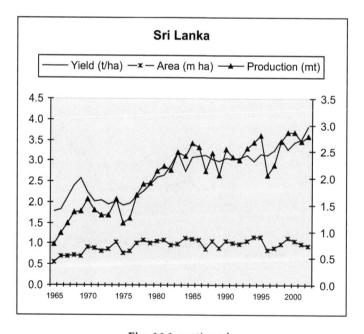

Fig. 16.1 continued.

501

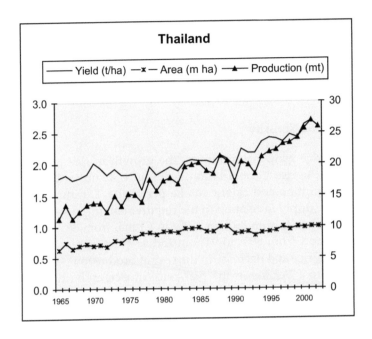

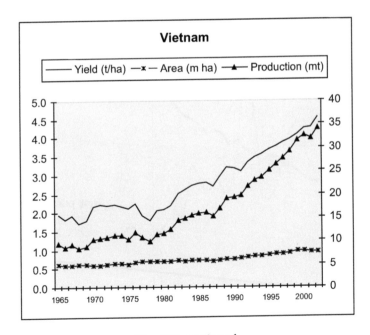

Fig. 16.1 continued.

from 2.1 to 4.0 tons per hectare (95%). World rice production increased from 257 million tons in 1966 to 600 million tons in 2000 (133%). Area planted to rice, average yields and total production in 12 Asian countries are shown in Fig. 16.1. Every country has had a marginal increase in area but dramatic increases in average yields and total production.[10]

Impact on Food Security

In many rice growing countries, the growth in rice production has outstripped the rise in population, leading to a substantial increase in cereal consumption and caloric intake per capita. During 1965-1990, the daily calorie supply in relation to the requirement improved from 81% to 120% in Indonesia, from 86% to 110% in China, from 82% to 99% in the Philippines and from 89% to 94% in India.[11] The increase in per capita availability of rice and decrease in the cost of production per ton of output contributed to a decline in the real price of rice, in both domestic and international markets. The unit cost of production is about 20-30% lower

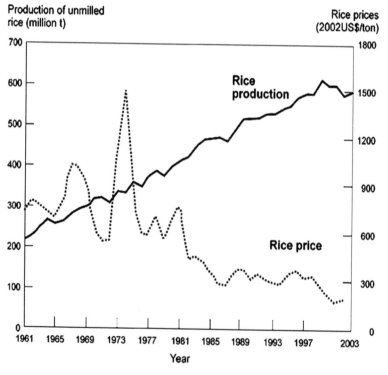

Fig. 16.2. Trends in world rice production and price (1961-2003).

for high yielding varieties than for traditional varieties of rice[12] and the price of rice adjusted for inflation is 40% lower than in mid 1960s (Fig. 16.2). The decline in food prices has benefited the urban poor and rural landless, who are not directly involved in food production but who spend more than one half of their income on food grains. As net consumers of grain, small and marginal farmers, who are dominant rice producers in most Asian countries, have also benefited from the downward trend in real prices of rice.

Impact on Landless Workers

The diffusion of high yielding varieties has also contributed to a growth in income for rural landless workers.[13] High-yielding varieties require more labor per unit of land because of increased intensive care in agricultural operations and harvesting of the larger output. The labor requirement has also increased because of the higher intensity of cropping, which has been made possible by the reduction in crop growth duration. As farm income increases, better-off farm households substitute leisure for family labor and hire more landless workers to do the work. The marketing of a larger volume of produce and an increased demand for non farm goods and services, resulting from higher farm income, have generated additional employment in rural trade, transport and construction activities. The economic miracle underway in Asia was triggered by the growth in agricultural income and its equitable distribution which helped expand the domestic market for non farm goods.[14]

Impact on Environmental Sustainability

In sharp contrast to rich countries where more of the environmental problems have been urban and industrial, the critical environmental problems in low-income developing countries are still rural, agricultural and poverty based. More than half of the world's poor live on lands that are environmentally fragile and they rely on natural resources over which they have little control. Land-hungry farmers resort to cultivating unsuitable areas such as erosion-prone hillsides and semiarid areas where soil degradation is rapid, as in tropical forests, where crop yields on cleared soils drop sharply after just a few years.

The widespread adoption of high-yielding varieties has helped most Asian countries meet their growing food needs from productive lands and thereby has reduced the pressure to open up more fragile lands for cultivation. If 1961 yields prevailed today, three times more land in China and two times more land in India would be needed to equal the 2000 rice harvest. If Asian countries attempted to produce a 2000 harvest at yield levels of 1960s, most of the forests, woodlands, pastures and range lands would have disappeared and mountainsides would be eroded, with disastrous consequences for upper watershed and productive lowlands, the extinction of wildlife habitat and the destruction of biodiversity. As an example, to produce the 2000 world rice production of 600 million tons at the yield levels of 1965, 135 million hectares more land would be required (Fig. 16.3).

The availability of rice varieties with multiple resistance to diseases and insects reduced the need for the application of insecticides and facilitated the adoption of integrated pest management (IPM) practices. Reduced insecticide use helps (1) enhance environmental quality, (2) improve human health in farming communities, (3) make more safer food available, and (4) protect useful fauna and flora.[15]

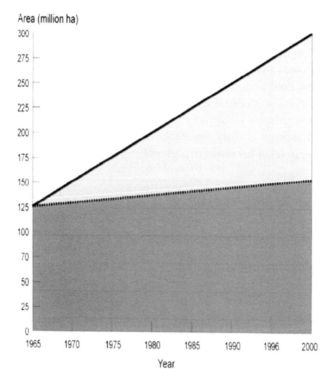

Fig. 16.3. Actual area planted to rice and the additional area which would have been required to produce 2000 level of production at the yield level of 1965.

FUTURE CHALLENGES

In spite of major increases in food production, 800 million people in the world go to bed hungry every night and many suffer from micronutrient deficiencies. According to UN estimates, the world population will grow from 6.5 billion now to 8.5 billion in 2030. Most of this increase (93%) will take place in developing world whose share of population is projected to increase from 78% in 1995 to 83% in 2020. Since 97% of the rice is produced and consumed in developing countries, it is estimated that rice production must increase by 30% to feed the rice consumers in 2030. This increased demand will have to be met from less land, with less water, less labor and fewer chemicals. To meet this challenge, we need rice varieties with higher yield potential, greater yield stability and higher micronutrient content.

Increasing the Yield Potential

Various strategies for increasing the yield potential of rice are being employed. These include: (1) conventional hybridization and selection procedures, (2) ideotype breeding, (3) hybrid breeding, (4) wide hybridization, and (5) genetic engineering.

Conventional Hybridization and Selection Procedures

Improvements in the yield potential of crops have been achieved through conventional hybridization and selection procedures. It is estimated that on the average about 1% increase has occurred per year in the yield potential of rice over a 35 year period since the development of first improved variety of rice IR8.[16] There is no reason to believe that such increase will not occur in the future if sufficient investment in research is made continuously.

Ideotype Breeding

Ideotype breeding aimed at modifying the plant architecture is a time tested strategy to achieve increases in yield potential. As discussed in an earlier section, breeding for short stature in rice resulted in doubling of yield potential. To increase the yield potential of rice further, a new plant type was conceptualized in 1990s. Modern short-statured rice varieties produce a large number of unproductive tillers and excessive leaf area which cause mutual shading and reduce canopy photosynthesis and sink size especially when they are grown under direct sowing conditions. To increase the yield potential of short-statured rice further, IRRI scientists proposed a new plant type (NPT) with the following characteristics.[17]

- Low tillering
- No unproductive tillers
- 200-250 grains per panicle
- Dark green and erect leaves
- Vigorous and deep root system

Breeding efforts to develop NPT were initiated in early 1990s. The objective was to develop improved germplasm with about 15% higher yield potential than that of existing high yielding varieties. Genetic resources for developing NPT were identified and hybridization and selection undertaken. Numerous breeding lines with desired ideotype were developed and shared with national rice improvement (NARS) programs. Three NPT lines have been released in China, two in Indonesia and one in Philippines. Other NARS are evaluating and further improving the NPT lines.

Hybrid Breeding

Rice hybrids with a yield advantage of 10-15% are now widely grown in China. Rice hybrids adapted to tropics and subtropics have been developed at IRRI and by the NARS. Various strategies are being employed to raise the level of heterosis. Efforts are underway to identify heterotic groups within the *indica* germplasm. Most of the hybrids grown to date are based on cytoplasmic-genic male sterility system. Sterile cytoplasm imposes a yield penalty of up to 5% in hybrids. Therefore many hybrid rice breeding programs are developing hybrids based on alternative genetic male sterility systems based on thermo or photo sensitivity. These genetic male sterility systems also allow wider choice of male parents in hybrid seed production. A limited area is planted to such so called two line hybrids in China.

Wide Hybridization

Crosses between crop cultivars and wild species, weedy races as well as intra-specific groups lead to widening of gene pools. Such gene pools are exploited for improving many traits including yield potential. Xiao *et al.*[18] reported that some backcross derivatives from a cross between an *Oryza rufipogon* accession from Malaysia and cultivated rice outyielded the recurrent parent by as much as 18%. They identified two QTL from wild rice with major contribution to yield increase. These QTL have been transferred to modern short-statured varieties and yield evaluation is underway. Molecular marker assisted backcrossing is a useful approach for bringing alleles for yield improvment from wild and primitive germplasm.

Genetic Engineering

Since protocols for rice transformation are well established,[19] it is now possible to introduce single alien genes that can selectively modify yield determining processes. In several crop species, incorporation of "stay green" trait or slower leaf senescence has been a major achievement of breeders in the past decade.[20] In some genotypes with slower senescence (stay green), the rubisco degradation is slower which results in longer duration of canopy photosynthesis and higher yields. The onset of senescence is controlled by a complement of external and internal factors. Plant hormones such as ethylene and abscisic acid promote senescence, while cytokinins are senescence antagonists. Therefore, over production of cytokinins can delay senescence. The *ipt* gene from *Agrobacterium tumefaciens* encoding an isopentenyl transferase[21] was fused with senescence specific promoter SAG12 and introduced into tobacco plants.[22] The leaf and floral senescence in transgenic plants was markedly delayed, biomass and seed yield was increased but other aspects of plant growth and development were normal. This approach appears to have great potential in improving canopy photosynthesis and increasing the yield potential of rice.

Breeding for Durable Resistance

Diseases and insects take serious toll of crop production. According to FAO estimates, diseases, insects and weeds cause as much as 25% yield losses annually in cereal crops. Similarly crop yields are reduced and fluctuate greatly as a result of biotic stresses such as drought, excess water (submergence), mineral deficiencies and toxicities and abnormal temperatures. Plant breeders have been improving the crops to withstand these biotic and abiotic stresses to impart yield stability.

Diverse sources of resistance to major diseases and insects have been identified and rice varieties with multiple resistance to disease and insects have been developed. Recent breakthroughs in cellular and molecular biology have provided tools to develop more durably resistant cultivars and to overcome the problem of lack of donors for resistance to some diseases and insects such as sheath blight and stemborers.

Yellow stemborer is widespread pest in Asia and causes substantial crop losses. Improved rice cultivars are either susceptible to the insect or have only partial resistance. Codon optimized *Bt* gene was introduced into rice and the transgenic rice showed excellent levels of resistance in the laboratory as well as in the field.[23] Bt rices have also been tested under field conditions in China[24] and showed excellent resistance. Besides *Bt* genes, other genes for insect resistance such as those for proteinase inhibitors, α-amylase inhibitors and lectins are also beginning to receive attention.

Two of the most serious and widespread diseases in rice production are rice blast caused by the fungus *Pyricularia oryzae* and bacterial blight caused by *Xanthomonas oryzae* pv. *oryzae*. Development of durable resistance to these diseases is the focus of coordinated effort at IRRI using molecular marker technology. Efforts to develop markers closely linked to bacterial blight resistance genes have taken advantage of the availability of near isogenic lines having single genes for resistance. Segregating populations were used to confirm cosegregation between RFLP markers and genes for resistance. RFLP markers were converted into PCR based markers and using these PCR based markers in marker assisted selection (MAS), several genes for bacterial blight resistance were pyramided. Thus *Xa4*, *xa5*, *xa13* and *Xa21* were combined into same breeding lines.[25] Pyramided lines showed a wider spectrum and higher level of resistance. Pyramided lines have been employed for moving genes into improved varieties grown in India.[26] *Xa21* has also been introduced into widely grown varieties through genetic engineering and transgenic lines are being evaluated under field conditions.

Breeding for Abiotic Stress Tolerance

The progress in developing crop cultivars for tolerance to abiotic stresses has been slow because of lack of knowledge of mechanism of tolerance, poor understanding of inheritance of resistance or tolerance, low heritability and lack of efficient techniques for screening of germplasm. Only a few cultivars with varying degrees of tolerance to abiotic stresses have been developed. Rainfed rice is planted to 40 million hectares worldwide and vast rainfed areas suffer from drought at some stage of growth cycle. QTL for various component traits for drought tolerance have been identified and are being introduced into improved cultivars. Genetic engineering techniques hold great promise for developing drought tolerant cultivars. Datta *et al.*[27] introduced *Dreb1A* gene in rice variety IR64 and transgenic plants showed good level of drought tolerance in greenhouse conditions. Large areas in river deltas of southeast and south Asia are submergence prone where water accumulates in fields after heavy rains for several days. Most rice varieties get killed after submergence of 3-4 days. However, rice variety FR13A can withstand submergence for up to 2 weeks. *Sub1* gene from FR13A has been bred into improved varieties such as Swarna,[28] BR11, Samba Mahsuri grown in submergence prone areas and these varieties have greater yield stability. *Sub1* gene was also cloned recently.[29]

Tackling the Hidden Hunger

In addition to protein-energy malnutrition, deficiencies of minerals and vitamins affect a high proportion of world's poor. Deficiencies of iron (Fe), Zinc (Zn) and vitamin A are most acute amongst poor rice consumers. Rice has a low amount of Fe and Zn and is completely devoid of vitamin A. A research project to develop improved varieties of rice with higher level of micronutrients was initiated at IRRI in 1992. Considerable variation for both Fe and Zn was observed in rice germplasm. A comparison of Fe and Zn contents of selected varieties such as Jalmagna and Juchen with widely grown varieties such as IR36 and IR64 indicated that former have twice as much Fe and 50% more Zn. Rice varieties with high Fe and Zn contents are tall, unimproved and low yielding and hence not suitable for modern agriculture. Efforts are underway to develop improved varieties with both high yield and higher levels of Fe and Zn. Crosses between these traditional varieties and high yielding varieties have produced progenies with high yield and high levels of these micronutrients. For example, an improved breeding line with short stature, IR68144-3B 2-2-3 from a cross of high yielding variety IR72 with tall traditional variety Zawa Bondy from India has high concentration of Fe in the grain, about 21 mg/kg in brown (i.e. unmilled) rice and its yield potential is comparable to improved varieties.

To study the genetics of iron and zinc content in the rice grains, two populations (backcross and F2) are under evaluation. Another accelerated breeding strategy, based on anther culture derived doubled haploids, is also being followed. More than 300 doubled haploids from 3 single crosses have been obtained and are under evaluation. We are also in the process of generating doubled haploid lines from 6 backcrosses. These doubled haploids would be field evaluated in replicated experiments and data on various agronomic traits and micronutrient content traits (iron and zinc) would be generated. These doubled haploid lines represent mapping populations for studying the genetics as well as for breeding. In addition, we are also in the process of developing single seed descent lines from the above single crosses and backcrosses. Germplasm screening to identify new donors with higher iron and zinc content is continuing.

β-Carotene, the precursor of vitamin A does not occur naturally in the rice endosperm. A genetic engineering project to introduce the biosynthetic pathway leading to production of β-carotene in rice endosperm was implemented by a team of Swiss and German scientists.[30] Two genes (*Psy* and *lcy*) from a plant (daffodil) and one (*crtI*) from a bacterium *Erwinia uredovora* were introduced into a rice variety Taipei 309. This resulted in the development of biosynthetic pathway leading to the development of beta-carotene in rice endosperm popularly called golden rice. Taipei 309 was used to introduce beta-carotene pathway genes as it is easy to transform. However, this variety is not cultivated due to its low yield and lack of adaptation to tropical and subtropical conditions. Subsequently new golden rice lines designated GR1 and GR2 in the background of two southern U.S. *japonica* type varieties Kaybonnet and Cocodrie respectively were developed by Syngenta. These US rice varieties are not adapted to the Asian rice growing conditions either Therefore, at IRRI efforts are underway to introgress the β-carotene loci from GR lines into popular Asian rice varieties, using marker aided backcrossing (MAB). Briefly the process involves, firstly, an event specific PCR based marker profiling to identify backcross plants possessing the target gene and usually referred to as 'foreground selection'. Subsequently, up to 70 SSR markers, providing genome wide coverage, are used to monitor recurrent genomic contribution, commonly called 'background selection'. Initially, 3 GR1 events (GR1-146, GR1-309 and GR1-652) as donor parents while 2 IRRI bred mega varieties (IR64 and IR36) and a popular Bangladeshi variety 'BR29' were used as recurrent parents. For one of the donor-recurrent parent combination (IR64 × GR1-309), BC2F4 seeds are available as of March 2008 for conducting an agronomic confined field trial. BC3F1 seeds have been obtained from three cross combinations (IR36 × GR1-309 and BR29 × GR1-309 and BR29 × GR1-146). Subsequently, we received 6 GR2 events (GR2-E, GR2-G, GR2-L, GR2-R, GR2-T and GR2-W). Four *indica* varieties, namely IR64, IR36, BR29 and PSB Rc 82 are being used as recurrent parents. Initially, two donor events namely GR2-E and GR2-G were chosen to precede further using marker aided backcrossing (MAB) and BC2F1 seeds have been produced. Marker aided backcrossing for reconstituting various recurrent parents is in progress.

FUTURE PROSPECTS

Selections from lowly *Oryza rufipogon* made by primitive men and women as well as varieties developed by numerous farmers during eight thousand years and in the last century by breeders have fed and nurtured vast population in Asia and elsewhere from times immemorial. Rise of Asian civilizations depended upon availability of abundant supplies of this precious grain. Interruption in adequate supplies of rice resulted in devastating famines as exemplified by great Bengal famine of 1943 and even larger famine in China during 1950s.

Continued political stability in the developing world and welfare of humanity at large is contingent upon adequate supplies of world's most important source of calories. Any perturbations in supplies and price fluctuations can lead to food riots as happened in several countries in mid 2008 when rice prices doubled in the domestic markets of these countries. Therefore, we must ensure continued adequate supplies of rice in the future and redouble our efforts to develop rice varieties and management practices to produce 30% more rice by 2030. As mentioned in an earlier section, we must raise the yield potential of rice, increase its durability to diseases and insects and tolerance to abiotic stresses. Low nitrogen use efficiency in rice production is a serious concern. Rice production consumes largest share of water used in agriculture. Dwindling water resources for agriculture is going to be a major constraint in future rice production. Therefore, major challenge for rice breeders is to develop varieties for higher nitrogen and water use efficiency. Fortunately, advances in rice genomics and breakthrough in molecular biology have provided tools and approaches to address these difficult problems of rice improvements. Cooperation between rice breeders and molecular biologists is essential to achieve these rice improvement objectives.

Yield potential of modern rice varieties is 10 tons per hectare. However, farmers harvest on the average only 5 tons per hectare. Agronomic practices must be fine-tuned to raise the average yields on farmer's fields.

References

1. Khush (1987).
2. Parthasarathy (1972).
3. Van Der Meulen (1951).
4. Parthasarathy (1972).
5. Paddock and Paddock (1967).

6. Khush (1999).
7. Khush (1995a).
8. Khush (1977).
9. Khush (1995a).
10. Khush and Virk (2005).
11. UNDP (1994).
12. Yap (1991).
13. Hyami *et al.* (1978), Hossain (1988).
14. Khush and Virk (2005).
15. Khush (1999).
16. Peng *et al.* (2000).
17. Khush (1995b).
18. Xiao *et al.* (1996).
19. Christou *et al.* (1991).
20. Evans (1993).
21. Akiyoshi *et al.* (1984).
22. Gan and Amasino (1995).
23. Datta *et al.* (1997).
24. Tu *et al.* (2000).
25. Huang *et al.* (1997).
26. Singh *et al.* (2001).
27. Datta *et al.* (2002).
28. Neerja *et al.* (2006).
29. Xu *et al.* (2006).
30. Ye *et al.* (2002).

References Cited

Preface

Barnhart, James (2003). The Empire of Rice. See website: www.BulletProofFilm. com.

Carney, Judith A. (2002). *Black Rice: The African Origins of Rice Cultivation in the Americas*. Harvard University Press.

Nanda, J.S. (2003). Antiquity and spread of rice cultivation. In Nanda, J.S, Sharma, S.D. (eds) *Monograph on Genus Oryza*. Science Publishers Inc, New Hampshire, USA.

Nilakanta Sastri, K.A. (1955). *A History of South India: From Prehistoric Times to the Fall of Vijayanagar*. Oxford Univrsity Press. Delhi.

Pakeman, S.A. (1964). *Ceylon*. Benn, London. Quoted by Toynbee, Arnold (1972) *A Study of History*. Strand Book Stall, Bombay.

Peñafiel, Samuel R. (2006). Ifugao Rice Terraces. Paper presented at the *International Forum on Globally Important Agricultural Heritage Systems (GIAHS): A Heritage for the Future*, FAO, Rome, Italy, October 24-26, 2006.

Porteres, R. (1956). Taxonomie agrobotanique des riz cultives: *O. sativa* Lin. et *O. glaberrima* Steudel. I-IV. *Journal d'Agriculture Tropicale et de Botanique Applique* 3:341-384, 541-580, 627-700, 821-856.

Robinson, Kim Stanley (2002). *The Years of Rice and Salt*. Bentam Books.

Tanaka, Junzo (2003). In Kansai International Public Relations Promotion Office News (KIPPO) 10:427, Wednesday, May 28, 2003. See website http://www.kansai.gr.jp

WARDA (2007). See website: http://www.warda.org.

Chapter 1: Domestication and Diaspora of Rice

Adshead, S.A.M. (1997). *Material Culture in Europe and China, 1400-1800: The Rise of Consumerism*. MacMillan.

Becker, L. and Diallo, R. (1996). The cultural diffusion of rice cropping in Cote d'Ivoire. *Geogr. Rev.* 86:505-526.

Bellwood, P. (1992). Southeast Asia before history. Pages 55-136 in Tarling, N. (ed), *The Cambridge History of Southeast Asia. Volume. From Early Times to c. 1800*. Cambridge Univ Press.

Bellwood, P. (2004). *First Farmers: The Origin of Agricultural Societies*. Blackwell Publishing.

Carney, Judith A. (2001). *Black Rice: The African Origins of Rice Cultivation in the Americas*. Harvard University Press, Cambridge.

Carpenter, A. (1978). The history of rice in Africa. Pages 3-10 in Buddenhagen I. and Persley, J. (eds), *Rice in Africa.* Academic Press, London.

Chang, T.T. (1976a). The origin, evolution, cultivation, dissemination and diversification of Asian and African rices. *Euphytica* 25:425-441.

Chang, T.T. (1976b). Rice. Pages 98-104 in Simmonds, N.W. (ed), *Evolution of Crop Plants.* Longman, London.

Chang, T.T. (1976c). The rice cultures. *Phil. Trans. Royal Society, London* B. 275:143-157.

Chang, T.T. (1983). The origins and early cultures of the cereal grains and food legumes. Pages 65-94 in Keightley, D.N. (ed), *The Origins of Chinese Civilization.* University of California Press, Los Angeles.

Chang, T.T. (1985). Crop history and genetic conservation: Rice—A case study. *Iowa State J. Res.* 59(4): 425-456.

Christensen, Peter (1993). *The Decline of Iranshahr: Irrigation and Environments in the History of Middle East 500 BC to AD 1500.* Museum Tusculenum Press, University of Copenhagen. (Translated from Danish).

Collins, R.O. and Burns, J.M. (2007). *A History of Sub-Saharan Africa.* Cambridge University Press.

Crawford, G.C. (2005). East Asian plant domestication. Pages 77-95 in Stark, Miriam Y. (ed), *Archaeology of Asia.* Blackwell Publishing.

Garris, A.J., Tai, T.H., Coburn, J., Kresovich, S. and McCouch, S. (2005). Genetic structure and diversity in *Oryza sativa* L. *Genetics* 169:1631-38.

Glaszmann, J.C. (1987). Isozymes and the classification of Asian rice varieties. *Theor. Appl. Genet.* 74:21-30.

Harlan, J.R. (1989). Wild-grass seed harvesting in the Sahara and Sub-Sahara of Africa. Pages 79-98 in Harris, D.R. and Hillman, G.C. (eds), *Foraging and Farming: The Evolution of Plant Exploitation.* Unwin Hyman, London.

Harlan, Jack R., de Wet, Jan M.J. and Stemler, Ann B.L. (eds) (1976). *Origins of African Plant Domestication.* Mouton Publishers, The Hague.

Higham, C.F.W. (1996). *The Bronze Age of Southeast Africa.* Cambridge Univ Press.

Higham, C.F.W. (1999). Archaeology, linguistics and expansion of the east and southeast Asian Neolithic. Pages 1003-1114 in Blench, R. and Spriggs, R. (eds), *Archaeology and Language. II. Archaeological Data and Linguistic Hypothesis.* Routledge, London.

Hudgens, J., Trillo, R. and Calonec, N. (2004). *The Guide to West Africa.* Rough Guides Ltd.

Inalcik, Halil, *et al.* (1994). *An Economic and Social History of the Ottoman Empire, 1300-1914.* Cambridge Univ Press.

Kiple, K.F. and Ornelas, K.C. (2000). *The Cambridge World History of Food.* Cambridge Univ Press.

Konishi, S., Ebana, K. and Izawa, T. (2008). Inference of the *japonica* rice domestication process from the distribution of six functional nucleotide polymorphisms of domestication-related genes in various landraces and modern cultivars. *Plant Cell Physiol.* 49(9):1283-93.

Li, C., Zhou, A. and Sang T. (2006). Rice domestication by reducing shattering. *Science* 311:1936-1939. Also see website: www.sciencemag.org/cgi/content/full/1123604/DC1

Linares, O.F. (2002). African rice (*Oryza glaberrima*): History and future potential. *PNAS* 99(25):16360-65.

Littelfield, D.C. (1981). *Rice and Slaves*. Louisiana State University Press, Baton Rouge, LA.

Londo, J.P., Chiang, Y.C., Hung, K.H., Chiang, T.Y. and Schaal, B.A. (2006). Phylogeography of Asian wild rice, *Oryza rufipogon*, reveals multiple independent domestications of cultivated rice, *Oryza sativa*. *PNAS* 103(25):9578-83.

MacDonald, K.C. (1997). Archaeology, language, and the peopling of West Africa. Pages 33-66 in Blench, R. and Spriggs, M. (eds), *Archaeology and Language. II. Archaeological Data and Linguistic Hypothesis*. Routledge, London.

Markus, Andrew (2002). *Building a New Community: Immigration and the Victorian Economy*. Allen and Unwin.

McIntosh, R.J. (2005). *Ancient Middle Niger: Urbanism and the Self Organising Landscape*. Cambridge University Press.

McNally, K.L., Bruskiewich, R., Mackill, D., Buell, C.R., Leach, J.E. and Hei Leung, H. (2006). Sequencing multiple and diverse rice varieties connecting whole-genome variation with phenotypes. *Plant Physiology* 141:26-31.

Misra, P.K. and Misro, B. (1969). Postulation of two subspecies in the African cultivated rice (*O. glaberrima* Steud.). *Indian J. Agric. Sci.* 39(10):966-970.

Morinaga, T. (1968). Origin and geographical distribution of Japanese rice. *Jap. Agric. Res. Quarterly* 3:1-5.

Morishima, H. (1984). Wild plant and domestication. Pages 3-30 in Tsunoda S. and Takahashi, N. (eds), *Biology of Rice*. Elsevier, Amsterdam.

Morishima, H. (2001). Evolution and domestication of rice. Pages 63-77 in Khush, G.S., Brar, D.S. and Hardy, B. (eds), *Rice Genetics. IV*. (Poceedings of the 4th International Rice Genetics Symposium, October 22-27, 2000). Science Publishers and IRRI, Manila, Philippines.

Murray, S.S. (2004). Searching for the origins of African rice domestication. *Antiquity* 78(300). Also see website: http://antiquity.ac.uk/projgall/murray/

Nakagahara, M. (1978). The differentiation, classification and center of genetic diversity of cultivated rice (*O. sativa* L.) by isozyme analysis. *Trop. Agric. Res. Ser.* 11:77-82.

Niane, D.T. and Ki-Zerbo, J. (1997). *General History of Africa. V. Africa from the Twelfth to the Sixteenth Century*. James Currey Publishers. (Prepared by UNESCO International Committee for the Drafting of a General History of Africa).

Ohnuki-Tierney, Emiko (1994). *Rice as Self: Japanese Identity through Rice*. Princeton University Press.

Oka, H.I. (1988). *Origin of Cultivated Rice*. Japan Scientific Society Press, Tokyo.

Oka, H.I. (1974). Experimental studies on the origin of cultivated rice. *Genetics* 78:475-486.

Oka, H.I. (1977). The ancestors of cultivated rice and their evolution. Pages 57-64 in *Meeting on African Rice Species*. January 25-26, 1977. IRAT-ORSTOM, Paris.

Olsen, K.M. and Purugganan, M.D. (2002). Molecular evidence on the origin and evolution of glutinous rice. *Genetics* 162:941-950.

Olson, J.S. (1996). *The People of Africa: An Ethnohistorical Dictionary.* Greenwood Publishing Group.

Pejros, Ilia and Shnirelman, V. (1999). Rice in Southeast Asia: A regional inter-disciplinary approach. Pages 379-389 in Blench, R. and Spriggs, R. (eds), *Archaeology and Language. II. Archaeological Data and Linguistic Hypothesis.* Routledge, London.

Portères, R. (1970). Primary cradles of agriculture in the African continent. Pages 43-58 in Fage, J. and Olivier, R. (eds), *Papers in African Prehistory.* Cambridge University Press, Cambridge.

Portères, R. (1976). African cereals: Eleusine, fonio, black fonio, teff, Brachiaria, Paspalum, Pennisetum and African rice. Pages 409-452 in Harlan, J.R., De Wet, J.M.J and Stemler, A.B.L. (eds), *Origins of African Plant Domestication.* Mouton, The Hague.

Potts, Daniel T. (1999). *The Archaeology of Elam: Formation and Transformation of an Ancient Iranian State.* Cambridge Univ Press.

Richards, Paul (1985). *Indigenous Agricultural Revolution: Ecology and Food Production in West Africa.* Hutchinson, London.

Ruskin, F.R. (1996). *Lost Crops of Africa. Vol. I. Grains.* National Academic Press, Washington.

Sang, T. and Ge, S. (2007). The puzzle of rice domestication. *Journal of Integrative Plant Biology* 49(6):760-768.

Sano, Y., Morishima, H. and Oka, H.I. (1980). Intermediate perennial-annual populations of *Oryza perennis* found in Thailand and their evolutionary significance. *Bot. Mag. (Tokyo)* 93:291-305.

Sato, T. (1997). *State of Rural Society in Medieval Islam: Sultans, Muqtas and Fallahun.* BRILL, Boston.

Saxena, A., Prasad, V., Singh, I.B., Chauhan, M.S. and Hasan, R. (2006). On the Holocene record of phytoliths of wild and cultivated rice from Ganga Plain: Evidence for rice-based agriculture. *Current Science* 90(11):1547-1551.

Schiedel, W. (2001). *Death on the Nile: Disease and Demography of Roman Egypt.* BRILL.

Second, G. (1982). Origin of the genic diversity of cultivated rice (*Oryza* sp): Study of the polymorphism scored at 40 isozyme loci. *Japan. J. Genet.* 57:25-57.

Semon, M., Nielsen, R., Jones, M.P. and McCouch, S.R. (2005). The population structure of African cultivated rice *Oryza glaberrima* (Steud.): Evidence for elevated levels of linkage disequilibrium caused by admixture with *O. sativa* and ecological adaptation. *Genetics* 169:1639-47.

Sharma, S.D. (2003). Origin of cultivated rices. Pages 311-329 in Nanda, J.S. and Sharma, S.D. (eds), *Monograh on Genus Oryza.* Science Publishers, Enfield, USA.

Sharma, S.D., Tripathy, S. and Biswal, J. (2000). Origin of *O. sativa* and its ecotypes. Pages 349-369 in Nanda, J.S. (ed), *Rice Breeding and Genetics: Research Priorities and Challenges.* Science Publishers, Enfield, USA.

Sweeney, Megan and McCouch, Susan (2007). The complex history of the domestication of rice. *Annals of Botany* 100(5):951-957.

Sweeney, M.T., Thomson, M.J., Cho, Y.G., *et al.* (2007). Global dissemination of a single mutation conferring white pericarp in rice. *PLoS Genet* 3(8):e133.

Tewari, R., Srivastava, R.K., Singh, K.K., *et al.* (2006). Second preliminary report of the excavations at Lahuradewa, District Sant Kabir Nagar, U.P.: 2002-2003-2004 & 2005-06. *Pragdhara* 16:35-68.

Vaughan, D.A., Lu, B.R. and Tomooka, N. (2008). The evolving story of rice evolution. *Plant Science* 174(4): 394-408.

Vitte, C., Ishii, T., Lamy, F., Brar, D. and Panaud, O. (2004). Genomic paleontology provides evidence for two distinct origins of Asian rice (*Oryza sativa* L.). *Mol. Genet. Genomics* 272:504-511.

Yamanaka, S., Nakamura, I., Nakai, H. and Sato, Y. (2003). Dual origin of the cultivated rice based on molecular markers of newly collected annual and perennial strains of wild rice species, *Oryza nivara* and *O. rufipogon*. *Genetic Resources and Crop Evolution* 50:529-538.

Yen, D.E. (1982). Ban Chiang pottery and rice. *Expedition* 24:51-64.

Chapter 2: Tai Participation

Bailey, Charles-James N. (1973). *Variation and Linguistic Theory. Center for Applied Linguistics*, Arlington, VA.

Condominas, Georges (1990). *From Lawa to Mon, from Saa' to Thai: Historical and Anthropological Aspects of Southeast Asian Social Spaces*. Department of Anthropology, Research School of Pacific Studies, Australian National University, Canberra, Australia.

Diamond, Jared (1997). *Guns, Germs and Steel: The Fates of Human Societies*. W.W. Norton and Company, New York.

Dinh-Hoa, Nguyen (1966). *Vietnamese-English Dictionary*. Charles E. Tuttle Co., Rutland, Vermont.

Edmondson, Jerrold and Gregerson, Kenneth (1997). Outling Kam-Tai: Notes on Ta Mit Laha. Paper presented at the 1997 SEALS meeting, University of Illinois, Urbana-Champaign.

Gedney, William J. (1972). A checklist for determining tones in Tai dialects. Pages 423-37 in Smith, M. Estellie (ed), *Studies in Linguistics in Honour of George L. Trager*. Mouton, The Hague, Netherlands.

Harris, Jimmy G. (1975). A comparative word list of three Tai Nua dialects. Pages 202-30 in Harris, Jimmy G. and Chamberlain, James R. (eds), *Studies in Tai Linguistics in Honour of William J. Gedney*. Allied Printers, Bangkok.

Harris, Jimmy G. and Chamberlain, James R. (1975). *Studies in Tai Linguistics in Honour of William J. Gedney*. Allied Printers, Bangkok.

Hartmann, John F. (1998). *Linguistic Evidence for the Origins of Tai Irrigated Rice Technology and Culture*. Paper presented at the International Conference on Sino-Tibetan Languages and Linguistics. Lund University, Sweden.

Hartmann, John F. (2004). Linguistic and historical continuities of the Tai Dam and Lao Phuan: Case studies in boundary crossings. Paper presented at the SEALS (Southeast Asian Linguistic Society) Conference, Bangkok, Thailand.

Judson, Adoniram (1955). *Judson's Burmese-English Dictionary*. Baptist Board of Publications, Rangoon.

Lansing, J.S. (1997). *Priests and Programmers: Technologies of Power in the Engineered Landscape of Bali*. Princeton University Press, Princeton.

Lass, Roger (1997). *Historical Linguistics and Language Change*. Cambridge University Press, Cambridge, UK.

Li, Fang-Kuei (1977). *A Handbook of Comparative Tai*. The University of Hawaii Press, Honolulu, Hawai'i.

Liu, J. (1985). Some observations on the archaeological site of Hemudu, Zhejiang Province, China. *Bulletin of the Indo-Pacific Prehistory Association* 60:40-45.

Londo, Jason P., Chiang, Y.C., Hung, K.H., Chiang, T.Y. and Schaal, Barbara A. (2006). Phylogeography of Asian wild rice, *Oryza rufipogon* reveals multiple independent domestication of cultivated rice, *Oryza sativa*. *PNAS* 103(25): 9578-83.

Luo, Wei and Hartmann John, F. (2000). GIS mapping and analysis of Tai linguistic and settlement patterns in southern China. *Geographic Information Science* 6(2):129-36.

O'Connor, Richard A. (1995). Agricultural change and ethnic succession in Southeast Asian states: A Case for regional anthropology. *Journal of Asian Studies* 54(4):968-96.

Panikar, S.N. and Gowda, M.K.M. (1976). On the origins of rice in India. *Science and Culture* 42:547-50.

Pearson, R. and Underhill, A. (1987). The Chinese Neolithic: Recent trends in research. *American Anthropologist* 89:807-22.

Renfrew, A.C. (1987). *Archaeology and Language: The Puzzle of Indo-European Origins*. Pilmco, London.

Sagart, Laurent (1999). *The Roots of Old Chinese*. Johns Benjamin Publishing Company, Amsterdam/Philadelphia.

Sharma S.D., Tripathy, S. and Biswal, J. (2000). Origin of *O. sativa* and its ecotypes. Pages 349-369 in Nanda, J.S. (ed), *Rice Breeding and Genetics: Research Priorities and Challenges*. Science Publication, Enfield.

Smith, Bruce D. (1995). *The Emergence of Agriculture*. Scientific American Library, New York.

Tan, Leshan (1994). The Tai before the Thirteenth Century: A Perspective from Chinese Records. *Journal of the Siam Society* 82(2):167-73.

Tanabe, Shigeru (1994). *Ecology and Practical Technology: Peasant Farming Systems in Thailand*. White Lotus, Bangkok.

Terweil, B.J. (1981). *The Tai of Assam and Ancient Tai Ritual. Vol. 2. Sacrifices and Time-Reckoning*. Centre for South East Asian Studies, Gaya.

Watabe, Tadayo (1978). The Development of Rice Cultivation in Thailand. Pages 15-39 in Ishii, Yoneo (ed), *Thailand: A Rice Growing Society*. University of Hawaii Press, Honolulu.

Chapter 3: Rice in Social and Cultural Life

Ahuja, S.C. and Ahuja, Uma (2006). Rice in religion and tradition. In *Souvenir, 2nd International Rice Congress*, Delhi, Oct 9-13, 2006.

Ahuja, Uma, Ahuja, S.C. and Thakrar, R. (1997). Rice in folklore of India. *Asian Agri-Hist.* 1(4):321-330.

Ahuja, Uma, Thakrar, R. and Ahuja, S.C. (2000). Fairs and festivals associated with rice cultivation. *Asian Agri-Hist.* 1(4):321-330.

Ahuja, Uma, Thakrar, R. and Ahuja, S.C. (2001). Alcoholic rice beverages. *Asian Agri-Hist.* 3(4):300-319.

Anonymous (1993). India Magazine. December 1993.

Bahadur, M. (2003). Rice culture in Manipur. Pages 89-100 in Hamilton, R.W. (ed), *The Art of Rice—Spirit and Sustenance in Asia.* UCLA Fowler Museum of Cultural History, Los Angeles.

Bhattacharyya, A. (1978). *Folklore of Bengal.* National Book Trust, New Delhi, India.

Chettiar, S.M.L.L. (1973). *Folklore of Tamil Nadu.* National Book Trust, New Delhi.

Crooke, W. (1896). *Religion and Folklore of Northern India.* (Revised Edition 1926). S. Chand & Co., New Delhi.

Das, J. (1972). *Folklore of Assam.* National Book Trust, New Delhi.

Das, K.B. and Mahapatra, L.K. (1979). *Folklore of Orissa.* National Book Trust, New Delhi.

Grist, D.H. (1986). *Rice.* Longman, Singapore.

Gupta, Shanti M. (1971). *Plant Myths and Traditions in India.* Leaden.

Hamilton, Roy W. (2003). *The Art of Rice—Spirit and Sustenance in Asia.* UCLA Fowler Museum of Cultural History, Los Angeles.

Hanchett, S. (1988). *Coloured Rice.* Hindustan Publishing Corporation, Delhi.

Huggan, R.R. (1995). Co-evolution of rice and humans. *Geo-Journal* 35(3):262-265.

Huke, Robert and Huke, Eleanor (1990). *Rice: Then and Now.* International Rice Research Institute, Manila, Philippines.

Ismani (1985). Rice culture viewed from myths, legends, rituals: Customs and artistic symbolism relating to rice cultivation in Indonesia. *East Asian Rice Studies (Tokyo)* 24(1-4):113-130.

Kling, Z. (1985). Rice field rituals in Malaysia. *East Asian Rice Studies (Tokyo)* 24 (1-4):131-154.

Kumar, T.T. (1988). *History of Rice in India.* Gian Publishing House, Delhi.

Majumdar, G.P. and Banerji, S.C. (trans & eds) (1960). *Krisi Pārāsara.* The Asiatic Society, Calcutta.

Panigarhi, Kusum Misra (1999). *Festivals of Diversity.* Navdanya, New Delhi.

Panikkar, K.N. (1991). *Folklore of Kerala.* National Book Trust, New Delhi, India.

Piper, Jacquilline M. (1993). *Rice in South East Asia.* Oxford University Press, Singapore.

Ponappa, K. (1978). Coorg Fare. *India Magazine* 8(6):60-65.

Raju, R.B. (1978). *Folklore of Andhra Pradesh.* National Book Trust, New Delhi.

Ray, A. (1993). *Mizoram.* National Book Trust, New Delhi.

Reid, Daniell (1984). *Rice in Chinese Rites.* Discovery, Nov. 1984:36.

Roy, S.C. (1912). *The Mundas and Their Country.* Calcutta.

Singh, K.M. (1992). Some common agricultural ceremonies and rites in Manipuri life. Pages 116-121 in *Agriculture in Ancient India.* Institute for Oriental Studies. Thane, India. 212 pp

Tan, S.K. (1985). Rituals, beliefs and art related to rice cultivation in the Philippines. *East Asian Rice Studies (Tokyo)* 24(1-4):155-160.

Taryo, Orayashi (1985). Myths of agricultural origin in Indo-pacific area: A cultural history approach. *East Asian Rice Studies (Tokyo)* 24(1-4):161-174.

Terwil, B.J. (1994). Legends in mainland Southeast Asia. Pages 5-36 in Walker, A.R. (ed), *Rice in Southeast Asia: Myth and Rituals. Contributions to Southeast Asian Ethnography-10*. Double Six Press, Singapore.

Tuan, D.T. (1985). Types of rice rultivation and its related civilization. *East Asian Rice Studies (Tokyo)* 24(1-4):54-56.

Walker, A.R. (ed), (1994). *Rice in Southeast Asia: Myth and Rituals. Contributions to Southeast Asian Ethnography-10*. Double Six Press, Singapore.

Watt, G. (1901). *A Dictionary of Economic Products of India. Reprint Edition (1972)*. Cosmo Publications, New Delhi, India.

Williams, C.A.S. (1941). *Outline of Chinese Symbolism and Art Motives (Third Revised Edition)*. Keity and Walsh Ltd, Shanghai.

Yuxiang, Yang (1994). Traditional rituals of the rice cultivation cycle among the ethnic minorities of Yunan Province, South West China. Pages 91-118 in Walker, A.R. (ed), *Rice in Southeast Asia: Myth and Rituals. Contributions to Southeast Asian Ethnography-10*. Double Six Press, Singapore.

Zefaniasy, R.B. (1985). Rice in Malagasy oral tradition. *East Asian Rice Studies (Tokyo)* 24(1-4):173-180.

Chapter 4: Domestication of Rice in China

Cai, H.W., Wang, X.K. and Pang, H.H. (1996). Isozyme studies on the *hsien-keng* differentiation of the common wild rice (*Oryza rufipogon*) in China. Pages 147-152 in Wang, X.K. (ed), *Origin and Differentiation of Chinese Cultivated Rice*. China Agri Univ Press, Beijing. (in Chinese)

Chang, T.T. (1983). The origins and early cultures of the cereal grain and food legumes. Pages 65-94 in Keightley, D.N. (ed), *The Origins of Chinese Civilization*. Asco Trade Typesetting Ltd, USA.

Ding, Y. (1957). The origin and evolution of Chinese cultivated rice. *Agricultural Journal* 8(3):243-260. (in Chinese)

Han, L.Z. and Cao, G.L. (2005). Status of collection, conservation and propagation of rice germplasm in China. *Journal of Plant Genetic Resources* 6(3):359-364. (in Chinese)

Huang, Y.H., Sun, C.Q. and Wang, X.K. (1996a). *Indica-japonica* differentiation of chloroplast DNA in Chinese common wild rice populations. Pages 166-170 in Wang, X.K. (ed), *Origin and Differentiation of Chinese Cultivated Rice*. China Agri Univ Press, Beijing. (in Chinese)

Huang, Y.H., Sun, X.L. and Wang, X.K. (1996b). Study on the center of genetic diversity of Chinese cultivated rice. Pages 85-91 in Wang, X.K. (ed), *Origin and Differentiation of Chinese Cultivated Rice*. China Agri Univ Press, Beijing. (in Chinese)

Liu, Z.M. (1975). Origin and development of Chinese cultivated rice. *Genetics* 2(1):23-30. (in Chinese)

Liu, Z.Y. (2003). The origin of rice cultivation in China from a comparative analysis between Yu-Chan-Rock and Niu-Lan Cave both Neolithic sites. *Agricultural Archaeology* 69:76-87. (in Chinese)

Morishima, H. and Gadrinab, L.U. (1987). Are the Asian common wild rices differentiated into the *indica* and *japonica* types? Pages 11-22 in *Crop Exploration and Utilization of Genetic Resources*. Proceedings of the International Symposium, Changhua, Taiwan. Taichung District Agricultural Improvement Station, Taiwan.

Peng, H.H. and Ying, C.S. (1993). Geographical distribution, research and utilization of wild rice species in China. Pages 17-27 in Ying, C.S. (ed), *Rice Germplasm Resources in China*. China Agricultural S&T Press, Beijing. (in Chinese)

Sano, R. and Morishima, H. (1992). *Indica-japonica* differentiation of rice cultivars viewed from variations in key characters and isozymes with special reference to landraces from the Himalayan hilly areas. *TAG* (84):266-274.

Sato, Y.I., Tang, S.X., Yang, L.J. and Tang, L.H. (1991). Wild rice seeds found in an oldest rice remain. *Rice Genetics Newsletter* (8):76-78.

Second, G. (1985). Evolutionary relationships in the *sativa* group of *Oryza* based on isozyme data. *Genet. Sel. Evol.* 17(1):89-114.

Sun, C.Q., Wang, X.K. and Yoshimura, A. (1996). Genetic differentiation of mitochondrial DNA in common wild rice (*O. rufipogon* Griff.) and cultivated rice (*O. sativa* L.). Pages 134-139 in Wang, X.K. (ed), *Origin and Differentiation of Chinese Cultivated Rice*. China Agri Univ Press, Beijing. (in Chinese)

Tang, L.H. (2002). Origin and genetic diversity of rice. Pages 1-11 in Lu, L.J. *et al.* (ed), *Rice Germplasm Resources*. Hubei S&T Press, Wuhan. (in Chinese)

Tang, S.X. (1996). The origin and evolution of cultivated rice. Pages 64-80 in Xiong, Z.M. *et al.* (eds), *Rice Breeding*. China Agricultural Press, Beijing. (in Chinese)

Tang, S.X. (2007). Rice origin and domestication. Pages 9-30 in Cheng, S.H and Li, J. (eds), *Modern Rice in China*. Jin Duen Press, Beijing. (in Chinese)

Tang, S.X., Min, S.K. and Sato, Y.I. (1993). Exploration on origin of *keng* rice (*japonica*) in China. *Chinese J. Rice Sci.* 7(3):129-136. (in Chinese)

Tang, S.X., Sato, Y.I. and Yu, W.J. (1994). Discovery of wild rice grains (*O. rufipogon*) from Hemudu ancient carbonized rice. *Agricultural Archaeology* 35:88-91. (in Chinese)

Tang, S.X., Wei, X.H., Jiang, Y.Z., Brar, D.S. and Khush, G.S. (2007). Genetic diversity based on allozyme alleles of Chinese cultivated rice. *Agricultural Sciences in China* 6(6):641-646.

Tang, S.X., Zhang, W.X. and Liu, J. (1999). The study on the bi-peak tubercle on lemma of Hemudu and Luojiajiao ancient excavated rice grains with electric scanning microscope. *Acta Agro. Sinica* 25(3):320-327. (in Chinese)

Wang, X.K., Zhang, J.Z., Chen, B.Z. and Zhou, H.Y. (1996). A new discovery about the origin of rice cultivation in China. Pages 8-13 in Wang, X.K. (ed), *Origin and Differentiation of Chinese Cultivated Rice*. China Agri. Univ Press, Beijing. (in Chinese)

Yan, W.M. (1989). Origin of Chinese rice agriculture (II). *Agricultural Archaeology* 18:72-83. (in Chinese)

You, X.L. (1987). Exploration on wild rice recorded in ancient writing in China. *Agriculture Ancient & Present* 1:1-6. (in Chinese)

You, X.L. (1992). Rice origin and its cultivation history. Pages 1-19 in Xiong, Z.M *et al.* (eds), *China Rice*. China Agricultural Press, Beijing. (in Chinese)

You, X.L. (2004). Relations of human migration, language evolution with the origin of agriculture. *Agri. History of China* 88:3-9. (in Chinese)

Chapter 5: History of Rice Culture in Korea

Anh, C.B. (1999). Carbonized Rice Excavated from the Site of Sonamri Bronze Age. Paper Presented at *in situ* Presentation Meeting. Sancheong-Sonamri, Gyeonnam Prov., Korea.

Anonymous (1976). Achievement of Japanese agricultural research in Korea. *Rept. Jap. Agr. Res.* 13:215-244.

Choi, H.C. (2005). Recognition of rice. II. Pages 154-190. Science and Horticulture Pub., Seoul, Korea.

Choi, H.O. (1978). Recent progress of rice breeding in Korea. *J. Korean Breeding Society* 14(3):201-238.

Crop Experiment Station (1990). *Rice Varietal Improvement in Korea*. Crop Experiment Station, RDA, Suweon. 52 pp.

Gweon-eop mobeom-jang (1913). *List of Rice Variety in Korea*. Gweon-eop mobeom-jang (Agricultural Demonstration Station), Suwon, Korea.

Hamada, K. and Umehara, S. (1920). *Report on the Studies on Excavation of Shell Mound in Kimhae, Korea*. Committee of Archological Investigation, Chosun-chongdokboo, Seoul, Korea.

Heu, H.M. *et al.* (1991). *Indica* rices grown in the ancient time in Korea. *Korean J. Crop Sci.* 36(3):241-248.

Heu, M.H. (1991). On the rough rice, milled rice and soybean grains enshrined in Buddha Image. *A Study of Relics Enshrined in Amita-Buddhist State Made in 1302*. Kyemong Cultural Foundation Research Paper Series No. 2:145-158. Onyang Falk Museum, Korea.

Heu, M.H. (1996). Rices cultivated in the ancient time in Korea. *Jour. Soc. of Prehistory and Antiquity* 7:19-29.

Heu, M.H. (1998). Excavated Grains at Suyanggae Historical Site. Reports on Suyanggae Excavations in 15th Century.

Heu, M.H. (2000). Botanical characteristics of ancient rice excavated at different places in Korea. Pages 65-85 in *International Symposium on Korean Prehistoric Rice Agriculture*.

Heu, M.H. *et al.* (1997). Grain excavated at Chodong-ri Bronze Age dwelling site. *Annual Report, Choongbook Univ. Museum* 8:159-163.

Hue, M.H. (1999). *Examinations of Carbonized Grains at Horokoru Historical Site. 2nd Report*. Kyunggi Province Land Museum.

Kim, W.Y. (1986). *Introduction of Korean Archeology*. Ilchisa Pub.

Kim, Y.K. and Suk, K.J. (1984). Studies on the Namkyung Historical Site. In *Scientific Encylopedia*. Pyungyang, N. Korea.

Kwangju Museum (1997). *Excavated Carbonized Rice Hulls at Wetland Historical Site in Kwangjoo*. National Kwangju Museum, Korea.

Lee, C.Y. and Park, T.S. (1979). Carbonized rice excavated at Songkook-ri dwelling site. *Songkook-ri*: 153-154. National Museum.

Lee, H.J. (1999). *Historical Remains Majunri Roonsam. Presentation Report*. Institute of Archaeological Ground Remains Study.

Lee, J.H. (1965). Research Report of Office of Rural Development 14:1-26.

Lee, S.K. (1997). Historical Remains at Daepyung Eown Plot 1. Paper presented at *in situ* presentation meeting at Daepyung, Gyeongna Univ. Museum.

Lee, S.K. (1999). Historical Remains at Okhyun Mookundong. Presentation Report. Kyung-nam Univ. Museum and Milyang Univ. Museum.

Lee, S.K. et al. (1992). *Transition of Rice Cultural Practices during the Chosun Dynasty.* Shinkoo Munhwasa Pub.

Lee, Y.J., Park, T.S. and Ha, M.S. (1994). Studies on the prehistoric rice culture in Korea. Pages 927-980 in *Songkok Treatise Collection.*

Lim, M.S., Kim, J.H. and Choi, H.O. (1983). Rice genetics and breeding technology. Pages 149-158 in *History of Agricultural Technology in Korea.*

Moon, H.P. (2005). Rice variety improvement and production technology in Korea. Pages 127-149 in Int'l Seminar: *Seed, Cultivation, Irrigation and Drainage for Increasing Food Production in the Northeast Asia (2005).*

Nakashima, N. (1995). Nabatake Ishekii. Pages 169-171 in *Genetical and Archeological Investigations on the Origin of Cultivated Rice and Ancient Rice Culture in East Asia.* Saga Univ., Japan.

National Museum of Korea (2000). *Rice in Korean History. New Millennium Special Exhibition.* Seoul, Korea.

Nishiyama, B. and Kumashiro, S. (Co-Trans.) (1957). *"Chi Min Yao Shu"* Translated in Japanese. The National Inst. of Agriculture, Ministry of Agriculture & Forestry, Japan.

Okajima, H. and Shida, Y. (1986). *"Fan Sheng-Chih Shu": An Agriculturist Book of China Written by Fan Sheng-Chih in The First Century BC.* Nobunkyo, Tokyo.

Rural Development Administration (2006). *Varietal Description of Major Crops.* 121 pp.

Shim, B.K. (1992). Paddy Field Site of First Excavation in Korea. The Historical Remains Site in Habookjung Yangsan. *Report of the Study on the Habookjung Historical Site.* Pusan Univ. Museum, Pusan, Korea.

Shu, H.S. (2003). *Characterization of Weedy Rice Germplasm.* Youngnam Univ. Gyeongsan, Korea.

Son, B.G. (1992). *Report on the Study of the Natural Environment and Archaeological Excavation in the Illsan Zone-1.* Institute of Korean Prehistory Culture, Seoul. Korea.

Takakura, H.A. (1995). Foreign exchange during introduction of rice culture in Japan. Pages 277-281 in *Genetical and Archaeological Investigations on the Origin of Cultivated Rice and Ancient Rice Culture in East Asia.* Saga Univ., Japan.

Wang, S.P. and Wang, C.C. (1995). Remains excavated at "Yang Chia Juan, Shi Ha". Pages 143-145 in *Genetical and Archaeological Investigations on The Origin of Cultivated Rice and Ancient Rice Culture in East Asia.* Saga Univ., Japan.

Wu, Q.Y. (1995). The examination and the studies on the carbonized rice excavated at the Dai Jui Zi remains. Pages 129-132 in *Genetical and Archaeological Investigations on the Origin of Cultivated Rice and Ancient Rice Culture in East Asia.* Saga Univ., Japan.

Yamazaki, S. (1995). Itatsuke Yiseki. Pages 174-175 in *Genetical and Archaeological*

Investigation on the Origin of Cultivated Rice and Ancient Rice Culture in East Asia. Saga Univ., Japan.

Yan, W.M. (1995). Rice farming during pre-historical period, Cinca. Pages 209-214 in *Genetical and Archaeological Investigation on the Origin of Cultivated Rice and Ancient Rice Culture in East Asia*. Saga Univ., Japan.

Yeom, J.S. (2007). Opening technology of plowing and weed control in the Chosun Dynasty. Pages 51-136 in *Symposium on Reillumination of Ancient Agricultural Technology in Korea*. The Korean Academy of Science and Technology, Seoul, Korea.

Yim, H.J. (1978). Excavated carbonized rice from the Heunam-ri historical dwelling site. "Henamri Site No. 4". *Archaeology and Anthropology (Seoul Natl. Univ.)* 8:30-38.

Yim, H.J. (1990). Studies on the archaeological excavation at Kimpo Semi-Peninsula Kyunggi Province, Korea. *Annual Report of Seoul National Univ.* 2: 1-22.

Chapter 6: History of Rice in Japan

Fujiwara, Hiroshi (1998). *Searching for the Origin of Rice Cultivation*. Iwanami Shinsho No. 554, 201 pp. (in Japanese)

Ishikawa, Rikinosuke (1901). *'Ine-shu Tokushitsu no Ben'*, republished with commentary in 1985 as Meiji Nosho Zenshu No. 2, p. 323-397, Rural Culture Association, Tokyo.

Kaneda, Chukichi (1995). Development of Rice Cultivation in Japan since Late 19th Century. Pages 15-23 in *Proc. Inter'l Scientific Symp. Asian Paddy Fields: Their Environmental, Historical, Cultural and Economic Aspects under Various Physical Conditions*. 29 May-3 June, 1995.

Matsuo, Takane (1952). Genecological studies on the cultivated rice. *Bulletin of the National Institute of Agricultural Sciences. Ser. D, No.* 2:1-111. (in Japanese with English discussion and summary)

Miyazaki, Antei (Yasusada) (1697). *'Nogyo Zensho'*, republished with commentary in 1978 as Nihon Nosho Zenshu No. 12 and 13. Rural Culture Association, Tokyo.

Nara, Senji (1881). *'Sinsen Beisaku Kairyoho'*, republished with commentary in 1985 as Meiji Nosho Zenshu No. 2, pp. 5-67. Rural Culture Association, Tokyo.

Okura, Nagatsune (1822). *'Nogu Benri-ron'*, republished with commentary in 1977 as Nihon Nosho Zenshu No. 15, pp. 119-315. Rural Culture Association, Tokyo.

Okura, Nagatsune (1826). *'Jokouroku'*, republished with commentary in 1977 as Nihon Nosho Zenshu. No. 15, pp. 3-118. Rural Culture Association, Tokyo.

Sato, Y-I., Fujiwara, H. and Udatsu, T. (1990). Morphological differences in silica body derived from motor cell of *indica* and *japonica* in rice. *Japan. J. Breed.* 40:495-504. (in Japanese with English summary)

Sato, You-Ichiro (2002). *'Ine no Nihon-shi' (History of Rice in Japan)*, pp. 197. *Kadokawa Senjo No. 337*, Tokyo, Japan. (in Japanese)

Sawada, Goichi (1926). *Numerical Studies on Public Welfare and Economy in the Nara Period (in Japanese)*, cited in Newspaper 'NOMIN', July 1, 2001.

Tomiyama, Kazuko (1993). *Rice in Japan: Environment and Culture Were Thus Created. Chuko Shinsho No. 1156*, Chuokoron Publ. 232 pp. (in Japanese)

Tsuchiya, Matasaburo (1707). 'Koka-shunju', republished with commentary in 1980 as Nihon Nosho Zenshu No. 4. Rural Culture Association, Tokyo. 387 pp.

Tsuchiya, Matasaburo (1717). 'Nogyo Zue', republished with commentary in 1983 as Nihon Nosho Zenshu No. 26, Rural Culture Association, Tokyo. 312 pp.

Yamaguchi, Jouji (2000). Wooden agricultural tools in the Yayoi period. Pages 587-609 in Hong-jong Lee (ed), Culture Change and Culture Contact of Ancient Society in Korea. (in Japanese)

Chapter 7: History of Rice in Southeast Asia and Australia

Alexander, J. and Coursey, D.G. (1969). The origins of yam cultivation. Pages 323-29 in Ucko, P.J. and Dimbleby, G.W. (eds), The Domestication and Exploitation of Plants and Animals. London.

Anonymous (1884). The French in Indo-China: With a Narrative of Garnier's Explorations in Cochin-China, Annam and Tonquin. Nelson and Sons, London.

Arbhabhirama, Anat; Phantumvanit, Dhira; Elkington, D. and Ingkasuwan, P. (1987). Thailand Natural Resources Profile. The National Environment Board and Thailand Development Research Institute, Bangkok.

Attwater, R. (1998). Livelihood, Risk and Common Property: Local Water Resource Development in Thailand. Tai Culture 3(2):71-77.

Brown, I. (1988). The Elite and the Economy of Siam c. 1890-1920. Oxford University Press, Singapore.

Chandraratna, M.F. (1964). Genetics and Breeding of Rice. Longmans, London.

Chang, T.T. (1988). The ethnobotany of rice in island Southeast Asia. Asian Perspectives 26:69-76.

Chang, T.T. and Bardenas, E. (1965). Present Knowledge of Rice Genetics and Cytogenetics. Technical Bulletin No. 4. International Rice Research Institute, Los Banos.

Changbao Li, Ailing Zhou and Sang Tao (2006). Rice domestication by reduced shattering. Science 311:1936-1939.

Chomchalow, N. (1993). Agricultural development in Thailand. In Penning de Vries et al. (eds), Systems Approaches for Agricultural Development. Kluwer, Netherlands.

Cohen, P.T. (1980). Irrigation and the Northern Thai State in the Nineteenth Century. In Davis, R.B. (in memory of) Patterns and Illusions: Thai History and Thought. Australian National University, Canberra.

Croll, E. and Parkin, D. (1992). Cultural understandings of the environment. In Croll, E. and Parkin, D. (eds), Bush Base: Forest Farm Culture, Environment and Development. Routledge, London.

de Campos, J. (1940). Early Portuguese Accounts of Thailand. Journal of the Siam Society 32:1-27.

De Datta, S.K. (1975). Upland rice around the world. In Major Research in Upland Rice. IRRI, Manila.

Donner, W. (1978). The Five Faces of Thailand: An Economic Geography. Institute of Asian Affairs, Hamburg and Hurst, London.

Dumarcay, J. and Smithies, M. (1995). Cultural Sights of Burma, Thailand and Cambodia. Oxford University Press, Kuala Lumpur.

Golomb, L. (1972). From Subsistence to Symbol: How Glutinous Rice Has Stuck.

Bound Monograph, Library of the Siam Society, Bangkok.

Grist, D.H. (1959). *Rice*. Longmans, London.

Groslier, B.P. (1962). *Indochina, Art in the Melting Pot of Races*. Methuen, London.

Guillon, E. (1999). *The Mon: A Civilisation of Southeast Asia. (Translated and edited by Di Crocco, J.V.)*. The Siam Society under the Patronage of His Majesty the King, Bangkok.

Gutkind, E.A. (1946). *Revolution of Environment*. Trobner and Co., London.

Haanant, Juanjai *et al.* (1987). *TDRI Natural Resources Profile*. Thailand Development Research Institute, Bangkok.

Hall, K.R. (1992). Economic History in Early South East Asia. In Tarling, N. (ed), *The Cambridge History of South East Asia. Volume 1: From Early Times to c. 1880*. Cambridge University Press, Cambridge.

Hanks, L.M. (1972). *Rice and Man: Agricultural Ecology in South East Asia*. Aldane, Atherton, Chicago.

Ho, P.T. (1959). *Studies on the Population of China: 1368-1953*. Harvard University Press, Cambridge, Massachusetts.

Hung, Ling-Chi (1793). See Ho, P.T. (1959).

Hutchinson, E.W. (1940). *Adventures in Siam in the 17th Century*. D.D. Books, London.

Ibrahim, M. (1972). *The Ship of Sulaiman*. O'Kane, J. (translator) *Persian Heritage Series No. 11*. Routledge and Kegan Paul, London.

Ingram, J.C. (1971). *Economic Change in Thailand: 1850-1970*. Stanford University Press, Stanford.

IRRI (1995). *Water: The Looming Crisis*. The International Rice Research Institute, Los Banos, Philippines.

Ishii, Y. (1978). History and Rice Growing. In Ishii, Yoneo (ed), *Thailand: A Rice Growing Society*. The University Press of Hawaii, Honolulu.

Judd, L. (1964). *Dry Rice Agriculture in Northern Thailand. South East Asia Data Paper No. 52*. Cornell University.

Jumsai, Sumet (1997). *Naga: Cultural Origins in Siam and the West Pacific*. D.K. Books, Bangkok.

Kato, K. (1998). Water irrigation and administration of the Tai of Sishuangpanna. *Tai Culture* 3(2):49-70.

Kulthong Kham, S. *et al.* (1964). *Rice Economy of Thailand*. Bangkok.

Kunstadter, P. and Chapman, E.C. (1978). Problems of shifting cultivation and economic development in Northern Thailand. In Kunstadter, P., Chapman, E.C. and Sabhasri, Sanga (eds), *Farmers in the Forest: Economic Development and Marginal Agriculture in Northern Thailand*. East West Centre, University Press of Hawaii, Honolulu.

Lamb, H.H. (1977). *Climate: Past, Present and Future*. Routledge, London.

Lourido, R.A. (1996). European trade between Macao and Siam, from its beginnings to 1663. *Journal of the Siam Society* 84:75-97.

Lu, J.J., and Chang, T.T. (1980). Rice in its temporal and spatial perspectives. Pages 1-74 in Luh, B.S. (ed), *Rice: Production and Utilization*. Westport, Conn.

Major, R.H. (1957). *India in the XVth Century: The Travels of Nicolo Conti*. Hakluyt Society.

Murray, S.O. (1996). *Angkor Life*. Bua Luang Books, Bangkok.

Office of Agricultural Economics (1998). *Trends in Agricultural Production for 1998, 1999*. Office of Agricultural Economics, Bangkok.

Owen, N.G. (1971). The Rice Industry of Mainland South East Asia 1850-1914. *Journal of the Siam Society* 59(2):75-143.

Pelzer, K.J. (1978). Swidden cultivation in South East Asia: Historical, ecological and economic perspectives. In Kunstadter, P., Chapman, E.C. and Sabhasri, Sanga (eds), *Farmers in the Forest: Economic Development and Marginal Agriculture in Northern Thailand*. East West Centre, University Press of Hawaii, Honolulu.

Penth, H. (1994). *A Brief History of Lan Na: Civilisations of North Thailand*. Silkworm Books, Chiang Mai, Thailand.

Phongpaichit, Pasuk and Baker, C. (1998). *Thailand's Boom and Bust*. Silkworm Books, Chiang Mai, Thailand.

Rogers, P. (1996). *North East Thailand from Prehistoric to Modern Times*. D.K. Books, Bangkok.

Saraya, D. (1989). State formation. In *'Culture and Environment in Thailand': A Symposium of the Siam Society*. Siam Society under Royal Patronage, Bangkok.

Sauer, C.O. (1952). *Agricultural Origins and Dispersals*. American Geographical Society, New York.

Setboonsarng, Suthad and Evenson, R. (1991). Technology, infrastructrure, output, supply and factor demand in Thailand's agriculture. In Evenson, R.E. and Pray, Carl (eds), *Research and Productivity in Asian Agriculture*. Cornell University Press, Ithaca.

Shoocongdea, Rasmi (1996). Rethinking the development of sedentary villages in western Thailand. In Bellwood, P. (ed), Indo-Pacific Pre-History: The Chiang Mai Papers. Volume 1. *Bulletin of the Indo Pacific Pre-History Association* 14:203-215.

Siribhadra, Smitthi, H.I. (1999). Some recently reopened 7th-13th century temples in Cambodia. Lecture Presented to the Siam Society under Royal Patronage, Bangkok.

Steinberg, D.J. (1987). *In Search of South East Asia: A Modern History*. University of Hawaii Press, Honolulu.

Suchitta, Pornchai (1989). Traditional technology as a culture-environment relationship in Thailand. In *Culture and Environment in Thailand: A Symposium of the Siam Society*. Siam Society under Royal Patronage, Bangkok.

Sukwong, Somsak (1989). Patterns of land use as influenced by forestry. In *Culture and Environment in Thailand: A Symposium of the Siam Society*. Siam Society under Royal Patronage, Bangkok.

Surareks, Vanpen (1998). The Müang Fai irrigation system of Northern Thailand: Historical development and management. *Tai Culture* 3(2):37-48.

Tanabe, S. (1978). Land reclamation in the Chao Phraya Delta. In Ishii, Yoneo (ed), *Thailand: A Rice Growing Society*. The University Press of Hawaii, Honolulu.

Tanabe, S. (1994). Sacrifice and the transformation of ritual: The Pu Sae Na Sae spirit cult in Northern Thailand. In Tamara, K. and Rajah, Ananda (eds), *Spirit Cults and Popular Knowledge in Southeast Asia*. Institute of Southeast Asian Studies, Singapore.

Tanaka, A., Kawana, K. and Yamaguchi, J. (1966). *Photosynthesis, Respiration and Plant Type of the Tropical Rice Plant. Technical Bulletin No. 7.* International Rice Research Institute, Los Banos.

Taylor, K.W. (1992). The early kingdom. In Tarling, N. (ed), *The Cambridge History of South East Asia Volume 1: From Early Times to c. 1880.* Cambridge University Press, Cambridge.

Thomson, V. (1967). *Thailand: The New Siam.* Paragon Books, New York.

Trebuil, G. (1984). A functional typology of farming systems in Sating Pra Area—Southern Thailand. In *Rural Thai Social Development of Thai Economy.* Thai Studies. Chulalongkorn University.

Vallibhotama, Srisakra (1989). Traditional Thai villages and cities: An overview. In *Culture and Environment in Thailand: A Symposium of the Siam Society.* Siam Society under Royal Patronage, Bangkok.

Van Beek, S. (1995). *The Chao Phya: River in Transition.* Oxford University Press, Kuala Lumpur.

van Liere, W.J. (1980). Traditional water management in lower Mekong basin. *World Archaeology* 11(3):274.

van Liere, W.J. (1989). Mon-Khmer approaches to the environment. In *Culture and Environment in Thailand: A Symposium of the Siam Society.* Siam Society under Royal Patronage, Bangkok.

Vavilov, N.I. (1930). The Problems of the Origins of Cultivated Plants and Domestic Animals as Conceived at the Present Time. In *Plant Breeding Abstracts (1930).*

Watabe, T. (1978). The development of rice cultivation. In Ishii, Y. (eds and trans by Hawkes, P. and S.), *Thailand: A Rice Growing Society.* The University Press of Hawaii, Honolulu.

Watabe, T. (1967). *Glutinous Rice in Northern Thailand.* Centre for South East Asian Studies, Kyoto University.

Wongthes, Pranee and Wongthes, Sujit (1989). Art, culture and environment of the Thai-Lao speaking groups. In *Culture and Environment in Thailand: A Symposium of the Siam Society.* Siam Society under Royal Patronage, Bangkok.

Wyatt, D.K. (1984). Laws and social order in early Thailand: An introduction to the Mangraisat. *Journal of South East Asian Studies* 15(2):245-252.

Wyatt, D.K. (1989). Discussion Leader's Comments. In *Culture and Environment in Thailand: A Symposium of the Siam Society.* Siam Society under Royal Patronage, Bangkok.

Yamchong, Cheah (1996). More thoughts on the ancient culture of the Tai people: The impact of the Hua Xia Culture. *Journal of the Siam Society* 84:29-48.

Yen, D.E. (1977). Hoabinhian Horticulture. In Allen, J., Golson, J. and Jones, R. (eds), Sunda and Sahul: prehistoric studies in Southeast Asia, Melanesia and Australia. Academic Press, London.

Yule, H. and Cordier, H. (1903). *The Travels of Marco Polo.* 276 pp.

Chapter 8: History of Rice in South Asia (Up to 1947)

Belfour, Betty (1899). *The History of Lord Lytton's Indian Administration, 1876-1880.* London.

Barbosa, Duarte (1918). The Book of Duarte Barbosa. I. Dames, M.L. (ed), London. Quoted in Tarafdar M.R. (1995). *Trade, Technology and Society in Medieval Bengal.* International Centre for Bengal Studies, Dhaka University, Dhaka.

Barker, R. and Molle, F. (2002). Perspectives on Asian irrigation. Paper Presented at the *Conference on Asian Irrigation in Transition: Responding to the Challenges Ahead.* Workshop held at Asian Institute of Technology, Bangkok, Thailand. April 22-23, 2002.

Barsamian, David (2001). Reflections of an Economist. India Together, September 2001. Alternative Radio, Boulder, USA.

Bedekar, V.V. (2002). Agriculture in Ancient India. Lecture delivered at the *Seminar on Agriculture in Ancient India* at the Institute of Oriental Studies, Thane on October 26, 2002.

Bellwood, B. (2004). *First Farmers: The Origin of Agricultural Societies.* Blackwell Publishing.

Bhatia, B.M. (1985). *Famines in India: A Study in Some Aspects of the Economic History of India with Special Reference to Food Problem.* Konark Publishers, Delhi.

Chakrabarti, D.K. (2001). *Archaeological Geography of the Ganga Plain: The Lower and the Middle Ganga.* Orient Longman.

Chauhan, M.S., Pokharia, A.K. and Singh, I.B. (2004). Pollen record of Holocene vegetation and climate change from Lahuradewa Lake. (Abstract) *Joint Annual Conference of IAS, ISPQS and IHCS and National Seminar on the Archaeology of the Ganga Plain*: 41. Lucknow.

Chibber, Vivek (1998). Breaching the Nadu: Lordship and Economic Development in Pre-Colonial South India. *Journal of Peasant Studies* 26(1):1-42.

Chopra, P.N. (ed) (2003). *The Gazetteer of India. Vol. II.* Publications Division, Ministry of Information and Broadcasting, Government of India, New Delhi.

Danielou, A. (1965). *Shilappadikaram (The Ankle Bracelet) by Ilango Adigal (An English Translation).* New Directions Publishing Co.

Danielou, A. and Kopalayyar, T.V. (1989). *Manimekhalai: The Dancer with a Magic Bowl by Merchant Prince Shattan Cattanar (An English Translation).* New Directions Publishing Co.

Das, Purnachandra (1993). *History of Orissa.* Sitaram Printers. Cuttack. (in Oriya).

Dash, K.C. (1997). Economic Life of Orissa under the Imperial Gangas. In: Patnaik, Nihar Ranjan (ed), *Economic History of Orissa.* Indus Publishing, New Delhi.

Davis, Mike (2001). *Late Victorian Holocausts: El Nino Famines and the Making of the Third World.* Verso.

Dutt, Romesh C. (1902). *Economic History of British India. Vol I.* Kegan Paul, Trench, Trubner, London.

Schultz, B., Thatte, C.D. and Fahlbusch, H. (2004). *The Indus Basin: History of Irrigation, Drainage and Flood Management.* International Commission on Irrigation and Drainage.

Fujiwara, H. (1996). *Origin of Rice Cultivation.* Iwanami Publishers Co. Ltd. (in Japanese).

Fuller, D.Q. (2002). Fifty years of archaeobotanical studies in India: Laying a solid foundation. Pages 247-364 in Settar, S. and Korisettar, Ravi (eds.), *Indian Archaeology in Retrospect: Archaeology and Interactive Disciplines. Vol. III.* Indian Council of Historical Research, New Delhi.

Gates, C. (2003). *Ancient Cities: The Archaeology of Urban Life in the Ancient Near East and Egypt, Greece and Rome.* Routledge.

Government of India (1928). *Report of the Royal Commission on Agriculture in India.* His Majesty's Stationery Office, London.

Grover, B.L. and Grover, S. (2004). *A New Look at Modern History.* S. Chand & Co., New Delhi.

Habib, Irfan (1964). Evidence for 16th Century Agrarian Conditions in Guru Granth Sahib. *The Indian Economic and Social History Review. I.* Quoted by Raychaudhury T. and Habib, Irfan (1982).

Harnetty, Peter (1991). De-industrialisation Re-visited: The Handloom Weavers of the Central Province of India *c.* 1800-1947. *Modern Asian Studies:* 25(3): 455-510.

Haroon Mohsini (nd). Sher Shah "Suri" and the Afghan Revival. See website: http://www.afghan-network.net/Culture/shershah.html

Harvey, E.L., Fuller, D.Q., Pal, J.N. and Gupta, M.C. (2005). Early agriculture of the Neolithic Vindhyas (North-Central India). *South Asian Archaeology* (2003):329-334.

Joshi, P.M., Kulakarni, A.R., Nayeem, M.A. and De Souza, T.R. (1996). *Mediaeval Deccan History: Commemoration Volume in Honour of Purushottam Mahadeo Joshi.* Popular Prakashan.

Kajale, M.D. (1989). Archaeobotanical investigations at Megalithic Bragimohari and its significance for ancient Indian agricultural system. *Man and Environment* 13:87-100.

Keay, J. (2001). *India: A History.* Grove Press.

Kharakwal, J.S., Rawat, Y.S. and Osada, T. (2007). Kanmer: A Harappan site in Kachchh, Gujarat, India. Pages 21-46 in Osada, T. (ed), *Linguistics, Archaeology and the Human Past, Occasional Paper 2.* Indus Project, Research Institute for Humanity and Nature, Kyoto.

Kumar, T. (1988). *History of Rice in India.* Gyan Publishing House, Delhi.

Kurasaki, Takashi (2006). Long-term agricultural growth and crop shifts in India and Pakistan. *Journal of International Economic Studies* 20:19-35.

Mehra, K.L. (2005). Rice in Indian History and Culture. Pages 1-31 in Sharma, S.D. and Nayak, B.C. (eds), *Rice in Indian Perspective.* Today and Tomorrow Printers and Publishers. New Delhi.

Mishra, Anupam (1993). *Aaj Bhi Khare Hain Talaab. (The Tanks Are There Even Today.)* Gandhi Peace Foundation, New Delhi. (in Hindi)

Mohan, R. (2007). Statistical system of India: Some reflections. Inaugural address by the Deputy Governor, Reserve Bank of India on the Statistics Day and Annual Conference on Financial Statistics at Department of Statistical Analysis and Computer Services, Reserve Bank of India, Mumbai on June 29, 2007.

Moreland, W.H. (1922). The Agricultural Statistics of Akbar's Empire. *Journal of the United Provinces Historical Society. II.* Quoted by Raychaudhury T. and Habib, Irfan (1982).

Nene, Y.L. (2005). Rice research in South Asia through ages. *Asian Agri-History* 9(2):85-106.

Nene, Y.L. (2006). Rice in the context of Indian agriculture. Pages 35-44 in *Souvenir of 2nd International Rice Congress—2006: Science, Technology and Trade for Peace and Prosperity.* October 9-13, 2006. New Delhi.

Nilakanta Sastri, K.A. (2004). *A History of South India from Pre-historic Times to the Fall of Vijayanagar. Fourth Edition.* Oxford University Press. New Delhi.

Pande, G.C. (1990). *Fundamentals of Indian Culture.* Motilal Banarsidass Publication.

Pokharia, A.K. (2006). Record of macrobotanical remains from the Aravalli Hill, Ojiyana, Rajasthan: Evidence for agriculture-based subsistence economy. *Current Science* 94(5):612-623.

Possehl, G.L. and Rissman, P.C. (1992). The chronology of prehistoric India: From earliest times to Iron Age. Pages 465-474 in Ehrich, R.W. (ed), *Chronologies in Old World Archaeology. I.* The University of Chicago Press. Chicago and London.

Ramiah, K. and Rao, M.B.V.N. (1953). *Rice Breeding and Genetics.* Indian Council of Agricultural Research, New Delhi.

Randhawa, M.S. (1979). *A History of the Indian Council of Agricultural Research.* Indian Council of Agricultural Research, New Delhi.

Raychaudhury, T. and Habib, Irfan (1982). *The Cambridge Economic History of India, Vol. I: 1200-1750.* Cambridge University Press, Cambridge, UK.

Richharia, R.H. and Govindaswami, S. (1990). *Rices of India.* Academy of Development Sciences, Kashele (Maharashtra).

Rothermund, Dietmar (1993). *An Economic History of India: From Pre-Colonial Times to 1991.* Routledge.

Saha, Pijushkanti (nd). Land tenure system in India: A historical perspective.

Sangwan, Satpal (2007). Level of agricultural technology in India (1757-1857). *Asian Agri-History* 11(1):5-25.

Saran, Mishi (2005). *Chasing the Monk's Shadow: A Journey in the Footsteps of Xuanzang.* Penguin, New Delhi.

Saraswat, K.S. and Pokharia, A.K. (2004). Plant economy at Lahuradewa: A preliminary contemplation. (Abstract). *Joint Annual Conference of IAS, ISPQS and IHCS and National Seminar on the Archaeology of the Ganga Plain, Lucknow,* 46.

Saraswat, K.S. (2004). Plant economy in ancient Malhar. *Pragdhara* 14:137-171.

Saraswat, K.S. (2005). Agricultural background of the early farming communities in the Middle Ganga Plain. *Pragdhara* 15:145-177.

Saxena, A., Prasad, V., Singh, I.B., Chauhan, M.S. and Hasan, R. (2006). On the Holocene record of phytoliths of wild and cultivated rice from Ganga Plain: Evidence for rice-based agriculture. *Current Science* 90(11):1547-1551.

Sen, Amartya (1981). *Poverty and Famines: An Essay on Entitlements and Deprivation.* Clarendon Press, Oxford, UK.

Sharma, G.R. (1985). From hunting and food gathering to domestication of plants and animals in the Belan and Ganga Valleys. Pages 369-371 in Misra,

V.N. and Bellwood, P. (eds), *Recent Advances in Indo-Pacific Prehistory.* Oxford & IBH, New Delhi.

Stein, B. (1998). *A History of India.* Blackwell Publishing.

Swaminathan, M.S. (2007). The crisis of Indian agriculture. *The Hindu* (A Daily Newspaper of India), Wednesday, August 15, 2007.

Tewari, R., Srivastava, R.K., Singh, K.K., Saraswat, K.S., Singh, I.B., Chauhan, M.S., Pokharia, A.K., Saxena, A., Prasad, V. and Sharma, M. (2006). Second preliminary report of the excavations at Lahuradewa, District Sant Kabir Nagar, U.P.: 2002-2003-2004 & 2005-06. *Pragdhara* 16:35-68.

Thakkar, Himanshu (1999). South Asia Network on Dams, Rivers and People, India. Assessment of Irrigation Options. Prepared for Thematic Review. IV.2. for World Commission on Dams.

Thapar, Romila (2003). *The Penguin History of India from the Origins to AD 1300.* Penguin Books, New Delhi.

Wallach, B. (2004). Agricultural development in British India. In *"Losing Asia, Modernization and the Culture of Development".* John Hopkins University Press.

Washbrook, David (2007). India in the early-modern world economy: Modes of production, reproduction and exchange. *Journal of Global History* (London School of Economics and Political Science, UK) 2:87-111.

Watt, G. (1891). *A Dictionary of the Economic Products of India.* Reprinted in 1972 by Cosmo Publications, Delhi.

Ziauddin, Barani (1357). *Tarikh-i-Firuz Shahi.* Quoted in: Raychaudhury T. and Habib, Irfan (1982).

Sri Lanka

Harischandra, B.W. (1998). *The Sacred City of Anuradhapura with Forty-Six Illus-trations.* (Reprint of 1985 Edition). Asian Educational Services, New Delhi.

De Silva, K.M. (1981). *A History of Sri Lanka.* C. Hurst & Co.

Newitt, M.D.D. (2005). *A History of Portuguese Overseas Expansion, 1400-1668.* Routledge.

Chapter 9: History of Rice in South Asia (1947-2007)

India

Ehrlich, Paul (1968). *The Population Bomb.* Ballantine. New York.

Government of India (2003). *Problems and Prospects of Rice Export from India.* Directorate of Rice Development, Patna.

Government of India (2006). *Report of the National Commission on Farmers.* Report submitted by M.S. Swaminathan, Chairman, National Commission on Farmers to the Government of India.

Government of India (2008). Economic Survey 2007-08. Ministry of Finance. See website: http://indiabudget.nic.in

Prasada Rao, U. (2004). Role of Directorate of Rice Research in genetic improve-ment of rice varieties of India. Pages 101-140 in Sharma, S.D. and Prasada

Rao, U. (eds), *Genetic Improvement of Rice Varieties of India*. Today and Tomorrow's Printers and Publishers, New Delhi

Saha *et al.* (1998). Long term fertilizer experiments. *Better Crops International* 16 (Special Supplement), May 2002.

Sen, A. (1981). *Poverty and Famines: An Essay on Entitlement and Deprivation*. Oxford University Press.

Shobha Rani, N., Prasad, G.S.V., *et al.* (2008). *High Yielding Rice Varieties of India. Technical Bulletin No 33*. Directorate of Rice Research, Hyderabad (India).

Sugihara, K. (2006). *Notes on Trade Statistics of British India*. Economics Faculty, Osaka University, Japan.

Tiwari, K.N. (2002). Rice production and nutrient management. *Better Crops International* 16 (Special Supplement), May 2002.

Wallach, B. (2004). Agricultural development in British India. In *Losing Asia: Modernization and the Culture of Development. Revised Edition*. Johns Hopkins University Press.

Bangladesh

Akmam, A. (1991). *Early Urban Centres in Bangladesh: An Archaeological Study*. Ph.D. dissertation, Calcutta University, Kolkata.

Alim, A. (1968). *Rice Improvement in East Pakistan*. Department of Agriculture, Government of East Pakistan. 134 p.

Banglapedia. (2004). http://banglapedia.search.com.bd/HT/F_0015.htm

Barua, B.M. (1934). The Old Brahmi Inscriptions of Mahasthan. *Indian Historical Quarterly* 10:57-66.

BBS (Bangladesh Bureau of Statistics). *Statistical Yearbooks: 1974, 1980, 2004*. BBS, Government of Bangladesh, Dhaka.

Bhandarkar, D.R. (1931). Mauryan Brahmi Inscription of Mahasthan. *Epigraphia Indica* 21:83-91.

BRRI (1978). *The Rice Situation in Bangladesh 1977*. BRRI, Joydebpur.

BRRI (2002). *(Three Decades of Success in Rice Research)*. BRRI, Joydebpur. (in Bengali)

Chakrabarti, A. (1991). *History of Bengal (c. 550 AD to c. 750 AD)*. The University of Burdwan, West Bengal, India.

Das, T. (2005). *Rices in Bangladesh*. Dr. Tulsi Das, 11B NAEM Road, Dhanmondi, Dhaka.

del Ninno, C., Dorosh, P.A. and Subbarao, K. (2005). Food Aid and Food Security in the Short and Long Run: Country Experiences from Asia and Sub-Saharan Africa. (PPT presentation). The World Bank.

Dorosh, P.A., del Ninno, C. and Shahabuddin, Q. (2004). *The 1998 Floods and Beyond: Towards Comprehensive Food Security in Bangladesh*. The University Press Limited, Dhaka and International Food Policy Research Institute (IFPRI), Washington, D.C.

Goletti, Francesco (1994). *The Changing Public Role in Rice Economy Approaching Self-suficiency: The Case of Bangladesh. Research Report 98*. IFPRI, Washington DC.

Halim, A., Rahman, M.L., Hossain, S.M.A., Hossain, M.A., Mamun, A.A. and Ali,

M.H. (1984). *Rice and Civilization: The Bangladesh Case*. Graduate Training Institute, Bangladesh Agricultural University, Mymensingh (Mimeo).

Hossain, S.M.A., Sattar, M. and Ahmed, J.U. (1981). *Bench Mark Survey, Kanhar Cropping Systems Research Site*. Graduate Training Institute and Department of Agronomy, Bangladesh Agricultural University, Mymensingh. (Mimeo)

Hossain, A. (1999). A second look at the 1974 famine: Additional insights and policy implications. *Journal of Bangladesh Studies*, Vol. 1, No. 1.

Hossain, S.M.A., Sattar, M. and Ahmed, J.U. (1983). *Component Technology Research*. CSRDP, Department of Agronomy, Bangladesh Agricultural University, Mymensingh. (Mimeo)

Islam, N. (2007). What was it about the 1974 famine? http://www.scholarsbangladesh.com/nurulislam1.php

Kabir, K.M.E. (ed) (1986). *Rice Research in Bangladesh*. IRRI, Philippines.

Malek, Q.M.A. (1973). *Rice cultivation in Comilla Kotwali thana: The role of cooperatives*. BIDS, Dhaka, Bangladesh (Mimeo).

MoA (Ministry of Agriculture, Government of Bangladesh) (2007). See website information on various aspects of rice. http://www.moa.gov.bd/index.htm

Niaz, M.S. and Miller, G.W. (1952). *Rice Situation*. Ministry of Food and Agriculture, Government of Pakistan, Karachi.

Osmany, S.R. (1987). *The Food Problem of Bangladesh*. UN WIDER Working Paper 29.

Samad, A. (1999). Private sector imports of fertilizer and pesticides in Bangladesh. In: Sidhu, S.S. and Mudahar, M.S. (eds), *Privatization and Deregulation: Needed Policy Reform for Agribusiness Development*. Kluwer Academic Publishers, The Netherlands.

Sattar, M. and Hossain, S.M.A. (1986). An evaluation of farmers' technology for seed production and post-harvest operations in *aus* rice. *Bangladesh. J. Exten. Edn.* 1(2) 1-12.

Sobhan, R. (1979). Politics of food and famine in Bangladesh. *Econ. & Political Weekly* 14(48):1973-80.

Pakistan

Blood, P.R. (1996). *Pakistan: A Country Study*. Diane Publishing.

Kayank, E. (1999). *Cross-national and Cross-cultural Issues in Food Marketing*. Haworth Press Inc.

Mahmood, M.A., Sheikh, A.D. and Kashif, M. (2007). Acreage supply response of rice in Punjab. *Pakistan J. Agric. Res.* 45(3):231-236.

Jaffrelot, C. and Beaumont, G. (2004). *A History of Pakistan and its Origins*. Anthem Press.

Haq I. (n.d.) Barrages and Dams in Pakistan. See website: www.waterinfo.net.pk/pdf/mbp.pdf

UNITAR (2004). Hiroshima Office for Asia and the Pacific, Series on Biodiversity Training Workshop on Wetlands, Biodiversity and Water: New Tools for the Ecosystem Management. Kushiro, Japan, 29 November to 3 December 2004, Kushiro International Wetland Centre. See website: http://www.unitar.org/hiroshima/programmes/kushiro04/team_presentations/team%204/indusbasin_summ.pdf

Sri Lanka

De Silva, K.M. (1981). *A History of Sri Lanka*. C. Hurst & Co.

Fernando, A. Denis N. (2002). The ancient hydraulic civilization of Sri Lanka. Summary of guest lecture addressed at the Diplomatic Training Institute BMICH on 18 September 2002.

Jayasuriya, G. (1985). Rice production in Sri Lanka. Pages 81-86 in *Impact of Science*. International Rice Research Institute, Manila.

Mottau, S.A.W. (1981). The Dutch in Ceylon. *Journal of the Dutch Burgher Union of Ceylon*, January-December 1981, Page 6-7. (Paper read before a group of visiting tourists from Holland at St Andrew's Hotel, Nawara Eliya on 9th July 1980.

Rafeek, M.I.M. and Samaratunga, P.A. (2000). Trade liberalisation and its impact on the rice sector of Sri Lanka. *Sri Lankan Journal of Agricultural Economics* 3(1):143-154.

Chapter 10: History of Rice in Western and Central Asia

Abdurahmanov, R. (1954). *Russko-uzbekskii slovar'*. Moscow.

Adams, R.M. (1962). Agriculture and urban Life in early Southwestern Iran. *Science* 136:109-122.

Adams, R.M. (1965). *Land behind Baghdad. A History of Settlement on the Diyala Plains*. University of Chicago Press, Chicago.

Adams, R.M. (1981). *Heartland of Cities: Surveys of Ancient Settlement and Land Use on the Central Floodplain of the Euphrates*. University of Chicago Press, Chicago.

Adams, R.M. and Nissen, H.J. (1975). *The Uruk Countryside. The Natural Setting of Urban Societies*. University of Chicago Press, Chicago.

Ahsan, M.M. (1979). *Social Life under the Abbasids 170-289 AH/786-902 AD*. Longman, London.

Andersson, M. and Svanberg, I. (1995). *Från Rio till Aral: om miljömord i Centralasien*. Brevskolan, Stockholm.

Ashtor, E. (1976). *A Social and Economic History of the Near East in the Middle Ages*. Collins, London.

Bacon, E.E. (1980). *Central Asians under Russian Rule*. Cornell University Press, Ithaca, NY.

Bailey, H.W. (1946). Gandhari. *Bulletin of the School of Oriental and African Studies* (University of London) 11:764-797.

Baskakov, N.A. (1967). *Russko-karakalpakskii slovar'*. Moscow.

Bawden, E. (1945). The Marsh Arabs of Iraq. *Geographical Magazine* 17:382-393.

Bazin, M., Bromberger, C., Balland, D. and Bāzargān, S. (1990). Berenj. Pages 147-163 in Yarshater, E. (ed), *Encyclopaedia Iranica*. Mazda, Costa Mesa, CA.

Beldiceanu, N. and Beldiceanu-Steinherr, I. (1978). Riziculture dans l'Empire Ottoman. *Turcica* 9-10:9-28.

Boardman, S. (1995). Archaeobotanical progress report, (In: 'International Merv Project. Preliminary Report on the Third Season (1994) by Herrmann G., Kurbansakhatov K. *et al.*) *Iran* 33:31-60.

Bolens, L. (1990). *La cuisine andalouse, un art de vivre X!e—XIIIe siècle*. Albin Michel, Paris.

Bosworth, C.E. (1999). *The History of al-Tabari 5. The Sasanids, the Byzantines, the Lakhmids and Yemen*. State University of New York Press, Albany, NY.

Boucharlat, R. (1993). Pottery in Susa during the Seleucid, Parthian and early Sasanian periods. Pages 41-57 in Finkbeiner, U. (ed), *Materialien zur Archäologie der Seleukiden- und Partherzeit im südlichen Babylonien und im Golfgebiet. Ergebnisse der Symposium 1987 und 1989 in Blaubeuren*. Wasmuth, Tübingen.

Brosh, M. (1986). The Diet of Palestine in the Roman Period—Introductory Notes. *Israel Museum Journal* 5:41-56.

Buxton, P.A. and Dowson, V.H.W. (1921). The Marsh Arabs of Lower Mesopotamia. *Indian Antiquary* 50:289-297.

Canard, M. (1959). Le riz dans le Proche Orient aux premiers siecles de l'Islam. *Arabica* 6:113-131.

Cappers, R.T.J. (2006). *Roman foodprints at Berenike: Archaeobotanical evidence of subsistence and trade in the Eastern Desert of Egypt*. Cotsen Institute of Archaeology, Los Angeles, CA.

Chang, C-Y. (1949). Land utilization and settlement possibilies in Sinkiang. *Geographical Review* 39:57-75.

Ch'ien, S. (1961). *Records of the Grand Historian of China*, 2. Colombia University Press, New York.

Christensen, A. (1936). *L'Iran sous les sassanides*. Ejnar Munksgaard, Copenhagen.

Clark, L.V. (1973). The Turkic and Mongol words in William of Rubruck's Journey. *Journal of the American Oriental Society* 93:559-572.

Costantini, L. (1979a). Plant remains at Pirak, Pakistan. Pages 326-333 in Jarrige, J.F. and Santoni, M. (eds), *Fouilles de Pirak*, 1. Boccard, Paris.

Costantini, L. (1979b). Notes on the Palaeoethnobotany of Protohistorical Swat. Pages 703-708 in Taddei, M. (ed), *South Asian Archaeology 1977*, Vol. II. Instituto Universitario Orientale, Naples.

Dalby, A. (2003a). *Flavours of Byzantium*. Prospect Books, Blackawton, Devon.

Dalby, A. (2003b). *Food in the ancient world from A to Z*. Routledge, London.

Davidson, A. (2006). Pilaf. Pages 607-607 in *The Oxford Companion to Food*. Oxford University Press, Oxford.

de Candolle, A. (1886). *Origin of cultivated plants. Second Edition*. Reprinted 1967 by Hafner Publishing Company, New York.

Doerfer, G. (1963). *Türkische und mongolische Elemente im Neupersischen* 1. Franz Steiner Verlag, Wiesbaden.

Dyer, S. (1980). Muslim Life in Soviet Russia: The Case of the Dungans. *Journal of Muslim Minority Affairs* 2:42-54.

Dyer, S., Tsiburzgin, V. and Shmakov, A. (1992). Karakunuz: An Early Settlement of the Chinese Muslims in Russia. *Asian Folklore Studies* 51:243-278.

Edmonds, C.J. (1958). The Marshmen of Southern Iraq: Review. *Geographical Journal* 124:92-94.

Faroqhi, S. (1977). Rural society in Anatolia and the Balkans during the Sixteenth Century. *Turcica* 9:161-195.

Feliks, J. (2006). Rice. Pages 283-284 in *Encyclopaedia Judaica. Second edition*, 17. Thomson Gale, Detroit, MI.

Ferdinand, K. (1959). Ris: træk af dens dyrkning og behandling i Østafghanistan. *Kuml: Årbog for jysk arkæologisk selskab* 1959:195-232.

Fernea, R. (1970). *Shaykh and Effendi. Changing Patterns of Authority among the El Shabbana of Southern Iraq.* Harvard University Press, Cambridge, MA.

Fieldhouse, P. (1986). *Food and Nutrition: Customs and Culture.* Chapman & Hall, London.

Fragner, B. (2000). From the Caucasus to the Roof of the World: A culinary adventure. Pages 49-62 in Zubaida, S. and Tapper, R. (eds), *A Taste of Thyme: Culinary Cultures of the Middle East.* Tauris Parke, London.

Fuller, D.Q. (2006). Agricultural Origins and Frontiers in South Asia: A Working Synthesis. *Journal of World Prehistory* 20:1-86.

Fuller, D.Q., Harvey, E. and Qin, L. (2006). Presumed domestication? Evidence for wild rice cultivation and domestication in the fifth millennium BC of the Lower Yangtze region. *Antiquity* 81:316-331.

Garris, A., Tai, T., Coburn, J., Kresovich, S. and McCouch, S. (2005). Genetic structure and diversity in *Oryza sativa* L. *Genetics* 169:1631-1638.

Gelb, M. (1995). An early Soviet ethnic deportation: The far-eastern Koreans. *Russian Review* 54:389-412.

Ghirshman, R. (1978). *Iran from the Earliest Times to the Islamic Conquest.* Penguin Books, Harmondsworth.

Glover, I.C. and Higham, C.F.W. (1996). New evidence for early rice cultivation in south, southeast and east Asia. Pages 413-444 in Harris, D.R. (ed), *The Origins and Spread of Agriculture and Pastoralism in Eurasia.* UCL Press, London.

Golomb, L. (1959). *Die Bodenkultur in Ost-Turkestan: Oasenwirtschaft und Nomadentum.* Anthropos, Freiburg.

Grant, M. (2000). *Galen on Food and Diet.* Routledge, London.

Greehan, J. (2007). *Everyday Life and Consumer Culture in Eighteenth-Century Damascus.* University of Washington Press, Seattle and London.

Guest, E. (1933). *Notes on Plants and Plant Products with their Colloquial Names in Iraq.* Baghdad.

Harris, M.V. (1989). Glimpses of an Iron Age landscape. *Expedition* 31(2/3):12-23.

Hedin, S. (1898). *En färd genom Asien 1893-97*, Vol. 1. Albert Bonnier, Stockholm.

Hill, S. and Bryer, A. (1995). Byzantine porridge: *tracta, trachanás* and *tarhana*. Pages 44-54 in Wilkins, J., Harvey, D. and Dobson, M. (eds), *Food in Antiquity.* University of Exeter Press, Exeter.

Hjelt, A. (1906). Pflanzennamen aus dem Hexaëmeron Jacob's von Edessa. Pages 571-579 in Bezold, C. (ed), *Orientalische Studien Theodor Nöldeke zum siebzigsten Geburtstag gewidmet.* I.A. Töpelmann, Giessen.

Hua, C., Nagamine, T., Kikuchi, F. and Fujimaki, H. (2004). Chuugoku shinkyō uiguru jichiku no ine hinshu no tokusei kaimei. *Ikushu kenkyuu* 6:117-123.

Inalcýk, H. (1982). Rice cultivation and the *celtükci-re'aya* system in the Ottoman Empire. *Turcica* 14:69-141.

Jarring, G. (1935). The Ordam-Padishah-System of Eastern Turkestan Shrines. Pages 348-354 in *Hyllningsskrift tillägnad Sven Hedin på hans 70-årsdag den 19*

Febr. 1935. Svenska Sällskapet för Antropologi och Geografi, Stockholm.

Jarring, G. (1951). *Materials to the Knowledge of Eastern Turki* 4. Gleerup, Lund.

Jarring, G. (1998). *Agriculture and Horticulture in Central Asia in the Early Years of the Twentieth Century with an Excursus on Fishing.* Almqvist & Wiksell International, Stockholm.

Jarring, G. (1964). *An Eastern Turki English Dialect Dictionary.* CWK Gleerup. Lund.

Johnson, J.C.A. (1940). The Kurds of Iraq. II. *Geographical Magazine* 11/1:50-59.

Krochmal, A. (1958). Rice Production in Afghanistan. *Economic Botany* 12:186-191.

Lagardere, V. (1996). La riziculture en al-Andalus (VIIIe-XVe siecles). *Studia Islamica* 83:71-87.

Laufer, B. (1919). *Sino-Iranica. Chinese Contributions to the History of Civilization in Ancient Iran.* Field Museum of Natural History, Chicago.

Le Strange, G. (1905). *The Lands of the Eastern Caliphate: Mesopotamia, Persia and Central Asia from the Moslem Conquest to the Time of Timur.* Cambridge University Press, Cambridge.

Liu, L., Lee, G.A., Jiang, L. and Zhang, J. (2007). Evidence for the early beginning (c. 9000 cal. BP) of rice domestication in China: A response. *Holocene* 17:1059-1068.

Londo, J.P., Chiang, Y.-C., Hung, K.-H., Chiang, T.-Y. and Schaal, B.A. (2006). Phylogeography of Asian wild rice, *Oryza rufipogon*, reveals multiple independent domestications of cultivated rice, *Oryza sativa. Proceedings of the National Academy of the United States of America* 103:9578-9583.

Mack. G.R. and Surina, A. (2005). *Food Culture in Russian and Central Asia.* Greenwood Press, Westport, Connecticut.

Makhmudov, K.M. and Salikhov, Sh.G. (1983). *Bljuda uzbekskoi kukhni.* Tashkent.

Merriam, G.P. (1926). The regional geography of Anatolia. *Economic Geography* 2:6-107.

Miller, N.F. (1981). Plant remains from Ville Royale II, Susa. *Cahiers de la D.A.F.I.* 12:137-142.

Miller, R.A. (1959). *Accounts of Western Nations in the History of the Northern Chou Dynasty.* University of California Press, Berkeley.

Miroschedji, P., Desse-Berset, N. and Kervran, M. (1987). Fouilles du chantier ville royale II a Suse (1975-1977). II. Niveaux d'époques achéménide, séleucide, parthe et islamique. *Cahiers de la D.A.F.I.* 15:11-143.

Morony, M.G. (1984). *Iraq after the Muslim Conquest.* Princeton.

Nakamura, G. (1993). The origins of rice cultivation: Problems associated with finds of rice and Ishibocho in the northern Indian subcontinent. *Bulletin of the Ancient Orient Museum* (Japan) 14:53-117.

Nasrallah, N. (2007). *Annals of the Caliphs Kitchens: Ibn Sayyar Al-warraq's Tenth-century Baghdadi Cookbook.* Brill, Leiden.

Nawal, N. (2007). *Annals of the Caliphs' Kitchens. Ibn Sayyār al-Warrāq's Tenth-Century Baghdadi Cookbook. English Translation with Introduction and Glossary.* Brill, Leiden.

Neely, J.A. (1974). Sassanian and early Islamic water-control and irrigation systems on the Deh Luran plain, Iran. Pages 21-42 in Downing, T.E. and

Gibson, M.G. (eds), *Irrigation's Impact on Society.* University of Arizona Press, Tucson.

Nesbitt, M. (1993). Archaeobotanical remains. In: The International Merv Project: Preliminary Report on the First Season (1992) by Herrmann, G., Masson, V.M., Kurbansakhatov, K. *et al. Iran* 31:39-62.

Nesbitt, M. (1994). Archaeobotanical Research in the Merv Oasis. In: The International Merv Project: Preliminary Report on the Second Season (1993) by G. Herrmann, K. Kurbansakhatov *et al. Iran* 32:53-75.

Nesbitt, M. and O'Hara, S. (2000). Irrigation agriculture in Central Asia: A long-term perspective from Turkmenistan. Pages 103-122 in Barker, G. and Gilbertson, D. (eds), *The Archaeology of Drylands: Living at the Margin.* Routledge, London.

Nesbitt, M. and Summers, G.D. (1988). Some recent discoveries of millet (*Panicum miliaceum* L. and *Setaria italica* (L.) P. Beauv.) at excavations in Turkey and Iran. *Anatolian Studies* 38:85-97.

Newman, J. (1932). *The Agricultural Life of the Jews in Babylonia between the Years 200 CE and 500 CE* Oxford University Press, Oxford.

Olufsen, O. (1911). *The Emir of Bokhara and His Country: Journeys and Studies in Bokhara.* Heinemann, Copenhagen.

Oppenheimer, A. (1983). *Babylonian Judaica in the Talmudic Period.* Ludwig Reichert Verlag, Wiesbaden.

Perry, C. (2007). *A Baghdad Cookery Book: The Book of Dishes* [Kitab al-Tabikh] by Muhammad b. al-Ḥasan b. Muhammad b. al-Karim. Prospect Books, Totnes.

Petrushevsky, I.P. (1968). The socio-economic condition of Iran under the Il-Khans. Pages 483-537 in Boyle, J.A. (ed), *Cambridge History of Iran,* 5. Cambridge University Press, Cambridge.

Potts, D.T. (1991). A note on rice cultivation in Mesopotamia and Susiana. *NABU (Nouvelles Assyriologiques Brèves et Utilitaires)* 1991:1-2.

Potts, D.T. (1994). Contributions to the agrarian history of eastern Arabia II. The cultivars. *Arabian Archaeology and Epigraphy* 5:236-275.

Potts, D.T. (1996). *Mesopotamian Civilization: The Material Foundations.* Ithaca, NY: Cornell University Press.

Rabin, C. (1966). Rice in the Bible, *Journal of Semitic Studies* 11:2-9.

Radloff, W. (1911). *Versuch eines Wörterbuch der Türk-Dialecte* 4. Académie impériale des sciences, Sankt Petersburg.

Rahimi-Laridjani, F. (1988). *Die Entwicklung de Bewässerungslandwirtschaft im Iran bis in sasanidisch-frühislamische Zeit.* Dr. Ludwig Reichert, Wiesbaden.

Rålamb, C. (1679). *Kort beskriffning om thet som wid then Constantinopolitaniske resan är föreluppit. 1658.* Henrich Keyser, Stockholm.

Räsänen, M. (1969). *Versuch eines etymologischen Wörterbuch der Türksprachen.* Suomalais-ugrilainen seura, Helsinki.

Redhouse (1974). *Redhouse English-Turkish Dictionary.* Redhouse yayaevi, Isbanbul.

Roden, C. (1986). *A New Book of Middle Eastern Food.* Penguin, Harmondsworth.

Roerich, G.N. (1931). *Trails to Inmost Asia. Five Years of Exploration with the Roerich Central Asian Expedition.* Yale University Press, New Haven.

Salim, S.M. (1962). *Marsh Dwellers of the Euphrates Delta.* Athlone Press, London.

Sallares, R. (1991). *The Ecology of the Ancient Greek World*. Duckworth, London.

Samuel, D. (2001). Archaeobotanical evidence and analysis. Pages 343-481 in Berthier, S. (ed), *Peuplement rural et aménagements hydroagricoles dans la moyenne vallée de l'Euphrate fin VIIe-XIXe siècle*. Institut français d'études arabes de Damas, Damascus.

Sang, T. and Ge, S. (2007). The puzzle of rice domestication. *Journal of Integrative Plant Biology* 49:760-768.

Schomburg, R.C.F. (1928). The oasis of Kelpin in Sinkiang. *The Geographical Journal* 71:381-382.

Schuyler, E. (1876). *Turkistan: Notes of a Journey in Russian Turkistan, Khokand, Bukhara and Kuldja*, 1. Sampson Low, Marston, Searle & Rivington, London.

Sencil, E.E. (1980). *The Famous Turkish Cookery*. Galleri Minyatür, Istanbul.

Simpson, St J. (2003). From Mesopotamia to Merv: Reconstructing patterns of consumption in Sasanian households. Pages 347-375 in Potts, T., Roaf, M. and Stein, D. (eds), *Culture through Objects: Ancient Near Eastern Studies in Honour of P.R.S. Moorey*. Griffith Institute, Oxford.

Simpson, St J. *et al.* (forthcoming). The small finds, in *Excavations at Kush* (Derek Kennet *et al.*). Indicopleustes forthcoming.

Smith, B.D. (1998). *The Emergence of Agriculture*. Scientific American Library, New York, NY.

Strabo (1932). *The Geography of Strabo* VII. Loeb Classic, London.

Svanberg, I. (1989). *Kazakh Refugees in the Republic of Turkey: A Study of Cultural Persistence and Social Change*. Almqvist & Wiksell International, Uppsala.

Sweeney, M. and McCouch, S. (2007). The complex history of the domestication of rice. *Annals of Botany* 100:951-957.

Täckholm, V. and Täckholm, G. (1941). *Flora of Egypt* 1. Fouad I University, Cairo.

Tengberg, M. (2003). Archaeobotany in the Oman peninsula and the role of eastern Arabia in the spread of African crops. Pages 229-237 in Neumann, K., Butler, A. and Kahlheber, S. (eds), *Food, Fuels and Fields: Progress in African Archaeobotany*. Heinrich Barth Institut, Köln.

Thesiger, W. (1967). *The Marsh Arabs*. Penguin, Harmondsworth.

Thompson, R.C. (1939). ꝕKURANGU and ꝕLAL(L)ANGU as possible 'rice' and 'indigo' in cuneiform. *Iraq* 6:180-183.

Tlemisov, K.A. (1990). *Nacional'naya kukhija Kazakhhov*. Kaynar. Alma-Ata.

Tosi, M. (1975). Hasanlu project 1974: Palaeobotanical survey. *Iran* 13:185-186.

Usenbayev, N.T., Yezhov, M.N., Zvantsov, A.B., Annarbayev, A., Zhoroyev, A.A. and Almerekov, K.S. (2006). Epidemicheskii iskhog malarii vivaks Kirgizstan. *Medicinskaia Parazitologiia i Parazitarnye bolezni* 1:17-20.

Usmanova, Z.I. (1963). 'Erk-Kala," *YuTAKE Reports* 12:20-94.

Venzke, M.L. (1987-92). Rice cultivation in the plain of Anticoh in the 16th century: the Ottoman fiscal practice. *Archivum Ottomanicum* 12:175-276.

Visson, L. (1999). *The Art of Uzbek Cooking*. Hippocrene Books, New York.

Vitte, C., Ishii, T., Lamy, F., Brar, D. and Panaud, O. (2004). Genomic paleontology provides evidence for two distinct origins of Asian rice (*Oryza sativa* L.). *Molecular Genetics and Genomics* 272:504-511.

Waines, D. (1995). al-Ruzz [rice]. Pages 652-653 in Bosworth, C.E., van Donzel, E., Heinrichs, W.P. and Lecomte, G. (eds), *The Encyclopaedia of Islam. New Edition*. E.J. Brill, Leiden.

Watson, A.M. (1983). *Agricultural Innovation in the Early Islamic World: The Diffusion of Crops and Farming Techniques, 700-1100.* Cambridge University Press, Cambridge.

Weber, S.A. (1990). Millets in South Asia: Rojdi as a Case Study. Pages 333-348 in Taddei, M. and Calleri, P. (eds), *South Asian Archaeology 1987*, I. Istituto Italiano per il Medio ed Estremo Oriente, Rome.

Weggel, O. (1985). *Xinjiang/Sinkiang: das zentralasiatische China.* Institut für Asienkunde, Hamburg.

Wenke, R.J. (1975-76). Imperial investments and agricultural developments in Parthian and Sasanian Khuzestan: 150 BC to AD 640. *Mesopotamia* 10/11:31-221.

Wenke, R.J. (1987). Western Iran in the Partho-Sasanian Period: The Imperial Transformation. Pages 251-281 in Hole, F. (ed), *The Archaeology of Western Iran. Settlement and Society from Prehistory to the Islamic Conquest.* Smithsonian Institution Press, Washington.

Werner, C.A. (1999). The dynamics of feasting and gift exchange in rural Kazakhstan. Pages 47-72 in Svanberg, I. (ed), in *Contemporary Kazakhs: Cultural and Social Perspectives.* Curzon, London.

Whitman, J. (1956). Turkestan Cotton in Imperial Russia. *American Slavic and East European Review* 15/2:190-205.

Wiens, H.J. (1967). Regional and seasonal water supply in the Tarim basin and its relation to cultivated land potentials. *Annals of the Association of American Geographers* 57:350-366.

Wilson, G.H. (1995). Minnith. Pages 371 in Bromiley, G.W. (ed), *The International Standard Bible Encyclopedia.* Eerdmans, Grand Rapids, MI.

Wulff, H.E. (1966). *The Traditional Crafts of Persia: Their Development, Technology and Influence on Eastern and Western Civilizations.* Harvard University Press, Cambridge, MA.

Xiao, Q.L., Xin, Y.Z., Zhang, H.B., Zhou, J., Shang, X. and Dodson, J. (2007). The record of cultivated rice from archaeobiological evidence in northwestern China 5000 years ago. *Chinese Science Bulletin* 52:1372-1378.

Yudakhin, K.K. (1957). *Russko-kirgizskij slovar'.* Moskva.

Zaouali, L. (2007). *Medieval Cuisine of the Islamic World. A Concise History with 174 Recipes.* University of California Press, Berkeley.

Ziuzhen, F. (2003). Rice-fish culture in China. *Aquaculture Asia* 8:4:44-46.

Zohary, D. (1998). The diffusion of south and east Asian and of African crops into the belt of Mediterranean agriculture. Pages 123-134 in Prendergast, H.D.V., Etkin, N.L., Harris, D.R. and Houghton, P.J. (eds), *Plants for Food and Medicine.* Royal Botanic Gardens, Kew.

Zubaida, S. (2000). Rice in the culinary cultures of the Middle East. Pages 93-104 in Zubaida, S. and Tapper, R. (eds), *A Taste of Thyme: Culinary Cultures of the Middle East.* Tauris Parke, London.

Chapter 11: History of Rice in Europe

Actis Caporale, A. (2005). Problematiche della coltivazione del riso nell'ottocento in Piemonte. *Annali dell'Accademia di Agricoltura di Torino* 246:33-46.

Aguilar Portero, M. (2001). *Cultivo del arroz en el sur de España*. Lince Artes Gràficas, Spain, 192 pp.

AIDAF-VC/BI (Associazione Interprovinciale dei dottori in Scienze agrarie e forestali di Vercelli and Biella) (2003). Il bilancio economico dell'azienda risicola in relazione alle proposte di riforma dell'O.C.M. riso. *Meeting Consorzio di Irrigazione e Bonifica Ovest Sesia-Baraggia, Vercelli Italia, 26 febbraio*.

Anagnostopoulos, H. (1954). La risicoltura greca. *Il riso* 7:14-16.

Angelini, F. (1936). *Il riso, tecnica ed economia della coltivazione*. Società anonima Arte della stampa, Roma, Italy. 268 pp.

Audebert, A. and Mendez del Vilar, P. (2007). Characterization of rice crop systems and rice sector organization in Camargue-France. Pages 185-225 in Ferrero, A., Vidotto, F. (eds), *Agro-economical traits of rice cultivation in Europe and India*. Edizioni Mercurio, Vercelli, Italy.

Barbaresi, S. and Gherardi, F. (2000). The invasion of the alien crayfish *Procambarus clarkii* in Europe with particular reference to Italy. *Biological Invasions* 2:259-264.

Barbier, J.M. and Mouret, J.C. (1991). La mise en valeur de la Camargue: histoire et place du riz. Proposition pour une adaptation de la riziculture au nouveau contexte économique et écologique. Pages 71-104 in *Territori dell'acqua e dell'agricoltura*.

Barbier, J.M. and Mouret, J.C. (1992). Le riz et la Camargue. *Inra mensuel* 64/65:39-51.

Barbier, J.M. and Mouret, J.C. (1993). La Camargue: une région de production avec des atouts mais aussi des contraintes. Pages 21-23 in *Riz-du débuché à la culture*. ITCF and CFR.

Barret, S.C.H. and Wilson, B.F. (1983). Colonizing ability in the *Echinochloa crusgalli* complex (barnyard grass). II. Seed biology. *Canadian Journal of Botany* 61:556-562.

Batalla, J.A. (1989). Malas hierbas y herbicidas en los arrozales españoles. *Phytoma* 8:36-42.

Bayer, D. and Hill, J. (1993). Weeds. Pages 32-55 in Flint, M. (ed), *Integrated Pest Management for Rice. Vol. 3280*. University of California and Division of Agricultural and Natural Resources.

Bayer, D.E. (1991). Weed management. Pages 267-309 in Luh, B.S. (ed), *Rice Production. Vol. I*. AVI Book, Van Nostrand Reinhold.

Betti, Z. (1783). *Memoria 2 aggiunta all'Agricoltore sperimentale di Cosimo Trinci, Gatti Venezia*. 266 pp.

Blandi, S. (1827). *Arriano Flavio, Periplo del Mar Rosso, da "Opere di Arriano Nicomediese-Tomo secondo-Opuscoli", volgarizzato da Spiridione Blandi, Tipi di Francesco Sonzogno q.m Gio. Batt.a, Milano. 1827*.

Bocchi, S. (1996). Principi di fertilizzazione del riso-1. L'azoto nell'agro-sistema della risaia sommersa. *Ente Nazionale Risi. Quaderno n. 13*.

Busi, R., Vidotto, F., Ferrero, A., Fischer, A.J., Osuna, M.D. and De Prado, R. (2004). Patterns of resistance to ALS-inhibitors in *Cyperus difformis* and *Schoenoplectus mucronatus* at whole plant level. Pages 27-31 in Ferrero, A., Vidotto, F. (eds), *Proceedings of the Conference "Challenges and Opportunities for Sustainable Rice-based Production Systems"*, 13-15 September 2004. Edizioni Mercurio, Vercelli, Italy.

Carreres, R. (2007). Characteristics of rice cultivation in the Valencia Region (Spain). Pages 147-184 in Ferrero, A., Vidotto, F. (eds), *Agro-economical traits of rice cultivation in Europe and India*. Edizioni Mercurio, Vercelli, Italy.

CEC (Commission of the European Communities) (2002). Rice, Markets, CMO and Medium Term Forecast. *Commission Staff Working Paper*. SEC (2002):788.

Chataigner, J. and Salmon, C. (1996). Performances compares de la riziculture européenne. *Cahiers Options Meéditerranéennes* 15(2):103-114.

Cinotto, S. (2007). Interviews with witness of the rice world—Valencia, Spain. Pages 223-270 in Cinotto, S. (ed), *And Then the Rice Fields Emptied Out: History, Memory and Representations of the Rice Society in the Great Transformation (1945-1965)*. Edizioni Mercurio, Vercelli, Italy.

Clampett, W.S. (2001). Major achievements in closing yield gaps of rice between research and farmers in Australia. *International Rice Commission Newsletter* 50:7-16.

Clavera, A.S. (2004). El cultivo del arroz de secano en cataluña (1778-1839). Una propuesta agronómica al problema del paludismo. *Asclepio* 51(2):169-196.

Crist, R.E. (1960). Rice culture in the Camargue. *Annals of the Association of American Geographers* 50(3):312.

Crocioni, A. (1973). *Meccanizzazione integrale delle aziende risicole*. Consiglio Nazionale delle Ricerche, Roma, Italy. 140 pp.

D'Agaro, E. and Sparacino, A.C. (2007). Modelling an alien crayfish population invading rice fields. Pages 348-349 in Bocchi, S., Ferrero, A., Porro, A. (eds), *Proceedings of the Fourth International Temperate Rice Conference, Novara, Italy, 25-28 June 2007*.

Della Chiesa, L. (1777). *Storia del Piemonte. Libro III*. 143 pp.

ENR (Ente Nazionale Risi), (2007). http://enterisi.it/

FAO (Food and Agriculture Organization of the United Nations) (1996). *Groups and Types of World Climates. Map*. FAO, Rome, Italy.

FAOSTAT, (2006). See website: http://faostat.fao.org/

FAOSTAT, (2007). See website: http://faostat.fao.org/

Faure, J. and Mazaud, G. (1996). Rice quality criteria and the European market. Page 27 *Proceedings of the 18th Session of the International Rice Commission, 5-9 September, 1996, Rome, Italy*.

Ferrero, A. (2005). Preface. In Ferrero, A., Scansetti, M. (eds), *Rice Landscapes of Life*. Edizioni Mercurio, Vercelli, Italy. 207 pp.

Ferrero, A. (2007). Rice scenario in the European Union. *Cahiers Agriculture* 16(4):272-277.

Ferrero, A. and Nguyen, V.N. (2004). The sustainable development of rice-based production systems in Europe. *International Rice Commission Newsletter—Special Edition* 53:115-124.

Ferrero, A. and Tabacchi, M. (2002). Agronomical constraints in rice culture: Are there any possible solutions from biotechnology? In *Proceedings of Rice Conference "Dissemination Conference of Current European Research on Rice", Turin (Italy), June 6-8, 2002*.

Ferrero, A., Tabacchi, M. and Vidotto, F. (2002). Italian rice field weeds and their control. Pages 535-544 in Hill, J.E. and Hardy, B. (eds), *Proceedings 2nd Temperate Rice Conference, 13-17 June 1999, Sacramento, CA, USA.* International Rice Research Institute, Los Baños (Philippines).

Ferrero, A. and Vidotto, F. (2007). Weeds and weed management in Italian rice fields. Pages 55-72 in Ferrero, A., Vidotto, F. (eds), *Agro-economical traits of rice cultivation in Europe and India.* Edizioni Mercurio, Vercelli, Italy.

Ferrero, A., Vidotto, F., Gennari, M. and Nègre, M. (2001). Behaviour of cinosulfuron in paddy surface waters, sediments, and ground water. *Journal of Environmental Quality* 30:131-140.

Finassi, A. (2001). A century of mechanization of rice growing in Italy. Pages 37-48 in Isolani, B. and Manachini, B. (eds), *Project for and education on rice: Italian wetlands.* ScientiArs Multimedia, Cologno Monzese, Italy.

Finassi, A. and Ferrero, A. (2004). Outline of the Italian farm structure. Pages 583-596 in Ferrero, A. and Vidotto, F. (eds), *Proceedings of the Conference "Challenges and Opportunities for Sustainable Rice-Based Production Systems".* Torino, Italy, 13-15 September 2004. Edizioni Mercurio, Vercelli, Italy.

Finassi, A. and Ferrero, A. (2007). Mechanization and labor organization of rice farms in the Vercelli area. Pages 101-113 in Ferrero, A. and Vidotto, F. (eds), *Agro-economical traits of rice cultivation in Europe and India.* Edizioni Mercurio, Vercelli, Italy.

Garrity, D.P., Oldeman, L.R., Morris, R.A. and Lenka, D. (1986). Rainfed lowland rice ecosystems: characterization and distribution. Pages 3-23 in *Progress in Rainfed Lowland Rice.* IRRI, Los Baños, Philippines.

Gherardi, F., Bertocchi, S., Brusconi, S., Quaglio, F., Colombo, F., D'Agaro, E. and Sparacino, A.C. (2007). The impact of an introduced crayfish (*Procambarus clarkii*) on rice fields. Pages 230-231 in Bocchi, S., Ferrero, A. and Porro, A. (eds), *Proceedings of the Fourth International Temperate Rice Conference, Novara, Italy, 25-28 June 2007.*

GPP (Gabinete de Planeamento e Políticas) (2007). *Agricultura, Silvicultura e Pesca—Indicadores 2007.* Ministério da Agricultura, do Desenvolvimento Rural e das Pescas, Lisboa, Portugal. 98 pp.

Grant, C.J. (1964). Soil characteristic associated with the wet cultivation of rice. In *The Mineral Nutrition of the Rice Plant.* Proceedings of a Symposium at the International Rice Research Institute. The Johns Hopkins Press, Baltimore, Maryland. 494 pp.

Greenland, D.J. (1997a). Rice farming today. Pages 43-74 in *The Sustainability of Rice Farming.* CAB International in Association with the International Rice Institute, Wallingford Oxon, UK.

Greenland, D.J. (1997b). The biophysical basis of the sustainability of rice farming. Pages 65-102 in *The Sustainability of Rice Farming.* CAB International in Association with the International Rice Institute, Wallingford Oxon, UK.

GSNSSG (General Secretariat of National Statistical Service of Greece) (2005). See website: http://www.statistics.gr

Herruzo, C., Zekri, S. and Velasco, A. (1992). *Recent Developments in Rice Production in Spain.* 24th Rice Technical Working Group. Little Rock. Arkansas. (USA).

Imbergamo, B. (2007). Militant, exploited and sexy: the images and discourse on rice weeders in Italy from the late eighteenth century to the second half of the twentieth. Pages 91-137 in Cinotto, S. (ed), *And Then the Rice Fields Emptied Out: History, Memory and Representations of the Rice Society in the Great Transformation (1945-1965)*. Edizioni Mercurio, Vercelli, Italy.

INE (Instituto Nacional de Estadística) (2007a). See website: http://www.ine.es/inebase/cgi/um

INE (Instituto Nacional de Estadística) (2007b). See website: http://www.ine.pt/index.htm

IRRI (International Rice Research Institute) (2007). World Rice Statistics. http://www.irri.org/science/ricestat/index.asp

ISTAT (Istituto Nazionale di Statistica) (2007). See website: http://www.istati.it

Lima, A. (1997). Current situation of rice production in Portugal and the main diseases that occur. *Cahiers Options Méditerranéennes* 15(3):29-40.

Linares, O.F. (2002). African rice (*Oryza glaberrima*): History and future potential. *Proceedings of the National Academy of Sciences of the United States of America* 99(25):16360-16365.

Lupi, D., Colombo, M., Giudici, M.L., Villa, B., Sparacino, A.C. and Ranghino, F. (2007). Present status of knowledge on *Lissorhoptrus oryzophilus* Kuschel (Rice Water Weevil) in Italy. Pages 138-139 in Bocchi, S., Ferrero, A. and Porro, A. (eds), *Proceedings of the Fourth International Temperate Rice Conference, Novara, Italy, 25-28 June 2007*.

Lupotto, E. (2004). Antifungal genes in rice as a strategy for crop protection. Pages 353-359 in Ferrero, A. and Vidotto, F. (eds), *Proceedings of the Conference "Challenges and Opportunities for Sustainable Rice-based Production Systems". Torino, Italy, 13-15 September 2004*. Edizioni Mercurio, Vercelli, Italy.

Luppi, G., Finassi, A. and Cavallero, A. (2000). Riso (*Oryza* sp. pl.). Pages 233-285 in Baldoni, R. and Giardini, L. (eds), *Coltivazioni erbacee. Cereali e proteaginose*. Patron Editore, Bologna, Italy.

MAPA (Ministerio de Agricultura, Pesca y Alimentación) (2006). *HECOS y cifras de la agricultura, la pesa y la alimentación en España*. MAPA, Madrid, Spain. 160 pp.

Marinone, N. (1992). *Il riso nell'anticità greca*. Pàtron Editore, Quarto Inferiore, Bologna, Italy. 156 pp.

Mastrangelo, N. (1988). *Il Riso (On rice)*. Agrimont. Calderoni Ed agricole, Bologna, Italy. 142 pp.

Morbello, T. (1952). Cenni storici sulla coltivazione. *Il riso* 2:19-20.

Motta, E. (1913). *La storia della coltura del riso in Lombardia*. Giornale di risicoltura. 110 pp.

Navarro, L., Díaz, J. and Rodríguez, M.J. (1997). Cultivo del arroz en Andalucía. Estructuras productivas. Pages 271-285 in Junta de Andalucía (ed), *Cultivo del arroz en clima mediterráneo*. Sevilla, Spain.

Nguyen, V.N. and Ferrero, A. (2006). Meeting the challenges of global rice production. *Paddy and Water Environment* 4:1-9.

Ntanos, D. (2001). Evolution of rice research and production in Greece. *International Rice Commission Newsletter* 50:43-48.

Oerke, E.C., Dehene, H.V., Schoenbeck, F. and Weber, A. (1994). Rice losses. *In*

Crop Production and Crop Protection. Estimated Losses in Major Food and Cash Crops. Elsevier Science B.V., Amsterdam. 808 pp.

Piqueras, J. and Boira, J.V. (2005). Spain. Rice in history. Pages 177-191 in Ferrero, A. and Scansetti, M. (eds), *Rice Landscapes of Life*. Edizioni Mercurio, Vercelli, Italy.

Pirola, A. (1968). *Heteranthera reniformis* Ruitz et Pavon (Pontederiaceae) avventizia delle risaie pavesi. *Il Riso* 4:15-21.

Tabacchi, M. and Ferrero, A. (2006). Morphological traits and molecular markers for classification of *Echinochloa* species from Italian rice fields. *Weed Science* 54(6):1086-1093.

Tinarelli, A. (2006). Il riso nella cultura d'occidente. Pages 17-36 in Giacosa, A., Rondanelli, M. and Tinarelli, A. (eds), *Chiccodoro, Il riso nutrizione e salute*. Printed by Torchio De' Ricci Certosa Di Pavia, Italy.

Van Tran, D. (2001). Progress in rice production in the Mediterranean region. *Medoryzae, FAO, CIHEAM* 9:5.

Vasconcellos, J. de C.E. (1953). *O arroz*. Ministério da Economia, Commissão Reguladora do Comércio de Arroz, Lisboa. 301 pp.

Chapter 12: History of Rice in Africa

North Africa

Balal, M.S. (1981). Overview of rice production and research in Egypt. *Proceedings of the First National Rice Institute Conference, 21-25 February 1981.*

Balal, M.S. (1989). Rice varietal improvement in Egypt. *Report on rice farming systems and new directions. Proceedings of an international symposium, 31 Jan-3 Feb 1987*, Rice Research and Training Center, Sakha, Egypt.

Crecelius, D. (1998). Egypt in the eighteenth century. Pages 59-86 in Daly, M.F., Petry, C.F. (eds), *The Cambridge History of Egypt, 640-1517*. Cambridge University Press.

Dankoff, R. (2004). *An Ottoman Mentality: The World of Evliya Celebi*. BRILL.

El-Tobgy, H.A. (1976). *Contemporary Egyptian Agriculture*. 2nd ed. 228 pp.

Ibrahim, M.A., El-Gohary, A.S.A., Willardson, L.S. and Sission, D.V. (1995). Irrigation interval effects on rice production in the Nile Delta. *Irrig. Sci.* 16(1):29-33.

Khush, G.S. (1993). *Breeding Rice for Sustainable Agricultural Systems*. International Rice Research Institute, Manila, Philippines/Crop Science Society of America, Madison, Wiscosin, USA.

Khush, G.S. (1990). Varietal needs for different environments and breeding strategies. Pages 68-75 in Muralidharan, K. and Siddiq, E.A. (eds), *New Frontiers in Rice Research*. Directorate of Rice Research. Hyderabad, India.

Momtaz, A. (1989). Research and management strategies for increased rice production in Egypt. *Report on Rice Farming Systems and New Directions. Proceedings of an International Symposium, 31 Jan-3 Feb 1987*. Rice Research and Training Center, Sakha, Egypt.

Nour, M.A., Abd El-Wahab, M.A.E. and Ghanem, S.A. (1994). Broadcast-seed rice as affected by different irrigation intervals. *Egypt. J. Appl. Sci.* 9(8):671-683.

Refaie, F.R. (1965). *Chemistry of Cereal Technology*. Cereal Technology Department. Ministry of Agriculture, Egypt. (In Arabic).

RRTC (Rice Research and Training Centre) (1996). *Proceeding of the First National Rice Research and Development Program Workshop. Final Results of 1996 Season.* Sakha. Kafr El-Sheikh, Egypt.

Sato, T. (2005). *State and Rural Society in Medieval Islam: Sultan, Muqtas and Fallahun.* BRILL.

Sayyid-Marsot, A.L. (1984). *Egypt in the Reign of Muhammad Ali.* Cambridge University Press. 300 pp

Scheidel, W. (2001). *Death on the Nile: Disease and Demography of Roman Egypt.* BRILL.

Sidky, A.R. (1998). *Major Achievements in Rice Production. The 6th National Rice Research Conference, 26-28 March 1988.* Proceedings-Rice Research and Training Center. Sakha, Egypt.

Schiedel, W. (2001). *Death on the Nile: Disease and Demography of Roman Egypt.* BRILL.

Taylor and Francis Group (2002). *Middle East and North Africa 2003.* Routledge.

Tucker, Judith E. (1985). *Women in Nineteenth Century Egypt.* Cambridge Univ Press.

Wally, Y.A (1989). *Rice Farming Systems—New Directions.* Proceedings of an International Symposium, 31 Jan-3 Feb 1987, Rice Research and Training Center, Sakha, Egypt.

West Africa

Africa Rice Center (WARDA) (2005-2006). Annual Report 2005-2006, 8-9 pp.

Africa Rice Center (WARDA) (2007). Africa Rice Trends: Overview of recent developments in the sub-Saharan Africa rice sector. Africa Rice Center (WARDA), Cotonou. See website: www.warda.org

Agboola, S.A. (1979). *An Agricultural Atlas of Nigeria.* Oxford University Press, London. 89 pp.

Campbell, G. (2005). *An Economic History of Imperial Madagascar, 1750-1895: The Rise and Fall of an Island Empire.* Cambridge: Cambridge University Press.

Chohin-Kuper, A., Kelly, V. and Mariko, D. (2001). 20 years of economic reform in Sub-Saharan Africa: How have farmers in Mali's Office du Niger irrigated rice system responded? (A Power Point Presentation).

Connah, G. (2004). *Forgotten Africa: An Introduction to its Archaeology.* Rutledge.

De Datta, S.K. (1981). *Principles and Practice of Rice Production.* John Wiley & Sons, Singapore.

Eltis, D., Morgan, P. and Richardson, D. (n.d.). The African Contribution to Rice Cultivation in the Americas. See website: www.uga.edu/colonialseminar/EltisEssay.pdf

Jiddah, I.M. (2002). Agriculture in Nupeland: Rice production and others. Page 334 in Idress, A.A., Ochefu, Y.A. (eds), *Studies in the History of Central Nigeria Area, Vol. 1.* CSS, Lagos.

Kwara State Ministry of Agriculture (n.d.). *Information Bulletin No. 147, Passim.*

Maclean, R., Insoll, T. (1991). The social context of food technology in Iron Age

Gao, Mali. *World Archaeology* 31(1):78-92.

McIntosh, Susan K. (n.d.). A tale of two floodplains: comparative perspectives on the emergence of complex societies and urbanism in the middle Niger and Senegal valleys. See website: http://www.arkeologi.uu.se/afr/projects/BOOK/Mcintosh/mcintosh.pdf

Ogot, B.A. (1999). *Africa from Sixteenth to Eighteenth Century*. James Currey Publishers.

Olumese, C.E. (1988). The Role of River Basin Development Authority in the agricultural development of Nigeria. Pages 33-341 in Sanda, A.O. (ed), *Corporate Strategy for Agricultural and Rural Development in Nigeria, Ile-Ife*. Les Shyraden (Nig) Ltd.

Portères, R. (1970). Primary cradles of agriculture in the African continent. Pages 43-58 in Fage, J. and Olivier, R. (eds), *Papers in African Prehistory*. Cambridge University Press.

Portères, R. (1976). African cereals: Eleusine, fonio, black fonio, teff, Brachiaria, Paspalum, Pennisetum and African rice. Pages 409-452 in Harlan, J.R., De Wet, J.M.J. and Stemler, A.B.L. (eds), *Origins of African Plant Domestication*. Mouton, The Hague.

East Africa

Ade Ajayi, J.F. (1998). *General History of Africa. VI. Africa in the Nineteenth Century until the 1880s*. James Currey Publications.

Africa Rice Center (WARDA) (2007). Africa Rice Trends: Overview of recent developments in the sub-Saharan Africa rice sector. Africa Rice Center (WARDA), Cotonou. See website: www.warda.org

Blench, R. (2006). Subsistence systems: Comparing crop names to trees and livestock. In Denham, T.P., Areso, J.V.I. and Vrydaghs, L. (eds), *Rethinking Agriculture: Archaeological and Ethnoarchaeological Perspectives*. Taylor and Francis Group.

Campbell, G. (2005). *An Economic History of Imperial Madagascar, 1750-1895: The Rise and Fall of an Island Empire*. Cambridge University Press.

Fremigacci, J. (1985). Madagascar. Pages 393-398 in Fage, J.D. (ed), *The Cambridge History of Africa. Vol. 7. c. 1905-c. 1940*. Cambridge Univ Press.

IRRI (1985). *Impact of Science on Rice*. International Rice Research Institute, Manila.

Maghimbi, S. (2007). Recent changes in crop patterns in the Kilimanjaro region of Tanzania: The decline of coffee and the rise of maize and rice. *African Study Monographs, Suppl.* 35:73-83.

Vaughan, D., Miyazaki, S. and Miyashita, K. (2004). The rice genepool and human migrations. Pages 1-13 in Werner, D. (ed), *Biological Resources and Migration*. Springer.

Chapter 13: The Rice Industry of the United States

Bleyhl, Norris Arthur (1955). *A History of the Production and Marketing of Rice in California*. Ph.D. dissertation, University of Minnesota, Minneapolis.

Carney, Judith A. (2001). *Black Rice: The African Origins of Rice Cultivation in the Americas*. Harvard University Press, Cambridge.

Chaplin, Joyce E. (1993). *An Anxious Pursuit: Agricultural Innovation and Modernity in the Lower South, 1730-1815*. University of North Carolina Press, Chapel Hill.

Coclanis, Peter A. (1989). *The Shadow of a Dream: Economic Life and Death in the South Carolina Low Country, 1670-1920*. Oxford University Press, New York and Oxford.

Coclanis, Peter A. (1993a). Distant thunder: The creation of a world market in rice and the transformations it wrought. *American Historical Review* 98: 1050-1078.

Coclanis, Peter A. (1993b). Southeast Asia's incorporation into the world rice market: A revisionist view. *Journal of Southeast Asian Studies* 24:251-267.

Coclanis, Peter A. (2000). How the Low Country Was Taken to Task: Slave-Labor Organization in Coastal South Carolina and Georgia. In Paquette, Robert Louis and Ferleger, Louis (eds), *Slavery, Secession and Southern History*. University Press of Virginia, Charlottesville.

Coclanis, Peter A. (2002). Review of Black Rice. *Journal of Economic History* 62: 247-248.

Coclanis, Peter A. (2006a). Rice. In McCusker, J. (ed), *Encyclopedia of World Trade Since 1450. 2 Vols*. Macmillan Reference, Detroit.

Coclanis, Peter A. (2006b). Rice. In Edgar, W.B. (ed), *The South Carolina Encyclopedia*. University of South Carolina Press, Columbia.

Coclanis, Peter A. (2008). Simple truths about rising food costs. *The Herald-Sun* [Durham, N.C.], July 15, 2008.

Coclanis, Peter A. (2009). Introduction. In Hollis, Margaret *et al.* (eds), *Heyward Family Letters, 1862-1871*. University of South Carolina Press, Columbia.

Coclanis, Peter A. and Komlos, John (1987). Time in the paddies: A comparison of rice production in the southeastern United States and lower Burma in the nineteenth century. *Social Science History* 11:343-354.

Coclanis, Peter A. and Marlow, J.C. (1998). Inland rice production in the South Atlantic States. *Agricultural History* 72 (Spring 1998):197-212.

Daniel, Peter (1985). *Breaking the Land: The Transformation of Cotton, Tobacco and Rice Cultures since 1880*. University of Illinois Press, Urbana.

Dethloff, Henry C. (1988). *A History of the American Rice Industry, 1685-1985*. Texas A&M University Press, College Station.

Edelson, S. Max (2006). *Plantation Enterprise in Colonial South Carolina*. Harvard University Press, Cambridge.

Eltis, David; Morgan, Philip D. and Richardson, David (2007). Agency and diaspora in Atlantic history: Reassessing the African contribution to rice cultivation in the Americas. *American Historical Review* 112: 1329-1358.

Gray, Lewis C. (1958). *History of Agriculture in the Southern United States to 1860. 2 Vols*. Peter Smith, Gloucester, Mass.

Griswold, Daniel (2006). *Grain Drain: The Hidden Cost of U.S. Rice Subsidies. Trade Briefing Paper No. 25*. Center for Trade Policy Studies, Cato Institute, Washington, D.C.

Hart, John Fraser (1991). *The Land That Feeds Us*. W.W. Norton, New York.

Hill, J.E., Roberts, S.R., Brandon, D.M., *et al.* (1992). *Rice Production in California. Publication 21498*, DANR Publications, Univ. of California, Davis, California.

Littlefield, Daniel C. (1981). *Rice and Slaves: Ethnicity and the Slave Trade in Colonial South Carolina.* Louisiana State University Press, Baton Rouge.

Maclean, J.L., Dawe, D.C., Hardy, B. and Hettel, G.P. (eds), (2002). *Rice Almanac: Source Book for the Most Important Economic Activity on Earth.* 3rd ed. International Rice Research Institute, Los Baños, Philippines.

MacPherson, David (1805). *Annals of Commerce, Manufactures, Fisheries, and Navigation, 4 Vols.* Nicols and Son, London.

Moore, John Robert (2000). *Grist for the Mill: An Entrepreneurial History of Louisiana State Rice Milling Company, 1911-1965, River Brand Rice Milling Company, 1946-1965 and Riviana Foods, 1965-1999.* The Center for Louisiana Studies, University of Louisiana at Lafayette, Lafayette, Louisiana.

Morgan, Philip D. (1998). *Slave Counterpoint: Black Culture in the Eighteenth-Century Chesapeake* and *Lowcountry.* University of North Carolina Press, Chapel Hill.

Phillips, Edward Hake (1954). The Gulf Coast rice industry. *Agricultural History* 25:91-96.

Post, Lauren C. (1940). The rice country of southwestern Louisiana. *Geographical Review* 30:574-590.

Pudup, Mary Beth and Watts, Michael J. (1987). Growing against the grain: Mechanized rice farming in the Sacramento Valley, California. Pages 345-384 in Turner, B.L. and Brush, Stephen B. (eds), *Comparative Farming Systems.* The Guilford Press, New York and London.

Scott, James C. (1998). *Seeing Like a State: How Certain Schemes to Improve the Human Condition Have Failed.* Yale University Press, New Haven.

Shortridge, James R. and Shortridge, Barbara G. (1983). Patterns of American rice consumption 1955 and 1980. *Geographical Review* 73:417-429.

Spicer, J.M. (1964). *Beginnings of the Rice Industry in Arkansas.* Arkansas Rice Promotion Association, Stuttgart.

Stewart, Mart A. (1996). *'What Nature Suffers to Groe': Life, Labor, and Landscape on the Georgia Coastal Plain, 1680-1920.* University of Georgia Press, Athens.

United States Department of Agriculture, Economic Research Service (2006). *Rice Backgrounder.* RCS-2006-01. Washington, D.C.

United States Department of Agriculture, Economic Research Service (2007). *Rice Situation and Outlook Yearbook.* RCS-2007. Washington, D.C.

United States Department of Commerce, Bureau of the Census (2007). *Statistical Abstract of the United States, 2008.* 127th edn. U.S. Government Printing Office, Washington, D.C.

United States Department of Commerce, Bureau of the Census (1964). *United States Census of Agriculture, Vol. II.* U.S. Government Printing Office, Washington, D.C.

Wood, Peter H. (1974). *Black Majority: Negroes in Colonial South Carolina from 1670 through the Stono Rebellion.* Knopf, New York.

Chapter 14: History of Rice in Hispanic America

Amaral, J.R. do (1923). *Ephemerides maranhenses: datas e factos mais notáveis da história do Maranhão (1499-1823).* Tip. Teixeira, São Luís. 174 pp.

Amaral, L. (1940). *História geral da agricultura brasileira*. Companhia Editora Nacional, São Paulo. 473 pp.

Anonymous (2006). História del arroz: Fedearroz: la fuerza del campo que cultiva el futuro. See website http://www.fedearroz.com.co.

Asociación Cultivadores de Arroz (2006). Historia del arroz en el Uruguay. See website: <http//: www.aca.com.uy>

Bastos, A. (1964). M. Efemérides. *Lavoura Arrozeira* 18(210): 43.

Bastos, A. (1966). M. Efemérides. *Lavoura Arrozeira* 20(230): 36-37.

Bernardes, B.C. (1947). Variedades de arroz cultivadas no Rio Grande do Sul. *Lavoura Arrozeira* 1(7): 17-21.

Beernardes, B.C. (1960). A pesquisa na Estação Experimental de Arroz de Gravataí e seus resultados. Pages 11-20 in *Rio Grande do Sul*. Secretaria da Agricultura Porto Alegre RS. Cultura do arroz.

Black, G.A. (1950). Os capins aquáticos da Amazônia. *IAN Boletim Técnico* 19: 53-94.

Boxer, C.R. (1981). O império colonial português (1415-1825). Edições 70, Lisboa. 406 pp.

Brandão, A.F. (1997). *Diálogos das grandezas do Brasil*. Editora Massangana, Recife. 242 pp.

Brasil, J.F. de A. (1910). *Cultura dos campos. 3. ed.* Mounier: Jeanbin, Paris. 377pp.

Brito, R.S. de (1966). Guiné Cabo Verde e São Tomé e Príncipe: alguns aspectos da terra e dos homens. Pages 13-46 in *Instituto Superior de Ciências Socias e Politicas Ultramarina Lisboa. Cabo Verde Guiné São Tomé e Príncipe: curso de extensão universitária ano lectivo de 1965-66*. Universidade Técnica de Lisboa, Lisboa.

Bruzzone, C. (2004). El mejoramiento de arroz en el Peru: actual y potencial. In: *Conferência e Taller Melhoramento Genético do Arroz na América Latina e Caribe Goiânia. Memórias... Goiânia: Embrapa Arroz e Feijão*. 1 CD-ROM. (Embrapa Arroz e Feijão. Documentos 160).

Carney, J. and Marin, R.A. (1999). Aportes dos escravos na história do cultivo do arroz africano nas Américas. *Estudos Sociedade e Agricultura. Rio de Janeiro* n. 12: 113-133.

Carney, J.A. (2001). *Black rice: The African Origins of Rice Cultivation in the Americas*. Harvard University Press, Harvard. 267 pp.

Carreira, A. (1988). *A Companhia Geral do Grão-Pará e Maranhão*. Editora Nacional, São Paulo. 344 pp.

Ciferri, R. (1960). *Lineamenti per una storia del riso in Itália*. Ente nazionale rise, Milan. 42 pp.

Coutinho, M. (2005). *Fidalgos e barões: uma história da nobiliarquia luso-maranhense*. Instituto Geia, São Luís. 483 pp.

D'Azevedo, J.L. (1901). *Os jesuítas no Grão-Pará: suas missões e a colonização*. Tavares Cardoso, Lisboa. 366 pp.

Dias, M.N. (1970). *Fomento e mercantilismo: a Companhia Geral do Grão-Pará e Maranhão (1755-1778)*. UFPA, Belém. 545 pp.

Duncan, T.B. (1072). *Atlantic islands: Madeira the Azores and the Cape Verdes in seventeenth-century commerce and navigation*. University of Chicago Press. 291 pp.

Echenique, S. da C. (1964). Voltando à vaca fria da história do arroz no Rio Grande do Sul. *Lavoura Arrozeira* 18(209):20-21.

Faculdade Zacarias de Goes (2005). Valença. See website: <http:// www.fazag. com.br/valenca-histórico.php>

Ferreira, A.R. (1855a). Diário da viagem philosophica pela Capitania de São José do Rio Negro com a informação do estado presente. *Revista Trimensal do Instituto Historico Geographico e Ethnographico do Brazil* 48:1-77.

Ferreira, A.R. (1855b). Memoria sobre a introdução do arroz branco no Estado do Gram-Pará. *Revista Trimensal do Instituto Historico Geographico e Ethnographico do Brazil* 48:79-84.

Freitas, J.C. de. (1919). *Relatório da Secretaria da Fazenda.* Tip. Teixeira, São Luís. 214 pp.

Gaioso, R.J. de S. (1970). *Compêndio histórico-político dos princípios da lavoura do Maranhão. Edição similar.* Livros de Mundo Inteiro, Rio de Janeiro. 337 pp.

Gandavo, P. de M. (2001). Tratado da terra do Brasil. See website: <http:// www.bn.br/>.

Germek, E. and Banzatto, N.V. (2002). *Melhoramento do arroz no Instituto Agronômico.* Instituto Agronômico Boletim 202. 56 pp.

Germek, E. and Banzatto, N.V. (1977/78). Participação da variedade paulista de arroz IAC 1246 na produção nacional. *O Agronômico* 29/30: 33-40.

Gonçalves, J.S., Souza, S.A.M. and Resende, J.V. de (1989). Pesquisa e produção de alimentos: o caso do arroz em São Paulo. *Agricultura em São Paulo* 36(2):171-199.

Gonçalves, P.A. (1964). O arroz na Itália. *Lavoura Arrozeira* 18(213):15.

Gonzáles, J. (1985). Origen taxonomia y anatomia da la planta de arroz (*Oryza sativa* L.). Pages 47-64 in Tascón, J.E. and Garcia, D.E. (eds), *Arroz: investigación y producción.* CIAT, Cali.

Granato, L. (1914). *O arroz.* Tip. Levi, São Paulo. 538 pp.

Guimarães, E.P., Sant'Ana, E.P. and Rangel, P.H.N. (1997). Embrapa e parceiros lançam 85 cultivares de arroz em 15 anos de pesquisa. *Embrapa Arroz e Feijão. Pesquisa em Foco* 4. 2 pp.

Holanda, S.B. de. (1947). O arroz em São Paulo na era colonial. *Digesto Econômico* 3(31):56-58.

Holanda, S.B. de. (1957). *Caminhos e fronteiras.* Livraria José Olympio, Rio de Janeiro. 334 pp.

Homma, A.K.O. (2003). *História da agricultura na Amazônia: da era pré-colombiana ao terceiro milênio.* Embrapa Informação Tecnológica, Brasília. 274 pp.

Iglésias, F. de A. (1958). *Caatingas e chapadões: notas impressões e reminiscências do Meio-Norte brasileiro (1912-1919). 2.ed.* Companhia Editora Nacional, São Paulo.

Instituto de Pesquisas e Experimentação Agropecuárias do Sul (1967). *Lavoura Arrozeira* 21(238):46-51.

IRRI. (2006). Rough rice yield by country and geographical region 1961-2004. See website: <http://www.irri.org/science/ricestat>

Jennings, P.R. (1961). Historia del cultivo del arroz en Colombia. *Agricultura Tropical* 17(2):79-89.

Luís, W. (1956). *Na Capitania de São Vicente.* Livraria Martins, São Paulo. 339 pp.

Madre de Deus Frei Gaspar da (1975). *Memórias para a história da Capitania de São Vicente.* Ed. Itatiaia, Belo Horizonte. 250 pp.

Marassi, M.A. (2004). *Mejoramiento de arroz en el Argentina: impactos alcanzados.* In Conferência e Taller Melhoramento Genético do Arroz na América Latina e Caribe Goiânia. Memórias... Goiânia: Embrapa Arroz e Feijão. 1 CD-ROM. (Embrapa Arroz e Feijão. Documentos 160).

Marques, C.A. (1970). *Dicionário histórico-geográfico da Província do Maranhão. 3.ed.* Cia. Editora Fon-Fon e Seleta, Rio de Janeiro. 634 pp.

Massera, E.J. (1983). As origens da rizicultura gaúcha. *Lavoura Arrozeira* 36(341): 12-18.

Melhoramentos da rizicultura no Rio Grande do Sul (1946). Secretaria da Agricultura Indústria e Comércio, Porto Alegre. 429 pp.

Milano, S. (2003). First came Balilla. *Slow Ark Bra* 6:50-51.

Moquete, C. (2004). *Melhoramento do arroz na República Dominicana: estado atual.* Conferência e Taller Melhoramento Genético do Arroz na América Latina e Caribe Goiânia. Memórias... Goiânia: Embrapa Arroz e Feijão. 1 CD-ROM. (Embrapa Arroz e Feijão. Documentos 160).

Morais, C. (1960). Ensinamentos da rizicultura européia. Pages 5-10 in *Rio Grande do Sul.* Secretaria da Agricultura. Cultura do arroz.

National Geographic (1994). *Rice the essential harvest.* National Geographic Society, Washington. 79 pp.

Nivia, N. de (1991). Arroceros que hacen historia. *Arroz en las Américas* 12(1): 13-14.

Novelli, N. de (1918). *Cultura do arroz no Rio Grande do Sul.* Ministério da Agricultura Indústria e Comércio, Rio de Janeiro. 40 pp.

Oliveira, J.T.M. de (2000). *Veneza e Portugal no século XVI: subsídios para a sua história.* Imprensa Nacional:Casa da Moeda, Lisboa. 382 pp.

Paxeco, M.F. (1923). *Geografia do Maranhão.* Tip. Teixeira, São Luís. 739 pp.

Pedroso, B.A. (1989). *Arroz irrigado: obtenção e manejo de cultivares. 3. ed.* Sagra, Porto Alegre. 179 pp.

Pereira, J.A. (2002a). *Cultura do arroz no Brasil: subsídios para a sua história.* Embrapa Meio-Norte. 226, Teresina. pp.

Pereira, J.A. (2002b). Contribuição dos Açores à colonização do Brasil nos séculos XVII e XVIII. *Boletim do Instituto Histórico da Ilha Terceira Angra do Heroísmo* 60: 261-279.

Pereira, J.A. (2004). *O arroz vermelho cultivado no Brasil.* Embrapa Meio-Norte, Teresina. 90 pp.

Pott, V.J. and Pott, A. (2000). *Plantas aquáticas do Pantanal.* Embrapa Comunicação para Transferência de Tecnologia, Brasília. 404 pp.

Ramon, N.D. de. (2004). *El mejoramiento genético del arroz en Venezuela: impactos alcanzados.* Conferência e Taller Melhoramento Genético do Arroz na América Latina e Caribe Goiânia. Memórias... Goiânia: Embrapa Arroz e Feijão. 1 CD-ROM. (Embrapa Arroz e Feijão. Documentos 160).

Rangel, P.H.N. (1998). *Origem e evolução do arroz. Goiânia:* EMBRAPA-CNPAF 1998. Palestra apresentada no I Curso Internacional de Melhoramento Genético de Arroz Goiânia.

Rasmussen, W.D. (1975). Jefferson Washington... and other farmers. In U.S.D.A.

That we may eat. Washington 15-22.

Resende, M.H. (1976). Origens do arroz em Pelotas. *Lavoura Arrozeira* 29(289):13-16.

Ribeiro, D. (1995). *O povo brasileiro: a formação e o sentido do Brasil*. Companhia das Letras, São Paulo. 476 pp.

Santos, C.M. dos (1979). Cultura indústria e comércio de arroz no Brasil colonial. *Lavoura Arrozeira* 32(315):6-20.

Silva, M.V. e (1950a). Elementos para a história do arroz no Brasil. *Lavoura Arrozeira* 4(39):11-16.

Silva, M.V. e (1950b). Elementos para a história do arroz no Brasil. *Lavoura Arrozeira* 4(40):19-23.

Silva, M.V. e (1955). *Elementos para a história do arroz em Portugal*. Coimbra: Grêmio da lavoura da beira Litoral. Boletim da Federação dos Grêmios da Lavoura da Beira Litoral. 72 pp.

Silva, M.V. e (1956). O melhoramento do arroz em Portugal. *Vida Agrícola* 19.

Sindicato Arrozeiro do Rio grande do Sul (1935). *A cultura do arroz no Rio Grande do Sul*. 152 pp.

Sousa, G.S. de (1974). *Notícia do Brasil*. Departamento de Assuntos Culturais do MEC, São Paulo. 489 pp.

Suárez, E., Alfonso, R., Hernández, J.L., Ávila, J., Puldón, V., Fuentes, J.L., Pérez, A.V., Pérez, N. and González, M.C. (2004). *Mejoramiento genético del arroz en Cuba: impactos alcanzados*. Conferência e Taller Melhoramento Genético do Arroz na América Latina e Caribe Goiânia. Memórias... Goiânia: Embrapa Arroz e Feijão. 1 CD-ROM. (Embrapa Arroz e Feijão. Documentos 160).

Sweeney, M.T., Thomson, M.J., Pfeil, B.E. and McCouch, S. (2006). Caught red-handed: Rc encodes a basic helix-loop-helix protein conditioning red pericarp in rice. *Plant Cell* 18: 283-294.

Varnhagen, F.A. de (1975). *História geral do Brasil. 8.ed.* Edições Melhoramentos / MEC, São Paulo.

Vasconcellos, J. de C. e (1949). *Melhoramento do arroz: normas a seguir no apuramento das novas formas de origem híbrida*. Comissão Reguladora do Comércio de Arroz, Lisboa. 12 pp.

Viveiros, J.F. de (1928). Cultura do arroz no Estado do Maranhão. *Boletim do Ministério da Agricultura Indústria e Comércio Rio de Janeiro* 2:201-205.

Viveiros, J. de (1992). *História do comércio do Maranhão (1612-1895) Edição facsimilar*. Associação Comercial do Maranhão, São Luís. 309 pp.

Viveiros, J. de (1999). *Alcântara no seu passado econômico social e político. 3.ed.* AML:Alumar, São Luís. 180 pp.

Chapter 15: History of Rice Marketing

Andrus, J. Russell (1936).Three economic systems clash in Burma. *Review of Economic Studies* 3(2):140-146.

Arasaratnam, S. (1988). The Rice Trade in Eastern India 1650-1740. *Modern Asian Studies* 22(3):531-549.

Becker, L. and Diallo, R. (1996). The cultural diffusion of rice cropping in Cote d'Ivoire. *Geographical Review* 86(4): 505-528.

Bullio, P. (1969). Problemi e geografia della risicoltura in Piemonte nei secoli XVII e XVIII. *Annuli della Fondazione Luigi Einaudi* 3:37-93

Carney, Judith A. (2001). African rice in the Columbian Exchange. *Journal of African History* 42(3):377-396.

Carpenter, A. (1978). The History of Rice in Africa. Pages 3-10 in Buddenhagen, I. and Persley, J. (eds), *Rice in Africa.* Academic Press, London.

Chandola, H.V. (2006). Basmati Rice: Geographical Indication or Mis-Indication. *Journal of World Intellectual Property* 9(2):166-188.

Chang, T.T. (1985). Crop history and genetic conservation: Rice—A case study. *Iowa State J. Res.* 59(4): 425-456.

Chevalier, A. (1932). Les céréales des régions subsahariennes et des oasis. *Revue de Botanique appliquée et d'Agriculture tropicale,* 742-59.

Coclanis, Peter A. (1993). Distant Thunder: The Creation of a World Market in Rice and the Transformations It Wrought. *American Historical Review* 98 (4):1050-1078.

Collinson, P. (1766). Of the introduction of rice and tar in our colonies. *Gentleman's Magazine* (June):278-80.

Cotton, H.J.S. (1874). The rice trade of the world. *Calcutta Review* 58:267-302.

Dasgupta, B. (2000). Trade in Pre-Colonial Bengal. *Social Scientist* 28 (5/6):47-76.

Dawe, D. (2001). The Changing Structure of the World Rice Market, 1950-2000. *Food Policy* 27:355-370.

Dawe, D. (2002). The Changing Structure of the World Rice Market, 1950-2000. *Food Policy* 27:355-370.

Dethloff, Henry C. (1982). The colonial rice trade. *Agricultural History* 56(1):231-243.

Dollinger, P. (1970). *The German Hansa.* Translated and Edited by Ault, D.S. and Steinberg, S.H. Stanford: California University Press.

Drayton (1808). *A View of South Carolina,* 113-17.

Falcon, Walter P. and Monke, Eric A. (1979/80). International Trade in Rice. *Food Research Institute Studies* 17(3):283-284

FAOSTAT. See website: http://faostat.fao.org/default.aspx

Gulati, A. and Narayanan, S. (2002). *Rice Trade Liberalization and Poverty.* MSSD Discussion Paper No. 51, International Food Policy Research Institute, Washington, D.C.

Harper, Lawrence A. (1935). *The English Navigation Laws: A Seventeenth Century Experiment in Social Engineering.* New York: Columbia University Press.

IBRD (1981). *Commodity Trade and Price Trends (1981).* International Board for Reconstruction and Development, Washington DC.

Johnson, Gale D. (1982). Grain insurance, reserves and trade: Contributions to food security for LDCs. Pages 255-286 in Alberto Valdes (ed), *Food Security for Developing Countries.* Westview Press, Boulder Colo.

Kratoska, Paul H. (1990). The British Empire and the Southeast Asian rice crisis of 1919-1921. *Modern Asian Studies* 24(1):115-146.

Latham, A.J.H. and Neal, L. (1983). The International market in rice and wheat, 1868-1914. *The Economic History Review* 36(2):260-280.

Lewicki, T. (1974). *West African Food in the Middle Ages.* Cambridge.

McIntosh, R.J. and McIntosh, S.K. (1981). The Inland Niger Delta before the

Empire of Mali: Evidence from Jenne-Jeno. *Journal of African History* 22:1-22.

Monke, E. and Pearson, S. (1991). The international rice market. In: Pearson, S., Falcon, W., Heytens, P., Monke, E. and Naylor, R. (eds), *Rice Policy in Indonesia*. Cornell University Press, Ithaca, NY.

Morgan, K. (1995). The organization of the colonial American rice trade. *The William and Mary Quarterly* 52(3):433-452.

Nash, R.C. (1992). South Carolina and the Atlantic economy in the late seventeenth and eighteenth centuries. *Economic History Review* 45(4):677-702.

OECD (Organisation for Economic Cooperation and Development) (2002). *Agricultural Policies in OECD Countries: Monitoring and Evaluation*. OECD, Paris.

Orden, D., Cheng, F., Nguyen, H., Grote, U., Thomas, M., Mullen, K. and Sun, D. (2007). *Agricultural Producer Support Estimates for Developing Countries*. Research Report 152. International Food Policy Research Institute, Washington DC.

Pounds, Norman J.G. (1973/1990). *An Historical Geography of Europe 450 BC-1330 AD*. Cambridge.

Siamwalla, A. and Haykin, S. (1983). *The World Rice Market: Structure, Conduct and Performance*. International Food Policy Research Institute, Washington, DC.

Thomas, P.J. (1935). India in the World Depression. *The Economic Journal* 45(179): 69-483.

Thompson, V. (1941). War and Further India's Rice. *Far Eastern Survey* 10(16): 183-188.

Timmer, C.P. and Walter, P.F. (1975). The political economy of rice production and trade in Asia. Pages 373-408 in Reynolds, L. (ed), *Agriculture in Development Theory*. Yale University Press, New Haven.

Tymowski, M. (1971). Les domaines des princes de Songhay (Soudan occidental): Comparaison avec la grande propriété foncière au début de l'époque féodal'. *Annales* 15:1637-43.

UNDP (1997). *Human Development Report*. Oxford University Press, New York.

Urickizer, V.D. and Bennett, M.K. (1941). *The Rice Economy of Monsoon Asia*. Stanford, California.

van der Eng, Pierre (2004). Productivity and comparative advantage in rice agriculture in Southeast Asia since 1870. *Asian Economic Journal* 18(4):345-370.

Warmington, E.H. (1974). *The Commerce between the Roman Empire and India*. South Asia Books, New York.

Wells, G.J. (1994). The GATT Agriculture Agreement—A Time to Face Market Forces. *Interpaks Digest* 2(1):1-3.

Yap, C. Ling (1996). Implications of the Uruguay round on the world rice economy. *Food Policy* 21(4/5):377-391.

Chapter 16: A Century of Rice Breeding

Akiyoshi, D.E., Klee, H., Amasino, R., Nestor, E.W. and Gordon, M. (1984). T-DNA *Agrobacterium tumefaciens* encodes an enzyme for cytokinin bio-synthesis. *Proc. Natl. Acad. Sci.* USA 81:5994-5998.

Christou, P., Ford, T.L. and Kofron, M. (1991). Production of transgenic rice (*Oryza sativa* L.) plants from agronomically important *indica* and *japonica* varieties via electric discharge particle acceleration of exogenous DNA into immature zygotic embryos. *Bio/Technology* 9:957-962.

Datta, S.K. (2002). Recent developments in transgenics for abiotic stress tolerance in rice. *JIRCAS Working Report* (2002):43-53.

Datta, S.K., Torrizo, L., Tu, J., Oliva, N. and Datta, K. (1997). *Production and molecular evaluation of transgenic rice plants.* IRRI Discussion Paper Series No. 21. International Rice Research Institute, Manila, Philippines.

Evans, L.T. (1993). Raising the ceiling to yield: Key role of synergism between agronomy and plant breeding. Pages 103-107 in Muralidharan, K. and Siddique, E.A. (eds), *New Frontiers in Rice Research.* Directorate of Rice Research, Hyderabad, India.

FAO (Food and Agriculture Organization) (1996). *Food Balance Sheets, 1992-1994 Average.* Rome, Italy.

Gan, S. and Amasino, R.A. (1995). Inhibition of leaf senescence by autoregulated production of cytokinin. *Science* 270:1986-1988.

Hayami, Y., Kikuchi, M., Moya, P., Bambo, L. and Marciano, E. (1978). *Anatomy of peasant economy: A rice village in Philippines.* International Rice Research Institute, Philippines.

Hossain, M. (1988). *Nature and impact of the green revolution in Bangladesh.* International Food Policy Research Institute, Research Report No. 67. Washington D.C. (USA).

Huang, N., Angeles, E.R., Domingo, J., Magpantay, G., Singh, G., Zhang, G., Kumaravadivel, N., Bennett, J. and Khush, G.S. (1997). Pyramiding of bacterial blight resistance genes in rice: Marker assisted selection using RFLP and PCR. *Theor. Appl. Genet.* 95:313-320.

Khush, G.S. (1977). Disease and insect resistance in rice. *Adv. Agron.* 29:265-341.

Khush, G.S. (1987). Rice breeding: Past, present and future. *J. Genet.* 66:195-216.

Khush, G.S. (1995a). Modern varieties—their real contribution to food supplies and equity. *Geo Journal:* 35:275-284.

Khush, G.S. (1995b). Breaking the yield barrier of rice. *Geo J.* 35:329-332.

Khush, G.S. (1999). Green Revolution: Preparing for the 21st century. *Genome* 42:646-655.

Khush, G.S. and Virk, P.S. (2005). *IR Varieties and Their Impact.* International Rice Research Institute, Los Banos, Laguna.

Neeraja, C., Maghirang-Rodriguez, R., Pamplona, A., Heuer, S., Collard, B., Septiningsih, E., Vergara, G., Sanchez, D., Xu, K., Ismail, A. and Mackill, D. (2007). A marker-assisted backcross approach for developing submergence-tolerant rice cultivars. *Theor. Appl. Genet.* 115:767-776.

Paddock, W. and Paddock, P. (1967). *Times of Famine.* Little Brown and Company, Boston.

Parthasarathy, N. (1972). Rice breeding up to 1960. In *Proceeding of the Symposium on Rice Breeding held at IRRI, Los Banos, Laguna.*

Peng, S., Laza, R.C., Visperas, R.M., Sanico, A.L., Cassman, K.G. and Khush, G.S.

(2000). Grain yield of rice cultivars and lines developed in Philippines since 1996. *Crop Sci.* 40:307-314.

Singh, S., Sidhu, J.S., Huang, N., Vikal, Y., Li, Z., Brar, D.S., Dhaliwal, H.S. and Khush, G.S. (2001). Pyramiding three bacterial blight resistance genes (*xa5, xa13* and *Xa21*) using marker-assisted selection into *indica* rice cultivar PR106. *Theor. Appl. Genet.* 102:1011-1015.

Tu, J., Zhang, G., Datta, K., Xu, C., He, Y., Zhang, Q., Khush, G.S. and Datta, S.K. (2000). Field performance of transgenic elite commercial hybrid rice expressing *Bacillus thuringiensis* endoprotein. *Nat. Biotechnol.* 18:1101-1104.

UNDP (United Nations Development Programme) (1994). *Human Development Report.* Oxford University Press, U.K.

Van Der Meullen, J.G.J. (1951). Rice improvement by hybridization and results. *Contrib. Gen. Agric. Res. Stn. Bogor* 116:1-38.

Xiao, J., Grandillo, S., Ahn, S.N., McCouch, S.R. and Tanksley, S.D. (1996). Genes from wild rice improve yield. *Nature* 384:123-124.

Xu, K., Xia, X., Fukao, T., Canlas, P., Maghirang-Rodriguez, R., Heuer, S., Ismail, A.I., Bailey-Serres, J., Ronald, P.C. and Mackill, D.J. (2006). *Sub1A* is an ethylene response factor-like gene that confers submergence tolerance to rice. *Nature* 442:705-708.

Yap, C.L. (1991). *A comparison of the cost of producing rice in selected countries. Economic and Social Development, Paper No. 101.* Food and Agriculture Organization (FAO), Rome (Italy).

Ye, X., Al-Babili, S., Kloti, A., Zhang, J., Lucca, P., Beyer, P. and Potrykus, I. (2000). Engineering the provitin A (β-carotene) biosynthetic pathway I into (carotenoid-free) rice endosperm. *Science* 287:303-305.

Index

Glutinous rice 11, 30, 99, 112, 202, 210, 211, 212
Gold rush 407
Golden rice 510
GPS technology 430
Grain fissuring 365
Grain quality 494
Great Bengal Famine 255
Great Depression 469
Greeks 457
Green Revolution 213, 214, 489
Guacari variety 433
Guangxi-Guizhou 28
Gundil 185

Han 28
Han Dynasty 96
Harappan civilization 8
Harappan culture 228
Harappan sites 321
Harlans 390
Harvesting festivals 56, 181
Hasanlu 332
Hasanlu Period III 322
Hausa 402
Hemudu 93, 97, 133
Hemudu site 87
Herbicides 353, 361, 364
Hidden Hunger 509
High Yielding Varieties (HYVs) 213, 275, 454
Higo-suki 174
Hispanic America 433
Hmong-Mien 9
Hokkaido 173
Horace 342
Huai 321
Huai River 98
Huen Tsang 234
Hybrid breeding 506
Hybrid rice 283, 366
Hybrid varieties 298
Hydraulic civilization 265
Hydraulic engineering 165
Hydraulic system 265
HYV 291, 295

Ibn Battuta 239, 391
ICAR 261, 280, 282
Ideotype breeding 505
Ideotype plants 104
Iguape 438
IITA 22
Inari 45
Incantations 181
Indian agriculture 262
Indica 2, 85, 89, 117, 122, 138, 139, 140, 320, 382
Indica-japonica differentiation 95
Indica-japonica hybridization 147, 489
Indonesia 216
Indus River 302
Indus Valley 321
Indus Valley Civilization 228
INGER 495
Inner Niger Delta 20
Insects 491
Intellectual Property Rights (TRIPS) 481, 482
International cooperation 494
International Rice Commission (IRC) 488
International trade 284
International Year of the Rice 336
IPEAS 448
IR varieties 491
IR36 491
IR8 275, 291, 436
Iran 318
IRAT 22, 408
Iron Age 131
Irrawaddy Delta 462
IRRI 275, 282, 291, 300, 304, 408, 434, 436, 449
Irrigated rice 30
Irrigation 258, 301, 304, 344, 377, 386, 397, 446, 449
Irrigation canals 165, 346
Irrigation channels 122
Irrigation system 204, 264
IRRI-Pak type 300
Ishikawa Rikinosuke 171
Isles de Loss 393
ITO 479

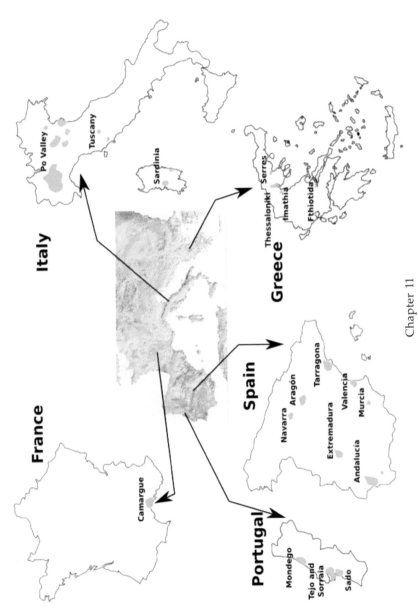

Chapter 11
Fig. 11.1. Major rice growing areas in Europe.

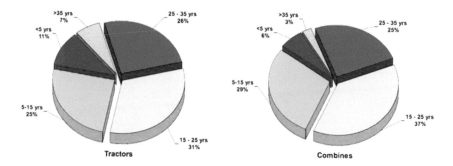

Chapter 11

Fig. 11.3. Age class distribution in the tractor and combine fleets in the Vercelli area, Italy.[92]

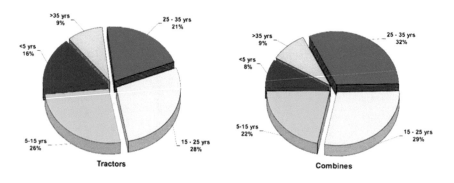

Chapter 11

Fig. 11.4. Age class distribution in the tractor and combine fleets in the Valencia area Spain.[93]

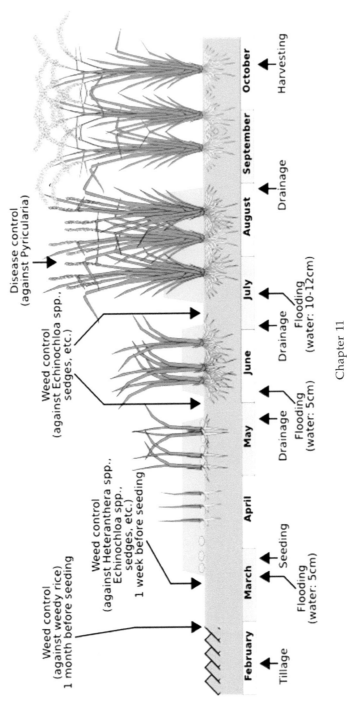

Chapter 11

Fig. 11.5. Scheme of main agronomical practices carried out during rice cultivation.

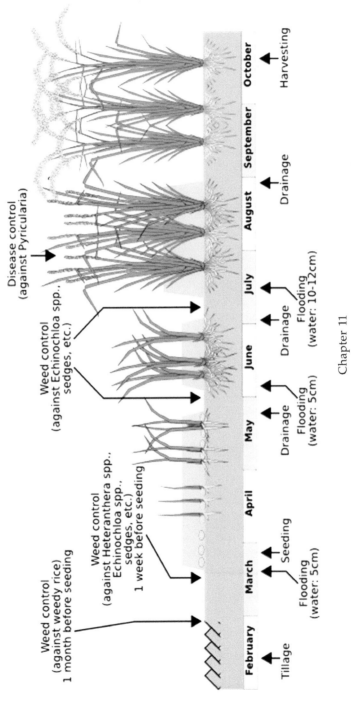

Chapter 11

Fig. 11.5. Scheme of main agronomical practices carried out during rice cultivation.

T - #0303 - 071024 - C22 - 229/152/26 - PB - 9780367383961 - Gloss Lamination